# ASTROPHYSICAL APPLICATIONS OF STELLAR PULSATION

## A SERIES OF BOOKS ON RECENT DEVELOPMENTS IN ASTRONOMY AND ASTROPHYSICS





Printed by BookCrafters, Inc.

First published 1995

Library of Congress Catalog Card Number: 95-77972
ISBN 1-866733-03-1
D. Harold McNamara, Managing Editor of Conference Series
Box 63 KMB
Brigham Young University
Provo, UT 84602
801-378-2298

pasp@astro.byu.edu
Fax 801-378-2265

# A SERIES OF BOOKS ON RECENT DEVELOPMENTS IN ASTRONOMY AND ASTROPHYSICS

Vol. 1-Progress and Opportunities in Southern Hemisphere Optical Astronomy: The CTIO 25th Anniversary Symposium
ed. V. M. Blanco and M. M. Phillips ISBN 0-937707-18-X

Vol. 2-Proceedings of a Workshop on Optical Surveys for Quasars
ed. P. S. Osmer, A. C. Porter, R. F. Green, and C. B. Foltz ISBN 0-937707-19-8

Vol. 3-Fiber Optics in Astronomy
ed. S. C. Barden ISBN 0-937707-20-1

Vol. 4-The Extragalactic Distance Scale: Proceedings of the ASP 100th Anniversary Symposium
ed. S. van den Bergh and C. J. Pritchet ISBN 0-937707-21-X

Vol. 5-The Minnesota Lectures on Clusters of Galaxies and Large-Scale Structure
ed. J. M. Dickey ISBN 0-937707-22-8

Vol. 6-Synthesis Imaging in Radio Astronomy: A Collection of Lectures from the Third NRAO Synthesis Imaging Summer School
ed. R. A. Perley, F. R. Schwab, and A. H. Bridle ISBN 0-937707-23-6

Vol. 7-Properties of Hot Luminous Stars: Boulder-Munich Workshop
ed. C. D. Garmany ISBN 0-937707-24-4

Vol. 8-CCDs in Astronomy
ed. G. H. Jacoby ISBN 0-937707-25-2

Vol. 9-Cool Stars, Stellar Systems, and the Sun. Sixth Cambridge Workshop
ed. G. Wallerstein ISBN 0-937707-27-9

Vol. 10-The Evolution of the Universe of Galaxies. The Edwin Hubble Centennial Symposium
ed. R. G. Kron ISBN 0-937707-28-7

Vol. 11-Confrontation Between Stellar Pulsation and Evolution
ed. C. Cacciari and G. Clementini ISBN 0-937707-30-9

Vol. 12-The Evolution of the Interstellar Medium
ed. L. Blitz ISBN 0-937707-31-7

Vol. 13-The Formation and Evolution of Star Clusters
ed. K. Janes ISBN 0-937707-32-5

Vol. 14-Astrophysics with Infrared Arrays
ed. R. Elston ISBN 0-937707-33-3

Vol. 15-Large-Scale Structures and Peculiar Motions in the Universe
ed. D. W. Latham and L. A. N. da Costa ISBN 0-937707-34-1

Vol. 16-Atoms, Ions and Molecules: New Results in Spectral Line Astrophysics
ed. A. D. Haschick and P. T. P. Ho ISBN 0-937707-35-X

Vol. 17-Light Pollution, Radio Interference, and Space Debris
ed. D. L. Crawford ISBN 0-937707-36-8

Vol. 18-The Interpretation of Modern Synthesis Observations of Spiral Galaxies
ed. N. Duric and P. C. Crane ISBN 0-937707-37-6

Vol. 19-Radio Interferometry: Theory, Techniques, and Application, IAU Colloquium 131
ed. T. J. Cornwell and R. A. Perley ISBN 0-937707-38-4

Vol. 20-Frontiers of Stellar Evolution, celebrating the 50th Anniversary of McDonald Observatory
ed. D. L. Lambert ISBN 0-937707-39-2

Vol. 21-The Space Distribution of Quasars
ed. D. Crampton ISBN 0-937707-40-6

Vol. 22-Nonisotropic and Variable Outflows from Stars
ed. L. Drissen, C. Leitherer, and A. Nota ISBN 0-937707-41-4

Vol. 23-Astronomical CCD Observing and Reduction Techniques
ed. S. B. Howell ISBN 0-937707-42-4

Vol. 24-Cosmology and Large-Scale Structure in the Universe
ed. R. R. de Carvalho ISBN 0-937707-43-0

Vol. 25-Astronomical Data Analysis Software and Systems I
ed. D. M. Worrall, C. Biemesderfer, and J. Barnes ISBN 0-937707-44-9

Vol. 26-Cool Stars, Stellar Systems, and the Sun, Seventh Cambridge Workshop
ed. M. S. Giampapa and J. A. Bookbinder ISBN 0-937707-45-7

Vol. 27-The Solar Cycle
ed. K. L. Harvey ISBN 0-937707-46-5

Vol. 28-Automated Telescopes for Photometry and Imaging
ed. S. J. Adelman, R. J. Dukes, Jr., and C. J. Adelman ISBN 0-937707-47-3

Vol. 29-Workshop on Cataclysmic Variable Stars
ed. N. Vogt ISBN 0-937707-48-1

Vol. 30-Variable Stars and Galaxies, in honor of M. S. Feast on his retirement
ed. B. Warner ISBN 0-937707-49-X

Vol. 31-Relationships Between Active Galactic Nuclei and Starburst Galaxies
ed. A. V. Filippenko ISBN 0-937707-50-3

Vol. 32-Complementary Approaches to Double and Multiple Star Research, IAU Collouquium 135
ed. H. A. McAlister and W. I. Hartkopf ISBN 0-937707-51-1

Vol. 33-Research Amateur Astronomy
ed. S. J. Edberg ISBN 0-937707-52-X

Vol. 34-Robotic Telescopes in the 1990s
ed. A. V. Filippenko ISBN 0-937707-53-8

Vol. 35-Massive Stars: Their Lives in the Interstellar Medium
ed. J. P. Cassinelli and E. B. Churchwell ISBN 0-937707-54-6

Vol. 36-Planets and Pulsars
ed. J. A. Phillips, S. E. Thorsett, and S. R. Kulkarni ISBN 0-937707-55-4

Vol. 37-Fiber Optics in Astronomy II
ed. P. M. Gray ISBN 0-937707-56-2

Vol. 38-New Frontiers in Binary Star Research
ed. K. C. Leung and I. S. Nha ISBN 0-937707-57-0

Vol. 39-The Minnesota Lectures on the Structure and Dynamics of the Milky Way
ed. Roberta M. Humphreys ISBN 0-937707-58-9

Vol. 40-Inside the Stars, IAU Colloquium 137
ed. Werner W. Weiss and Annie Baglin ISBN 0-937707-59-7

Vol. 41-Astronomical Infrared Spectroscopy: Future Observational Directions
ed. Sun Kwok ISBN 0-937707-60-0

Vol. 42-GONG 1992: Seismic Investigation of the Sun and Stars
ed. Timothy M. Brown ISBN 0-937707-61-9

Vol. 43-Sky Surveys: Protostars to Protogalaxies
ed. B. T. Soifer ISBN 0-937707-62-7

Vol. 44-Peculiar Versus Normal Phenomena in A-Type and Related Stars
ed. M. M. Dworetsky, F. Castelli, and R. Faraggiana ISBN 0-937707-63-5

Vol. 45-Luminous High-Latitude Stars
ed. D. D. Sasselov ISBN 0-937707-64-3

Vol. 46-The Magnetic and Velocity Fields of Solar Active Regions, IAU Colloquium 141
ed. H. Zirin, G. Ai, and H. Wang ISBN 0-937707-65-1

Vol. 47-Third Decinnial US-USSR Conference on SETI
ed. G. Seth Shostak ISBN 0-937707-66-X

Vol. 48-The Globular Cluster-Galaxy Connection
ed. Graeme H. Smith and Jean P. Brodie ISBN 0-937707-67-8

Vol. 49-Galaxy Evolution: The Milky Way Perspective
ed. Steven R. Majewski ISBN 0-937707-68-6

Vol. 50-Structure and Dynamics of Globular Clusters
ed. S. G. Djorgovski and G. Meylan ISBN 0-937707-69-4

Vol. 51-Observational Cosmology
ed. G. Chincarini, A. Iovino, T. Maccacaro, and D. Maccagni ISBN 0-937707-70-8

Vol. 52-Astronomical Data Analysis Software and Systems II
ed. R. J. Hanisch, J. V. Brissenden, and Jeannette Barnes ISBN 0-937707-71-6

Vol. 53-Blue Stragglers
ed. Rex A. Saffer ISBN 0-937707-72-4

Vol. 54-The First Stromlo Symposium: The Physics of Active Galaxies
ed. Geoffrey V. Bicknell, Michael A. Dopita, and Peter J. Quinn ISBN 0-937707-73-2

Vol. 55-Optical Astronomy from the Earth and Moon
ed. Diane M. Pyper and Ronald J. Angione ISBN 0-937707-74-0

Vol. 56-Interacting Binary Stars
ed. Allen W. Shafter ISBN 0-937707-75-9

Vol. 57-Stellar and Circumstellar Astrophysics
ed. George Wallerstein and Alberto Noriega-Crespo ISBN 0-937707-76-7

Vol. 58-The First Symposium on the Infrared Cirrus and Diffuse Interstellar Clouds
ed. Roc M. Cutri and William B. Latter ISBN 0-937707-77-5

Vol. 59-Astronomy with Millimeter and Submillimeter Wave Interferometry
ed. M. Ishiguro and Wm. J. Welch ISBN 0-937707-78-3

Vol. 60-The MK Process at 50 Years: A Powerful Tool for Astrophysical Insight
ed. C. J. Corbally, R. O. Gray, and R. F. Garrison ISBN 0-937707-79-1

Vol. 61-Astronomical Data Analysis Software and Systems III
ed. Dennis R. Crabtree, R. J. Hanisch, and Jeannette Barnes ISBN 0-937707-80-5

Vol. 62-The Nature and Evolutionary Status of Herbig Ae / Be Stars
ed. P. S. Thé, M. R. Pérez, and E. P. J. van den Heuvel ISBN 0-937707-81-3

Vol. 63-Seventy-Five Years of Hirayama Asteroid Families: The role of Collisions in the Solar System History
ed. R. Binzel, Y. Kozai, and T. Hirayama ISBN 0-937707-82-1

Vol. 64-Cool Stars, Stellar Systems, and the Sun, Eighth Cambridge Workshop
ed. Jean-Pierre Caillault ISBN 0-937707-83-X

Vol. 65-Clouds, Cores, and Low Mass Stars
ed. Dan P. Clemens and Richard Barvainis ISBN 0-937707-84-8

Vol. 66- Physics of the Gaseous and Stellar Disks of the Galaxy
ed. Ivan R. King ISBN 0-937707-85-6

Vol. 67-Unveiling Large-Scale Structures Behind the Milky Way
ed. C. Balkowski and R. C. Kraan-Korteweg ISBN 0-937707-86-4

Vol. 68-Solar Active Region Evolution: Comparing Models with Observations
ed. K. S. Balasubramaniam and George W. Simon ISBN 0-937707-87-2

Vol. 69-Reverberation Mapping of the Broad-Line Region in Active Galactic Nuclei
ed. P. M. Gondhalekar, K. Horne, and B. M. Peterson ISBN 0-937707-88-0

Vol. 70-Groups of Galaxies
ed. Otto G. Richter and Kirk Borne ISBN 0-937707-89-9

Vol. 71-Tridimensional Optical Spectroscopic Methods in Astrophysics
ed. G. Comte and M. Marcelin ISBN 0-937707-90-2

Vol. 72-Millisecond Pulsars—A Decade of Surprise, ed. A. A. Fruchter
M. Tavani, and D. C. Backer ISBN 0-937707-91-0

Vol. 73-Airborne Astronomy Symposium on the Galactic Ecosystem: From Gas to Stars to Dust
ed. M. R. Haas, J. A. Davidson, and E. F. Erickson ISBN 0-937707-92-9

Vol. 74-Progress in the Search for Extraterrestrial Life,
ed. G. Seth Shostak ISBN 0-937707-93-7

Vol. 75-Multi-Feed Systems for Radio Telescopes
ed. D. T. Emerson and J. M. Payne ISBN 0-937707-94-5

Vol. 76-GONG '94: Helio- and Astero-Seismology from the Earth and Space
ed. Roger K. Ulrich, Edward J. Rhodes, Jr., and Werner Däppen ISBN 0-937707-95-3

Vol. 77-Astronomical Data Analysis Software and Systems IV
ed. R. A. Shaw, H. E. Payne, and J. J. E. Hayes ISBN 0-937707-96-1

Vol. 78-Astrophysical Applications of Powerful New Databases
ed. S. J. Adelman and W. L. Wiese ISBN 0-937707-97-X

Vol. 79-Robotic Telescopes: Current Capabilities, Present Developments, and Future Prospects for Automated Astronomy
ed. Gregory W. Henry and Joel A. Eaton ISBN 0-937707-98-8

Vol. 80-The Physics of the Interstellar Medium and Intergalactic Medium
ed. A. Ferrara, C. F. McKee, C. Heiles, and P. R. Shapiro ISBN 0-937707-99-6

Vol. 81-Laboratory and Astronomical High Resolution Spectra
ed. A. J. Sauval, R. Blomme, and N. Grevesse ISBN 1-886733-01-5

Vol. 82-Very Long Baseline Interferometry and the VLBA,
ed. J. A. Zensus, P. K. Diamond, and P. J. Napier ISBN 1-886733-02-3

Vol. 83-Astrophysical Applications of Stellar Pulsation, IAU Colloquium 155
ed. R. S. Stobie and P. A. Whitelock ISBN 1-886733-03-1

Inquiries concerning these volumes should be directed to the:
Astronomical Society of the Pacific
CONFERENCE SERIES
390 Ashton Avenue
San Francisco, CA 94112-1722
415-337-1100
e-mail asp @ stars.sfsu.edu

ASTRONOMICAL SOCIETY OF THE PACIFIC
CONFERENCE SERIES

Volume 83

# ASTROPHYSICAL APPLICATIONS OF STELLAR PULSATION

Proceedings of IAU Colloquium 155
held in Cape Town, 6-10 February 1995



Edited by
R. S. Stobie and P. A. Whitelock

ASTRONOMICAL SOCIETY OF THE PACIFIC
CONFERENCE SERIES

Volume 83

## Table of Contents

## Part 2. Poster Papers

# Preface

Every 2 or 3 years, astronomers interested in stellar pulsation and its many applications have organised a meeting in a different part of the world. This was the first time that such a meeting had been held in South Africa. The topic of the conference, Astrophysical Applications of Stellar Pulsation, was highly appropriate as South African astronomers have made seminal contributions to the study of pulsating stars over many years. Many new and exciting results were announced during the conference and there was much stimulating discussion.

In total 152 astronomers from 26 countries attended this conference in Cape Town from 6-10 February 1995. The conference, which was hosted by the South African Astronomical Observatory, was unique for a number of reasons. First, it was the largest gathering of astronomers ever to take place in South Africa. Secondly, it was the first time the International Astronomical Union had supported a conference on the African continent. Thirdly, a unique experience was provided for participants who were accommodated in an old prison (Breakwater Lodge), and attended the scientific sessions in a mental institution (Valkenberg Hospital), which happens to have a superb conference venue adjacent to SAAO. We hope that all astronomers got out again!

Prior to the conference we took advantage of the presence of international astronomers interested in "astronomy in science education" to hold a one-day workshop on this subject (organised by Dr P A Whitelock and Prof J Percy). The aim was to show teachers how they might use astronomy to stimulate an interest in science, and to provide exciting illustrations of basic physical principles. The workshop was attended by over 100 teachers and trainee-teachers and was highly successful, judging by the comments of the teachers.

Returning to the pulsation conference, we would like to take this opportunity to thank all the people who contributed to the success of the pulsation meeting: the Scientific Organising Committee for generating a stimulating scientific programme; the Local Organising Committee for their effort in ensuring the smooth running of the conference; the Sponsors for the financial support; and last, but not least, the participants for ensuring a very successful conference.

## Sponsors

Department of Arts, Culture, Science and Technology, Republic of South Africa
International Astronomical Union
International Science Foundation
SANLAM (South African Life and Mutual)
South African Astronomical Observatory
Captour, Fair Lady
Kirstenbosch National Botanical Gardens
Prof John Percy

*R. S. Stobie and P. A. Whitelock*
May 1995

## SCIENTIFIC ORGANISING COMMITTEE

M. Breger, Institut für Astronomie, Universität Wien, Austria
C. Cacciari, Osservatorio Astronomico, Bologna, Italy
P. L. Cottrell, Mount John University Observatory, New Zealand
A. N. Cox, Los Alamos Astrophysics, USA
W. Gieren, Pontifica Universidad Católica de Chile, Chile
M. Jerzykiewicz, Wroclaw University Observatory, Poland
G. Kovacs, Konkoly Observatory, Hungary
A. J. Longmore, Royal Observatory, Edinburgh, UK
J. M. Matthews, University of British Columbia, Canada
J. R. Percy (Chairman), University of Toronto, Canada
H. Saio, Tohoku University, Japan
R. S. Stobie, South African Astronomical Observatory, South Africa
C. Waelkens, Instituut voor Sterrekunde, Universiteit Leuven, Belgium
P. A. Whitelock, South African Astronomical Observatory, South Africa

## LOCAL ORGANISING COMMITTEE

L. A. Balona, J. A. R. Caldwell, A. Hultzer, D. O'Donoghue
R. S. Stobie (Chairman), B. Warner

# WILLIAM H. WEHLAU (1926 - 1995)

The astronomical world lost an accomplished but humble scientist when Professor Emeritus William (Bill) H. Wehlau passed away in Cape Town soon after the close of this Colloquium. Although a tragic footnote to the meeting, it seems fitting that Bill's final scientific endeavours were at a conference on stellar pulsation, since he was a genuine pioneer in the field on several fronts.

Bill laid the foundation for many of the astronomical period-searching techniques we take for granted today. His 1964 paper with Kam-Ching Leung (ApJ 139, 843) on the multiple periods of $\delta$ Delphini was the first published use of Fourier integral techniques to interpret stellar light curves. The quality and quantity of data (25 nights of three-colour photometry) and sophistication of the fit (a sum of 6 cosines) would not be out of place in the modern literature on asteroseismology of $\delta$ Scuti variables (see his paper in these Proceedings). In the next decade, Bill again was a pioneer; this time in stellar surface imaging through inversion of spectral line-profile variations into abundance and temperature maps of Ap stars. Such techniques are now widely used to analyse high-degree NRP in rapid rotators. By the 1980s, Bill's diverse interests in pulsation and spectral peculiarity converged in the rapidly oscillating Ap stars. He played key roles in the first detection of velocity oscillations in an roAp star and in translating optical-IR photometry of roAp pulsations into the first empirical atmospheric temperature profile for a star other than the Sun.

Bill's modesty meant he never considered himself a pioneer, but even he might have found it hard to dismiss an impressive legacy which includes: the Astronomy Department and Observatory he founded in 1966 at the University of Western Ontario and headed for 25 years; the success of the Canada-France-Hawaii Telescope which he helped guide in its early years as Chair of its Board and Advisory Council; and several powerful analytical tools which are still advancing our knowledge of stars.

*J. Matthews*

## ADRIAAN JAN WESSELINK (1909 - 1995)

A.J. Wesselink made important contributions to the study of a wide variety of variable stars. However, it is for his work on the determination of the radii of pulsating stars (the "Baade-Wesselink" method) that he is perhaps best known to astronomers generally and his name will continue to be mentioned in this connection.'Wes' had considerable independence of mind and wide interests. Thus he wrote a classic paper on the nature of the surface of the moon as derived from lunar eclipse observations and he was intrigued by the possibility that some forms of stellar variability might be explained by relaxation oscillations.

He spent 18 years in mid-career in South Africa, first as astronomer in charge of the Leiden southern station, Johannesburg (1946-50) and then as chief assistant (i.e., assistant director) at the new Radcliffe observatory Pretoria (1950-64). Amongst his many investigations with the Radcliffe 1.9m reflector the most notable was probably his discovery with Thackeray of RR Lyraes in the Magellanic Clouds; work which was of such crucial importance for a revision of the extragalactic distance scale.

His friends remember him for his cheerfulness, his enthusiasm for astronomy and his devotion to his family.

*M. W. Feast*

## List of Participants

| | |
|---|---|
| Conny Aerts | kgafa01@cc2.kuleuven.ac.be |
| Michael Albrow | mda@saao.ac.za |
| Elio Antonello | elio@astmim.mi.astro.it |
| Shimon Asida | shimon@bilbo.fiz.huji.ac.il |
| Nathalie Audard | audard@obs.aau.dk |
| Dominique Ballereau | ballereau@mesioa.obspm.fr |
| Luis Balona | lab@saao.ac.za |
| Katalin Barlai | barlai@rmki.kfki.hu |
| Thomas G. Barnes | tgb@astro.as.utexas.edu |
| Jean-Philippe Beaulieu | beaulieu@iap.fr |
| Tim Bedding | tbedding@hq.eso.org |
| Juan Antonio Belmonte | jba@iac.es |
| Gaetano Belvedere | gbelvedere@astrct.ct.astro.it |
| Leonid Berdnikov | berdnik@sai.msu.su or @sai.msk.su |
| David Bersier | bersier@scsun.unige.ch |
| Giuseppe Bono | bono@oat.ts.astro.it |
| Paul A Bradley | pbradley@lanl.gov |
| Michel Breger | breger@astro.ast.univie.ac.at |
| Eduardo Brieva | dc33035@unalcol.unal.edu.co |
| David Buckley | dibnob@saao.ac.za |
| Gilbert Burki | burki@scsun.unige.ch |
| Maurizio Busso | busso@astto2.to.astro.it |
| R. Paul Butler | paul@further.berkeley.edu |
| Carla Cacciari | cacciari@astbo3.bo.astro.it |
| John Caldwell | jac@saao.ac.za |
| Jessica Chapman | jmc@aaoepp.aao.gov.au |
| Cesare Chiosi | chiosi@astrpd.pd.astro.it |
| Jörgen Christensen-Dalsgaard | jcd@obs.aau.dk |
| Geoff Clayton | gclayton@fenway.colorado.edu |
| Gisella Clementini | gisella@astbo3.bo.astro.it |
| Kem H. Cook | kcook@llnl.gov |
| Peter Cottrell | p.cottrell@csc.canterbury.ac.nz |
| A W J Cousins | post@saao.ac.za |
| Jan Cuypers | jan@astro.oma.be |
| Cecil G Davis | cgj@lanl.gov |
| Patrick de Laverny | laverny@graal.univ-montp2.fr |
| Wojciech Dziembowski | wd@camk.edu.pl |
| Chris Engelbrecht | engelca@risc2.unisa.ac.za |
| Nancy Remage Evans | evans@nereid.sal.ists.ca |
| Michael Feast | mwf@uctvax.uct.ac.za |
| Don Fernie | fernie@astro.utoronto.ca |
| Søren Frandsen | srf@obs.aau.dk |
| Janos Gál | j.gal@physx.u-szeged.hu |
| Alfred Gautschy | gautschy@astro.unibas.ch |

| | |
|---|---|
| Mike Gaylard | mike@bootes.hartrao.ac.za |
| Wolfgang Gieren | wgieren@astro.puc.cl |
| Ron Gilliland | gillil@stsci.edu |
| Ian Glass | isg@saao.ac.za |
| Peter Goldreich | pmg@nicholas.caltech.edu |
| Philip Goode | pgoode@tesla.njit.edu |
| Marie Jo Goupil | goupil@mesiob.obspm.fr |
| Martin Groenewegen | groen@iap.fr |
| Joyce Guzik | joy@beta.lanl.gov |
| Harm Habing | habing@strw.leidenuniv.nl |
| Gerald Handler | handler@astro.ast.univie.ac.at |
| Chris Haniff | cah@mrao.cam.ac.uk |
| Tony Hearn | ahearn@fys.ruu.nl |
| Jan Hers | jhers@docnet.infolink.co.za |
| Ken Hinkle | hinkle@noao.edu |
| Scott Horner | horner@astro.psu.edu |
| Shaun Hughes | hughes@mail.ast.cam.ac.uk |
| Mikołaj Jerzykiewicz | mjerz@ii.uni.wroc.pl |
| Eiji Kambe | kambe@astro.ubc.ca |
| Shashi M Kanbur | shashi@astro.gla.ac.uk |
| Lex Kaper | lkaper@eso.org |
| Steven Kawaler | sdk@iastate.edu |
| Franz Kerschbaum | kerschbaum@astro.ast.univie.ac.at |
| Dave Kilkenny | dmk@saao.ac.za |
| Chris Koen | ck@saao.ac.za |
| Geza Kovács | kovacs@buda.konkoly.hu |
| Kevin Krisciunas | kevin@jach.hawaii.edu |
| Jerzy Krzesinski | sfkrzesi@cyf-kr.edu.pl |
| Friedrich Kupka | kupka@galileo.ast.univie.ac.at |
| Don Kurtz | dkurtz@uctvax.uct.ac.za |
| Rainer Kuschnig | kuschnig@pluto.ast.univie.ac.at |
| Henny Lamers | hennyl@sron.ruu.nl |
| Dave Laney | cdl@saao.ac.za |
| Warrick Lawson | wal@phadfa.ph.adfa.oz.au |
| Umin Lee | lumin@physique.ens-lyon.fr |
| Holger Lehmann | vhl@rz.uni-jena.de |
| Tom Lloyd Evans | tle@saao.ac.za |
| Jan Lub | lub@strw.leidenuniv.nl |
| Tony Lynas-Gray | aelg@astro.ox.ac.uk |
| Andre Maeder | maeder@scsun.unige.ch |
| Luciano Mantegazza | mantegazza@axppv.pv.infn.it |
| Peter Martinez | peter@uctvax.uct.ac.za |
| Janet A Mattei | aavso@cfa.harvard.edu |
| Jaymie Matthews | matthews@astro.ubc.ca |
| Rodney Medupe | mdprod01@uctvax.uct.ac.za |
| Marie-Odile Mennessier | menes@graal.univ-montp2.fr |
| John Menzies | jwm@saao.ac.za |
| Siobahn Morgan | smorgan@nitro9.earth.uni.edu |
| Paweł Moskalik | pam@alfa.camk.edu.pl |

| | |
|---|---|
| Emma Nasi | nasi@astrpd.pd.astro.it |
| James M Nemec | nemec@dao.nrc.ca |
| Darragh O'Donoghue | dod@uctvax.uct.ac.za |
| Gabriel Pajdosz | sfpajdos@cyf-kr.edu.pl |
| Jan Willem Pel | pel@ksw.rug.nl |
| John Percy | jpercy@credit.erin.utoronto.ca |
| Jørgen Petersen | oz@astro.ku.dk |
| Jörg Pfleiderer | joerg.pfleiderer@uibk.ac.at |
| Karen Pollard | krp@saao.ac.za |
| Frederic Pont | pont@scsun.unige.ch |
| Ennio Poretti | poretti@astmim.mi.astro.it |
| Andy Reid | ahnr@star.ucl.ac.uk |
| Paul Retief | fissjpr@pukrs1.puk.ac.za |
| Eloy Rodríguez | eloy@iaa.es |
| Ian Roxburgh | i.w.roxburgh@qmw.ac.uk |
| Georgij Rudnitskij | gmr@sai.msk.su |
| Dimitar Sasselov | sasselov@cfa.harvard.edu |
| Andries Schutte | schutte@unizul1.uzulu.ac.za |
| Hiromoto Shibahashi | shibahashi@dept.astron.s.u-tokyo.ac.jp |
| Giovanni Silvestro | silvestro@to.infn.it |
| Myron Smith | msmith@iuegtc.gsfc.nasa.gov |
| Michael Snowden | michael@serendib.im.lk |
| Fatma Soufi | soufi@evris1.obspm.fr |
| Jürgen Staude | jstaude@aip.de |
| Stanislav Štefl | sstefl@sunstel.asu.cas.cz |
| Martin Stift | stift@astro.ast.univie.ac.at |
| Bob Stobie | rss@saao.ac.za |
| Jesper Storm | jstorm@eso.org |
| Denis Sullivan | denis.sullivan@vuw.ac.nz |
| Laszlo Szabados | szabados@ogyalla.konkoly.hu |
| Karoly Szatmary | k.szatmary@physx.u-szeged.hu |
| Terry Teays | teays@iuegtc.gsfc.nasa.gov |
| Olga Tsiopa | tsiopa@gaoran.spb.su |
| Yitzchak Tuchman | tuchma@vax.fiz.huji.ac.il |
| David Turner | turner@hercules.stmarys.ca |
| Catherine Turon | turon@obspm.fr |
| Igor Usenko | root@astro.odessa.ua |
| Tim van Hoolst | kgafa06@cc2.kuleuven.ac.be |
| Jozsef Vinko | vinko@physx.u-szeged.hu |
| Sergei Vorontsov | s.v.vorontsov@qmw.ac.uk |
| Christoffel Waelkens | fgafa01@cc1.kuleuven.ac.be |
| Alistair Walker | walker@ctiow3.ctio.noao.edu |
| Brian Warner | warner@physci.uct.ac.za |
| Amelia Wehlau | afwehlau@uwo.ca |
| William Wehlau | whwehlau@uwovax.uwo.ca |
| Werner W Weiss | weiss@astro.ast.ac.at |
| Douglas L Welch | welch@physics.mcmaster.ca |
| Klaus Werner | supas075@astrophysik.uni-kiel.d400.de |

| | |
|---|---|
| Marion West | marion@bootes.hartrao.ac.za |
| Patricia Whitelock | paw@saao.ac.za |
| Peter Wood | wood@mso.anu.edu.au |
| Atara Yaari | atara@bilbo.fiz.huji.ac.il |
| Albert Zijlstra | azijlstr@eso.org |
| E J Zita | zita@ac.grin.edu |
| Endre Zsoldos | zsoldos@ogyalla.konkoly.hu |
| Stanislaw Zola | sfzola@cyf-kr.edu.pl or zola@oa.uj.edu.pl |

*Astrophysical Applications of Stellar Pulsation*
*ASP Conference Series, Vol. 83, 1995*
*R. S. Stobie and P. A. Whitelock, eds.*

# Population I Stellar Structure and Evolution: facing the lingering difficulties to make a step forward

André Maeder

*Geneva Observatory, CH–1290 Sauverny, Switzerland*

## 1. Introduction

The understanding of stellar structure, evolution and pulsations is a key piece of the proper understanding of the Universe. The interpretation of the luminosity and spectra of galaxies, the properties of star populations and of starbursts, the calibration of distances, the nucleosynthetic prescriptions etc... are all requiring good models of stellar structure and evolution. The many interesting and new results in the field have been presented in several recent reviews by Maeder (1991), Chiosi et al. (1992), Maeder & Conti (1994). Table 1 below presents a list of the recent grids of stellar models. The various columns give the author's name, then the mass range, the metallicity Z range, the kind of convective assumption adopted, where "Schw." means Schwarzschild's criterion, "oversh" means with convective overshooting, "all" means that different assumptions have been used. In the last column, a few remarks are given. It is interesting to note that the models mentioned above often arrive to very different conclusions about the kind of mixing supported by the observations. Authors of models with overshooting generally conclude that overshooting is best supported, authors of models without overshooting favour the absence of overshooting, while authors who consider various convective assumptions generally favour semiconvection (Stothers & Chin 1991; Mowlawi & Forestini 1994).

Rather than reviewing all the arguments for and against some specific convective assumptions I shall examine here the various discrepancies between stellar models and observations. Hopefully, from a good appraisal of the existing discrepancies we may learn more about the future improvements needed in model physics, in particular regarding mixing. This is not of minor consequences since all model outputs depend on specific mixing assumptions.

## 2. Extended cluster main sequences

It has been known for a long time that the upper main sequences of clusters in the age range from that of Pleiades to M67 cannot be well fitted with stellar models using Schwarzschild's criterion (cf. Maeder & Meynet 1989). The observed sequence goes to a larger distance from the ZAMS than predicted. A comparison of the HR diagrams of 30 clusters with isochrones (cf. Meynet et al. 1993) well confirms the need of some core extension to properly account for the upper main sequences of clusters. Some cases, like that of NGC 4651, are beautiful illustrations of the necessity of enlarged cores. For very young clusters, such an effect also seems present (see §7); however the sparse nature of the sequence does not allow us to be very affirmative about the top of the main sequence.

The extended cluster sequences have generally been interpreted in terms of convective cores extended by overshooting. An increase of the core by about 0.20 $H_p$ is necessary, where $H_p$ is the local pressure height. It is also quite clear that, even with the new OPAL opacities, some core extension is necessary (cf. Meynet et al. 1992). Other features in the HR diagrams of clusters also support the same conclusion (cf. Maeder & Meynet 1989; Lattanzio et al. 1991; Brocato et al. 1994); however, they are generally less straightforward in interpretation and could be partly attributed to other effects.

Actually one can be rather confident that some small core increase with respect to what is predicted by Schwarzschild's criterion is necessary. However, the origin is certainly uncertain, it could be overshooting, but any additional mixing process in the deep interior would do the same.

## 3. N/C and $^{13}C/^{12}C$ enrichments on the Red Giant Branch

The internal distribution of chemical elements built during MS evolution is reached by the deep external convective zone in the red giant branch and this produces the first dredge-up of chemical elements. The changes of N/C as a function of luminosity on the red giant branch have been calculated for stars of different initial masses by Charbonnel (1994). Similar curves are obtained, for example, for the abundance ratio $^{13}C/^{12}C$. One notices that in principle, for sufficiently accurate measurements, one would be able to have some mass estimate of red giants in the field.

Various observational comparisons (cf. Sweigart et al. 1989; Gilroy & Brown 1991; Charbonnel 1994) well support the theoretical results for the first dredge-up: the luminosity and amplitude of the changes in N/C and $^{13}C/^{12}C$ are correctly predicted. Incidentally, this shows that there is no need of a large overshooting below the convective envelopes as assumed in some models. Also, the results on helioseismology (cf. Gilliland, this meeting) show a good correspondence of the calculated and "observed" lower boundary of the solar convective zone. Comparisons for low-mass stars ($M < 2\ M_\odot$) on the upper red giant branch show, however, significant discrepancies: N/C and $^{13}C/^{12}C$ are higher than predicted by about a factor of 3 (see ref. above). This suggests (cf. Charbonnel 1994) the occurrence of some extra-mixing appearing on the upper red giant branch. Clues are given that the extra-mixing needed to account for the CNO observations is only efficient when the hydrogen-burning shell reaches the chemical discontinuity created by the convective envelope at its maximum extent.

## 4. Abundances in Cepheids and AGB stars

There are not enough data on abundances in Cepheids to allow a correct comparison of observations and models. However, it is interesting to emphasize that the Cepheid models predict different surface He-contents and N/C according to metallicity Z and masses (cf. Schaller et al. 1992). In general, the higher the mass, the larger the enrichments since convective dredge-up is more efficient.

Fig. 1 shows the HR diagram for Z = 0.001 with indications of the Cepheid instability strip and of the regions of slow He-burning for Z = 0.001 and Z = 0.020. One notices that the intersections of the lines of slow He-burning with the instability strip occur at different L, a well known fact which is responsible for the different distributions of luminosities and periods according to the metallicity of the galaxy (cf. Iben & Tuggle 1975). However, the new point is that the surface He-content $Y_s$ is expected to be much lower at lower Z: $Y_s$ = 0.243 at Z = 0.001 compared to $Y_s$ = 0.33 at Z = 0.020. This is rather different from the value $Y_s$ = .29 found by Iben & Tuggle for Cepheids at various Z. Simultaneously, the N/C ratio should be of the order of 0.7 to 1 in Cepheids compared to a value of 4 for the cosmic abundances. The large helium differences result from the difference of the dredge-up as well as from the initial values. The difference in helium content should have some consequences for the driving mechanism and thus for the amplitudes of Cepheid pulsations at different Z. Such differences of amplitude distributions between the Galaxy, LMC and SMC are known to exist (cf. van Genderen 1978) and we may suspect a relation with the different helium contents.

Table 1. Recent grids of models (only the most recent reference in each series is given)

| GRIDS | M | Z | mix. | REMARKS |
|---|---|---|---|---|
| **Brocato & Castellani** 1993 | 0.6 – 20 $M_\odot$ | .002 – .02 | SCHW | • Laol Opacities<br>• Concl.: no overshoot |
| **Claret & Gimenez** 1992 | 1 – 40 $M_\odot$ | .01 – .03 | OVERSH | • Structure constants for binaries |
| **Geneva (Meynet et al.)** 1994 | 0.8 – 120 $M_\odot$ | .001 – 0.04 | OVERSH | • Isochrones<br>• Various $\dot{M}$; WR<br>• Chemistry; pulsation |
| **Lattanzio** 1991 | 1 – 5 $M_\odot$ | .001 – .02 | SCHW | • Laol Opacities<br>• From ZAMS, ZAHB |
| **Mowlawi & Forestini** 1994 | 2.5 – 20 $M_\odot$ | .02 | ALL | • Pre – MS evolution<br>• Concl.: semiconvection |
| **Padova (Bertelli et al.)** 1994 | 0.6 – 120 $M_\odot$ | .0004 – .10 | OVERSH | • Isochrones<br>• Population models<br>• Chemistry |
| **Stothers & Chin** 1992 | 1.5 – 60 $M_\odot$ | .02 – .03 | ALL | • Tests<br>• Concl.: no or small overshooting |
| **VandenBerg** 1992 | 0.15 – 1.25 $M_\odot$ | .0001 – .007 | SCHW | • Laol Opacities<br>• O – enhanced for low Z |
| **Vassiliadis & Wood** 1993 | 0.89 – 5.0 $M_\odot$ | .001 – .016 | SCHW | • Laol Opacities<br>• Through TP – AGB phase |

Recent models of TP-AGB stars by Vassiliadis & Wood (1993) use a new algorithm for the mass loss rates $\dot{M}$, based on the observed relation between $\dot{M}$

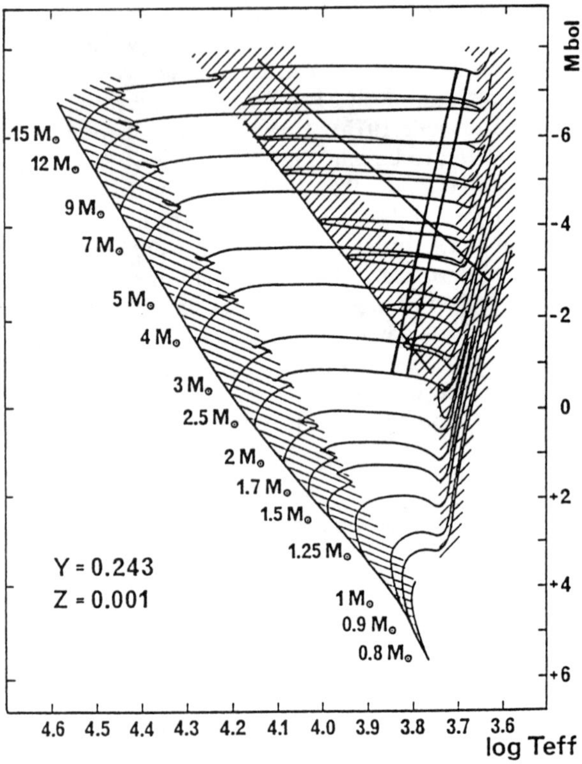


Figure 1. Grids of models at Z=0.001 and Y=.234 (from Schaller et al. 1992) with indication of the Cepheid instability strip. The line connecting the extremity of the blue loops for models with Z=0.02 and Y=0.28 is also given. The surface helium content for stars on the blue loops are very different: $Y_s$=0.243 at Z=0.001 and $Y_s$=0.33 at Z=0.02.

and the pulsation periods of Miras. These models correctly predict the luminosity of the RGB tip as well as the final vs. initial mass relation. However, the long standing difficulty of the models to produce C-stars at a luminosity as low as that observed is still unsolved. Of course, some tricks such as changing the mixing length ratio or Z may help, but the problem of an earlier efficient transport of C to the stellar surface remains.

## 5. He and mass discrepancy in O-stars

The problem of rotational mixing has a long history (cf. Tassoul 1978; Zahn 1994). In this context, the recent finding of Herrero et al. (1992) that all fast rotators among O-stars show surface He-enrichments is a very important result. It shows that rotationally induced mixing is strong enough to transport to the surface the newly processed elements during a fraction of the MS lifetimes.

This implies that all stellar properties are influenced by this effect: lifetimes, luminosities, nucleosynthesis, etc... In §10 below we try to explain why such effects preferentially occur in the most massive stars. We may also mention the so-called ON-stars. These are O-type stars, often close to the zero-age sequence and which show He- and N-enrichments (Walborn 1988; Howarth & Prinja 1989). A substantial fraction (about 50%) of them are binaries and some are also fast rotators. Their location in the HR diagram suggests (cf. Maeder 1987) the possibility of homogeneous evolution as a result of tidal or rotational mixing.

Another discrepancy was also claimed for O-stars: the so-called mass discrepancy (cf. Kudritzki et al. 1992; Herrero et al. 1992). The claimed effects were very large: a 50 $M_\odot$ star with a mass value assigned from the luminosity (evolutionary mass) might have a spectroscopic mass of 20 $M_\odot$. Doubts were cast on the reality of this discrepancy by Lamers & Leitherer (1993) and more recently by Schaerer (1995) who showed that the mass discrepancy was mainly due to the fact that the spectroscopic masses were based on atmosphere models, which did not account for the spherical extension of stars. This explains why the discrepancy was found to be larger for supergiants and Of stars than for MS stars. We may also notice that if there is some He-enrichment in O-stars, there should normally be some change in the M-L relation. However, in view of the observed He-enrichments, a 45 $M_\odot$ star, say, could possibly shine like a standard model of 50 $M_\odot$, but a 20 $M_\odot$ star not like one of 50 $M_\odot$.

## 6. He- and N-excesses in OBA supergiants and in SN 1987 A

For the time being the most important result in the field is still that by Walborn (1976, 1988) who showed that ordinary OB supergiants have He- and N-enrichments as a result of CNO processing, and that it is only the small group of peculiar OBC stars that have the normal cosmic abundance ratios. Several other recent studies fully confirm the above conclusions (cf. Howarth & Prinja 1989; Herrero et al. 1992; Gies & Lambert 1992). Also, B-supergiants in the LMC and SMC generally show He- and N-enrichments. A high-dispersion study of LMC B-supergiants by Fitzpatrick & Bohannan (1993, and reference therein) found among 62 B0.7 to B3 supergiants only seven OBC stars, while the remainder showed evidence of strongly contaminated surfaces by CNO processing. The progenitor of SN 1987 A, which was a B-supergiant, also had N/C and N/O ratios larger than solar values by factors 37 and 12 respectively (Fransson et al. 1989); however, in this special case the B-supergiant was obviously at a more advanced stage than most B-supergiants on the average. Anyhow, there are enough results to show that the general rule among B-supergiants is He- and N-enrichments, and this places very severe constraints on the models.

There are two alternatives to explain the He- and N-enrichments in supergiants and both bring about new problems.

I) The first alternative is that the blue supergiants are on the blue loops after a first red supergiant stage, where dredge-up has occurred producing the observed surface enrichments. The problems related to this hypothesis are the following: 1. At Z = 0.02, current grids of models (Schaller et al. 1992; Alongi et al. 1993; Arnett 1991; Brocato & Castellani 1993) only predict blue loops for masses equal or lower than 15 $M_\odot$. At higher masses there are no blue loops and no predicted enrichments. 2. The blue loops, when they exist (for $M \leq 15\ M_\odot$), do not go leftwards enough to reach continuity with the main sequence and this creates the so-called blue Hertzsprung gap (see below). 3. Even on the blue loops the predicted enrichments do not seem high enough to account for the observations. This was shown for three A-supergiants (Venn 1993), for four F-supergiants (Luck & Lambert 1985) and also in the case of SN 1987 A. II) A possibility is that some blue supergiants are still on the MS. This is not predicted by current stellar models. However, we may assume that the effects producing enriched O-stars (see §5) are also active in lighter lower mass stars and that they lead to surface enrichments in supergiants resulting from a larger range of initial masses.

This case of supergiants shows that we still need to understand their evolutionary stage and that, in any case, additional mixing is necessary to explain their surface abundances.

## 7. The blue Hertzsprung gap and the ledge

As mentioned above, the observed distribution of massive stars in the HR diagram shows a continuous distribution from the MS to the blue and yellow supergiants, with no sign of the predicted gap at the end of the MS (Stothers & Chin 1977; Chiosi et al. 1978; Meylan & Maeder 1982; Chiosi et al. 1992). Several different possible explanations have been proposed in terms of enhanced opacities (Nasi & Forieri 1990), of scatter in mass loss rates (Meynet et al. 1994), extended atmospheres and binaries (Tuchmann & Wheeler 1989, 1990). Our feeling is that no good solution has been found for this problem yet and that new evolutionary tracks are needed to account for this important fact.

A more or less visible drop-off in the number of LMC supergiants beyond a line somehow parallel to the MS, but shifted by $\sim$ 0.6 dex in log $T_{eff}$ was claimed by Fitzpatrick & Garmany (1990) who called it "the ledge". I am not convinced that this is a very constraining effect since most models predict a rather fast evolution of supergiants between log $T_{eff} = 3.9$ and 3.6; thus the ledge is a rather unclear and not constraining feature.

## 8. B/R ratios in galaxies

The number ratio B/R of blue to red supergiants is an increasing function of the metallicity Z in galaxies, typically increasing by a factor of 10 from the SMC to the solar neighbourhood (cf. Humphreys & McElroy 1984; Langer & Maeder 1995). Indeed, there is generally no problem for a given set of stellar models to fit the observed B/R at a given Z, since B/R at 25-40 $M_\odot$ typically changes from infinity in case of no mass loss to zero in case of high mass loss. However, it is much more difficult to reproduce the observed decrease of B/R with decreasing Z. As shown by Langer & Maeder (1995), no assumption in stellar model physics explored so far is able to cope with this trend. All models do the opposite! In general, models with Schwarzschild's criterion can more or less reproduce the observed B/R at $Z = 0.02$, but they really fail to do it at low Z, as for example in the SMC. There they predict no or few red supergiants, while many are observed. On the other hand, some models with Ledoux criterion correctly reproduce the observed B/R in the SMC (cf. Stothers & Chin 1992; Brocato & Castellani 1993), but they generally fail at Z = 0.01 or higher (see also Arnett 1991).

For supergiants there is often a stage close to neutrality between a blue and a red location and this is why any change in model assumptions can easily make the star shift from red to blue and vice-versa. Thus the test is a highly sensitive one. In that respect we notice that a restricted scheme for mixing (Ledoux) looks generally better at low Z, while an extended scheme might be better at high Z. This suggests some connection of the B/R problem with internal mixing.

## 9. The WR stages

Wolf-Rayet (WR) stars are nowadays considered as bare cores left over from massive stars by mass loss. Recent works have succeeded in explaining the large observed changes of the number ratios WR/O, WC/WN in galaxies of different metallicities (cf. Maeder & Meynet 1994). These changes are due to the larger mass loss rates at higher Z. Also, the models were able to explain the observed trend of WC9, WC8 and WC7 to be located in inner regions of the Galaxy. The interesting case of starbursts was also discussed (cf. Maeder & Conti 1994).

The sequence of Wolf-Rayet stars nicely corresponds to a sequence of more advanced chemical processing becoming visible at the stellar surfaces (L means late, E early).

| WNL | WNE | WCL | WCE | WO |
|---|---|---|---|---|
| *CNO equilibrium H present* | *CNO equilibrium* | *early He burning* | *advanced He-burning* | *more advanced* $O > C$ |

The standard models (cf. Schaller et al. 1992) predict an abrupt transition from WN to WC stars, since the He-core is growing and building up a steep chemical discontinuity at its outer edge. Thus the standard models predict almost no (< 1%) stars with intermediate characteristics of WN and WC stars. However, there are about 4% of such intermediate stars (Conti & Massey 1989). It has been shown by Langer (1991) that a mild mixing like that produced by semiconvective diffusion can account for these stars. The existence of WN/WC stars undoubtedly shows that an extra-mixing process is at work, even during the short He-burning phases.

## 10. Implications for stellar models

The various comparisons above tend to favour some additional mixing in stars which leads to more He and N brought to the stellar surface during evolution. These abundance excesses generally occur in post-MS evolution (red giants, C-star, supergiants, WR stars); however, in the case of rotating O-stars, the abundance excesses even appear during MS evolution. Indeed, all the above facts plead for a profound renewal in the physics of stellar models leading to a new generation of evolutionary grids.

The long standing debate between models with or without overshooting, with or without semiconvection and Ledoux criterion had led to no clear conclusions, some results supporting core overshooting while others support on the contrary restricted cores. The problem has been debated for about 25 years with no generally accepted conclusions.

The effects of mixing due to rotation have also to be considered. This is the case in recent works by Zahn (1992), where the effects of shears are found to be the most important ones. The horizontal shear generates a turbulence, which tends to limit the size of the horizontal shear and to generate a law of angular rotation velocity $\Omega = \Omega(r)$, thus 1-D models may still be meaningful. Secondly, this horizontal turbulence strongly limits the effect of meridional circulation, so that the main mixing effect results from the vertical shear $\frac{\partial \Omega}{\partial r}$. The vertical mixing occurs when $\frac{\partial \Omega}{\partial r}$ is larger than some critical value defined by Richardson's criterion. A new, more consistent form of this criterion has recently been given (Maeder 1995). For sufficiently high $\frac{\partial \Omega}{\partial r}$ the mixing effect can be treated as a diffusion with a coefficient

$$D_{shear} \simeq \frac{2}{5} K \left( \frac{\Omega}{N} \frac{\partial \ell n \Omega}{\partial \ell n r} \right)^2$$

Mixing thus grows with rotation, where $K$ is the radiative diffusivity $K = \frac{4acT^3}{3\kappa \rho^2 C p}$ and $N$ the Brunt-Väisälä frequency. At the same time a discussion of semiconvection and rotation (Maeder 1995) indicates that semiconvection can only exist for shears $\frac{\partial \Omega}{\partial r}$ smaller than that given by Richardson's criterion. Thus the following picture emerges:

**Low initial $\Omega$ and low $\partial\Omega/\partial r$ during evolution**

— Ledoux criterion, restricted convection with a mild semiconvective diffusion given by

$$D_{SC} = \frac{1}{4} K \frac{(\nabla_{rad} - \nabla_{ad})}{(\nabla_{ad} - \nabla + \frac{\Gamma+1}{\Gamma} \frac{\varphi}{\delta} \nabla_\mu)}$$

where $\nabla_{rad}$ and $\nabla_{ad}$ are the radiative and adiabatic gradients respectively, $\Gamma$ the Peclet number ($\lesssim 0.1$), $\varphi$ and $\delta$ thermodynamic coefficients and $\nabla_\mu = (\frac{\partial \ell n \mu}{\partial \ell n P})_{P,T}$.

**High initial $\Omega$ and high $\partial\Omega/\partial r$ during evolution**

— Core extended by the effect of $D_{shear}$ and surrounded by a region of mild mixing, depending on the degree of shear. To some extent the strong shear mixing at the edge of the core may mimic some effects of overshooting. To a first order, in the case of no magnetic coupling between the star and the matter lost in stellar winds, Zahn's results (1992) indicate a local conservation of the angular momentum. Thus the degree of shear $\partial\Omega/\partial r$ during evolution directly results from the value of the initial angular velocity $\Omega$.

Thus the above scheme suggests a change (growth) of the diffusion coefficient with the value of angular velocity. This influences the distribution of elements inside the stars and thus one has

$$TRACK = function\ (M,\ Z,\ \Omega)$$

It is likely that for given mass $M$ and metallicity $Z$ one no longer has a single evolutionary track (and related results), but a family of tracks depending on $\Omega$. The steep growth of $K$ with stellar mass suggests that such effects will be larger for more massive stars. The observed distributions of stars in the HR diagram and of surface abundances will result from a mixture of tracks of different rotation velocities. Such a mixture of tracks might also well explain the absence of the blue Hertzsprung gap (cf. §7).

On the whole we conclude that the collection of lingering discrepancies shown here demands a new generation of stellar evolutionary models, in which more internal mixing is present. The account for rotational effect is a good line to be explored.

## References

Alongi, M., Bertelli, G., Bressan, A., Chiosi, C., Fagotto, F., et al. 1993, A&AS, 97, 851

Arnett, D. 1991, ApJ, 383, 295

Bertelli, G., Bressan, A., Chiosi, C., Fagotto, F., & Nasi, E. 1994, A&AS, 106, 275

Brocato, E., & Castellani, V. 1993, ApJ, 410, 99

Brocato, E., Castellani, V., & Piersimoni, A.M. 1994, A&A, 290, 59

Charbonnel, C. 1994, A&A, 282, 811

Chiosi, C., Bertelli, G., & Bressan, A. 1992, ARA&A, 30, 235

Chiosi, C., Nasi, E., & Sreenivasan, S.R. 1978, A&A, 63, 103

Claret, A., & Gimenez, A. 1992, A&AS, 96, 255

Conti, P., & Massey, P. 1989, ApJ, 337, 251

Fitzpatrick, E.L., & Bohannan, B. 1993, ApJ, 404, 734

Fitzpatrick, E.L., & Garmany, C.D. 1990, ApJ, 363, 119

Fransson, C., Cassatella, A., Gilmozzi, R., Kirshner, R.P., Panagia, N., et al. 1989, ApJ, 336, 429

Gies, D.R., & Lambert, D.L. 1992, ApJ, 387, 673

Gilroy, K.K., & Brown, J.A. 1991, ApJ, 371, 578

Herrero, A., Kudritzki, R.P., Vilchez, J.M., Kunze, D., Butler, K., & Haser, S. 1992, A&A, 261, 209

Howarth, I.D., & Prinja, R.K. 1989, ApJS, 69, 527

Humphreys, R.M., & McElroy, D.B. 1984, ApJ, 284, 565

Iben, I., & Tuggle, R.S. 1975, ApJ, 197, 39

Kudritzki, R.P., Hummer, D.G., Pauldrach, A.W.A., Puls, J., Najarro, F., & Imhoff, J. 1992, A&A, 257, 655

Lamers, H.J.G.L.M., & Leitherer, C. 1993, ApJ, 412, 771

Langer, N. 1991, A&A, 248, 531

Langer, N., & Maeder A. 1995, A&A, in press

Lattanzio, J.C. 1991, ApJS, 76, 215

Lattanzio, J.C., Vallenari, A., Bertelli, G., & Chiosi, C. 1991, A&A, 250, 340
Lennon, D.J. 1994, Space Sci. Rev., 66, 127
Luck, R.E., & Lambert, D.L. 1985, ApJ, 298, 782
Maeder, A. 1987, A&A, 178, 159
Maeder, A. 1991, QJRAS, 32, 217
Maeder, A. 1995, A&A, in press
Maeder, A., & Conti, P. 1994, ARA&A, 32, 227
Maeder, A., & Meynet, G. 1989, A&A, 210, 155
Maeder, A., & Meynet, G. 1994, A&A, 287, 803
Meylan, G., & Maeder, A. 1982, A&A, 108, 148
Meynet, G., Maeder, A., Schaller, G., Schaerer, D., & Charbonnel, C. 1994, A&A, 281, 638
Meynet, G., Mermilliod, J.C., & Maeder, A. 1993, A&AS, 98, 477
Mowlawi, N., & Forestini, M. 1994, A&A, 282, 843
Nasi, E., & Forieri, C. 1990, Ap&SS, 166, 229
Schaerer, D. 1995, Thesis, Univ. Geneva
Schaller, G., Schaerer, D., Meynet, G., & Maeder, A. 1992, A&AS, 96, 269
Stothers, R.B., & Chin, C.W. 1977, ApJ, 211, 189
Stothers, R.B., & Chin, C.W. 1991, ApJ, 374, 288
Stothers, R.B., & Chin, C.W. 1992, ApJ, 390, 136
Sweigart, A.V., Greggio, L., & Renzini, A. 1989, ApJS, 229, 624
Tassoul, J.L. 1978, Theory of rotating stars, Princeton Series in Astroph., p. 3
Tuchman, J., & Wheeler, J.C. 1989, ApJ, 344, 835
Tuchman, J., & Wheeler, J.C. 1990, ApJ, 363, 255
VandenBerg, Don A. 1992, ApJ, 391, 685
van Genderen, A.M. 1978, A&A, 65, 147
Vassiliadis, E., & Wood, P.R. 1993, ApJ, 413, 641
Venn, K.A. 1993, ApJ, 414, 316
Walborn, N.R. 1976, ApJ, 205, 419
Walborn, N.R. 1988, in Atmospheric Diagnostics of Stellar Evolution, IAU Coll. 108, Ed. K. Nomoto, Springer Verlag, Berlin, p. 70
Zahn, J.P. 1992, A&A, 265, 115
Zahn, J.P. 1994, Space Sci. Rev., 66, 285

## Discussion

*Roxburgh*: First, I should stress that the initial magnetic field is also an independent parameter which along with mass, composition and rotation determine the evolution of stars. Next, I wish to stress that the use of the mixing length model to model stellar convective zones is a substantial source of error - since it is not a theory but a model with tunable parameters - and we do not know how to tune the parameters except by either fitting the Sun and applying the same value over a wide range of surface parameters - or by adjusting the mixing length to fit each individual observation. But there is at least some hope that numerical simulations can give us better insight into both surface convective zones and convective overshoot in the central regions.

*Evans*: I might add to your list of problems which the inclusion of rotation may solve. At present up to half the binary Cepheids cannot be fit with isochrones (without rotation). These are binaries with orbital periods longer than a year, so single star evolution should be appropriate.

*Maeder*: You are quite right, the fact that some binaries, Cepheids or not, do not fit the same isochrone was long recognised as a problem. This is really a point which deserves close attention in future.

*Turner*: What has changed in recent years to make the treatment of rotation in stellar evolutionary models more tractable (in comparison, say, with your published results of about 10 years ago)?

*Maeder*: In the previous work of 1987, it was considered that the horizontal turbulence is the main source of vertical diffusion of chemical elements. Now, more recent works by Zahn (1992) show that the mixing effects of the vertical shears are even more important and that they add, to the first order approximation, their effects to those of horizontal turbulence.

*Nemec*: For many years there has been a problem understanding why in some globular clusters there are no double-mode RR Lyrae (RRd) stars (e.g., $\omega$ Cen, M5) whereas in other clusters there are many (e.g., IC4499, M15). Faster rotation in $\omega$ Cen and M5 has been advocated as a possible reason for this. Can you comment on this?

*Maeder*: The interpretation you are mentioning implies that there would be systematic differences in the distributions of rotational velocities (both the initial ones and those in RR Lyrae) for stars in globular clusters. At present there are no observations which permit one to decide on these matters.

*Astrophysical Applications of Stellar Pulsation*
*ASP Conference Series, Vol. 83, 1995*
*R. S. Stobie and P. A. Whitelock, eds.*

# Population II Stellar Structure and Evolution: A Few Selected Topics

Cesare Chiosi

*Astronomy Department, University of Padua, Vicolo Osservatorio 5, 35122 Padua, Italy*

**Abstract.** In this review we discuss in some detail the evolution of low mass stars, and other related topics, limiting ourselves to the class of stars with solar and super-solar metal content that likely constitute the basic component of elliptical galaxies and bulges. In particular, we deal with the stars responsible of the UV flux observed in elliptical galaxies and the role played by the chemical composition and mass loss by stellar wind in producing hot progenies with effective temperatures in the range indicated by the UV data. In this context, we briefly report on recent chemo - spectro - photometric models for elliptical galaxies. In addition to this, we discuss the subject of galaxy ages by means of the ultraviolet properties and narrow band indices. Finally, we briefly report on a new formulation of diffusive mixing in stellar interiors and its effects on the structure and evolution of intermediate- and high-mass stars.

## 1. Introduction

In this review, no attempt is made to cover all the topics that could be addressed by a report on the progress made in understanding the structure and evolution of population II stars. Recent reviews on the subject are by Iben & Renzini (1983, 1984), Hesser (1988), Renzini & Fusi-Pecci (1988), Iben (1991), VandenBerg (1991), Fusi-Pecci & Cacciari (1991), and Chiosi et al. (1992) to whom the reader should refer for all details. Rather we have selected a few topics on which most effort has been concentrated over the past few years. These are all related to the evolution of old, low-mass stars of high metal content. Stars of this type are likely to be the major constituents of bulges in nearby galaxies and elliptical galaxies in general, in addition to well known cases of metal-rich clusters in the solar vicinity and Galactic Bulge. The peculiarity of the high metallicity objects is that they do not follow the classical scheme in which the core He-burning phase (HB stars) gets redder and redder at increasing metallicity. Indeed, if a threshold metallicity is passed the core He-burning phase splits in two parts, one near the RGB and another one at high effective temperature (cf. Horch et al. 1992, Fagotto et al. 1994a,b,c). These stars are the right candidates to be the dominant source of UV radiation observed in elliptical galaxies (Burstein et al. 1988).

## 2. Observational hints on metallicities and ages

Direct determinations of metallicities in elliptical galaxies are still rather limited. Most of the information comes from metal line-strength indices and their radial variations (Carollo et al. 1993, Carollo & Danziger 1994, Davies et al. 1993) or colour gradients (Schombert et al. 1993). The evidence arises that the metallicities for elliptical galaxies are solar or larger. Furthermore, for bright ellipticals there are also indications that the $\alpha$-elements (O, Mg, Si, etc.) are enhanced with respect to Fe. The average [Mg/Fe], in particular, exceeds that of the most metal rich stars in the solar vicinity by about 0.2-0.3 dex, and the ratio [Mg/Fe] is expected to increase with the galactic mass up to this value. Due to their proximity, the metallicity of stars in the Galactic Bulge is less of a problem, because this is currently measured by spectroscopic methods (cf. Rich 1990, McWilliam & Rich 1994). The mean metallicity is slightly lower than solar and long tails towards the metal-poor ($[Fe/H] = -1$) and metal-rich end ($[Fe/H] = 1$) are also present. High metallicities seem also to be indicated by the color-magnitude diagrams (CMD) of stellar populations in the bulges of nearby Galaxies (M31 and M32). In the case of M32 (Freedman 1989, 1992), these stars are most likely in the RGB and AGB phases and metallicities as high as [M/H]=0.1 are possible (cf. Bica et al. 1990, 1991). Concerning the age, from the magnitude of the brightest AGB stars in M32, Freedman (1989, 1992) concludes that a fraction of these could be as young as 5 Gyr. This is also supported by the new calibration for the maximum AGB star luminosity as a function of the age and metallicity by Bressan et al. (1994, BCF). Similar conclusions would be obtained looking at the CMDs for M32 by Elston & Silva (1992) and Davidge & Jones (1992). Finally, deep photometry and CMDs of selected areas of the Galactic Bulge have been presented by Terndrup (1988), Frogel et al. (1990), Udalski et al. (1993a,b), Holtzman et al. (1993), Ng (1994), and Ng et al. (1995), and of clusters and companion fields by Ortolani et al. (1990, 1992, 1993). Since all these regions are observed across the Galactic Disk, a significant contamination by disk stars is present. Therefore, deciphering the information contained in those CMDs requires models for the star density distribution and laws of extinction along the line of sight. Despite the difficulty, Bertelli et al. (1995) studying the region named Terzan 1 ($l = 357.5; b = 1.0$), argue that stars with a large range of metallicity ($0.008 < Z < 0.08$) and ages from 15 to 12.5 Gyr are present, and more important that stars of the type named H-HB (see below) are perhaps visible.

## 3. UV excess in elliptical galaxies: the data

Much of the available information is from the study of Burstein et al. (1988). The main points are the following:
(1) All studied ellipticals have detectable flux short-ward of about 2000 Å.
(2) There are large galaxy to galaxy differences in the level of the UV flux. The intensity of the UV emission is measured by the colour (1550-V).
(3) The colour (1550-V) correlates with the index Mg2, the velocity dispersion $\log \Sigma$ and the luminosity (mass) of the galaxy. The few galaxies (e.g., NGC 205) in which active star formation is seen do not obey these relations.

(4) An important constraint is posed by the HUT observations by Ferguson et al. (1991) and Ferguson & Davidsen (1993) of the UV excess in the bulge of M31. In this case the UV emission shows a drop-off short-ward of about 1000 Å whose interpretation requires that the temperature of the emitting source must be about 25,000 K. Only a small percentage of the $912 \leq \lambda \leq 1200$ Å flux can be coming from stars hotter than 30,000 K and cooler than 20,000 K.

## 4. UV Excess in elliptical galaxies: the sources

Excluding ongoing star formation, the UV excess owes its origin to an old component that gets hot enough to power the integrated spectral energy distribution (ISED) of a galaxy in the far UV regions. Three possible candidates are envisaged (cf. Greggio & Renzini 1990, BCF):
(1) The classical post asymptotic giant branch (P-AGB) stars (see Bruzual 1992, Bruzual & Charlot 1993, Charlot & Bruzual 1991). However, even if a small drop-off can be predicted just long-ward of 912 Å they cannot be the sole source of UV flux because of their high mean temperature (about 100,000 K). Nevertheless, these stars have been detected with HST observations in the nucleus of M31 (Bertola et al. 1995) and can contribute as much as 50% to the UV light.
(2) Very blue horizontal branch (HB) stars of extremely low metallicity (Lee 1994). These stars have effective temperatures hotter than about 15,000 K but much cooler than those of the P-AGB stars. Therefore, depending on their actual effective temperature, they can generate ISEDs in agreement with the observational data. However, most likely they are not the dominant source of the UV flux because the analysis by BCF clarifies that in the wavelength interval $2000 < \lambda < 3000$ Å the ISEDs of the bulge of M31 and of elliptical galaxies like NGC 4649 are fully consistent with the notion that virtually no stars with metallicity lower than $Z = 0.008$ ought to exist in the mix of these stellar populations.
(3) Finally, the hot horizontal branch (H-HB) and AGB-manqué stars of very high metallicity (say $Z>0.07$) which are expected to be present albeit in small percentages in the stellar content of bulges and elliptical galaxies in general. Indeed, these stars have effective temperatures in the right interval and generate ISEDs whose intensity drops short-ward of about 1000 Å by the amount indicated by the observational data. (Ferguson et al. 1991, Ferguson & Davidsen 1993).

On the side of the stellar evolution theory, the appearance of the various types of UV sources is governed by several important physical factors, each of which is affected by a certain degree of uncertainty still far from being fully assessed. These are the efficiency of mass loss during the RGB and AGB phases, the enrichment law $\Delta Y/\Delta Z$, and finally for the specific case of P-AGB stars the detailed relation between the initial and final mass of the stars at the end of the AGB phase.

P-AGB stars are always present in the stellar mix of a galaxy. The major problem with these stars is the high effective temperature on which we have already reported, and the precise relation between their mass and the turn-off mass (and hence age). The initial-final mass relationship is not firmly established. The most popular empirical determination is by Weidemann (1987). Matching this relation depends on the efficiency of mass loss during the AGB phase and other details of model structure (for instance the initial size of the carbon-oxygen core). In general, relations steeper than Weidemann (1987) would predict more massive P-AGB stars and because of their lifetime smaller UV-fluxes at given turn-off mass. The UV flux increases as the stellar models approach the Weidemann (1987) relation. The response of the UV flux to details of initial-final mass relation (for instance its dependence on the metallicity) is so strong that firm conclusions cannot yet be reached. Suffice to recall that for ages older than about 10 Gyr, the whole problem is driven by the initial-final mass relation in the mass range 0.8 to 1.0 $M_\odot$.

The formation of H-HB and AGB-manqué with metallicity $Z \gg Z_\odot$ is possible either with low values of the enrichment ratio ($\Delta Y/\Delta Z \simeq 1$) and strong dependences of the mass-loss rates on the metallicity (Greggio & Renzini 1990) or even with canonical mass-loss rates and suitable enrichment laws ($\Delta Y/\Delta Z \simeq 2.5$) as in Horch et al. (1992) and Fagotto et al. (1994a,b,c). These topics are briefly examined below.

### 4.1. Does the mass loss rate from RGB stars increase with the metallicity ?

Studying the stars in the red clump of the old open cluster M67, Tripicco et al. (1993) argue that the rate of mass loss along the RGB increases with the metallicity above the value holding for Globular Clusters and predicted by the classical Reimers (1975) relation. Assuming the current estimates of metallicity (nearly solar), age (about 5 Gyr), and distance modulus ($(m-M)_o$=9.55) for M67, Tripicco et al. (1993) conclude that the mass of the giant stars in the red clump is about 0.7 $M_\odot$. Since at the above age, the masses at the turn-off and tip of the RGB are about 1.2 $M_\odot$ and 1.3 $M_\odot$, respectively, it follows that in clusters with nearly solar metallicity, much more mass ought to be lost by the red giant stars than classically predicted. However, Carraro et al. (1995) pointed out that in order to derive the mass of the red stars in the clump from their position in the CMD, the color difference $\Delta(B-V)$ between the turn-off and the base of the RGB should be known with a precision smaller than $\delta[\Delta(B-V)] = 0.01$ mag and, at the same time, the distance modulus should be determined with a precision higher than 0.025 mag. The analysis of the problem clarifies that in the global isochrone fitting technique, both the distance modulus and mass of the clump stars depend on the above colour difference according to the relations $\partial \mathrm{M_V^{TO}}/\partial\Delta(B-V) \simeq -9.2$ and $\partial\ \mathrm{M^{HB}}/\partial\Delta(B-V) \simeq -22\ \mathrm{M_\odot/mag}$. Therefore, an uncertainty in $\Delta(B-V)$ as small as $\delta[\Delta(B-V)] = 0.01$ mag implies an uncertainty in the distance modulus of $\delta(m-M)_o = 0.1$ mag and, even more important, an uncertainty in the mass of the clump stars of 0.22 $M_\odot$. Fits of the CMD of M67 are possible, in which the classical value of the mass for the stars in question is recovered. Therefore, the claim that the rate of mass loss from RGB stars increases with the metallicity is not supported by the present day data for M67. Similar remarks apply to the companion paper by Liebert et al. (1994) on NGC 6791.

### 4.2. Metal-poor, low-mass stars of very old age

Low-mass, metal-poor stars suffering normal mass loss during the RGB phase, at ages older than about 20 Gyr can undergo the core He-burning phase in the required range of effective temperatures (Lee 1994). The main difficulty with this possibility is the age itself which is much older than commonly assigned to globular clusters (13 ± 2 Gyr, Fusi-Pecci & Cacciari 1991). In addition to this, if the UV source resides in the HB stars of low metallicity, the question arises whether their luminosity and relative frequency are compatible with the observed ISED of elliptical galaxies in the range 2000 < $\lambda$3500 Å. Indeed, the analysis of BCF pointed out that the observed ISEDs hint that only very few stars with metallicity lower than about Z=0.008 ought to exist (the analogue of the G-Dwarf Problem in the solar vicinity). In contrast, as the luminosity of the HB stars of low metallicity is significantly fainter than the corresponding one of H-HB and AGB manqué stars (see Fagotto et al. 1994a,b,c), the relative number of such stars required to produce the total amount of UV flux could exceed the limit set by the ISED.

### 4.3. Enrichment Law $\Delta Y/\Delta Z$

The law of enrichment $\Delta Y/\Delta Z$ is highly uncertain (Pagel 1989, Pagel et al. 1992). Equally, for its temporal behaviour. With $\Delta Y/\Delta Z = 2.5$ (the solar case) and normal rates of mass loss for the RGB phase ($\eta \simeq 0.35 - 0.45$), the appearance of the first generation of H-HB and AGB manqué stars in single stellar population occurs at the age of about 7.6 Gyr. This occurs at even younger ages if $\Delta Y/\Delta Z$ is higher than 2.5 (see Fagotto 1994 for details).

## 5. New chemo-spectro-photometric models of elliptical galaxies

BCF and Tantalo et al. (1995, TCBF) constructed new chemo - spectro - photometric models of elliptical galaxies particularly designed to match the colour-magnitude relation (CMR), cf. Bower et al. (1992a,b), and to provide a robust explanation for the UV flux and its dependence on the galactic luminosity (and hence mass), the index $Mg_2$, and $\log \Sigma$. No details of these models are given here for the sake of brevity. They can be found in BCF and TCBF. Suffice to recall that BCF made use of the closed-box approximation, whereas TCBF adopted the infall description. In both cases the enrichment law is $\Delta Y/\Delta Z = 2.5$ and the mass-loss rates are normal ($\eta = 0.35$ in BCF and $\eta = 0.45$ in TCBF). Finally, the models allow for galactic winds halting star formation. Galactic winds are at the base of the current interpretation of the CMR for elliptical galaxies (Bower et al. 1992a,b).

The models by BCF and TCBF show that the H-HB and AGB-manqué stars are the most plausible candidates as dominant source of UV radiation. Their occurrence is possible in galaxies of sufficiently high mass, which in virtue of the very efficient star formation rate possess higher values both for the mean and peak metallicity. According to the BCF and TCBF models, a few percent of the mass in stars stored in the highest metallicity bins can generate the desired amount of UV flux. In contrast galaxies of lower mass and lower mean and peak metallicity in turn, are not able to generate this type of stars and their UV flux is simply produced by their PN component. A natural result of this scheme is that the correlation of the UV flux with the $Mg_2$ index and the velocity dispersion can be easily accounted for (see BCF and TCBF).

## 6. The UV Excess as an Age Indicator

BCF have suggested that the uprise in the UV flux at the onset of the H-HB and AGB manqué stars can be used as probe of galaxy ages. Since these stars are expected to appear at the rest-frame age of about 7.6 Gyr for $\Delta Y/\Delta Z = 2.5$, a sudden change in the (1550-V) colour is expected to be observable at relatively low red-shifts, perhaps reachable with the present day space instrumentation, thus allowing us to date a galaxy from the stellar clock running in the constituent stellar populations. The precise value of the red-shift corresponding to the above age depends on the particular model of the Universe in usage, namely the choice made for the Hubble constant $H_0$, the deceleration parameter $q_0$, and finally the red-shift of galaxy formation $z_{for}$. See BCF and TCBF for all details.

## 7. Is there a real age range among elliptical galaxies ?

The merge of the recent work on population synthesis in narrow band photometry ($H_\beta$, $Mg_2$, $< MgFe >$ indices) by Gonzales (1993), Worthey (1992), and Worthey et al. (1994) and the models for elliptical galaxies in broad band photometry by BCF and TCBF allows us to tackle the question whether or not elliptical galaxies span a large range of ages. This problem has been addressed by Gonzales (1993) by means of the evolutionary path of single stellar populations (SSP) in the so-called $H_\beta$ versus $< MgFe >$ plane. Galaxies are seen to fall in a rather narrow band comprised between evolutionary paths of SSPs with nearly solar metallicity (narrow range of $< MgFe >$) and intersect lines of different age. For the sake of illustration, M32 is at the young age side, while NGC4649 is at the old age side. Since the analysis was made using SSPs of given composition instead of galaxy models with ample range of metallicities, the above finding was subjected to the obvious criticism that in real galaxies it would not be possible to disentangle age from metallicity effects.

Bressan et al. (1995, BCT) included the narrow band photometry in the models for elliptical galaxies of BCF and TCBF and found that the galaxy paths run almost parallel to those of the SSPs and that all galaxies of the Gonzales (1993) sample are matched by models whose stellar contents have mean metallicity slightly above solar and may contain peak values up to Z=0.08. Analyzing the properties of the sample galaxies in the space of the parameters $H_\beta$, $< MgFe >$, (1550-V), $\log \Sigma$, and broad band colours such as (U-V), (B-V), (V-K), etc., BCT reached the following conclusions:

(1) The observed distribution of galaxies in the $H_\beta$ versus $< MgFe >$ plane is not a sequence of metallicity with bluer galaxies significantly more metal-poor than the red ones. Metallicity effects are expected to broaden the distribution mainly along the $< MgFe >$ axis.

(2) The observed distribution of galaxies in the $H_\beta$ versus $< MgFe >$ plane is not a mere sequence of age with bluer galaxies significantly younger than the red ones. Although some scatter in the age is possible, it is not likely to be as

large as indicated by the formal match of data with isochrones in this diagram. M32 in particular according to its $H_\beta$ should have an age of about 3 Gyr, in disagreement with its (1550-V)=4.5 for which much older ages are needed even considering all the uncertainties in the final-initial mass relationship.
(3) The observed distribution of galaxies in the $H_\beta$ versus $< MgFe >$ plane does not agree with the expectation from the CMR of Bower et al. (1992a,b) on the notion that it represents a mass-metallicity sequence of nearly coeval objects, in which star formation took place in an initial episode. Most likely, the history of star formation was more complicated than this simple scheme.
(4) The observed (1550-V) colours are not compatible with ages younger than about 7 Gyr. UV excesses stronger than (1550-V)=5.5 at younger ages are possible only if the age is younger that 1- 2 Gyr. However, in such a case the remaining UBV colours would be too blue compared with the observational data. Most likely, all galaxies in the sample are older than at least 7 Gyr.
(5) For the subgroup of galaxies whose (1550-V) colours are compatible with old ages (say about 13-15 Gyr), there seems to be a certain spread in age which roughly amounts to about 3 Gyr consistently with the small scatter of the CMR.
(6) With the kind of modelling in use, i.e., initial episode of star formation interrupted by galactic, the difficulties encountered with the interpretation of the $H_\beta$, $< MgFe >$, (1550-V), and $\log \Sigma$ data can be removed by invoking later episodes of star formation in a sort of burst mode. However, the major problem with this suggestion and the notion of coevality, is that some mechanism synchronizing the bursts of star formation should be required in order to dislocate galaxies from the locus in the $H_\beta$ versus $< MgFe >$ plane consistent with the CMR of coeval objects (CMR baseline) to their actual position. Indeed, even if the statistics is limited, no galaxies are seen along the CMR baseline but for those at the red end.
(7) There is a interesting scenario emerging from this analysis. In brief, galaxies of high velocity dispersion (and hence mass) tend to be confined to the red end of the $H_\beta$ versus $< MgFe >$ distribution. Their properties are consistent with being old objects (some scatter in the age is perhaps possible) that are able to exhaust the star forming process at very early epochs with no need for later episodes of stellar activity. In contrast, galaxies of lower velocity dispersion (and hence mass) have more contrasting properties. In these systems it seems as if star formation, following the initial activity most likely at epochs as old as in other galaxies, later went through a series of episodes taking place at different epochs that vary from galaxy to galaxy. Equivalently, it could be that star formation (in the nucleus) continued perhaps at minimal levels over significantly longer periods of time. This would mimic a sort of age sequence.
(8) The idea that the overall duration of the star forming activity is inversely proportional to the galaxy mass does not contradict the current information of abundance ratios in elliptical galaxies (Carollo et al. 1993, Carollo & Danziger 1994, Matteucci 1994). Indeed it would be consistent with the expected trend for the ratio [Mg/Fe] inferred from the narrow band indices according to which the average [Mg/Fe] in giant elliptical galaxies exceeds that of the most metal rich stars in the solar vicinity by about 0.2-0.3 dex. Furthermore, the ratio [Mg/Fe] is expected to increase with the galactic mass up to this value. Matteucci (1994) argues that a possible explanation of this trend could an efficiency

of star formation increasing with the galactic mass. As a consequence of this, massive ellipticals should have formed on shorter time scales than smaller ellipticals as suggested by the analysis of the parameters $H_\beta$, $< MgFe >$, (1550-V), and $\log \Sigma$.

## 8. A new model of diffusive mixing in stellar models

In this section we briefly report on a new formulation of diffusive mixing in stellar interiors proposed by Deng (1994) and Deng et al. (1995a,b). It is aimed to cast light on the kind of mixing that should take place in the so-called overshoot regions surrounding fully convective zones (inner cores, external envelopes and intermediate shells). In these layers, instead of assuming instantaneous and fully efficient mixing, a diffusive scheme is adopted. The algorithm makes use of the so-called *scale length most effective for mixing* expressed as $l_d = P_{dif} \times 10^{-5} l_0$, where $l_0$ is the largest scale in the unstable region in units of the local pressure scale height $H_P$, and finally $P_{dif}$ is a fine tuning parameter of the order of unity. Furthermore, it takes two important phenomena of turbulence into account. They are intermittence and stirring, whose occurrence is known from laboratory fluid dynamics (see Deng et al. 1995a,b for more details). The major difference with respect to standard stellar models is that, while fully unstable regions turn out to be completely homogenized by the diffusive algorithm in use, this is not the case for the overshoot regions which undergo partial mixing and build up smooth chemical profiles.

The analysis by Deng et al. (1994a) shows that $P_{dif} = 0.4$ leads to stellar models able to match a number of properties of massive and intermediate mass stars. The key feature of these models is that they possess at the same time evolutionary characteristics that are separately typical of models calculated with different schemes of mixing. In other words, they share the same properties of models with standard overshoot, namely a wider main sequence band, higher luminosity, and longer lifetimes, but also perform extended loops that are the main signature of the semiconvective description of convection at the border of the core.

With the aid of these stellar models, Deng et al. (1995b) analyze the CMDs of supergiant stars in the LMC (Fitzpatrick & Garmany 1990) and the Milky Way (Blaha & Humphreys 1989). Particular attention is paid to the star counts across the CMD, the many stars in the Hertzsprung-Russell gap and the related suspected widening of the main sequence band, the so-called *Ledge*, the ratio of blue to red supergiants $N_B/N_R$, the location of Wolf-Rayet stars (WR), and the blue progenitor of the SN1987A. It is found that taking into account a plausible scatter in the chemical composition of the supergiant stars in the samples, the many stars in the gap and the Ledge can be easily accounted for. However, problems remain as far as the number frequencies of stars among the various spectral types, the possibility that the main sequence band can extend to spectral types up to B3, the ratio $N_B/N_R$, and finally the location of WR stars in the CMD. It is suggested that a distinct evolutionary scenario must be invoked for this type of stars (Deng et al. 1995c). Finally, the blue progenitor of SN1987A is not matched by the models, even if clues are found within the same evolutionary scheme for a possible interpretation.

**Acknowledgments.** This study has been financed by the Italian Ministry of University, Scientific Research and Technology (MURST) and the Italian Space Agency (ASI).

## References

Bertelli, G., Bressan, A., Chiosi, C., et al. 1995, A&A, in press

Bertola, F., Bressan, A., Burstein, D., et al. 1995, ApJ, 438, 680

Bica, E., Alloin, D., & Schmidt, A. 1990, A&A, 228, 23

Bica, E., Barbuy, B., & Ortolani, S. 1991, ApJ, 382, L15

Blaha, C., & Humphreys, R. M. 1989, AJ, 98, 1598

Bower, R. G., Lucey, J. R., & Ellis, R. S. 1992a, MNRAS, 254, 589

Bower, R. G., Lucey, J. R., & Ellis, R. S. 1992b, MNRAS, 254, 601

Bressan, A., Chiosi, C., Fagotto, F. 1994, ApJS, 94, 63, BCF

Bressan, A., Chiosi, C., Tantalo, R. 1995, A&A, submitted, BCT

Bruzual, G. 1992, in The Stellar Populations of Galaxies, B. Barbuy & A. Renzini, Dordrecht: Kluwer Academic Publishers, 311

Bruzual, G., & Charlot, S. 1993, ApJ, 405, 538

Burstein, D., Bertola, F., Buson, L. M., Faber, S. M., & Lauer, T.R. 1988, ApJ, 328, 440

Carraro, G., Girardi, L., Bressan, A., & Chiosi, C. 1995, A&A, submitted

Carollo, C. M., & Danziger, I. J. 1994, ESO preprint no. 1003

Carollo, C. M., Danziger, I. J., & Buson, L. 1993, MNRAS, 265, 553

Charlot, S., & Bruzual, G. 1991, ApJ, 367, 126

Chiosi, C., Bertelli, G., & Bressan, A. 1992, ARA&A, 30, 235

Davidge, T. J., & Jones J. H. 1992, AJ, 104, 1365

Davies, R. L., Sadler, E. M., & Peletier, R. F. 1993, MNRAS, 262, 650

Deng, L., Bressan, A., & Chiosi, C. 1995a, A&A, submitted

Deng, L., Bressan, A., & Chiosi, C. 1995b, A&A, submitted

Deng, L., Bressan, A., & Chiosi, C. 1995c, A&A, submitted

Elston, R., & Silva, D. R. 1992, AJ, 104, 1360

Fagotto, F. 1994, in Stellar Populations, M.S.A.It., in press

Fagotto, F., Bressan, A., Bertelli, G., & Chiosi, C. 1994a, A&AS, 100, 647

Fagotto, F., Bressan, A., Bertelli, G., & Chiosi, C. 1994b, A&AS, 104, 365

Fagotto, F., Bressan, A., Bertelli, G., & Chiosi, C. 1994c, A&AS, 105, 39

Ferguson, H. C., & Davidsen, A. F. 1993, ApJ, 408, 92

Ferguson, H. C., Davidsen A. F., Kriss G. A., et al. 1991, ApJ, 382, L69

Fitzpatrick, E. L., & Garmany, C. D. 1990, ApJ, 363, 119

Fusi-Pecci, F., & Cacciari, C. 1991, in New Windows to the Universe, F. Sanchez & M. Vasquez, Cambridge: Cambridge Univ. Press, 364

Freedman, W. L. 1989, AJ, 98, 1285

Freedman, W. L. 1992, AJ, 104, 1349

Frogel, J. A., Terndrup, D. M., Blanco, V. M., & Whitford, A. E. 1990, ApJ, 353, 494

Gonzales, J. J. 1993, Ph.D. Thesis, Univ. California, Santa Cruz

Greggio, L., & Renzini, A. 1990, ApJ, 364, 35

Hesser, J. E. 1988, in Progress and Opportunity in Southern Hemisphere Optical Astronomy, V. M. Blanco & M. M. Philip, Astron. Soc. Pac. Conf. Ser. vol. 1, Provo: Brigham Young Univ., 161

Holtzman, J. A., Light, R. M., Baum, W. A., et al. 1993, AJ, 106, 1826

Horch, E., Demarque, P., & Pinsonneault, M. 1992, ApJ, 388, L53

Iben, I. Jr. 1991, ApJS, 76, 55

Iben, I. Jr., & Renzini, A. 1983, ARA&A, 21, 271

Iben, I. Jr., & Renzini, A. 1984, Phys. Rep. 105 (6), 329

Lee, Y-W. 1994, ApJ, 423, 248

Liebert, J., Saffer R. A., & Green, E. M. 1994, AJ, 107, 1408

Matteucci, F. 1994, A&A, 288, 57

McWilliam, A., & Rich, R. M. 1994, ApJS, 91, 749

Ng, Y. K. 1994, Ph.D. Thesis, University of Leiden, the Netherlands

Ng, Y. K., Bertelli, G., Bressan, A., & Chiosi, C. 1995, A&A, in press

Ortolani, S., Barbuy, B., & Bica, E. 1990, A&A, 236, 362

Ortolani, S., Bica, E., & Barbuy, B. 1992, A&AS, 92, 441

Ortolani, S., Bica, E., & Barbuy, B. 1993, ApJ, 408, L29

Pagel, B. E. J. 1989, in Evolutionary Phenomena in Galaxies, J. Beckmam, & B. E. J. Pagel, Cambridge: Cambridge University Press, 201

Pagel, B. E. J., Simonson, E. A., Terlevich, R. J., & Edmunds, M.G. 1992, MNRAS, 255, 325

Reimers, D. 1975, Mém. Soc. Roy. Sci. Liege, 6$^e$ Ser., 8, 369

Renzini, A., Fusi-Pecci, F. 1988, ARA&A, 26, 199

Schombert, J. M., Halan, P. C., Barsony, M., & Rakos, K. D. 1993, AJ, 106, 923

Terndrup, D. M. 1988, AJ, 96, 884

Tantalo, R., Chiosi, C., Bressan, A., & Fagotto, F. 1995, A&A, submitted, TCBF

Tripicco, M. J., Dorman B., & Bel, R. A. 1993, AJ, 106, 618

Udalski, A., Szymański, M., Kałużny, J., et al. 1993a, Acta Astron., 43, 289

Udalski, A., Szymański, M., Kałużny, J., et al. 1993b, Acta Astron., 43, 69

VandenBerg, D. A. 1991, in The Formation and Evolution of Star Clusters, A.S.P. Conf. Ser. vol. 13, Provo: Brigham Young Univ., 183

Weidemann, V. 1987, A&A, 188, 74

Worthey, G. 1992, Ph.D. Thesis, Univ. California, Santa Cruz

Worthey, G, Faber, S. M., Gonzales, J. J., & Burstein D. 1994, ApJS, 94, 687

## Discussion

*Roxburgh*: I didn't quite understand the model you have of convective diffusion. What one seems to get from numerical simulation is an overshoot region with conditions little different from the unstable region then a very rapid decay of convective velocities.

*Astrophysical Applications of Stellar Pulsation*
*ASP Conference Series, Vol. 83, 1995*
*R. S. Stobie and P. A. Whitelock, eds.*

# Pulsation and Evolution. An observer's viewpoint

C. Waelkens

*Instituut voor Sterrenkunde, Katholieke Universiteit Leuven, Celestijnenlaan 200 B, B-3001 Heverlee, Belgium*

**Abstract.** We review observational evidence on the interaction between stellar pulsation and evolution. We discuss to what extent observations of pulsating stars with variable amplitudes and pulsation periods have implications on our understanding of stellar structure and evolution. The probable link between mass loss and pulsation in AGB stars and in hot luminous stars appears to be the strongest way in which pulsations affect evolution. We point out the possibility that forced oscillations in the components of binaries may have important consequences on evolution, that could offer an explanation for some classes of peculiar evolved objects.

## 1. Introduction

*Evolution* is the central paradigm that characterizes most astrophysical research, if not natural sciences in general. Confronting stellar pulsations with stellar evolution theory is therefore a key issue, both from the theoretical and observational points of view. Because, however, evolution and pulsation timescales differ so widely, direct confrontation of both is not an easy matter, and most evidence is of an indirect nature.

In this review, we will first address the question how *evolution affects pulsation.* Obvious aspects are observations of stars entering or leaving the various instability strips in the HR diagram, and the variations of pulsation periods. While the available evidence is rather consonant with stellar evolution theory, it is a striking fact that it probably tells us more about stellar structure than about evolution. Since observed stellar pulsations most often do not significantly affect the deep interior of stars, it is rare that *pulsation affects evolution.* An important issue is, however, pulsation-induced mass loss, for which evidence exists in hot massive stars as well as in evolved low-mass stars.

The last part of this review is devoted to the possible role resonant forced oscillations may play in the evolution of components of binaries. There is no hard direct evidence for such phenomena, but some indirect arguments can be put forward that they may be significant.

## 2. Onset and decay of pulsation

The very existence of well delimited instability strips in the HR diagram evidently is the best proof that evolution affects pulsation. A useful check is the statistics of variable stars versus constant stars. The population of the Cepheid strips is now well understood in terms of the excursions massive supergiants make from and back to the red supergiant stage. As far as Miras are concerned, the fact that these stars outnumber the planetary nebulae only by a factor 3 indicates that the duration of the Mira stage is rather short, i.e., of the order of $6\,10^4$ years only (Wood 1990), which may be seen as an indication of the effect of pulsation on evolution (see below).

The observational record of stars *entering* an instability strip and so starting pulsation, is very poor. Of course, no observer will commit himself to the monitoring of a constant star, waiting for its incipient pulsation. In these proceedings, Barlai presents evidence for an RR Lyrae star in M15 that may be in the process of starting its pulsations. The prospects of detecting the onset of Mira-type pulsations are much better now: comparing numbers and timescales, one finds that this should happen in the LMC once in 40 years, a fact which should prompt the EROS and MACHO teams to preserve their database for future generations.

Some well documented cases exist of stars with declining pulsation amplitudes, eventually stopping pulsating completely, e.g., the Cepheids Y Oph (Fernie 1990) and Polaris (Fernie et al. 1993), and the $\beta$ Cephei star $\alpha$ Virginis. All cases, however, concern objects that are rather centrally situated in the respective instability strips, and it seems that the evanescent nature of the pulsations is an effect of changing internal structure rather than of evolution. A better case for a star leaving the instability strip may be RU Cam (Teays 1990).

A remarkable case is the peculiar Cepheid HR 7308, the amplitude of which undergoes a modulation on a timescale of some 1000 days. A new explanation for this object has been put forward by Van Hoolst & Waelkens (1995), who interpret its behaviour as a resonant interaction between the second radial overtone and a non-radial mode with approximately the same period. This interpretation involves a 1-1 resonance in which energy is cyclically exchanged between a non-radial mode, which is pulsationally stable by itself, and an unstable radial mode. Such a 1-1 resonance can only happen for the second and higher overtones; it would be of interest to find out whether similar behaviour is present in some of the second overtone pulsators discovered by the MACHO project in the Magellanic Clouds (Welch, this conference).

## 3. Period variations

The change of the internal structure of a star during its crossing of an instability strip must affect its period. Such period changes have been searched for and found in several classes of pulsating variables. However, in most cases, it appears that the period variations caused by internal structural adjustments of another nature dominate the small effects of evolution, so that the diagnostic value of these results for stellar evolution purposes is somewhat deceiving.

Not unexpectedly, the clearest results have been found for Cepheids, for which a long record exists and which evolve on a relatively short timescale. A general result which emerges, is that the period increases and decreases allow one to locate Cepheids on the various paths through the instability strip, the statistics of which are in good qualitative agreement with stellar evolution (Fernie 1984). For RR Lyrae stars, random changes due to internal adjustments often dominate (Sweigart & Renzini 1979; Stothers 1980). On the other hand, Lee (1991) found a significant evolutionary increase in the periods of RR Lyrae stars in Oosterhoff II clusters (but see also Barlai, this conference).

For Mira's period changes are dominated by random effects (Percy et al. 1990), a view consistent with Whitelock's (1990) suggestion that hardly any evolutionary period change occurs for these stars. This suggests that the period increase due to a luminosity increase is compensated by the decrease due to mass loss, and fits well into a picture of a short-lived Mira phase. For RV Tauri stars, which according to Jura (1986) are rapidly evolving post-AGB stars, the situation is more confusing. While for R Scuti a period decrease consistent with rapid evolution is observed (Percy et al. 1991), the agreement is less good for U Mon, and the period of AC Her even displays an oscillatory behavior (Zsoldos 1988). An interesting result is that the absence of rapid period variations in 89 Her pleads against rapid post-AGB evolution for this star (Fernie & Sasselov 1989), in good agreement with the fact that this object has a Flamsteed number!

For main-sequence pulsators, the expected evolutionary changes are low, and observations show that other effects dominate. In some $\beta$ Cephei and some $\delta$ Scuti stars period increases and decreases are observed simultaneously for different modes. There is no disagreement with evolution theory, but the diagnostic value of period variations for stellar evolution is low. A notable exception may be BW Vul, for which Sterken & Jerzykiewicz (1990) claimed that the rapid period increase points to a post-core-hydrogen-burning evolution, which would put the exceptionally large pulsation amplitude of this star into a new perspective. Another most interesting result is the systematic presence of period *decreases* for evolved $\delta$ Scuti stars (Breger, this conference).

Extensive and detailed observations exist on the period variations of pulsating degenerate objects, where again structural and evolutionary effects interfere. The interpretation of the observations of white-dwarf pulsations is treated elsewhere in this volume by Kawaler.

We are then tempted to conclude that the observations of period changes in pulsating stars are useful though most often not conclusive for stellar evolution purposes. This conclusion should not discourage us to continue such observations, because of the wealth of other information they contain. *Pulsation period variations may be the best diagnostic for probing the random structural changes inside stars*, and have often led to the very important discovery of orbital motion of pulsating stars.

## 4. Pulsations and mass loss

Stars at the luminous tip of the AGB all are large-amplitude pulsating variables and all undergo severe mass loss. Both the large pulsation amplitudes and the mass loss are linked to the low gravity of these stars. A crucial question is whether a more direct link exists between pulsation and mass loss, i.e., is the mass loss induced by the pulsations? If not, the coexistence between mass loss and pulsation on the tip of the AGB would just be a coincidence, and the mass loss may be triggered by other mechanisms, such as thermal pulses or hydrogen recombination instabilities (Wagenhuber & Weiss 1994).

According to theoretical computations (Bowen 1988; Bowen & Willson 1991; Feuchtinger et al. 1993) the shocks associated with Mira pulsations increase the scale height of the atmosphere of an AGB star so that the density in its external, low-temperature, layers is sufficiently high for dust to form; radiation pressure then accelerates the dust which drags the gas with it.

An observational difficulty for the Bowen scenario is the fact that an extended post-shock region would produce molecular emission, which is not observed (Bessell et al. 1989). Second, direct observations of detached circumstellar shells and accurate modeling of energy distributions suggest that the mass loss is not a continuous process. Episodic mass loss is then probably related to thermal pulse phases. This is not necessarily an argument against the pulsation-induced-mass-loss hypothesis, since also pulsation amplitudes vary with the changing interpulse luminosity. On the other hand, various observations suggest a causal link between pulsation and mass loss. The fact that a large fraction of dust-enshrouded AGB stars are Miras suggests more than just a coincidence. Furthermore, a clear correlation exists between pulsation amplitude and mass loss rates (Whitelock 1990). In my opinion, the strongest argument is the short duration of the Mira phase, as it is computed from statistics in our Galaxy and the LMC: once a star becomes a Mira, its days on the AGB are numbered.

Some relation between pulsation and mass loss has also since long been suspected for hot luminous stars. We do not have to enter the discussion of pulsations of Be stars, since in these objects the mass loss hardly affects the evolution. More important for stellar evolution is the long-standing mystery of the origin of the very high mass loss rates of Wolf-Rayet stars. Maeder (1985) proposed that WR stars are vibrationally unstable. From the finding by Glatzel et al. (1993) that stars on the helium main sequence are violently unstable, Langer et al. (1994) elaborated on the effects of these pulsations on mass loss. The pulsations would lift the mass loss beyond the sonic point, where the radiation-driven wind would take over, in a way somewhat reminiscent of what is seen in AGB stars.

From the direct observational point of view, the situation with WR pulsations is somewhat frustrating. Blecha et al. (1993) announced the first direct observation of pulsations in WR40, but these observations could never be reproduced, not by themselves, nor by any other group. An observational caveat clearly is that the photospheres of these stars (if they can be defined) are hardly ever seen because of the dense wind. By all means, if WR pulsations have been observed once, they should be confirmed soon rather than late.

Violent pulsational instabilities are now also found for evolved hydrogen burning very massive and luminous stars (Glatzel & Kiriakidis 1993a, 1993b). It appears that the incorporation of the new opacities to stellar models implies that pulsational instabilities are inescapable in the upper part of the HR diagram, where the Cepheid and $\beta$ Cephei instability strips converge in the region where luminous blue variables are observed (Dziembowski 1992). According to Langer et al. (1994), the regions in the HR diagram of largest pulsational instability and of highest mass loss rates are coincident. The present situation thus holds much promise that the long-standing problem of the high mass loss rates of very massive stars may be reaching a solution, in which pulsations would play a major role.

## 5. Forced stellar oscillations

Among late-type giants and supergiants, a diverse zoo exists of objects the peculiarities of which can only be explained by mass transfer in a binary system. The *Barium stars* consist of an s-process enhanced star orbiting a white-dwarf. Since these objects have survived as fairly wide systems, it is excluded that the star that presently is a white dwarf has ever filled its Roche lobe, since then fast mass transfer and spiral-in would have occurred. However, some systems presently are so close that no Mira would fit into the Roche lobe of the white dwarf.

Tout & Eggleton (1988) have proposed that such close systems can be explained if one assumes that heavy mass transfer, induced by the tidal interactions, has occurred well before the AGB star filled its Roche lobe. This mass loss would then have impeded the star to grow to the size of typical single AGB stars, so that the decrease of the size of the system due to mass transfer never led to catastrophic mass transfer, and relatively wide systems with a white-dwarf component could survive. There is no clear understanding, however, of the way tidal interactions in well detached systems can increase mass loss rates by two orders of magnitude, as proposed by Tout & Eggleton.

A similar problem is encountered in some objects assumed to be post-AGB stars, such as HR 4049 and HD 44179, the central star of the Red Rectangle nebula. The latter object is particularly interesting, because it displays a remarkable bipolar nebula. That previous AGB evolution is involved for these objects is attested by the carbon richness of their circumstellar environment. Both objects occur in binaries with orbital periods that are much too short (434 and 298 days) to have accommodated a terminal AGB star of normal size (Van Winckel et al. 1995). Also the RV Tauri star 89 Herculis occurs in a binary, with an orbital period of 288 days (Waters et al. 1993).

Since observational evidence is accumulating that the high mass loss rates of AGB stars are linked to pulsations, it is worthwhile investigating this hypothesis for these peculiar binaries. In no case can a Mira fit into the system without destroying it. However, we propose that stellar oscillations may have occurred in these objects in an earlier phase, as *forced oscillations* by the companion. Indeed, since an AGB star undergoes substantial variations of size in a relatively short period, it appears hardly escapable that at some point a resonance develops between a free oscillation mode of the AGB star and the orbital motion. If during this resonance orbital energy is transfered to the oscillation mode, it could make the latter grow to macroscopic amplitudes, and cause pulsation-induced mass loss.

Ruymaekers & Smeyers (1994) have developed a theoretical framework for the passage through resonance of a dynamic tide in a rapidly evolving star. They find that such a passage indeed does affect the radial component of the tidal displacement. It would certainly be interesting to apply this theory to high-order g-modes in evolving AGB stars. If a positive answer is found, it would provide an independent way of probing the relation between pulsation and mass loss.

An interesting aspect in both HR 4049 and HD 44179 is that the present orbits are fairly eccentric. At first sight, a large eccentricity complicates the picture still further, in the sense that mass transfer and tidal interactions usually tend to circularize stellar orbits. However, it is not yet clear how the forced oscillations affect the eccentricity of the orbit. Moreover, observations by Matthieu (1992) and theoretical developments by Artymovicz et al. (1991) have shown that binary stars probably get their eccentricity before the main sequence, from tidal interactions with the circumsystem disk. Tidal interactions with the post-AGB circumsystem disk which is presently observed around both HR 4049 and HD 44179 may then also have been responsible for the large present eccentricity of these systems.

## References


Artymovicz, P., Clarke, C., Lubow, S.H., & Pringle, J.E. 1991, ApJ, 370, L35

Bessell, M.S., Brett, J.M., Scholz, M., & Wood, P.B. 1989, A&A, 213, 209

Blecha, A., Schaller, G., & Maeder, A. 1993, Nature 360, 320

Bowen, G.H. 1988, ApJ, 329, 299

Bowen, G.H., & Willson, L.A. 1991, ApJ, 375, L53

Dziembowski, W.A. 1992, in: *Pulsation, Rotation and Mass Loss in Early-Type Stars*, Proc. IAU Symposium No. 162, eds. L.A. Balona, H.F. Henrichs & J.M. Le Contel, p.55

Fernie, J.D. 1990, PASP, 102, 905

Fernie, J.D. 1994, in: *Observational Tests of the Stellar Evolution Theory*, Proc. IAU Symposium No. 105, eds. A. Maeder & A. Renzini, Reidel, Dordrecht, p. 441

Fernie, J.D., Kamper, K.W., & Seager, S. 1993, ApJ, 416, 820

Fernie, J.D., & Sasselov, D.D. 1989, PASP, 101, 513

Feuchtinger, M.U., Dorfi, E.A., & Höfner, S. 1993, A&A, 273, 513

Glatzel, W., & Kiriakidis, M. 1993a, MNRAS, 262, 85

Glatzel, W., & Kiriakidis, M. 1993b, MNRAS, 263, 375

Glatzel, W., Kiriakidis, M., & Fricke, K.J. 1993, MNRAS, 262, L7

Jura, M. 1986, ApJ, 309, 732

Langer, N., Hamann, W.R., Lennon, M., Najarro, F., Pauldrach, A.W.A., & Puls, J. 1994, A&A, 290, 819

Lee, Y.W. 1991, ApJ, 367, 524

Maeder, A. 1985, A&A, 147, 300

Matthieu, R.D. 1992, in: *Binaries as Tracers of Stellar Formation*, eds. A. Duquennoy & M. Mayor, Cambridge University Press, p.155

Percy, J.R., Colivas, T., Sloan, W.B., & Mattei, J.A. 1990, in: *Confrontation between Stellar Pulsation and Evolution*, eds. C. Cacciari & G. Clementini, ASP Conf. Series, p.446

Percy, J.R., Sasselov, D.D., Alfred, A., & Scott, G. 1991, ApJ, 375, 691

Ruymaekers, E., & Smeyers, P. 1994, A&AS, 104, 401

Sterken, C., & Jerzykiewicz, M. 1990, in: *Confrontation between Stellar Pulsation and Evolution*, eds. C. Cacciari & G. Clementini, ASP Conf. Series, p.236

Stothers, R. 1980, PASP, 92, 475

Sweigart, A.V., & Renzini, A. 1979, A&A, 71, 66

Teays, T.J. 1990, in: *Confrontation between Stellar Pulsation and Evolution*, eds. C. Cacciari & G. Clementini, ASP Conf. Series, p.128

Tout, C.A., & Eggleton, P.P. 1988, MNRAS, 231, 823

Van Hoolst, T., & Waelkens, C. 1995, A&A, 295, 361

Van Winckel, H., Waelkens, C., & Waters, L.B.F.M. 1995, A&A, 293, L25

Wagenhuber, J., & Weiss, A. 1994, A&A, 286, 121

Waters, L.B.F.M., Waelkens, C., Mayor, M., & Trams, N.R. 1993, A&A, 269, 242

Wood, P.R. 1990, in: *Confrontation between Stellar Pulsation and Evolution*, eds. C. Cacciari & G. Clementini, ASP Conf. Series, p.355

Whitelock, P.A. 1990, in: *Confrontation between Stellar Pulsation and Evolution*, eds. C. Cacciari & G. Clementini, ASP Conf. Series, p.365

Zsoldos, E. 1988, Inf. Bull. Variable Stars 3192

## Discussion

*Habing*: Estimating the number of PNe versus the number of Miras is dangerous: in our Galaxy the number of PNe is uncertain by a factor of 10.

*Waelkens*: I believe that the statistics of Miras are more reliable than that. So, if there are much more planetary nebulae, that would shorten Mira lifetimes still more. On the other hand, the census in the LMC should be fairly complete, and it is in agreement with the present statistics for our Galaxy.

*Whitelock*: I was interested in your comments about binary and pulsation interaction as a cause for mass ejection in AGB/post-AGB stars. It has worried me for some while that what we refer to as post-AGB phenomena may be associated with binary interaction. Do you know of any 'post-AGB' stars that are clearly solitary?

*Waelkens*: Yes. It seems that single stars have IR excesses only at IRAS wavelengths, while binaries also have near-IR excesses. Good candidate single post-AGB stars are HD 161796, HD 187885 and a few others

*Feast*: One would not expect a large amplitude Mira to start up suddenly. The evidence (e.g., globular clusters) shows that Miras are part of a wider instability strip which includes semi-regular (lower amplitude) variables as a prior evolutionary phase.

*Waelkens*: I cannot predict how Mira pulsations would start. My point is that the extensive coverage of the LMC by the MACHO and EROS projects may show us how it happens, once every 50 years or so.

*Mattei*: In the AAVSO Observing Program there are about 50 semi-regular variables that had large amplitudes in the past and presumably have much smaller amplitudes now. Bob Cadmus from Grinnell College is monitoring them photoelectrically. These stars deserve attention for they may evolve into something else other than being long period variables. Also I want to report that the 75-year AAVSO data on maxima and minima dates of long period variables, used extensively to search for period changes, will soon be extended to 90 years with the addition of data from 19755 to 1990.

*Dziembowski*: What could be the non-evolutionary structural changes causing period changes?

*Waelkens*: We can only guess. Earlier this morning we were told that mixing processes should be important. Such processes may also occur in discontinuous events. Remember that mass loss was first considered to be a continuous process, but now there is plenty of evidence that mass loss occurs in bursts.

*Burki*: Concerning HR 7308, I was in favour of a star evolving at the border of the instability strip, because the number of cycles between successive extrema of amplitude is in good agreement with the theoretical predictions by J. and A. Cox. Of course, in this case, we need an interaction between evolution and pulsation. In your proposed explanation, have I correctly understood that there is not an addition of two modes and that they cannot be observationally, separately, detected?

*Waelkens*: The basic idea is that there is one unstable radial mode which transfers energy to another, normally stable, non-radial mode with about the same period. As the amplitude of the radial mode declines and that of the non-radial one grows, damping of the non-radial mode starts and the radial one takes over again. Such a one-to-one resonance can only occur from the second radial overtone on, but then it is not unlikely, since the non-radial mode spectrum is rich in the vicinity of the second radial overtone. You have convincingly shown that HR 7308 is a second overtone pulsator.

*Percy*: (i) With regard to RV Tauri stars: the apparent period changes may be due to random fluctuations, as with the Mira stars. Endre Zsoldos and I have begun a project to test this hypothesis. (ii) For the Mira stars: because we have data on about 400 stars over more than 75 years, we *may* be able to average out the random fluctuations and detect the evolutionary changes.

*Waelkens*: Both projects are very important and underscore the interest of the AAVSO database. If no evolutionary period changes are found for the RV Tauri stars, or a least for some of them, that would have serious implications for our understanding of these stars in terms of stellar evolution.

*Astrophysical Applications of Stellar Pulsation*
*ASP Conference Series, Vol. 83, 1995*
*R. S. Stobie and P. A. Whitelock, eds.*

# Recent Developments in Stellar Pulsation Theory

A. Gautschy

*Astronomisches Institut der Universität Basel, Venusstr. 7, CH-4102 Binningen, Switzerland*

**Abstract.** A major driving force behind the efforts invested in stellar pulsation calculations in the recent years was certainly the success story of the new generation of opacity data. The new opacities did not only solve old problems but they also increased the complexity of quantitative stellar pulsation studies. Besides surveying the consequences of the new generation of stellar opacities for pulsation theory we also address more formal aspects of dealing with stellar oscillations. The section reviewing recent literature on the application of stellar pulsation theory to different classes of pulsating variables is strongly biased towards the high-luminosity part of the Hertzsprung-Russell (HR) diagram. We finish by mentioning recent changes in our beliefs concerning the envelope structure of oscillating white dwarfs.

## 1. Introduction

Computing stellar pulsation problems was considerably revived in recent years after the release of a new generation of stellar opacity data. The dramatic increase of the Rosseland opacity mean in the range of a few hundred thousand Kelvin allowed outstanding problems in stellar pulsation theory to be solved. But not only old problems were revisited with new microphysical data and mature computer codes also new features, that are to be attributed to the improved opacities, in the pulsational behaviour of stars were encountered.

The Hertzsprung-Russell diagram in Fig. 1 shows the distribution of a small subset of all pulsating variables. Essentially all major evolutionary stages over a large range of stellar masses encounter pulsational instabilities. Hence, pulsations may be potentially important tools for probing aspects of stellar constitution over a wide range of the physical parameter space. Recently, Brown et al. (1994) addressed the prospected gain in accuracy to constrain global stellar quantities – e.g., mass, luminosity, chemical abundances, age – from solar-like oscillations in distant stars. The results were rather disillusioning, however, in particular when compared with the achievements on the sun. The main reason for the reduced information gain is obviously the constraint of whole-disk observations only in distant stars.

Nevertheless, high-quality stellar oscillation data can play a role in a new era of quantitative stellar physics. Observational efforts and instrumental developments reached a level that allow the detection and the monitoring of variability at amplitudes of the order of $10^{-3}$ mag as a matter of routine. This achieve-

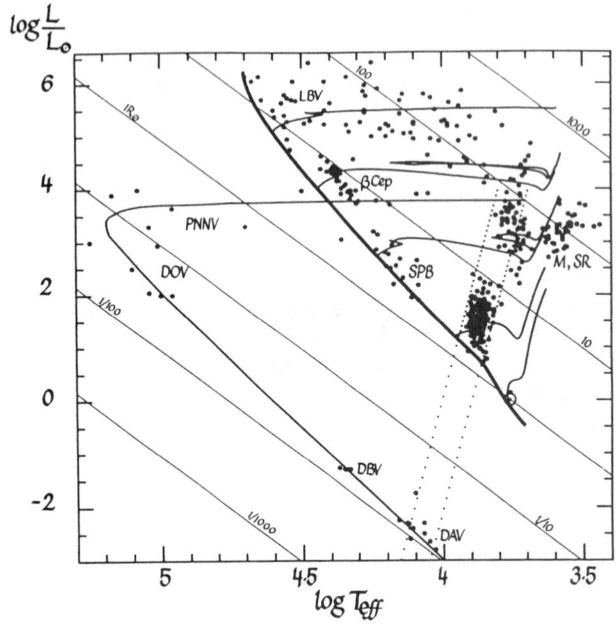


Figure 1. Distribution of some observed pulsating variable stars (dots) on the HR diagram. For convenience some guiding lines from stellar evolution are included. The heavy line passing from upper left to lower right shows the ZAMS. Evolutionary paths for some stellar masses, labeled at the tracks, are included. Common abbreviations are added to the figure indicating the location of some major families pulsating variable stars.

ment went along with the identification of a multitude of modes in stars that were believed to be mono-mode pulsators or even stable stars hitherto. Therefore, future observational approaches to stellar pulsations will conceivably shift from monitoring secular changes to long time-series monitoring of multi-mode pulsational variability – most of them typically on very low amplitude levels.

The following review must be considered as a biased survey of the recent literature on the theory of stellar pulsations. Particular emphasis is put on the impact of the new generation of opacity data, on pulsational instabilities in luminous stars, and on recent challenges of our beliefs concerning white-dwarf envelopes.

## 2. Methodical aspects of stellar pulsation theory

Many of the results in the more formal topics of stellar pulsation theory were driven by the requirement to interpret the high-quality data from the solar oscillation monitoring. In particular, second-order expansions for the coupling of nonradial oscillations with stellar rotation and a possible magnetic field were derived (Gough & Thompson 1990, Dziembowski & Goode 1992, Goode & Thompson 1992). Takata & Shibahashi (1994) suggested using directly vector spherical harmonics for an easier derivation of the selection rules for perturbations to nonradial eigenfunctions in rotating, magnetized stars. They aimed at explaining the observed frequency splittings in roAp stars, in particular the septuplet claimed for HR3831 (Kurtz et al. 1993).

Pulsation – convection coupling still defies a satisfactory amalgamation. Motivated by solar oscillations, Balmforth (1992a, b) discussed comprehensively the effects on radial oscillation modes in the sun of his time-dependent, non-local mixing-length approach. In the third paper of the series, Balmforth (1992c) criticizes the lack of predictive power of the excitation theories of solar modes based on stochastic driving. Xiong (1991, and references therein) is an advocate of a statistical approach to convection. He proposed a scheme for pulsation – convection coupling, but only the troublesome local convection theory found some applications to stars (Xiong 1981). Stellingwerf, on the other hand applied his non-local, time-dependent approach extensively to RR Lyrae stars (Stellingwerf 1984, and references therein).

Despite the powerful and user-friendly computers presently available, we are still far from being able to simulate nonlinear, nonradial oscillations of stars. An alternative approach is the use of nonlinear mode coupling equations. In this approach, the temporal evolution of the amplitudes of a preselected set of modes is described by a system of nonlinearly coupled equations. The coefficients in the equations are typically a combination of linear eigenmodes and of stellar structure quantities. Very recently, Goupil & Buchler (1994) and Van Hoolst (1994) presented the first, rather intricate, derivations of nonlinear, nonadiabatic, nonradial amplitude equations. The complexity of the problem is considerable and we eagerly await the first predictions for frequency splittings and amplitude distributions of multiplet components due to nonlinear interaction.

Results from adaptive-mesh radiation hydrodynamic codes were published in the last few years. Gehmeyr (1992, 1993) studied RR Lyrae stars including a time-dependent convection treatment which is not far from the Stellingwerf approach. Feuchtinger & Dorfi (1994) presented very smooth lightcurves for purely radiative RR Lyare star models. In both approaches, it seems that a *fully* adaptive mesh is not applicable due to numerical error propagation, in particular due to those picked up in the deep interior if the mesh is moving too unconstrained and which are then advected outwards. Hence, the mesh is freely adaptive in the exterior only, close to the driving region but it remains Lagrangian in the deep interior. Such a treatment is reminiscent of the Castor et al. (1977) method.

Also in the neutron star community, stellar stability problems led to innovative developments. To investigate sectorial oscillation modes in rapidly rotating neutron star models (in the Newtonian limit), Ipser & Lindblom (1990) made use of and adapted a two-potential approach to hydrodynamics that was originally proposed by Poincaré. First-order post-Newtonian corrections to the neutron star oscillation frequencies were provided by Cutler (1991) and Cutler & Lindblom (1992). Kokkotas & Schutz (1992) discussed the mathematical aspects of *quasi-normal modes* that occur when not only the fluid configuration but also the space-time metric is perturbed.

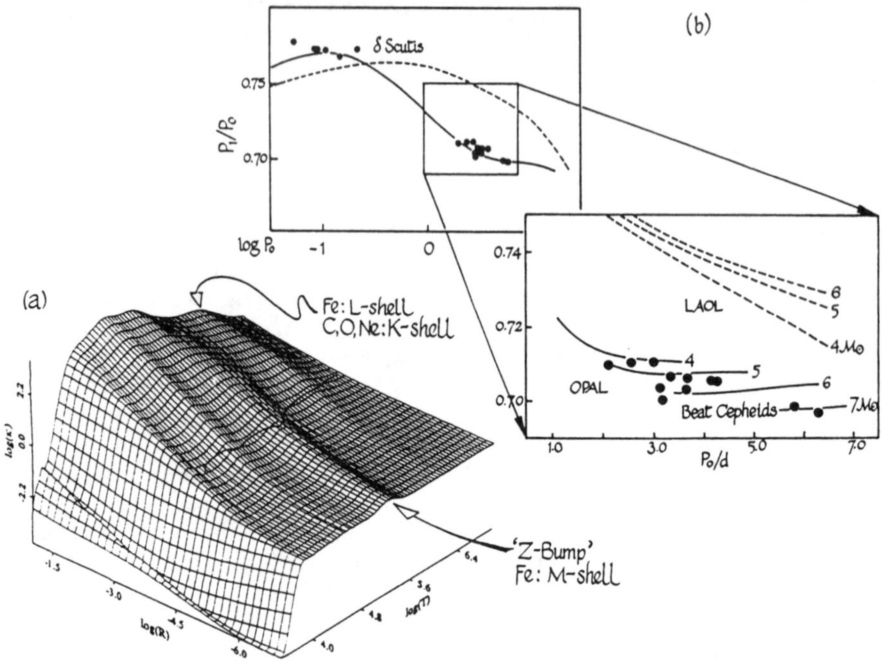


Figure 2. (a) The "opacity mountain" as obtained from OP calculations for a X = 0.7, Z = 0.02 composition. The two most prominent new features (ridges) found in the new opacity data are indicated. The solid line meandering along the slope of the $\kappa$-mountain marks the path traced out by the interior of a $12M_{\odot}$ ZAMS model. (b) Petersen diagram ($P_1/P_0$ vs. $P_0$) for population I variables ($\delta$ Sct and double-mode Cepheids). Dots indicate observed period data. Dashed lines clearly display the disagreement with the old LAOL data. Solid lines show the results of Moskalik et al. (1992) (lower right) and of Christensen-Dalsgaard (1993)(upper left), both based on OPAL opacity data.

## 3. Applications to classes of pulsating variables

Without doubt, the new generation of stellar Rosseland-mean opacity data boosted considerably the activities in the computational branch of the stellar pulsation community. Accounting for $\Delta n = 0$ transitions in the atomic calculations increased the Rosseland means by up to a factor of 3 or 4 (depending on the density range) at temperatures around 200 000 K. This bump in the opacity is often referred to as the "Z-bump" since it is caused mainly by inter – M-shell transitions of Fe; but also more subtle contributions from Ni, Mn, and Cr are seen (Seaton et al. 1994). New stellar opacities are now available from two independently working groups using also different approaches to the microphysics: OPAL (Rogers & Iglesias 1992; Igelsias et al. 1992) and OP (Seaton et al. 1994). Comparisons of the results of these two groups are very encouraging.

A second, smaller bump – with an increase of the Rosseland mean by about 10% over the old LAOL data – is found around $10^6$ K. In contrast to the Z-bump, this hotter bump , essentially caused by L-shell transition in Fe and by K-shell transitions in C, O, Ne, is "seen" by some solar oscillation modes. The envelope convection zone stops at temperatures below that of the hotter bump so that the sound speed is modified by its presence. The new opacities were identified as an important contributor to an improved solar "standard model" (Guenther et al. 1992).

For the period ratios of the different radial overtones of Cepheids and other variables in the classical instability strip, it was the Z-bump that had a major impact. The mass discrepancy for the double-mode and for the bump Cepheids is now considered to be essentially resolved (Moskalik et al. 1992). Also the RR Lyrae variables – despite their having Population II heavy element admixtures – showed modified period ratios. The combination of an improved treatment of hydrogen in a non-ideal stellar plasma – leading to a reduced opacity below $\approx 100\,000$ K – together with the reduced Z-bump are considered to be responsible for the particular behaviour. In contrast to the Cepheids, the new RR Lyrae period ratios do not improve the situation. Calculations by Kovács et al. (1992a) reveal considerable disagreement between double-mode masses for both Oosterhoff-type double-mode RR Lyrae stars and canonical horizontal-branch stellar evolution theory. A possible solution coming from non-solar admixtures of elements (such as C, N, O, Na, Mg) was suggested (Kovács et al. 1992b).

Double-mode pulsators – Cepheids as well as RR Lyrae stars – can still not be simulated in a convincing way. For RR Lyrae stars at least, numerical models pulsating simultaneously in two modes, with both of them being present for a long times, were obtained recently. Kovács & Buchler (1992) mentioned that purely radiative models can reproduce correct, i.e., observed, period ratios but they do not match the observed amplitude ratios. It was also realized that viscosity – numerical or physical – plays a crucial role for the final outcome. Bono & Stellingwerf (1993) announced their success in producing double-mode RR Lyrae models including convection. They noticed that their double mode models satisfy the resonance condition $3\omega_0 - \omega_1 = \omega_2$ (with $\omega_i \equiv 1/P_i$). On the other hand, Kovács & Buchler (1988) observed a $2\omega_0 = \omega_3$ resonance to obtain for their double-mode candidates. If, however, resonances are really needed or play a crucial role for the explanation of the double-mode phenomenon remains to be shown.

The OPAL/OP tables had a another important impact on pulsation theory. The Z-bump causes pulsational driving for stars in favourable temperature and luminosity ranges. The instability region running more or less parallel to the upper main sequence was eventually identified as an iron instability strip. Hence, the basic destabilisation mechanism of the $\beta$ Cepheids is classical $\kappa$-mechanism. The Z-bump is found to drive one or two radial (fundamental and/or first overtone) and a few low-order, low-degree $p$-modes above roughly $8M_\odot$. At the low-mass end of the $\beta$ Cep strip some additional unstable $g$-modes of high spherical degree and high radial order were found numerically (Dziembowski et al. 1993). The $g$-modes are excited by being trapped such that radiative dissipation in the deep interior is considerably reduced. For the unstable $p$-modes it is the classical argument (e.g., Cox 1980) of the correspondence of the thermal time-scale of the envelope overlying the driving region and of the pulsation period that obtains. A more detailed discussion of B-type pulsators is presented by Moskalik (these proceedings).

The driving and damping contributions to the total work integral are nearly equal for the B-star pulsators. Small changes in the form and the location of the Z-bump have, therefore, considerable influence on the position and on the extension of the instability region. The presence or absence of $\beta$ Cepheids in stellar systems are being considered as rough metallicity indicators. Furthermore, the Warsaw group found (see Moskalik, these proceedings) that the boundaries of the instability strip react also on the particular handling of Ni, Cr, Mn in the opacity calculations. It is the different treatment of these elements by the OP and the OPAL group that allows the pulsation theorists to investigate the influence of these subtle aspects on the pulsation properties of stars.

Towards lower masses, below about $7-8M_\odot$, to about $3M_\odot$ the slowly pulsating B (SPB) stars are observed having long periods between about 0.4 – 3 days. These stars are now understood as a natural extension of the pulsation mechanism of the $\beta$ Cepheids into the $g$-mode domain (Gautschy & Saio 1993, Dziembowski et al. 1993). For cooler stars the characteristic thermal time-scale of the envelope lying above the driving region of the Z-bump increases and tends to destabilize longer periods, in this case this is equivalent with higher-order $g$-modes. At the low-mass boundary of the SPB instability region the radial orders of the overtones favoured for destabilisation are so high that radiative dissipation due to the short spatial length-scales of the eigenfunctions eventually wins over the Z-bump driving. For further discussion see also Moskalik (these proceedings). The essence of this explanation is that the SPB stars can, physically spoken, be understood in the same manner as the $\beta$ Cepheids. In this respect, they are the same.

Also massive-star models are influenced by the Z-bump. The question of the upper-mass limit for stable main sequence models was readdressed after the new opacities became available. Stothers (1992) claimed that for a Z = 0.03 sequence of stellar models, only models exceeding $148M_\odot$ turned unstable by the $\epsilon$-mechanism. Glatzel & Kiriakidis (1993), on the other hand, found that the fundamental radial mode became pulsationally overstable already at $79M_\odot$. The contribution of the nuclear burning to the instability was, however, of minor importance. In the calculations of Glatzel & Kiriakidis, the envelope was the main contributor to the driving. They identified the Z-bump to induce strange modes (for a discussion see Gautschy & Glatzel 1990, Glatzel 1994) that are responsible for strong pulsational instabilities in the upper main sequence star models. Even down to $58M_\odot$ models, strange mode instabilities were encountered at high frequencies intermingled in higher overtone regular pulsation modes. If pulsational instabilities have any influence on the structure and on the evolution of very massive stars, then the strange modes, and not nuclearly unstable modes, dictate the outcome.

Kiriakidis et al. (1993) analyzed a large number of stellar evolution models with respect to their pulsational properties. Very broad domains of pulsational instabilities due to strange modes (see Fig. 3) were encountered. As strange modes, induced by the Z-bump, dominate in the hot domain of the HR diagram, i.e., close to the main sequence for the massive stars, their instabilities

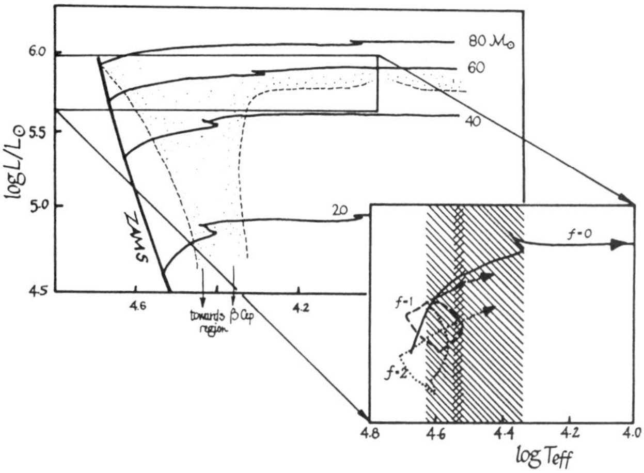


Figure 3. High luminosity part of the HR diagram. The evolutionary tracks indicate the stellar mass domain analyzed towards pulsational instabilities by Kiriakidis et al. (1993). The instability domain (for Z = 0.02 is outlined by the dotted region. The inlet at the lower right shows stellar evolution calculations accounting for mass-loss parameterized by the growth rates of the pulsations (Langer et al. 1994). Details are given in the text.

are again sensitive to the heavy element abundances. Kiriakidis et al. (1993) claimed that they might have uncovered the origin of the Humphreys – Davidson limit (Humphreys & Davidson 1984). Langer et al. (1994) used the results of the pulsation analyses of Kiriakidis et al. (1993) for their massive-star evolution calculations. The basic new hypothesis was that the pulsations induce mass loss that can be parameterized by the unusually large growth rates of the most unstable mode. The simplified parameterization introduced by Langer et al. was calibrated with the result of one particular non-linear simulation of Kiriakidis (1992). The result of the stellar evolution, assuming that the pulsations themselves do not react on the strong mass loss, are shown in the inset at the lower right of Fig. 3. The different values of $f$ indicate the effect a multiplicative factor in the mass-loss rate formula on the evolutionary track. The assumption that the unstable pulsation modes do not react on the modified envelope structure due to the substantial mass loss is certainly too naive. Nevertheless, the experiment of Langer et al. (1994) is a first step in the right direction of coupling pulsation properties with stellar evolution.

Upon the rumour of short-period pulsations in Wolf-Rayet stars (Blecha et al. 1992), Glatzel et al. (1993) investigated the stability properties of helium main-sequence models. Such models were considered reasonable approximations to observed Wolf-Rayet stars. Glatzel et al. (1993) were confronted, as already before in the case of hydrogen main-sequence stars, with pulsational instabilities due to strange modes at masses above $4M_\odot$. The periods of unstable modes ranged from about one to 25 minutes. The pulsation mode destabilised by $\epsilon$-mechanism at masses above $14M_\odot$ has a period of about 30 minutes, in contrast.

Again, the growth rates of the unstable strange modes were orders of magnitude higher than the weak instability of the $\epsilon$-destabilised mode.

Some subclasses of hydrogen-deficient stars – very luminous, low-mass stars – are known to be pulsating. In recent years it was realized that the very broad instability domain of these stars can be understood with unstable strange modes (e.g., Gautschy & Glatzel 1990). A few stars in this peculiar class of pulsators defied an explanation, however. One of them is V645 Her, the low luminosity was insufficient to induce unstable strange modes in its envelope. Applying the new opacity data, Saio (1993) was able to show that this star (and the same applies for LSS 3184) are pulsationally destabilised by the Z-bump in the same manner as the $\beta$ Cepheids. Hence, instability mechanism for the low luminosity H-deficient stars is different from that of the luminous ones. Saio found the iron strip to extend and widen towards high luminosities and eventually to merge with the very broad instability domain of the strange modes.

Based on the results for H-deficient stars, Gautschy (1993) studied post-AGB stars that are suitable candidates for cool central stars of planetary nebulae (CSPNe), with regard to possible pulsational instabilities. For masses exceeding roughly $0.75 M_{\odot}$, unstable strange modes could indeed be identified again. The coolest CSPNe are predicted to show periods around half a day with sharply decreasing periods as the stars get hotter. Very similar results were obtained for Wolf-Rayet – type CSPNe with their peculiar chemical abundances (Gautschy 1995). Observations, in particular newly available data from old Sonneberg monitoring plates, confirmed that the peculiar star FG Sge was pulsating already in the 1930s (Van Genderen & Gautschy 1995). This supports the theoretical findings of very broad pulsational instability domains in the upper HR diagram for stars with high $L/M$ ratios.

In the domain of oscillating white dwarfs the WET consortium in particular continues to challenge stellar pulsation theory (see the contributions of Kawaler and Bradley in these Proceedings). In their approach, all conclusions drawn from the high-resolution oscillation-frequency spectra are based on the framework of adiabatic pulsation theory. Besides the limitations of perturbation approaches to account for rotation and magnetic fields, adiabatic theory is probably on safe grounds. Presently, the main goal is to understand and to self-consistently embed the "seismological" results in the theory of stellar structure and evolution.

Nonadiabatic theory, on the other hand, is less well understood. For seismological work this is, however, not crucial. Nevertheless, it is important to understand the physical mechanism behind destabilizing white dwarf oscillation modes. We will content ourselves here with considering the family of DA variables only.

For more than a decade, the ZZ Ceti variables were advocated to admit thin H envelopes with $M_{\rm H}/M_{*} < 10^{-8}$ only (see e.g., Bradley et al. 1989, and references therein). White dwarf models with thicker envelopes always turned out to be pulsationally stable in the numerical calculations. Only very recently, with more careful modeling and more sophisticated computer codes, it was recognized (Bradley & Winget 1994, Fontaine et al. 1994) that the stability properties of DAVs are actually rather insensitive to the hydrogen layer thickness, at least for H envelope masses below $10^{-4} M_{*}$. Furthermore, mode identification in variable DA white dwarfs was hinting at thick hydrogen envelopes (Fontaine et al.

1992 for G226-29, Bergeron et al. 1993 for GD 165 and the review by Clemens 1993) for some time. Members of the stellar evolution community are certainly delighted about these recent developments. For the first time the results from oscillation theory of DAVs are consistent with predictions from canonical stellar evolution theory concerning the expected amount of hydrogen-rich material floating on top of white dwarfs. It is the nuclear burning during the pre – white-dwarf evolution that essentially fixes that thickness. Very thin envelope hydrogen layers could be explained only if significant mass loss or some mixing mechanism could be identified.

The extension of the convection zone in ZZ Ceti variables is another source of uncertainty. This aspect is important for the understanding of the $g$-mode instability since convection – pulsation interaction seems to play a yet incompletely understood role. To match the theoretical ZZ Ceti blue edge with observed temperatures of the hottest ZZ Ceti star (the present candidate is G117-B15A), the efficiency – and hence the mixing-length – of the convection in the outermost layers of the DAVs is adjusted in an ad hoc way. The spectroscopic calibration of these white dwarfs is also far from trivial and hence far from unanimously accepted. Estimates of $T_{\rm eff}$ for G117-B15A, and hence of the ZZ Cetis' blue edge, decreased by roughly 800 K within the last decade (Weidemann & Koester 1984, Koester et al. 1995). In any case, pulsation theorists postulated rather efficient and hence thick convection zones, based on modified mixing-length approaches, to comply with the observed blue boundary of the instability domain. Recent results from 2-D radiation hydrodynamic simulations (Ludwig et al. 1994) do not support the view of deep convection zones. Heuristic fits of mixing-length models to the full dynamical simulations seem to favour a mixing length of 1.5 pressure scale heights. This applies at least to the locations of vanishing entropy gradients. The thermal structure of the hydrodynamically simulated convection zone, however, differs considerably from one based on a mixing-length approach. Obviously, even the simplest class of pulsating white dwarfs might require some more work before we can claim with certainty to have understood them.

**Acknowledgments.** Financial support from the Swiss National Science Foundation and from the IAU is gratefully acknowledged.

## References

Balmforth, N.J. 1992a, MNRAS, 255, 603

Balmforth, N.J. 1992b, MNRAS, 255, 632

Balmforth, N.J. 1992c, MNRAS, 255, 639

Bergeron, P., et al. 1993, AJ, 106, 1987

Blecha, A., Schaller, G., & Maeder, A. 1992, Nature, 360, 320

Bono G.,& Stellingwerf, R.F. 1993 in IAU Colloq. 139, New Perspectives on Stellar Pulsation and Pulsating Variable Stars, eds. J. M. Nemec & J. M. Matthews, (Cambridge:University Press), p. 275

Bradley, P.A., & Winget, D. E. 1994, ApJ, 421, 236

Bradley, P.A., Winget, D.E., & Wood, M.A. 1989, in IAU Colloq. 114, White Dwarfs, ed. G. Wegner (New York:Springer), p.286

Brown, T.M., et al. 1994, ApJ, 427, 1013

Castor, J.I., Davies, C.G., Davison, D.K. 1977, Los Alamos Rep. LA-6664

Clemens, J.C. 1993, Baltic Astronomy, 2, 467

Cox, J.P. 1980, Theory of Stellar Pulsation, (New Jersey:Princeton Press)

Christensen-Dalsgaard, J. 1993, in IAU Colloq. 137, Inside the Stars, eds. W.W. Weiss & A. Baglin, (ASP: San Francisco), p. 483

Cutler, C. 1991, ApJ, 374, 248

Cutler, C., & Lindblom, L. 1992, ApJ, 385, 630

Dziembowski, W.A., & Goode, P.R. 1992, ApJ, 394, 670

Dziembowski, W.A., Moskalik P., Pamyatnykh, A.A. 1993, MNRAS, 265, 588

Feuchtinger, M., Dorfi, E.A. 1994, A&A, 291, 209

Fontaine, G., Brassard, P., & Wesemael F. 1994, ApJ, 428, L61

Fontaine, G., Brassard, P., Bergeron, P., & Wesemael, F. 1992, ApJ, 399, L91

Gautschy, A. 1993, MNRAS, 265, 340

Gautschy, A. 1995, A&A, in press

Gautschy, A., & Glatzel, W. 1990, MNRAS, 245, 597

Gautschy, A., & Saio, H. 1993, MNRAS, 262, 213

Gehmyer, M. 1992, ApJ, 399, 265

Gehmyer, M. 1993, ApJ, 412, 341

Van Genderen, A., & Gautschy, A. 1995, A&A, 294, 453

Glatzel, W. 1994, MNRAS, 271, 66

Glatzel, W., & Kiriakidis M. 1993, MNRAS, 262, 85

Glatzel, W., Kiriakidis M., & Fricke K.J. 1993, MNRAS, 262, L7

Goode, P.R., & Thompson, M.J. 1992, ApJ, 395, 307

Gough, D.O., & Thompson, M.J. 1992, MNRAS, 242, 25

Goupil M.-J., & Buchler J.R. 1994, A&A, 291, 481

Guenther, D.B., Demarque, P., Kim, Y.-C., & Pinsonneault, M.H. 1992, ApJ, 387, 372

Van Hoolst, T. 1994, A&A, 292, 471

Humphreys, R.M., & Davidson, K. 1984, Science, 223, 243

Iglesias, C.A., Rogers, F.J., & Wilson, B.G. 1992, ApJ, 397, 717

Ipser, J.R., & Lindblom L. 1990, ApJ, 355, 226

Kiriakidis, M. 1992, PhD thesis, Göttingen

Kiriakidis, M., Fricke, K.J., & Glatzel, W. 1993, MNRAS, 264, 50

Koester, D., Allard, N., & Vauclair G. 1995, A&A, in press

Kokkotas, K.D., Schutz B.F. 1992, MNRAS, 255, 119

Kovács, G., & Buchler, J.R. 1988, ApJ324, 1026

Kovács, G., & Buchler, J.R. 1992, ApJ, 404, 765

Kovács, G., Buchler, J.R., & Marom 1992a, A&A, 252, L27

Kovács, G., Buchler, J.R., Marom, A., Iglesias, C.A., & Rogers, F.J. 1992b, A&A, 259, L46

Kurtz, D.W., Kanaan, A., & Martinez, P. 1993, MNRAS, 260, 343

Langer, N., Hammann, W.-R., Lennon, M., Najarro F., Pauldrach, A.W.A., & Puls, J. 1994, A&A, 290, 819

Ludwig, H.-G., Jordan, S., & Steffen, M. 1992, A&A, 284, 105

Moskalik, P., Buchler, J.R., & Marom, A. 1992, ApJ, 385, 685

Rogers, F.J.,& Iglesias, C.A. 1992, ApJS, 79, 507

Saio, H. 1993, MNRAS, 260, 465

Seaton, M.J., Yu Yan, Mihalas D., & Pradhan, A.K. 1994, MNRAS, 266, 805

Stellingwerf, R.,F. 1984, ApJ, 284, 712

Stothers, R.B. 1992, ApJ, 392, 706

Takata, M., & Shibahashi, H. 1994, PASJ, 46, 301

Weidemann, V., & Koester D. 1984, A&A, 132, 195

Xiong, D.R. 1981, Scientia Sinica, 24, 1406

Xiong, D.R. 1991, Proc. ASA, 9, 26

## Discussion

*Kawaler*: Are you sure that the $ds/dr < 0$ line for the hydrodynamic models of white dwarf envelopes is correct? It is within the photosphere ($\sim 10^{-16} M_\odot$) and looks to be where the top of the convection zone is in the interior models of Tassoul et al. (1990). Also, where was the bottom of the hydrodynamic model grid within the stellar model?

*Gautschy*: The line *is* correct. Indeed, the disagreement in depth is rather frightening. The bottom of the computational box of the hydrodynamic simulations lies at least at about $10^{-13}M_*$.

*Christensen-Dalsgaard*: Have you considered using the Gough-Balmforth theory for time-dependent convection in the white-dwarf case? It appears to work quite well in the Sun and solar-like stars.

*Gautschy*: If you give me more time to work on, then I certainly will try to include a nonlocal time-dependent treatment in the oscillation code.

*Goldreich*: Don't you have a big problem handling dynamic convection when you calculate growth or decay rates for $g$-modes of white dwarfs?

*Gautschy*: The full treatment of convection dynamics and its interaction with the oscillation is at the moment out of reach. It should be included particularly when the convection zone is deep. If it turns out to be shallow, convection might adapt to the particular pulsation phase on times shorter than the period of the pulsation. The state at the moment is that I just did the the simplest thing - and it did not work.

*Astrophysical Applications of Stellar Pulsation*
*ASP Conference Series, Vol. 83, 1995*
*R. S. Stobie and P. A. Whitelock, eds.*

# Oblique Pulsator Model for the Blazhko Effect of RR Lyrae Stars

Hiromoto Shibahashi and Masao Takata

*Department of Astronomy, University of Tokyo, Tokyo, Japan*

**Abstract.** By assuming that the RR Lyrae stars have fairly strong dipole magnetic fields with the symmetry axis oblique to the rotation axis of the star, we show that the oscillation mode which would be a pure radial oscillation in absence of the magnetic field has a quadrupole component, which is axisymmetric with respect to the magnetic axis. The aspect angle of the quadrupole component changes due to the stellar rotation, and this apparent variation is interpreted as the Blazhko effect in RR Lyrae stars.

About one third of RR Lyrae stars show long-term amplitude modulation, which was first noticed by Blazhko (1907), and this phenomenon is called the Blazhko effect after his name. This phenomenon has been a puzzle for a long time (see, e.g., Teays 1993), and it still remains to be solved. The period of amplitude modulation is typically in the range of 30 days to 100 days. Some possible explanations have been proposed for the Blazhko effect: (i) the modulation in terms of beats among multiple modes; (ii) the internal resonances between the radial fundamental mode and the third harmonic (e.g., Moskalik 1986); (iii) the effect of a magnetic field (e.g., Cousens 1983). The existence of strong magnetic fields in the RR Lyrae stars is still controversial. In the following, we assume that RR Lyrae stars have substantial magnetic fields and that the magnetic symmetry axis is inclined to the rotation axis of the star, and we propose the oblique pulsator model for the Blazhko effect (see Shibahashi 1994).

We assume the Lorentz force is stronger than the Coriolis force and ignore the effect of the centrifugal force. We treat these forces as perturbations. We assume the magnetic field is a poloidal field and expand it in terms of Legendre harmonics, and assume that the magnetic axis is inclined to the rotation axis of the star. The effects of the Lorentz force and the Coriolis force appear both in the eigenfrequency shift and in deformation of the eigenfunction. Our interpretation of the Blazhko effect is as follows: The radial eigenfunction excited by the kappa mechanism is deformed mainly by the Lorentz force to have additional non-radial components, whose symmetry axis coincides with the magnetic axis which is assumed to be inclined to the rotation axis of the star. As the star rotates, the aspect angle of the non-radial components varies and then these components manifest themselves as the long-term modulation of the luminosity variation of RR Lyrae stars. This model is essentially the same as the oblique pulsator model for the rapid oscillations of the Ap stars (Kurtz 1982), but a little bit different from Cousens' (1983) model since we adopt the regular perturbation method.

Takata and Shibahashi (1994) derived the selection rules concerning the perturbation to the eigenfunctions. According to their result, in the case of a pure dipole magnetic field, the selection rules say the allowed perturbations to eigenfunctions of the degree $l = 0$ are only the spheroidal components with the $l = 2$ and $m = 0$ and the toroidal components with the $l = 1$ and $m = 0$. Since the dominant component of the oscillations of RR Lyrae stars is the radial mode, the amplitude is almost constant over the rotational phase of the star. However, the aspect angle of the quadrupole oscillation component caused by the magnetic field varies with the rotation of the star, and hence the contribution to the apparent luminosity variation from this component synchronously varies with the rotation of the star. We interpret the Blazhko effect as being a manifestation of this periodic variation. According to this model, if one performs a frequency spectrum analysis with a long time span, one will get a quintuplet fine structure in the frequency spectrum. The relative height of each peak of the quintuplet is dependent on the magnetic field strength and the geometrical configuration among the rotation axis of the star, the magnetic axis, and the line-of-sight. The separation between the adjacent peaks is equal to the rotation frequency of the star. At certain geometrical configurations, a pair of peaks in the quintuplet fine structure have so small amplitudes that they cannot be detected. In such a case, the fine structure should appear practically as a triplet [see observations by Smith et al. (1994) and Kovacs (1994)], and the separation between the adjacent peaks may be either the rotation frequency of the star or twice the rotation frequency. In the case that the magnetic axis passes through the stellar center, the fine structure is symmetric with respect to the central component. The off-axis components of the magnetic fields lead to asymmetry. According to this model, the Blazhko period must be equal to the rotational period of the star or twice that. Though the rotation periods of RR Lyrae stars have not been observationally determined, they are thought to be several ten days and seem consistent with the observed range of Blazhko periods. We performed numerical simulations, in which the magnetic field strength and the geometrical configurations were treated as free parameters, and we found that a magnetic field of order one kilogauss is needed to explain the observed Blazhko effect of RR Lyrae stars.

## References


Blazhko, S. 1907, Astron. Nachr., 175, 325

Cousens, A. 1983, MNRAS, 203, 1171

Kovacs, G. 1994, A&A, preprint

Kurtz, D. W. 1982, MNRAS, 200, 807

Moskalik, P. 1986, Acta Astron., 36, 333

Shibahashi, H. 1994, in: Helio- and Asteroseismology from the Earth and Space: GONG '94, ed. R. K. Ulrich, ASP Conf. Series, in press

Smith, H., Matthews, J. M., Lee, K. M., Williams, J., Silberman, N. A., & Bolte, M. 1994, AJ, 107, 679

Takata, M., & Shibahashi, H. 1994, PASJ, 46, 301.

Teays, T. J. 1993, in IAU Coll. No. 139, eds. J. M. Nemec and J. M. Matthews (Cambridge Univ. Press), p. 410

*Astrophysical Applications of Stellar Pulsation*
*ASP Conference Series, Vol. 83, 1995*
*R. S. Stobie and P. A. Whitelock, eds.*

# New Results on Pulsating OB Stars

Paweł Moskalik

*Copernicus Astronomical Center, Polish Academy of Sciences, ul. Bartycka 18, 00–716 Warsaw, Poland*

**Abstract.** Until very recently the physical mechanism driving oscillations in $\beta$ Cep and other early type stars has been a mystery. The breakthrough came with the publication of new OPAL and OP opacity data. Model calculations with the new opacities have demonstrated that the pulsations are driven by the familiar $\kappa$-mechanism, acting in the metal opacity bump at $T \approx 2 \times 10^5$K. The mechanism excites the low order $p$- and $g$-modes in the upper part of the instability strip and the high order $g$-modes in the lower part of the strip. The theoretical instability domains agree well with the observed domains of the $\beta$ Cep and the SPB stars. In this review I present these recent theoretical results and discuss their consequences for our understanding of B stars pulsations.

## 1. Introduction

Pulsational variability is a widely spread phenomenon among O and B stars. The first group of pulsators to be identified in this part of the H–R diagram were the $\beta$ Cep stars, discovered in the beginning of the century (Frost 1902; Albrecht 1908). Today 58 confirmed and 81 suspected members of this group are known (Sterken & Jerzykiewicz 1993). These early B stars display single- or multiperiodic variability of their brightness and radial velocity, with periods of $0.1 - 0.3$ day. In terms of pulsation theory such periods imply oscillations in low order $p$-modes. Several other groups of O and B pulsators have been discovered in the recent years. In 1976 Smith & Karp (see also Smith 1977) have identified the group of 53 Per stars, which show multiperiodic line profile variations with periods of the order of $1 - 2$ day. These variations are interpreted as resulting from nonradial oscillations of low spherical harmonic degree $\ell$. Photometric variability on similar time scales has been detected in several mid to late B stars by Waelkens (1991), who calls his objects the *Slowly Pulsating B-type Stars*, or the SPB stars. Both 53 Per and the SPB stars are slow rotators and most likely they are closely related. 53 Per itself (the only one of its group to have adequate data) is photometrically identical to the SPB stars (Smith et al. 1984; Huang et al. 1994). Line profile variations of *high* order ($\ell \geq 4$) have been discovered in $\zeta$ Oph (Walker, Yang & Fahlman 1979) and in several other early type stars. In contrast to the SPB and 53 Per-type stars, all members of this group are fast rotators. Also in this case the variability is interpreted in terms of nonradial pulsations (Vogt & Penrod 1983). Another group of OB stars whose variability is believed to be due to pulsations are the hot supergiants (Maeder 1980). Finally, nonradial pulsations have also been suggested as an explanation of the periodic brightness changes observed in many Be stars (the so called $\lambda$ Eri-type stars), although the nature of this variability is still subject of a heated debate (Baade & Balona 1994). For the detailed discussion of the observational status of the OB variables the reader is referred to, e.g., Balona (1995).

The origin of pulsation in the $\beta$ Cep and other OB type variables has been a long standing puzzle, challenging our understanding of stellar structure and stability. Already in the sixties it was realised that the opacity mechanism acting on the $He^+$ ionization zone, which is the cause of pulsation of Cepheids and RR Lyr stars (Baker & Kippenhahn 1962), cannot work in much hotter OB variables. The correct solution was first suggested by Simon (1982), who postulated that the metal opacities should be augmented by factor of $2-3$. He pointed out that such a modification resolves the period ratio discrepancy in Cepheids, which has been another long-standing problem, and at the same time provides a driving mechanism for the $\beta$ Cep stars. Simon's suggestion has been initially met with scepticism by atomic physisists (e.g., Magee, Merts & Huebner 1984), but nevertheless inspired new efforts in computing stellar opacities. As a result, two new, independently calculated, sets of opacity tables have been recently published: those of the OPAL project (Iglesias, Rogers & Wilson 1992) and those of the OP project (Seaton et al. 1994). Both sets of calculations show a large local opacity maximum at $T \approx 2 \times 10^5$K, which has been absent in the older Los Alamos tables. This new feaure is due to the combined effect of millions of metal lines and is therefore called a "metal" opacity bump.

The publication of new opacities had an immediate impact on the problem of the OB star pulsations. Already the first model computations with the OPAL tables (Cox et al. 1992; Moskalik & Dziembowski 1992; Kiriakidis et al. 1992) demonstrated that the excitation of the $p$-mode oscillations is possible in the $\beta$ Cep-type stellar models. These early results were not fully satisfactory though. The models were unstable only in the fundamental mode (contrary to observations) and only for a somewhat high metallicity of $Z \geq 0.03$. In this early work the first version of the OPAL tables was used, in which the spin-orbit coupling in the iron atoms was not taken into account. In the subseqent release of the OPAL tables (Iglesias et al. 1992) this effect was included, which led to a further enhancement of the metal bump. The improved tables were used in the next set of $\beta$ Cep and SPB pulsation models, computed by Dziembowski & Pamyatnykh (1993), Gautschy & Saio (1993) and Dziembowski, Moskalik & Pamyatnykh (1993). The new models, with a standard metallicity of $Z = 0.02$, show a remarkable overall agreement with the observations. The calculations performed with the OP tables give very similar results (Pamyatnykh et al. 1994). These two facts make us believe that the long sought mechanism exciting pulsations in OB star has been finally identified. In the following, I will discuss these recent theoretical results, with particular attention given to the morphology of the instability strip and to the physics of the driving. The latest review of this subject has been given by Dziembowski (1995). The new opacities have also been employed in the modeling of pulsations in the early type supergiants. In this context it is in order to mention the papers by Glatzel & Kiriakidis (1993) and by Kiriakidis et al. (1993). Unfortunately, because of limited space, this very interesting work has to stay beyond the scope of this review.

## 2. B stars instability strip

The main result of the stability surveys of stellar models built with the new opacities is the discovery of a theoretical instability strip of the B stars. In Fig. 1 we show the new instability domain in the H–R diagram. Only modes of $\ell \leq 2$, i.e., modes which are most likely to be observed *photometrically*, have been considered in determining the boundaries of the domain. The models displayed in the figure are calculated with the OP tables, assuming standard composition of $X = 0.7$ and $Z = 0.02$ (Pamyatnykh et al., in preparation). The influence of the particular choice of the opacity data (OPAL vs. OP) is discussed in Pamyatnykh et al. (1994). For comparison we also display the cluster $\beta$ Cep variables as well as the field SPB stars for which luminosities and effective temperatures are known.

The B stars instability strip begins at masses of about $3M_{\odot}$ (spectral type B9) and extends to masses well above $50M_{\odot}$, covering the whole range of spectral types B and O. Between this strip and the $\delta$ Sct instability strip there is only a narrow mass range, centered on $2.5M_{\odot}$, where the models are pulsationally stable. The $\beta$ Cep variables occupy only a small region of the strip. Nevertheless, by analogy with the *Cepheid instability strip*, we propose to call the *whole* new instability domain the *$\beta$ Cephei instability strip.*

The instability strip splits into two, partially overlapping, but otherwise distinct regions, where physically different types of oscillatory modes are driven.

### 2.1. $\beta$ Cep domain

In this domain the instability affects the low order $p$- and $g$-modes. In a typical model we find instability of the fundamental and of the first overtone $p$-modes ($p_1$ and $p_2$) as well as of the lowest order gravity mode $g_1$. For metallicity slightly above $Z = 0.02$ the second overtone ($p_3$) might also be excited. The instability extends from $\ell = 0$ to at least $\ell = 8$. The excitation of $g_1$ mode is of particular interest, because its frequency can be a sensitive probe of the evolutionary changes in the deep stellar interior, specifically in the $\mu$-gradient zone (Dziembowski & Pamyatnykh 1991). The periods of the unstable modes range from 0.1 to 0.3 day, which is in very good agreement with the observed period range of the $\beta$ Cep stars.

The Blue Edge of the $\beta$ Cep instability domain is very sensitive to the adopted metallicity. For $Z = 0.02$ (as in Fig. 1), the instability at lower masses ($9 - 10M_{\odot}$) sets in already on ZAMS, but with the increasing mass the strip becomes narrower. This trend continues up to $20M_{\odot}$. For even higher mass the strip widens in both directions, finally reaching ZAMS again. Calculations of Kiriakidis et al. (1993) show that for $\log(L/L_{\odot}) \geq 5.8$ the $\beta$ Cep instability domain and the Cepheid instability strip merge.

The instability of the low order p-modes extends beyond the Main Sequence phase. The post-MS evolution is, however, very fast. Consequently, the observed Red Edge of the instability strip is effectively determined by the Terminal Age Main Sequence (or TAMS). Such an *Effective Red Edge* is plotted with the dashed line.

The theoretical instability strip encompasses all the observed $\beta$ Cep variables (both in clusters and in the field). The observational points do not cover

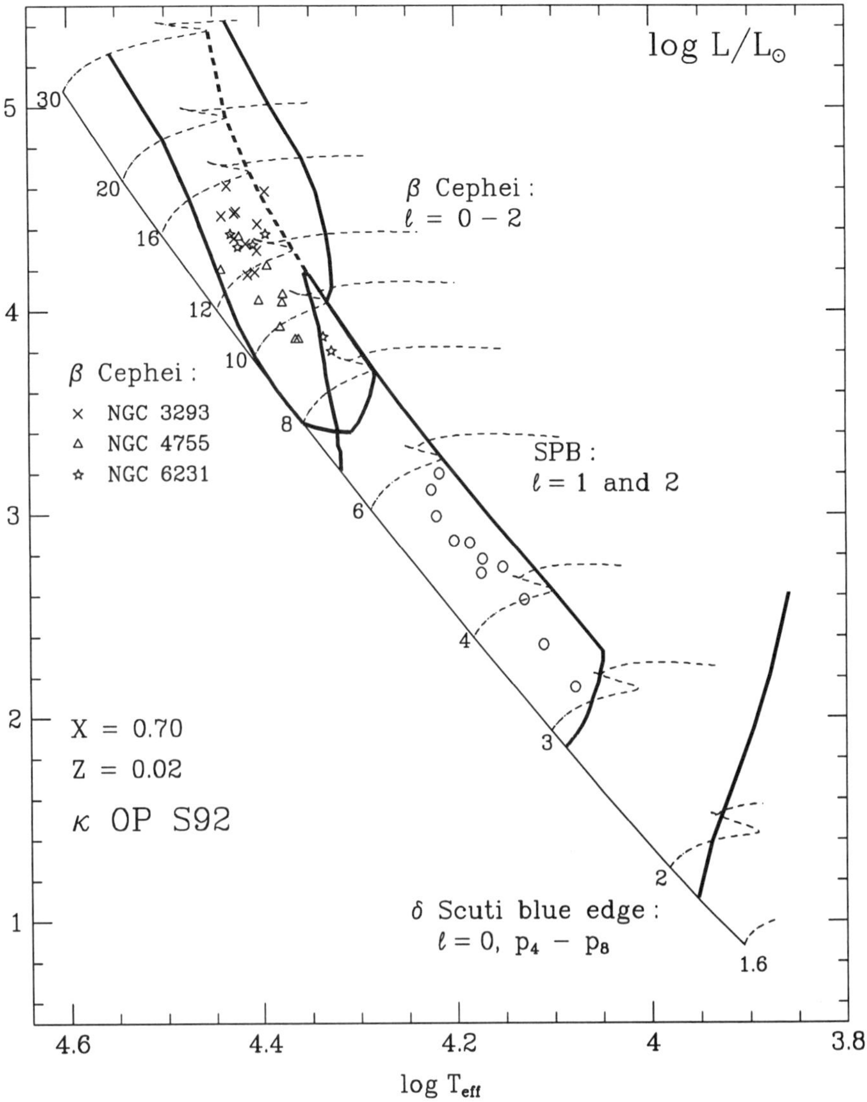


Figure 1. Theoretical instability domains of the $\beta$ Cep and the Slowly Pulsating B-type Stars in the H–R diagram. Only modes of $\ell \leq 2$ are considered. The models are computed with OP tables (S92 mixture, see Seaton et al. 1994) for standard composition of $X = 0.7$, $Z = 0.02$. For comparison we show $\beta$ Cep stars observed in young open clusters NGC 3293, NGC 4755 and NGC 6231 (Balona 1994; Balona & Koen 1994; Heynderickx 1991) and the field SPB stars (Waelkens 1994). From Pamyatnykh et al. (in preparation).

the whole theoretical domain, though. There is a significant lack of variables with $M < 9M_\odot$. The high luminosity part of the theoretical strip also remains unpopulated, with the exception of two field stars V986 Oph and HD 34656 (Balona 1995). We do not have a good explanation for this pattern; perhaps in these parts of the H–R diagram the stars preferentially choose to pulsate in modes of high $\ell$. Such modes cannot be detected photometrically.

### 2.2. SPB domain

In this domain the instability mechanism excites the high order $g$-modes. The instability sets in at ZAMS and continues until TAMS, but not beyond. The high temperature limit of the domain (in Fig. 1 located at $T_{\rm eff} \approx 4.35$) is sensitive to the metallicity and to the choice of the metal mixture. In a given model several (from 5 to 30) consecutive $g$-modes are unstable at each $\ell$. The total number and the radial order of the excited modes increases as the evolution of the star progresses. The modes are approximately equally spaced in period, as predicted by the asymptotic theory of pulsation, with the spacing depending on the mass and the age of the star. The small periodic departures from the equidistant pattern are caused by partial trapping of the modes in the $\mu$-gradient zone. In that sense the SPB stars are very similar to the DOV and DBV white dwarfs (Kawaler, these proceedings) and can be studied seismologically with the same theoretical tools. The unstable modes have periods between 0.5 and 4.1 day, which agree very well with periods observed in the SPB stars.

The instability is not limited to $\ell \leq 2$. Already for $4M_\odot$ it extends to $\ell = 7$. With each increasing $\ell$ the instability domain shifts higher up the Main Sequence band, and for $\ell \geq 6$ it extends well into the $\beta$ Cep region. The excitation of these relatively high-$\ell$ modes in more massive models may be relevant to pulsations of the $\zeta$ Oph variables.

All known SPB stars fall within the theoretical instability domain displayed in Fig. 1. They populate the middle and lower parts of the domain. The 53 Per variables (not plotted) would appear in the upper part of the domain and above it. Recalling that the 53 Per stars have been detected through the line profile variations (more sensitive to higher $\ell$), while the SPB stars have been detected photometrically, it is tempting to speculate that the 53 Per stars are just the extension of the SPB phenomenon to higher $\ell$ ($\ell = 3, 4$). We would like to caution the reader, though, that with the exception of 53 Per itself, the variables of this group are not well observed. Much more observational work is needed before any final conclusion on the SPB – 53 Per connection can be drawn.

## 3. The instability mechanism in B stars

Pulsational instability in B stars is driven by the familiar opacity mechanism ($\kappa$-mechanism), i.e., the same mechanism which operates in Cepheids and RR Lyr stars. However, in B stars the mechanism acts on a different opacity maximum.

For further discussion it is convenient to write down the work integral, $W$, which represents the net gain of energy by an oscillatory mode during one cycle:

$$W = -\int d^3x\,\nabla_{\rm ad} \oint dt\,\frac{\delta P}{P}\,\delta{\rm div}\mathbf{F_r} \tag{1}$$

where $\delta$ denotes the Lagrangean perturbation and other symbols have their usual meaning. Eq. (1) neglects the perturbations of the nuclear reaction rate and of the convective flux. The former effect is not important at all in B stars. The effects of convection are not very important either, since in the driving zone the convection carries very little (if any) of the total energy flux. Adopting the diffusion approximation, $\delta{\rm div}\mathbf{F_r}$ can be expressed as

$$\delta{\rm div}\mathbf{F_r} = \frac{1}{4\pi r^2}\,\frac{d\delta L_r}{dr} \tag{2}$$

where

$$\frac{d\delta L}{dr} = \frac{dr}{d\log T}\,\frac{d}{dr}\Big(\frac{\delta T}{T}\Big) - \frac{\delta\kappa}{\kappa} + 4\Big(\frac{\delta T}{T} + \frac{\delta r}{r}\Big) \tag{3}$$

The eqs. (2) and (3) are strictly valid for radial modes and approximately valid for nonradial modes (in the latter case every perturbed quantity contains the angular factor $Y_l^m(\theta,\phi)$). The first term in eq. (3) describes radiative dissipation, which is the main damping effect stabilizing the modes. The $\kappa$-effect driving is due to the second term. Inserting eq. (2) and (3) into eq. (1) we obtain

$$W = \ldots + \int d^3x \frac{\nabla_{\rm ad}}{4\pi r^2} \oint dt\,\frac{\delta P}{P}\,\frac{d}{dr}\Big(\frac{\delta\kappa}{\kappa}\Big) + \ldots \tag{4}$$

Eq. (4) shows, that the opacity perturbation contributes locally to the driving in those zones where during compression the derivative $\frac{d}{dr}(\frac{\delta\kappa}{\kappa})$ is positive. Noticing that $\delta P/P$, $\delta T/T$ and $\delta\rho/\rho$ are almost in phase and assuming that these perturbations vary slowly throughout the driving region, we find that the $\kappa$-effect *driving* occurs in places where $\kappa_T = (\frac{\partial\log\kappa}{\partial\log T})_\rho$ *increases with* $r$. The same effect will locally *damp* the oscillations if $\kappa_T$ *decreases with* $r$.

The proper behaviour of $\kappa_T$ is a necessary, but not a sufficient condition for the $\kappa$-effect to work. The other requirement is that the thermal timescale $\tau_{th}$ of the zone, defined as

$$\tau_{th} = \frac{1}{L}\int_{surface}^{M(T)} T c_p\, dM \tag{5}$$

is comparable to or longer than the period of the mode in question. If this condition is violated, then the zone re-adjusts quickly to the thermal equilibrium during the pulsation cycle, and consequently ${\rm div}\mathbf{F_r} \approx 0$.

In Fig. 2 we show two models, which are representative of the $\beta$ Cep and the SPB stars, respectively. All quantities are plotted as a function of the temperature in the model, so that the important features in the opacities occur in the same position. In both cases the opacity patterns are similar. The two main features are the $He^+$ bump at $\log T \approx 4.6$ and the metal bump at $\log T \approx 5.3$ (the H ionization bump is absent because the surface temperature is too high). The

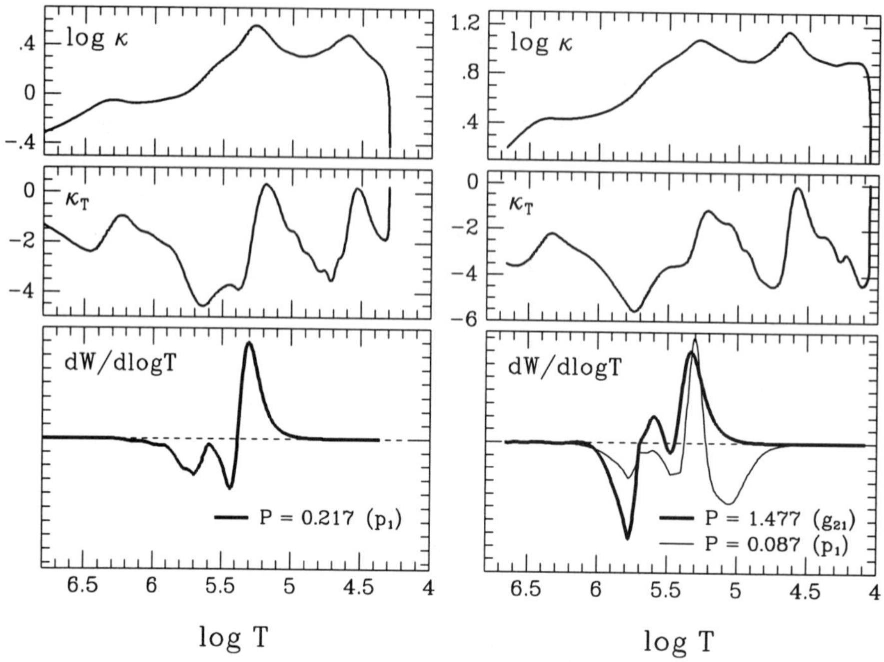


Figure 2. Opacity $\kappa$, opacity derivative $\kappa_T$, and differential work integral $dW/d\log T$ (positive in driving zone) for selected pulsation models, plotted vs. temperature. *Left*: $\beta$ Cep star model ($M = 12M_\odot$, $\log L/L_\odot = 4.27$, $\log T_{\rm eff} = 4.378$, $X_c = 0.13$). *Right*: SPB-star model ($M = 4M_\odot$, $\log L/L_\odot = 2.51$, $\log T_{\rm eff} = 4.142$, $X_c = 0.37$). Both models have initial composition of $X = 0.7$, $Z = 0.02$. The modes plotted correspond to $\ell = 1$.

metal bump is a new feature of the OPAL and OP opacities, which was absent in the older Los Alamos tables. In the lowest panel of the plot we show for selected modes the contribution to the work integral from different zones. The $He^+$ bump, which drives the pulsation in the classical variable stars (Cepheids, RR Lyr, $\delta$ Sct), plays no role in B stars. This is so, because the thermal timescale $\tau_{th}$ at the position of this bump is much too short. On the other hand, the timescale requirement can be satisfied at the position of the metal bump. Both $\beta$ Cep and the SPB variables are destabilized by the $\kappa$-mechanism operating on this feature.

Fig. 2 explains why the same mechanism excites different types of modes in $\beta$ Cep and in the SPB stars. In the $12M_\odot$ model $\tau_{th}$ at the zone of the metal bump is comparable to the periods of the low order $p$-modes. For these modes the local $\kappa$-effect driving is most efficient and the modes become unstable. The high order $g$-modes have considerably longer periods. Consequently, the driving at $\log T \approx 5.3$ is much less efficient and cannot overcome the radiative damping occuring deeper in the model. Therefore, these modes in the $\beta$ Cep-type models are stable. In the case of the $4M_\odot$ model, the thermal timescale at the metal bump is about an order of magnitude longer. The difference is caused by the lower surface temperature (the zone of $\log T = 5.3$ is deeper) and the lower luminosity of the model. The $\tau_{th}$ at the zone of the metal bump is now comparable to periods of the high order $g$-modes and these are the modes which become unstable. On the other hand, for the low order $p$-modes the timescale condition is satisfied already at $\log T \approx 5.0$. As a result, the zone where $d\kappa_T/dr < 0$ becomes "activated". The strong damping effect of this zone stabilizes the $p$-modes in the SPB stars, as can be seen in Fig. 2.

Finally, we need to explain why the opacity mechanism excites in the B stars either short period $p$-modes or long period $g$-modes, but never the modes with the *intermediate* periods. The thermal timescale condition at the metal bump is satisfied for such modes somewhere between $4M_\odot$ and $12M_\odot$, so why are these modes never destabilized? The answer lies in the shape of their eigenfunction $\delta P/P$. For these modes, the eigenfunction has very small amplitude at the driving zone and considerably larger amlitude in the deep interior. As a result, the driving is too weak to overcome the radiative damping in the inner zones and the mode remains stable.

## 4. A few remarks

The discovery of the metal opacity bump has provided a very simple, and fully satisfactory explanation of the B stars pulsations. The $\kappa$-mechanism operating in $\beta$ Cep and the SPB stars is fundamentally very similar to the mechanism driving pulsations of Cepheids or $\delta$ Sct stars. In contrast to the classical variables, though, our understanding of B stars pulsations is not hampered by the lack of a time dependent-theory of convection, since the convective transport plays no, or very little role in the driving region. There is a usual convection zone associated with the $He^+$ ionization zone, but this region is neutral for the driving or damping of the oscillations. This favourable situation allows us to put much more confidence in the prediction of the linear nonadiabatic pulsation theory. It also makes it possible to calculate reliably *all* boundaries of the B stars instability domain.

Since the driving mechanism relies on the metal opacity bump, it is quite natural that the size of the instability strip depends on the metallicity $Z$. The most affected is the Blue Edge of the strip, which becomes cooler with the decreasing $Z$. More detailed calculations show that the shape of the strip depends also on the choice of the metal mixture (Pamyatnykh et al. 1994; Pamyatnykh et al., in preparation). For the OP opacities with S92 mixture, the domain of $\beta$ Cep pulsations vanishes for $Z < 0.012$. The SPB domain is more robust and can survive at lower metallicities. It vanishes only for $Z < 0.006$. Thus, it might be possible to see the SPB variables in the moderately metal deficient stellar populations, which are already lacking the $\beta$ Cep variables. Another consequence of the $Z$-sensitivity of the driving mechanism is that the Blue Edge of observed instability strip of the field $\beta$ Cep stars should be somewhat diffuse, because of the scatter in $Z$ among the Galactic variables.

Our theory of the B star pulsations is currently limited to the linear approach. The first attempt at the nonlinear modeling has been undertaken only recently (Moskalik & Buchler 1994). Nonlinear theory is necessary to predict the amplitudes at which the instability will saturate and to understand the mode selection process. We note in passing, that in a typical $\beta$ Cep star there are 20–30 linearly unstable modes of $\ell \leq 2$ (counting the $m$-splitting), for the SPB star the respective number is well above 100. Only a small fraction of these modes reach detectable amplitudes and can be observed.

## 5. Applications

With our newly gained understanding of the B star pulsations we can use these objects as sensitive tools of stellar diagnostic. The presence of the $\beta$ Cep variables in the stellar cluster can be used to infer the metallicity of the population and to constrain the amount of the convective overshooting (Pamyatnykh et al., this conference). For an individual $\beta$ Cep star, the value of $\ell$ for each mode and the order for the radial modes can be derived from the multicolour photometry (Cugier et al. 1994). The detailed fit of the pulsation model to the observed frequency spectrum can then be performed, yielding parameters of the star. Such an attempt has been already undertaken for 16 Lac (Dziembowski & Jerzykiewicz, in preparation). Also the SPB stars, with their rich $g$-mode spectrum, can become very fruitful targets for seismology when adequate data become available. The $g$-modes penetrate the whole radiative interior of the star and their frequencies are sensitive to the structure of the $\mu$-grandient zone. These stars will provide a very strong test to the convective overshooting models. The $m$-splitting of the pulsation frequencies should also enable the measurement of the internal rotation rate, which will constrain the theory of the angular momentum evolution.

**Acknowledgments.** This work has been supported in part by KBN (Poland) through grant No 2–P304–013–07. Generous travel grant from IAU is also acknowledged. I am grateful to W. Dziembowski and A. Pamyatnykh for many enlightening discussions. Finally, it is my privilege to thank the organizers of the meeting for their warm hospitality during the conference.

## References

Albrecht, S. 1908, Lick Observatory Bull., 5, 62

Baade, D. & Balona, L. A. 1994, in IAU Symp. 162, Pulsation, Rotation and Mass Loss in Early-Type Stars, eds. L. A. Balona, H. F. Henrichs & J. M. Le Contel, Dordrecht: Kluwer, p. 311

Baker, N. & Kippenhahn, R. 1062, Zeit. Astrophys., 54, 114

Balona, L. A. 1994, MNRAS, 267, 1060

Balona, L. A. 1995, Ap&SS(in press)

Balona, L. A. & Koen, C. 1994, MNRAS, 267, 1071

Cox, A. N., Morgan, S. M., Rogers, F. J. & Iglesias, C. A. 1992, ApJ, 393, 272

Cugier, H., Dziembowski, W. A. & Pamyatnykh, A. A. 1994, A&A, 291, 143

Dziembowski, W. A. 1995, in Astrophysical Applications of Powerful New Atomic Databases, eds. S. J. Adelman & W. L. Wiese, ASP Conference Series (in press)

Dziembowski, W. A., Moskalik, P. & Pamyatnykh, A. A. 1993, MNRAS, 265, 588

Dziembowski, W. A. & Pamyatnykh, A. A. 1991, A&A, 248, L11

Dziembowski, W. A. & Pamyatnykh, A. A. 1993, MNRAS, 262, 204

Frost, E. B. 1902, ApJ, 15, 340

Gautschy, A. & Saio, H. 1993, MNRAS, 262, 213

Glatzel, W. & Kiriakidis, M. 1993, MNRAS, 262, 85

Heynderckx, D. 1991, PhD Thesis, Katholieke Universiteit Leuven

Huang, L., Guo, Z., Hao, J., Percy, J. R., Fieldus, M. S., Fried, R., Palovski, R. K., Bozić, H., Ruzić, Z., Paparó, M., Vetö, B & Dukes, R. 1994, in IAU Symp. No 162, Pulsation, Rotation and Mass Loss in Early-Type Stars, eds. L. A. Balona, H. F. Henrichs & J. M. Le Contel, Dordrecht: Kluwer, p. 33

Iglesias, C. A., Rogers, F. J. & Wilson, B. G. 1992, ApJ, 397, 717

Kiriakidis, M. El Eid, M. F. & Glatzel, W. 1992, MNRAS, 255, 1p

Kiriakidis, M., Fricke, K. J. & Glatzel, W. 1993, MNRAS, 264, 50

Maeder, A. 1980, A&A, 90, 311

Magee, N. H., Merts, A. L. & Huebner, W. F. 1984, ApJ, 283, 264

Moskalik, P. & Buchler, J. R. 1994, in IAU Symp. No 162, Pulsation, Rotation and Mass Loss in Early-Type Stars, eds. L. A. Balona, H. F. Henrichs & J. M. Le Contel, Dordrecht: Kluwer, p. 19

Moskalik, P. & Dziembowski, W. A. 1992, A&A, 256, L5

Pamyatnykh, A. A., Dziembowski, W. A., Moskalik, P. & Seaton, M. J. 1994, in IAU Symp. 162, Pulsation, Rotation and Mass Loss in Early-Type Stars, eds. L. A. Balona, H. F. Henrichs & J. M. Le Contel, Dordrecht: Kluwer, p. 70

Seaton, M. J., Yu Yan, Mihalas, D. & Pradhan, A. K. 1994, MNRAS, 266, 805

Simon, N. R. 1982, ApJ, 260, L87

Smith. M. A. 1977, ApJ, 215, 574

Smith, M. A. & Karp, A. H. 1976 in Proc. Solar and Stellar Pulsation Conf., eds. A. N. Cox & R. G. Deupree, Los Alamos, p. 289

Smith, M. A., Fitch, W. S., Africano, J. L., Goodrich, B. D., Halbedel, W., Palmer, L. H.& Henry, G. W. 1984, ApJ, 282, 226

Sterken, C. & Jerzykiewicz, M. 1993, Space Sci. Rev., 62, 95

Vogt, S. S. & Penrod, G. D. 1983, ApJ, 275, 661

Waelkens, C. 1991, A&A, 246, 453

Waelkens, C. 1994, in The Impact of Long-Term Monitoring on Variable Star Research, eds. C. Sterken & M. de Groot, NATO ASI Ser., Vol. 436, Dordrecht: Kluwer, p. 255

Walker, G. A. H., Yang, S. & Fahlmman, G. G. 1979, ApJ, 233, 199

## Discussion

*Maeder*: In principle, as you showed, the driving of $\beta$ Cep pulsations depends very much on metallicity. Is there any evidence of differences in the amplitudes or so, related to the galactic locations of $\beta$ Cep stars?

*Moskalik*: Pulsation amplitude is not directly connected with the growth rate (it depends on other factors as well) and cannot be used as an indicator of the strength of driving. On the other hand, with the increasing metallicity the blue edge of the instability strip becomes hotter and the $\beta$ Cep population should be on average bluer. This effect might be already observed by Waelkens et al. (A&A, 251, 69), who notice that $\beta$ Cep stars close to the centre of the Galaxy are all fairly blue.

*Bedding*: How firm is your statement that there should be no $\beta$ Cep stars in the LMC? Kjeldsen & Baade (IAU Symp. 162, p. 29) found tentative evidence for *low-amplitude* $\beta$ Cep pulsation in the young LMC cluster NGC 371.

*Moskalik*: As we decrease the metallicity of the models the instability strip becomes narrower and shifts progressively up the H–R diagram and finally disappears. The exact value of $Z$ at which the strip vanishes depends somewhat on the metal mixture and also on the choice of the opacity tables. With OP tables and S92 mixture we find $Z = 0.012$ as the threshold value. The metallicity of the LMC is on average $Z \approx 0.014$ (1.4 times less than solar) and in a young cluster it might be slightly higher. According to these numbers, the $\beta$ Cep stars could be present in the LMC, but they should be very rare. On the other hand, the SMC is more metal deficient ($Z \approx 0.005$) and there should be no $\beta$ Cep stars there, unless the stellar population is locally enriched in the iron-group elements.

*Smith*: How do you solve the problem in your picture of avoiding a plethora of high overtone modes being excited with long periods in 53 Per (SPB) stars?

*Moskalik*: I don't. The results, which I have presented are obtained with the linear pulsation theory, which only tells us which modes are excited. The questions of mode selection and of the final pulsation amplitudes lie in the domain of a nonlinear theory. We have no credible nonlinear theory for these stars. The problem you mentioned is not specific to the SPB stars; for all the multimode pulsators ($\beta$ Cep, SPB, $\delta$ Sct, white dwarfs) we have many modes which are theoretically excited and only a small fraction of them is observed.

*Smith*: Suppose it were the case, as many of us in the B star community have suggested, that there are long period nonradial modes in late-O/early-B 53 Per stars and/or in Be stars. Would you agree in this case that an additional excitation mechanism was necessary?

*Moskalik*: Our current theory describes nonrotating or slowly rotating stars and as such cannot be directly applied to the Be stars. Whether metal bump driving would work for these fast rotating objects is an open question. The 53 Per/SPB stars are all slow rotators, so the theory does apply here. The models predict excitation of modes with periods of 0.5 – 4.0 day. The longer period modes cannot be driven because the thermal timescale in the zone of the metal bump is too short for them. If you observe modes with periods significantly longer than 4.0 day, then indeed an additional driving mechanism would be necessary.

*Jerzykiewicz*: You cannot explain 53 Per stars by invoking high $\ell$ because they were defined by Myron Smith as $\ell = 2$ pulsators. The high $\ell$ late O and early B pulsators are called $\zeta$ Oph stars.

*Moskalik*: I had in mind the modes of $\ell$ only slightly higher, say $\ell = 3, 4$. Such modes will still cause the *low order* line profile variability, yet they will be practically undetectable photometrically, because of very strong cancellation effect.

*Bono*: What is the physical mechanism which provides the quenching of the pulsation close to the red edge?

*Moskalik*: The $\beta$ Cep pulsations are driven by $\kappa$-mechanism acting on the metal opacity bump at $\log T \approx 5.3$. At lower $T_{\rm eff}$ the metal bump is located deeper in the envelope and the thermal timescale $\tau_{th}$ of the zone becomes longer. (The periods become longer as well, but this effect is smaller.) As we approach the red edge of the strip, the timescale condition $\tau_{th} \approx P$ is satisfied already at $\log T < 5.3$. Consequently, the damping zone around $\log T = 5.0$ is activated and the mode becomes linearly stable. This is the same reason for which $p$-modes are stable in the SPB stars (see Fig. 2).

The answer is different for the $g$-modes in the SPB stars. In this case the red edge of the instability strip is given by TAMS. As the evolution progresses beyond TAMS, the star develops a wide $\mu$-gradient zone left behind the shrinking core. The high value of the Brunt-Väisälä frequency causes the $g$-mode eigenfunctions to have large amplitudes and short wavelengths here. The resulting very strong local radiative damping overcomes the envelope $\kappa$-mechanism driving.

*Astrophysical Applications of Stellar Pulsation*
*ASP Conference Series, Vol. 83, 1995*
*R. S. Stobie and P. A. Whitelock, eds.*

# Mode Identification of the $\beta$ Cephei star BW Vulpeculae [1]

C. Aerts[2], T. Van Hoolst, and P. Mathias[3]

*Instituut voor Sterrenkunde, Katholieke Universiteit Leuven, Celestijnenlaan 200 B, B-3001 Heverlee, Belgium*

**Abstract.** We identify the pulsation mode of BW Vul by means of both the moment method and the method of photometric amplitudes and find a radial pulsation. We briefly study the non-linear behaviour of BW Vul.

## 1. Introduction

BW Vul (HD 199140) is a $\beta$ Cephei star of which the radial-velocity (RV) curve has an extremely large amplitude (200 km/s peak-to-peak) and shows a velocity still-stand. A detailed spectroscopic analysis of BW Vul is presented by Furenlid et al. (1987), while Moskalik & Buchler (1994) studied hydrodynamical models for BW Vul and found that the star pulsates in a fundamental mode. Both studies are based on the *assumption* that BW Vul exhibits a radial pulsation, although existing mode identifications point towards a non-radial mode ($\ell = 2, m = -1$: Kubiak 1978, $\ell = -m = 2$: Odell 1981).

## 2. Observations

Our 97 spectra were obtained with the CFHT f8.2 Coudé Spectrograph on July 10-13, 1992. A resolving power of 30,000, a S/N ratio of 250, and a time resolution of P/30 (P=4h49min) was obtained. The region of the spectrum centred around the Si III triplet ($\lambda\lambda$ 4553, 4568, 4575) was observed. Twenty-four multicolour observations in the Geneva system, spread over 10 years, were kindly put at our disposal by the Observatoire de Genève.

## 3. Mode identification

The question has arisen if the harmonics in the light-curve of BW Vul (van der Linden & Sterken 1987) are due to a non-linear effect of one mode or to the presence of different modes. Both the multicolour photometry and the variation

[1]Based on observations obtained with the CFHT, Hawaii

[2]Senior Research Assistant, Belgian National Fund for Scientific Research

[3]Present Address: Observatoire de la Côte d'Azur, Département Fresnel, BP 229, F-06304 Nice Cedex 4, France

of the line parameters (RV, EW, FWHM) indicate that the relation between the amplitudes of the variations with $f = 4.974\,c/d$ and with its harmonics is the same. This is not what we expect if different modes would be present, since they would have a different signature in, e.g., the RV compared to the FWHM, and also in the colours. We thus prefer a mono-periodic pulsation model for BW Vul.

The mode is identified by means of the moment method (Aerts et al. 1992), from which we find a radial pulsation with an amplitude of 90 km/s. This result is confirmed by an analysis with the method of photometric amplitudes (Heynderickx et al. 1994), from which we also find that non-adiabatic effects are important in BW Vul. We thus contradict earlier identifications and find a result that is fully consistent with the studies by Furenlid et al. (1987) and by Moskalik & Buchler (1994).

## 4. Non-linear pulsations

In order to quantify the non-linear effects of BW Vul, we have calculated a RV curve for a non-linear radial pulsation in case of BW Vul by means of amplitude equations. Our theoretical curve shows no velocity still-stand, and the amplitudes of the terms varying with the harmonics are much smaller than the corresponding observed ones. Possible explanations are that we have used an adiabatic theory, that shocks are not taken into account with amplitude equations, and that the radiative transfer in the atmospheric layers was not taken into account.

## 5. Conclusions

We have found that BW Vul pulsates in a radial mode, that its non-adiabatic effects are important, and that its non-linear effects are large and cannot be treated by means of amplitude equations in the adiabatic approximation. We plan to study our line profiles in more detail by means of a non-adiabatic, non-linear pulsation theory in combination with hydrodynamical models.

Full details of this study will be published elsewhere (submitted to A&A).

## References

Aerts, C., De Pauw, M., & Waelkens, C. 1992, A&A, 266, 294

Furenlid, I., et al., 1987, ApJ, 319, 264

Heynderickx, D., Waelkens, C., & Smeyers, P. 1994, A&AS, 105, 447

Kubiak, M. 1978, Acta Astron., 28, 153

Moskalik, P., & Buchler, J.R. 1994, in *Pulsation, Rotation and Mass Loss in Early-Type Stars*, eds. L.A. Balona, H.F. Henrichs & J.M. LeContel, Kluwer Academisch Publishers, p.19

Odell, A.P. 1981, ApJ, 246, L77

van der Linden, D., & Sterken, C. 1987, A&A, 186, 129

*Astrophysical Applications of Stellar Pulsation*
*ASP Conference Series, Vol. 83, 1995*
*R. S. Stobie and P. A. Whitelock, eds.*

# Rapid Pulsations in Ap Stars

Peter Martinez and D.W. Kurtz
*Department of Astronomy, University of Cape Town, Rondebosch 7700, South Africa*

**Abstract.** The rapidly oscillating Ap (roAp) stars are upper main-sequence stars that pulsate in non-radial $p$ modes of high overtone. We present an observer's overview of the roAp phenomenon and discuss significant developments since the last major observational reviews of Kurtz (1990a) and Matthews (1991).

## 1. Introduction

The Ap (for 'A peculiar') stars are magnetic, chemically peculiar main-sequence stars spanning a temperature range that starts just below the cool edge of the instability strip and extends well beyond the blue edge to hotter temperatures (Wolff 1983). The magnetic fields in these stars are predominantly global dipolar fields with typical effective field strengths of 1–2 kG. The Ap stars exhibit spectrum, magnetic, photometric and radial velocity variability, all with the same period in a given star. The oblique rotator model (Stibbs 1950, Wolff 1983) explains the variations by postulating that the spectral peculiarities originate in spots of anomalous abundance on the stellar surface and that the magnetic axis is inclined with respect to the rotation axis. In this model the variations are a consequence of seeing the spots and the magnetic geometry from variable aspect as the star rotates. The coolest Ap stars, those with anomalously strong lines of Sr, Cr, Eu and other rare earths, overlap the instability strip where it crosses the main sequence. Kurtz's (1982) discovery of rapid pulsations in a number of cool Ap stars has elicited much interest because of the potential for using the pulsations to probe these stars' magnetic structures, atmospheres and interiors. Much of the early work on these rapidly oscillating Ap (roAp) stars concentrated on characterizing the roAp phenomenon and searching for new members of the class. Table 1 lists the 27 roAp stars known as of this writing. The Cape roAp star survey, the most prolific source of new roAp stars, is described by Martinez elsewhere in these proceedings. This review begins with an overview of our current understanding of the phenomenology of roAp star light curves. This is followed by a discussion of the physics of the roAp phenomenon and the oblique pulsator model. In the second half of the review, we focus on recent observational developments. This review builds on recent reviews by Kurtz (1990a), Matthews (1991) and Shibahashi (1991). Kurtz and Matthews emphasize the observations while Shibahashi emphasizes theoretical developments.

Table 1. The rapidly oscillating Ap stars.

| HD<br>*HR* | $V$ | Spectral Type | Period[a] (min) | $\Delta B$[b] | $\Delta\nu$[c] ($\mu$Hz) | $P_{rot}$ (day) | $B_{eff}$ (G) |
|---|---|---|---|---|---|---|---|
| 6532 | 8.4 | Ap SrCrEu | 7.1 | 5 | 47 | 1.9448 | |
| 9289 | 9.4 | Ap SrEu | 10.5 | 3 | | | |
| 12932 | 10.2 | Ap SrEuCr | 11.6 | 4 | | | |
| 19918 | 9.3 | Ap SrEuCr | 14.5 | 2 | | | |
| 24712<br>*1217* | 6.0 | Ap SrEu(Cr) | 6.2 | 10 | 68 | 12.4572 | +400 to +1300 |
| 42659 | 6.8 | Ap SrCrEu | 9.7 | 0.8 | 52 | | |
| 60435 | 8.9 | Ap Sr(Eu) | 12.0[d] | 16 | | 7.6793 | < 1000 |
| 80316 | 7.8 | Ap Sr(Eu) | 7.4 | 2 | | 2.1? | |
| 83368<br>*3831* | 6.2 | Ap SrEuCr | 11.6 | 10 | | 2.851962 | -700 to +700 |
| 84041 | 9.3 | Ap SrEuCr | 15.0 | 6 | 60 | 3.69 | |
| 86181 | 9.3 | Ap Sr | 6.2 | 5 | | | |
| 101065 | 8.0 | Controversial | 12.1 | 16 | 58 | 3.94? | -2200 |
| 119027 | 10.0 | Ap SrEu(Cr) | 8.7 | 2 | 52 | | |
| 128898<br>*5463* | 3.2 | Ap SrEu(Cr) | 6.8 | 5 | 50 | 4.46 | -300 |
| 134214 | 7.5 | Ap SrEu(Cr) | 5.6 | 7 | | | |
| 137949 | 6.7 | Ap SrEuCr | 8.3 | 3 | 40 | | +1400 to +1800 |
| 150562 | 9.8 | A/F(p Eu) | 10.8 | 0.8 | | | |
| 161459 | 10.3 | Ap EuSrCr | 12.0 | 1.3 | | | |
| 166473 | 7.9 | Ap SrEuCr | 8.8 | 2 | 68 | | |
| 176232<br>*7167* | 5.9 | F0p SrEu | 11.6 | 0.6 | 51 | | |
| 190290 | 9.9 | Ap EuSr | 7.3 | 2 | 58 | | |
| 193756 | 9.2 | Ap SrCrEu | 13.0 | 0.9 | | | |
| 196470 | 9.7 | Ap SrEu(Cr) | 10.8 | 0.7 | | | |
| 201601<br>*8097* | 4.7 | F0p | 12.4 | 3 | 58 | | -800 to +500 |
| 203932 | 8.8 | Ap SrEu | 5.9 | 2 | 66 | | |
| 217522 | 7.5 | Ap (Si)Cr | 13.9 | 4 | 31 | | |
| 218495 | 9.4 | Ap EuSr | 7.4 | 1 | | | |

[a] In the case of multi-periodic stars, the period listed is that with the highest amplitude.

[b] $\Delta B_{max}$, the amplitudes listed are peak-to-peak amplitudes, in mmag.

[c] This column lists the asymptotic $p$-mode spacing inferred from measurements of the frequency spacing in the multi-periodic stars.

[d] This is a representative value for the most commonly detected periods in this star.

## 2. The phenomenology of roAp star light curves

1. The pulsation periods of the roAp stars are in the range 5.6 to 15.0 min. The amplitudes are all $\leq$ 16 millimagnitudes (mmag) peak to peak in Johnson $B$ light. Because of the shortness of these periods, roAp stars are detected and studied using non-differential high-speed photometry. Experience has shown that it is possible to distinguish between real variations in the star and variations caused by sky transparency. Nonetheless, stable, well maintained equipment and an excellent photometric site are required for roAp star photometry. The observational techniques and common pitfalls have been discussed in detail by Martinez (1993a).

2. The light curves of many roAp stars show complex changes in amplitude on time-scales of several hours. This is the signature of beating among several excited modes. In the Fourier frequency domain, the signal content of the oscillations manifests itself as a region of power containing a series of equally spaced peaks. This is reminiscent of the Fourier spectrum of the solar 5-min oscillations. There is a significant difference though; in the Sun, the envelope of excited frequencies contains all the modes expected, whereas in the roAp stars only a few modes are excited to observable amplitudes.

3. In about $\frac{1}{3}$ of the roAp stars the harmonics of the principal oscillation frequencies are seen in the Fourier spectra of the light curves. What this means is that the pulse shape in these stars is non-sinusoidal, implying that the oscillations are non-linear. These non-linearities are not understood; $\delta$ Scuti stars which oscillate with amplitudes of a few tenths of a magnitude do not show harmonics, yet roAp stars which oscillate with amplitudes of 1–2 mmag do.

4. In addition to short-term amplitude modulation caused by beating of independent pulsation frequencies, the oscillations also undergo amplitude modulation on time scales of several days, or longer. For those roAp stars with known magnetic periods, the period of the amplitude modulation is equal to the rotation period. The bottom panel in Fig. 1 illustrates this for HR 3831. The pulsation amplitude is maximum at magnetic extremum (i.e., when one of the magnetic poles is viewed directly) and minimum at quadrature. This amplitude modulation manifests itself in the Fourier spectrum of the oscillations as an equally spaced frequency multiplet centered on the pulsation frequency. The spacing of the frequencies in the multiplet is equal to the rotation frequency.

5. For HR 3831 Kurtz et al. (1992b) have shown that the period of the pulsation amplitude modulation is equal to the period of the mean-light photometric variation caused by oblique rotation. They use this to show that an interpretation of the pulsation amplitude modulation as rotationally perturbed $m$ modes requires that the Ledoux (1951) rotational splitting constant $C_{n\ell} \leq 2 \times 10^{-5}$ at the $3\sigma$ confidence level, making that interpretation highly unlikely.

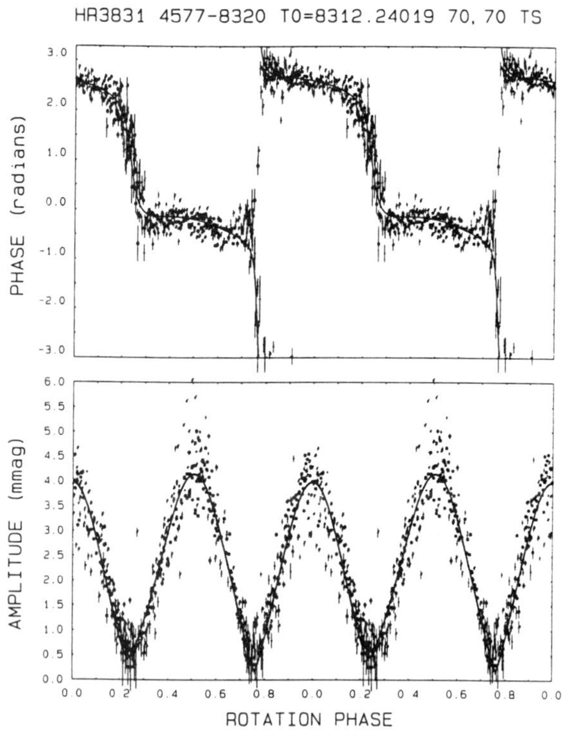


Figure 1. The observed phase and amplitude of the pulsations in HR 3831 as a function of rotation phase. Phase 0.0 is defined to be magnetic maximum.

6. Kurtz et al. (1992b) showed for HR 3831 that the time of magnetic extremum coincides with the time of pulsation-amplitude maximum but differs from the mean-light extremum at the $5\sigma$ confidence level. In HR 1217, a similar result obtains at the $3\sigma$ level (Kurtz et al. 1989). These findings indicate a deviation from cylindrical symmetry for the magnetic field geometry and abundance distributions in those two stars.

7. The phase behaviour of the rapid oscillations has been studied in detail for two roAp stars, namely HD 6532 and HR 3831. Between quadratures, the phase is approximately constant. At quadratures, the phase undergoes a $\pi$-radian jump. This variation is shown for HR 3831 in Figure 1.

8. The oscillation spectra of the roAp stars are not stationary. In some multi-periodic stars the distribution of amplitudes among the different frequencies changes on a time-scale shorter than the rotation period. In other stars a given frequency pattern is constant for a few rotation cycles. The topic of mode changes and lifetimes forms the subject of Section 5.

9. It was recently discovered that the frequency of the oscillation in the singly-periodic star HR 3831 varies on a time-scale of several hundred days in a fashion not yet well characterized. Similar, though less extensive results have been obtained for a few other roAp stars and it appears that such frequency variations may be the norm, rather than the exception. This topic is discussed fully in Section 6.

## 3. The physics of the roAp phenomenon

The shortness of the pulsation periods in the roAp stars implies that they pulsate in high-order $p$ modes similar to those of the well-studied solar 5-minute oscillations. The very fact that we observe the oscillations photometrically implies that low-order modes are excited. The asymptotic theory of low-degree, high-overtone ($n \gg \ell$) $p$-mode pulsation (Tassoul 1980) predicts that the eigenfrequencies of the modes are given by

$$\nu_{n\ell} \simeq \Delta\nu(n + \ell/2 + \epsilon) + \delta\nu, \tag{1}$$

where $\epsilon$ is a constant dependent on the equilibrium structure of the star and $\delta\nu$ is a second-order term. To first order, the eigenfrequencies have a uniform spacing $\Delta\nu$ given by

$$\Delta\nu = \left(2\int_0^R \frac{dr}{c(r)}\right)^{-1}, \tag{2}$$

where $c(r)$ is the sound speed. This spacing $\Delta\nu$ is the reciprocal of the time it takes for a sound wave to cross the star. For the Sun $\Delta\nu = 135\,\mu$Hz, which translates to a sound crossing time of 2.06 hr. The roAp stars have $\Delta\nu$ values between 30 and 70 $\mu$Hz, corresponding to sound crossing times of 9.3–4.0 hr, consistent with their larger radii. When $\Delta\nu$ is expressed in terms of structural parameters by $\Delta\nu \propto \sqrt(GM/R^3)$, one may readily show, using the mass-luminosity relation, that the loci of constant $\Delta\nu$ are essentially lines of constant $R$ in a theoretical H-R diagram. These loci have been calculated by a variety of people modelling $p$-mode oscillations in A-type stars, *viz.* Gabriel et al. (1985), Shibahashi & Saio (1985) and Heller & Kawaler (1988).

One of the properties of eq. (1) is that, to first order, it is degenerate for modes with the same value of $(n + \ell/2)$. Thus, consecutive overtones $(n, n+1)$ of modes with exclusively even or odd $\ell$ are spaced by $\Delta\nu$, whereas alternating even and odd $\ell$ modes are spaced by $\frac{1}{2}\Delta\nu$. The problem is to decide whether the observed spacing in a given star corresponds to $\Delta\nu$ or $\frac{1}{2}\Delta\nu$. The various A-star models mentioned above all indicate that $\Delta\nu = 50$–$80\,\mu$Hz for a slightly evolved $2M_\odot$ star. In several roAp stars a frequency spacing of $\sim 25$–$30\,\mu$Hz is observed. In such cases, the theoretical overtone spacing, and the fact that Ap stars are not highly evolved objects, are used to argue that alternating even and odd $\ell$ modes are present. However, such conclusions are only secure if an independent luminosity estimate is available for the star. Matthews (1989) has noted that the ratio $\frac{\Delta\nu}{<\nu>}$ falls around the preferred values 0.040 and 0.025 for the roAp stars. This result, if confirmed, could be used to remove the ambiguity in determining $\Delta\nu$. Another way suggested to determine $\Delta\nu$ is to measure the rate of frequency change due to evolution (Heller & Kawaler 1988). A more evolved star has a smaller value of $\Delta\nu$ and a higher rate of frequency change. In practice this is probably not possible for reasons discussed in Section 6.

We now turn to a discussion of the second-order term, $\delta\nu$. We have already noted that Tassoul's relation is degenerate for modes $(n, \ell)$ and $(n \pm 1, \ell \mp 2)$. The term $\delta\nu$ lifts this degeneracy by introducing a slight frequency separation $\delta\nu \ll \Delta\nu$ between such modes. Physically, what happens is that modes $(n, \ell)$ and $(n \pm 1, \ell \mp 2)$ have very similar mode structures outside the core but slightly dissimilar mode structures in the centre. This difference in central mode structures is a measure of the sound speed gradient in the stellar core. Since this gradient changes as nuclear burning changes the molecular weight distribution in the core, the small frequency separation contains information about the star's evolutionary state. Christensen-Dalsgaard (1988) has created a theoretical asteroseismological H-R diagram by plotting $\Delta\nu$ against $\delta\nu$ for a grid of models. By using these two observable quantities to plot stars in such a diagram, it should be possible to obtain asteroseismological mass and age estimates. Much observational work in recent years has been aimed at providing stars to calibrate such a diagram. At present, the Sun is the only star which can be plotted on this diagram with certainty.

## 4. The Oblique Pulsator Model

Much of the phenomenology of roAp star light curves can be understood in terms of the oblique pulsator model (Kurtz 1982, 1990a). This model postulates that a roAp star pulsates non-radially and that the pulsation axis coincides with the magnetic axis, which is itself inclined with respect to the rotation axis. This very simple model can explain the observed amplitude modulation of the rapid oscillations and the oscillation phase flips at magnetic crossover (Fig. 1) as natural consequences of viewing the pulsations from varying aspect as the star rotates. The oblique pulsator model predicts that a frequency of mode $\ell$ is split by rotation into a multiplet of $(2\ell + 1)$ frequencies. The multiplet is centered on the pulsation frequency in the rest frame of the star, and the multiplet frequency spacing is the rotation frequency, $\Omega_{rot}$. The amplitudes of the side-band components are determined by the relative strength of the Coriolis force ($\propto \Omega_{rot}$) to the Lorentz force ($\propto B^2$). In the limiting case where magnetic effects dominate, the pulsation axis is locked to the magnetic axis and the resulting frequency multiplet is symmetric. As rotation effects increase, the pattern of side-lobes becomes increasingly asymmetrical.

When it was first proposed, this model seemed to offer a substantially adequate explanation for the observation of an equally spaced asymmetric frequency triplet in the roAp star HR 3831. However, in the last few years it has become apparent that the situation is a good deal more complicated than was previously thought. Kurtz et al. (1993) have discovered that the fine structure in HR 3831 originally thought to be a triplet is in fact part of a septuplet. Indeed, they showed that it is possible to express the oscillations as a sum of axisymmetric spherical harmonics of degree $\ell = 0$, 1, 2 and 3 using the spherical harmonic decomposition technique developed by Kurtz (1992a). The solid lines in Fig. 1 show the fits they obtained. Although the observations were well reproduced by this procedure, there was no physical justification for using it. To explain the septuplet in HR 3831, Shibahashi & Takata (1993) refined the treatment of the effects of the Lorentz and Coriolis forces on non-radial oscillations in a rotating magnetic oblique pulsator. Using their refined model, they showed that the effect of the magnetic field on an axisymmetric dipole mode is to distort the dipole oscillation pattern in such a way that axisymmetric octupole components arise; this leads to an asymmetric equally spaced frequency septuplet instead of a triplet. This distortion of the basic dipole pattern led Shibahashi & Takata to christen their model the 'distorted oblique pulsator model.' [1]

## 5. Mode changes and mode lifetimes

Several roAp stars are observed to switch their principal pulsation frequency from time to time. Such frequency shifts are usually commensurate with some integer multiple of the $p$-mode spacing, $\Delta\nu$, and they are interpreted as shifts to a nearby overtone. An example of a star where such frequency shifts have been noted is HD 217522 (Kreidl et al. 1991). Observations in 1982 showed pulsation with a frequency $\nu_1 = 1215.29\,\mu$Hz whereas observations in 1989 showed that the principal pulsation frequency had shifted to $\nu_1 = 1199.9\,\mu$Hz, a shift of $15.4 \pm 0.1\,\mu$Hz. Moreover, the 1989 data showed a new frequency at $\nu_2 = 2017.4\,\mu$Hz at such a significance level that there was no chance it could have been missed in the 1982 observations. Such a frequency shift is orders of magnitude too high to arise from evolutionary changes in HD 217522. Kreidl et al. have interpreted the frequency shift as a mode change and have used the size of the shift to estimate the mass and luminosity of HD 217522.

In HD 217522, the mode change was not actually observed, but for the roAp star HD 60435 a transition from one pulsation mode to another was well observed. HD 60435 has the richest $p$-mode spectrum of the roAp stars. It pulsates with many frequencies having a basic spacing of $25.8\,\mu$Hz. This spacing is interpreted as $\Delta\nu/2$, the spacing of alternating even and odd $\ell$ modes. During a multi-site observing campaign from Chile and South Africa in 1985 (Matthews et al. 1987) this star was observed to undergo several mode transitions. In fact, so many mode changes were observed that one cannot meaningfully talk about a principal oscillation frequency for HD 60435. The lifetimes of some of the modes seem to be as short as a few days. A rotation period of $7.6793 \pm 0.0006$ days has been determined for HD 60435 by Kurtz et al. (1990b) from a study of its mean light variations. In many cases the modes have lifetimes shorter than one rotation cycle. These conclusions are secure. They were obtained from over 400 hr of contemporaneous multi-site data and there is no chance that they are artefacts produced by aliasing and frequency resolution problems.

There are other roAp stars with mode lifetimes on time-scales on days to weeks. Examples of such stars are HD 203932 (Martinez et al. 1990b), HR 1217

[1] We do not review the Spotted Pulsator Model of Mathys (1985) here as it has not been developed further since last reviewed by Kurtz (1990a). We note that although the oblique pulsator model is more successful at explaining the phenomenology of roAp light curves, the physics invoked by the spotted pulsator model may yet prove to be necessary for a detailed, complete description of the oscillations.

(Kurtz et al. 1989), HD 84041 (Martinez et al 1993b), HD 201601 (Libbrecht 1988) and HD 119027 (Martinez et al 1993c). These stars are not as well studied as HD 60435, but there is strong evidence that they pulsate in several modes and that their behaviour is qualitatively similar to that of HD 60435.

Kurtz (1990a) has pointed out that the pulsation mode lifetimes appear to be longer in those stars which pulsate in a few modes and shorter in stars with rich $p$-mode spectra. An example of a mode-stable star is HR 3831, which is singly periodic with harmonics. The principal dipole oscillation is amplitude-stable over more than 10 yr. There are slight frequency changes on time-scales of several hundred days which are discussed in Section 6. Another example of a well studied mode-stable star is HD 101065 which pulsates with one large amplitude mode and at least one other much smaller amplitude mode (Martinez & Kurtz 1990a). The large-amplitude mode seems to suffer some amplitude or phase instability on time-scales of 10 yr but is stable on shorter time-scales. HR 1217 pulsates with 6 modes (Kurtz et al. 1989) that are stable over one rotation cycle, but variable on a time-scale of 4 rotation cycles.

## 6. Frequency changes

From an asteroseismological standpoint, the richer the oscillation spectrum of the star, the more information it contains. However, there is one very important parameter which is best studied in singly periodic stars — it is the rate of frequency change. The way in which a frequency changes is an important clue to the cause of the change. A frequency change can be gradual or abrupt; it can be secular or periodic. Two of the most probable causes of gradual frequency change in a pulsating star are evolution and orbital motion. It is usually possible to identify orbital motion since the Doppler frequency shifts are periodic and sinusoidal (for circular orbits). Intrinsic frequency variations pose a tougher problem because the long-term secular frequency changes expected from evolution are confounded with comparatively short-term variations which are not yet understood.

Heller & Kawaler (1988) have investigated evolutionary period changes theoretically for a series of A star models which they allowed to evolve off the main sequence. Their models yield values of $|d\ln\nu/dt| \sim 10^{-16}$–$10^{-17}\,\mathrm{s}^{-1}$. For a given $T_{eff}$, models with a smaller $p$-mode spacing have a higher luminosity and a higher $|d\nu/dt|$. By using the measured values of $d\nu/dt$ it should be possible to resolve ambiguities in the determination of $\Delta\nu$. In HR 1217, for example, there are two possible values of $\Delta\nu$, *viz.* $\Delta\nu = 34\,\mu$Hz and $68\,\mu$Hz (Kurtz et al. 1989). According to Heller & Kawaler's models, $d\nu/dt$ should be detectable with the 6 yr of observations available if $\Delta\nu = 34\,\mu$Hz, whereas it should not be detectable if $\Delta\nu = 68\,\mu$Hz. Unfortunately, Kurtz et al. (1989) find that alias ambiguities and the interaction of the many frequencies in HR 1217 preclude this. This illustrates the importance of the remark made in the opening paragraph of this section that $d\nu/dt$ is more readily determined in singly periodic stars. It also demonstrates the practical difficulty of using $d\nu/dt$ measurements to resolve ambiguities in the determination of $\Delta\nu$.

Several other roAp stars with less complicated oscillation spectra than HR 1217 have data sets long enough to permit a determination of $d\nu/dt$. In

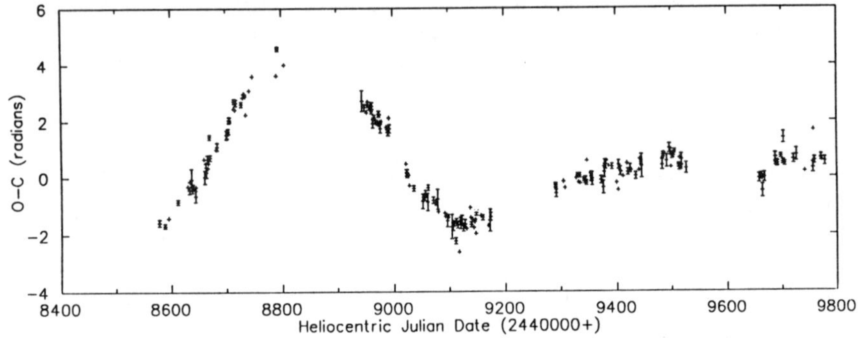


Figure 2. The long-term behaviour of the pulsation phase in HR 3831.

HD 101065, which oscillates with three frequencies, Martinez & Kurtz (1990a) find a rate of frequency change 10–100 times higher, and of the opposite sign, than expected for a hydrogen-core-burning A star. HD 137949 is another star that shows a period change several orders of magnitude larger than theoretical expectations for evolutionary effects (Kurtz, 1991).

Intrigued by these large frequency shifts, three years ago we started monitoring the pulsation frequency of HR 3831 on every possible night throughout its entire observing season (Fig. 2). These data show that the frequency of the principal pulsation mode varies cyclically by about 0.12 $\mu$Hz on a time-scale of about two years. Note that by 'cyclic' we do not mean 'periodic' - the exact nature of the variations remains to be characterized. Similar, though less extensive, results have been obtained for $\alpha$ Cir (Kurtz et al. 1994b), HD 134214 (Kreidl et al. 1994) and HD 12932 (Martinez et al 1994). In fact, there are no roAp stars which have been monitored over several seasons for which a single, constant frequency fits the data. Frequency variability in these stars seems to be the norm. In a thorough analysis of 12 yr of HR 3831 data, Kurtz et al. (1994a) were able to exclude abrupt phase changes, a drift of the pulsation pole with respect to the magnetic pole, and orbital motion as the cause of the frequency variations. They also excluded evolutionary changes because $|d\nu/dt|$ is orders of magnitude higher than expected and its sign changes from season to season, contrary to all expectations of the way an evolutionary frequency change should behave.

So what is the cause of these frequency variations? For a possible answer to this question, we turn to helioseismology. One of the interesting recent results of helioseismology is that the frequencies of the solar $p$ modes vary with the solar cycle (Elsworth et al. (1990), Libbrecht & Woodard (1990) and Bachmann & Brown (1993)). The range of that variation is about 0.5 $\mu$Hz. This motivated Kurtz et al. (1994a) to suggest that the frequency variations in HR 3831 may be caused by variations in the stellar magnetic field, possibly in a stellar magnetic cycle. A direct observational test of this suggestion is not possible at present since inference from the Sun suggests that the these conjectured magnetic variations are orders of magnitude lower than can be detected with conventional magnetic measurements. Recently, Soon et al. (1993) suggested that the square of the ratio of the chromospheric oscillatory time-scale to the rotation period,

$(P_{cyc}/P_{rot})^2$, is a measure of the theoretical dynamo number for cool, chromospherically active stars. Mindful of the danger of applying this suggestion to A stars, if we identify $(P_{cyc})$ with the cycle time of the frequency variation in HR 3831 we find $(P_{cyc}/P_{rot})^2 = 4.6$, which fits the extrapolation of Soon et al.'s relationship to higher temperature very well. Is this agreement purely a coincidence? Further monitoring of the principal pulsation frequency in HR 3831 and other roAp stars will allow us to test this.

## References


Bachmann, T.K., & Brown, T.M. 1993, ApJ, 411, L45

Christensen-Dalsgaard, J. 1988, Proc. IAU Symp. No. 123, p. 295, (ed.) Christensen-Dalsgaard, J. & Frandsen, S., Reidel, Dordrecht, Holland

Elsworth, Y., Howe, R., Isaak, G.R., McLeod, C.P. & New, R. 1990, Nature 345, 322

Gabriel, M., Noels, A., Scuflaire, R. & Mathys, G. 1985, A&A, 143, 206

Heller, C.H., & Kawaler, S.D. 1988, ApJ, 329, L43

Kreidl, T.J., et al. 1991, MNRAS, 250, 477

Kreidl, T.J., et al. 1994, MNRAS, 270 115

Kurtz, D.W. 1982, MNRAS, 200, 807

Kurtz, D.W. et al. 1989, MNRAS, 240, 881

Kurtz, D.W. 1990a, ARA&A, 28, 607

Kurtz, D.W., van Wyk, F., & Marang, F. 1990b, MNRAS, 243, 289

Kurtz, D.W. 1991, MNRAS, 249, 468

Kurtz, D.W. 1992a, MNRAS, 259, 700

Kurtz, D.W., Kanaan, A., Martinez, P., & Tripe, P., 1992b, MNRAS, 255, 289

Kurtz, D.W., Kanaan, A., & Martinez, P. 1993, MNRAS, 260, 343

Kurtz, D.W., Martinez, P., van Wyk, F., Marang, F., & Roberts, G. 1994a, MNRAS, 268, 641

Kurtz, D.W., Sullivan, D.J., Martinez, P., & Tripe, P. 1994b, MNRAS, 270, 674

Ledoux, P. 1951, ApJ, 114, 373

Libbrecht, K.G. 1988, ApJ, 330, L51

Libbrecht, K.G., & Woodard, M.F. 1990, Nature, 345, 779

Martinez, P., & Kurtz, D.W. 1990a, MNRAS, 242, 636

Martinez, P., Kurtz, D.W., & Heller, C.H. 1990b, MNRAS, 246, 699

Martinez, P. 1993a, in: D. Kilkenny, E. Lastovica, E. & J.W. Menzies (eds.), *Precision Photometry*, South African Astronomical Observatory, Cape Town, p. 134

Martinez, P., et al. 1993b, MNRAS, 263, 273

Martinez, P., Kurtz, D.W., & Meintjes, P. 1993c, MNRAS, 260, 9

Martinez, P., Kurtz, D.W., & van Wyk, F. 1994, MNRAS, 271, 305

Mathys, G. 1985, A&A, 151, 315

Matthews, J.M., Kurtz, D.W., & Wehlau, W.H. 1987, ApJ, 313, 782

Matthews, J.M. 1989, In: Seismology of the Sun and Sun-like stars, ESA Special Publication 286, p. 547

Matthews, J.M. 1991, PASP, 103, 5

Shibahashi, H., & Saio, H. 1985, PASJ, 37, 245

Shibahashi, H. 1991, In: Challenges to theories of the structure of moderate-mass stars, ed. Gough, D.O. & Toomre, J., Springer, Berlin, p. 393

Shibahashi, H., & Takata, M. 1993, PASJ, 45, 617

Soon W.H., Baliunas, S.L., & Zhang, Q. 1993, ApJ, 414, L33

Stibbs, D.W.N. 1950, MNRAS, 110, 395

Tassoul, M. 1980, ApJS, 43, 469

Wolff, S.C. 1983, The A-Type Stars: Problems and Perspectives, NASA Special Publication 463

## Discussion

*Roxburgh*: First I would like to take issue with your remark that Ap stars do not have convective regions. I would expect they have both convective core and a thin convective envelope. Next it is perfectly possible for a dynamo acting in one (or both) of these convective regions to drive a steady dynamo. It is not necessarily the case that dynamo driven fields have to be oscillating. Thus the magnetic field could be dynamo driven rather than fossil – or possibly have a dynamo component in addition to a fossil field. Since the Alfven travel time from centre to surface could be of the order of 1 year, this could produce surface variations which could be responsible for the observed variations in frequency.

*Waelkens*: Is there any possibility that your O-C variations are caused by orbiting planets?

*Kurtz*: No. The O-C diagrams do not repeat exactly from cycle to cycle as would be expected for a Doppler shift in frequency caused by orbital rotation.

*Maeder*: You mentioned these very interesting slow changes of the Ap stars frequencies. Would you please comment a bit more on their possible physical origin?

*Kurtz*: We have shown that the changes in the pulsation frequencies are intrinsic to the star, so this indicates a change in the sound speed or a change in the size of the acoustic cavity. We have guessed that these physical changes might be associated with a magnetic cycle, since the amplitude and time-scale of the frequency variations are similar to that seen in the Sun. The problem with the suggestion is that the magnetic fields in the Ap stars are thought to be fossil in origin, not dynamo driven, hence there is no theory to explain magnetic any magnetic cycles. So far as we know, there have been no other speculations about the physical origin of the frequency variations.

*Shibahashi*: You have concentrated in your talk mainly on the photometric observations. Some components with degree $\ell > 1$ may be substantially smeared out in the case of the luminosity variations integrated over the entire surface of the star. However, those components and the toroidal components, which might be induced by the magnetic fields, may contribute to the line profile variation. Could you say something about what has been done and what will be done in the spectroscopic observations to detect those components?

*Kurtz*: A few roAp stars have observed radial velocity variations: e.g., HR1217 was observed by Matthews, Wehlau, Walker & Young; $\gamma$ Equ by Matthews (these proceedings). No one has yet detected line profile variations.

*Jerzykiewicz*: As you know, RRc variables show pulsation frequency variations similar to those of roAp stars. Would you care to comment on this?

*Kurtz*: No.

*Smith*: Regarding your plans to invert the $\Delta m_\lambda(t)$ to obtain $T(\tau)$ gradients in the atmosphere through limb darkening, it might also be productive to attempt this in the UV with Voyager. This effect works very well for NRP B stars (53 Per itself) in which differential increments in line blanketing provide some very sensitive tests of limb darkening-dependent parameters (including the nonradial modes themselves).

*Kurtz*: We have considered the feasibility of using the GHRS on HST to get the observations. For the A stars, though, there is little flux beyond the Balmer jump. We plan to work from the ground in the visible and infrared in the near future to see what we can learn.

*Matthews*: To apply the limb darkening analysis, we must assume the flux variations *un*weighted by limb darkening. In the UV, the spectral distribution is *so* distorted by line blanketing and the Balmer jump that one must assume a model atmosphere, which is dangerous (and circular) at this stage.

*Goldreich*: Solar $p$-mode frequency changes are associated with variations of order 200 G in rms magnetic field at the level of the photosphere. Would such variations be observable in Ap stars?

*Kurtz*: The errors in magnetic field measurements for Ap stars are generally 200 G or worse. Given a good reason, though, some improvement on this could be obtained for the brighter stars with an intensive observing effort.

*Goldreich*: Similar frequency variations could result from few percent variations of entropy in the photosphere. This might result from an interplay between convection and diffusive separation of elements.

*Astrophysical Applications of Stellar Pulsation*
*ASP Conference Series, Vol. 83, 1995*
*R. S. Stobie and P. A. Whitelock, eds.*

# Astrophysical Applications of Delta Scuti Stars

Michel Breger

*Institut für Astronomie, Universität Wien, Türkenschanzstr. 17, A-1180 Wien, Austria*

**Abstract.** The $\delta$ Scuti stars pulsate with a large number of low-order radial and nonradial $p$-modes. There also exists some evidence for the presence of $g$-modes.

The pulsational variability outside the classical $\delta$ Scuti star instability strip (A1V - F0V on the main sequence) is examined. New measurements do not confirm the short-period variability of the hot B9p star ET And. However, a number of cool, early-F stars shows variability around one day: the $\gamma$ Dor variables. An explanation involving nonradial $g$ modes appears the most promising.

An examination of available information on the multi-periodicity of $\delta$ Scuti stars shows no systematic behaviour in which pulsation modes are excited to an observable level with luminosity and temperature. The asteroseismological potential of $\delta$ Scuti variables is briefly examined.

So far, the period changes of five luminous Pop. I $\delta$ Scuti variables have been studied. While stellar evolution predicts period increases, decreases have been observed and reported for all five stars. The statistical significance of this result and alternative explanations are briefly examined.

## 1. Introduction

The extension of the Cepheid Instability Strip crosses the main sequence among the A and F-type stars in a 1000K wide band. The $\delta$ Scuti variables are short-period pulsators situated on and near the main sequence inside the strip with periods from one half to eight hours. Among stars of Pop. I, the amplitudes of pulsation are usually small with visual amplitudes of a few mmag, but can be as large as 800 mmag.

The position of the $\delta$ Scuti stars of Pop. I in the Hertzsprung-Russell Diagram is shown in Fig. 1. Stars with uncertain absolute magnitudes or uncertain variability were rejected for this plot. The edges of the instability strip are drawn to represent the observed distribution of stars with photometrically detectable short-period variability.

While the Sun shows thousands of detectable pulsation modes, the maximum number of modes detected so far for individual $\delta$ Scuti stars is only around 15. An attractive aspect of the study of $\delta$ Scuti stars is the wide range of the stellar evolutionary stages of these stars, ranging from pre-main-sequence through hydrogen burning on the main sequence stage to the giant stage.

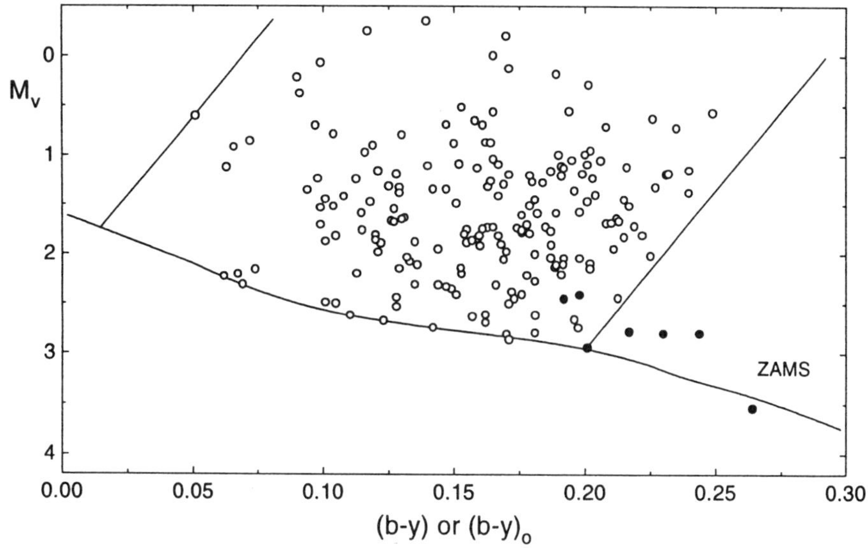


Figure 1. Position of the Pop. I $\delta$ Scuti (open circles) and the related $\gamma$ Dor variables (closed circles) in the HR diagram.

## 2. Pulsation outside the classical borders of the instability strip

The Lower Instability Strip contains several different related types of pulsators: the Pop. I $\delta$ Scuti stars (radial and nonradial low-order $p$-modes), the Pop. II $\delta$ Scuti stars (also called SX Phe stars, radial modes with large amplitudes) and the roAp stars (magnetic stars, high-order $p$-modes).

### 2.1. ET And and the Maia variables

Pulsation beyond the hot instability strip border, in the region bordered by the $\delta$ Scuti stars and the 53 Per/SPB variables (probably $g$ modes) has been suspected many times (e.g., Struve 1955, Lehmann et al. 1995) but not confirmed. Some inconclusive studies of these mythical Maia variables are available. If the reported variability were real, the period lengths would indicate radial or nonradial $p$-mode pulsation, as in the $\delta$ Scuti stars.

In recent years, the most often quoted example of a short-period, hot pulsator has been the B9p(Si) star ET And. This star was suspected by Hildebrandt, Schöneich & Lange (1985) and Kuschnig et al. (1990) to show $\delta$ Scuti-like variability of 0.10 d and a peak-to-peak amplitude of about 4 mmag in $B$. This variability would be superposed on the 1.6 d variations caused by the rotation of this peculiar Si star. An interpretation of the rapid variations in terms of $p$-modes would suggest that the hot border of the instability strip would need to be shifted considerably towards hotter temperatures of 11 500 K. However, a two-week multi-site campaign undertaken during 1994 September coordinated by R. Kuschnig and W. Weiss showed the short-period constancy of ET And and identified a previously used comparison star as the variable. With this discovery of constancy, the Maia variables return to their mythical status.

### 2.2. The $\gamma$ Dor variables

The variability of a number of stars on or beyond the cool border of the traditional $\delta$ Scuti instability strip (see Fig. 1) is more convincing. This variability is unusual enough so that a different type of variable might be involved: the periods are too long for $\delta$ Scuti-like $p$-mode pulsation.

Burke, Burke & Lady (1977) noticed that the F2V star, HD 164615, varied photometrically with a period of 0.815 d and an amplitude of 0.05 mag. Abt, Bollinger & Burke (1983) call this star a probable spotted single F-type star. The variability would be produced by the spot rotating with the star. They rule out ellipsoidal variability on the basis of constant radial velocity (measured scatter of $\pm 1.6$ km s$^{-1}$ compared to an error of $\pm$ 2.0 km s$^{-1}$).

A number of other variable early F stars with characteristic periods of variation ~1 d have been proposed: the star $\gamma$ Dor (Cousins 1966, Balona, Krisciunas & Cousins 1994), 9 Aur (Krisciunas & Guinan 1990), HD 224638 and HD 224945 (Mantegazza, Poretti & Zerbi 1994), GM Com (Handler 1994), and possibly HR 1459 (Krisciunas, private communication). The star HD 96008 (P = 0.31 d, Matthews 1990) probably is an ellipsoidal variable (Mantegazza, Poretti & Zerbi 1994) and is omitted here. While the determination of periods in the 1 d range is difficult at best, the situation is made even more complicated by the possible existence of double waves in the light curves.

Three explanations have been proposed:

1. Close double stars: this explanation is unlikely because of the need to match the observed $v \sin i$ values with the radial velocity amplitudes. The derived mass functions are improbably small. We refer to the discussions by Balona, Krisciunas & Cousins (1994) concerning the star $\gamma$ Dor and Abt, Bollinger & Burke (1983) for HD 164615.

2. Surface spots: this explanation is attractive because it can explain the small radial velocity changes, the double-wave behaviour of the light curve (two or more spots) and even the existence of two close periods (differential rotation). The remaining light curve irregularities could be explained by the formation and destruction of the spots and/or by their migration. The periods of variation are compatible with the expected rotation periods of the stars. However, the origin of such spots in such relatively hot stars is difficult to explain, especially if one considers the dynamo mechanism which requires a deep convection layer.

3. Nonradial $g$-mode pulsation: The discovery of three periods in $\gamma$ Dor (Balona, Krisciunas & Cousins 1994) of 0.757, 0.733 and 0.678 d, respectively, as well as the line-profile variations, are best explained by nonradial pulsation. These relatively long periods cannot be matched by radial or $p$-mode pulsation and would require g modes. Detailed models for $g$-mode pulsators should test whether the observed small or zero radial velocity changes observed for some stars might be a problem.

At this point, explanations involving *g*-mode pulsations seem to be the most attractive. However, the study of these $\gamma$ Dor variables is still in its infancy and may offer many surprises in the future.

## 3. Asteroseismology of $\delta$ Scuti stars

The rich mixture of simultaneously excited radial and nonradial *p*-modes in many Pop. I $\delta$ Scuti stars makes them an ideal tool for asteroseismological studies. The excited *p*-modes are of low order, viz. $0 \leq \mathrm{n} \leq 7$. Modes of higher orders have not yet been found, e.g., a WET/DSN campaign to look for these modes in the star FG Vir could not photometrically detect any (Breger et al. 1995). For low-order modes, the asymptotic theory of frequency spacing does not hold, so that we have to compare the observed frequency spacings with pulsation models computed specifically for the star studied.

One of the most challenging and promising aspects of $\delta$ Scuti star asteroseismology concerns the modes with mixed character: low-order *p*-modes may develop *g*-mode type behaviour in the interior (see Dziembowski & Krolikowska 1990, Dziembowski & Goode 1992). This means that the study of the observed *p*-mode spectrum is a probe of the stellar interior. The trapped modes may even provide an asteroseismological test for convective overshooting theories (Dziembowski & Pamyatnykh 1991). An additional probe of the stellar structure is provided by observing the frequency spacing of rotationally split modes:

$$\sigma_{\mathrm{m}} = \sigma_0 + (\mathrm{C_L} - 1)\ \mathrm{m}\Omega + \mathrm{D_L m^2}\Omega^2/\sigma_0,$$

where $\sigma$ represents the frequency of pulsation of the mode with azimuthal quantum number m, $\Omega$ is the rotational frequency, while $\mathrm{C_L}$, $\mathrm{D_L}$ are coefficients depending on the model and pulsation mode (see Dziembowski & Goode 1992).

An example of the methods which can be applied can be found in Goupil et al. (1993), who match the five observed frequencies of GX Peg with models and examine whether the star rotates as a solid body.

Not all m values are excited at the same time to an observable amplitude, so that the observed frequency spectra of $\delta$ Scuti stars usually do not show complete patterns of rotational splitting of successive radial orders. The missing frequencies make the identification of the observed frequencies with pulsation modes difficult. A large number of reliable mode identifications for each star are usually required to put constraints on models of stellar structure and evolution. Consequently, we urge that the highest priority should be given to increase the number and reliability of the frequencies for individual stars. Some of the modes studied have amplitudes just below 1 mmag: lengthy multi-site campaigns involving high-precision photometry are needed.

Several astronomical teams are concentrating on multi-site campaigns of individual stars. Some of these are:

1. Delta Scuti Network (DSN), e.g., Breger et al. (1989),
2. a group at Merate Observatory, e.g., Mantegazza & Poretti (1992),

3. STEPHI Network, e.g., Belmonte et al. (1994),
4. STACC, e.g., Frandsen et al. (1994),
5. Whole Earth Telescope (WET), e.g., Handler (1995).

What is the status of our knowledge of multiple pulsation modes of $\delta$ Scuti stars? Very few stars have been studied in detail and we have selected the most promising candidates for asteroseismology from data of hundreds of stars in the literature. The criteria by inclusion in the present group are:

1. The number of reliably detected frequencies must be $\geq 4$,
2. Information of reasonable quality on the basic stellar parameters such as $T_{\text{eff}}$, $\log g$ and $M_V$ must exist.
3. Stars with variable amplitudes were included only when the variability was judged to be intrinsic to the star rather than caused by incomplete multi-frequency solutions. For variable amplitudes an average amplitude was calculated for each frequency.

The results are presented in Figs. 2 and 3, where the stars have been ordered in decreasing temperature. To correct for the different stellar radii, the frequency scales for the individual stars have been normalised by the value of the pulsation constant, $Q$. This makes it possible to show the predicted values for the four low-order radial modes. The observationally determined $Q$ values are uncertain to about 18% even for stars with excellent $uvby\beta$ values (see Breger 1990a). Consequently, small horizontal shifts are still possible.

Most of the stars shown in the diagrams are not radial pulsators. Nevertheless, since the frequency separations between successive radial orders of nonradial modes are similar to those of radial modes, the lines drawn can be an important diagnostic tool for mode identification.

The following stars were selected: CD -24 7599 , FM Com (Paparo et al. 1993, Bax, Wehlau & Sharpe 1995), $\theta^2$ Tau (Breger et al. 1989), GX Peg (Michel et al. 1992), BU Cnc (Breger et al. 1993, Belmonte et al. 1994), BN Cnc (Belmonte et al. 1994, Kovacs 1981), UU Ari (Ostermann et al. 1991), FG Vir (Breger et al. 1995), 21 Mon (Stobie et al. 1977), AI Vel (Walraven, Walraven & Balona 1992), HD 18878 (Mantegazza & Poretti 1993), HD 224639 (Mantegazza, Poretti & Zerbi 1995), BI CMi (Mantegazza & Poretti 1994), X Cae (Mantegazza & Poretti 1992), 4 CVn (Breger 1990b), 44 Tau (Poretti et al. 1993).

Surprisingly, Figs. 2 and 3 do not show any systematic behaviour of the excited pulsation modes with temperature and luminosity. The astrophysical reasons for mode selection by the stars must be regarded as presently unknown.

## 4. Period variability and evolution

The short periods of $\delta$ Scuti stars allow the observational available determination of the period changes expected from stellar evolution within a few decades. The

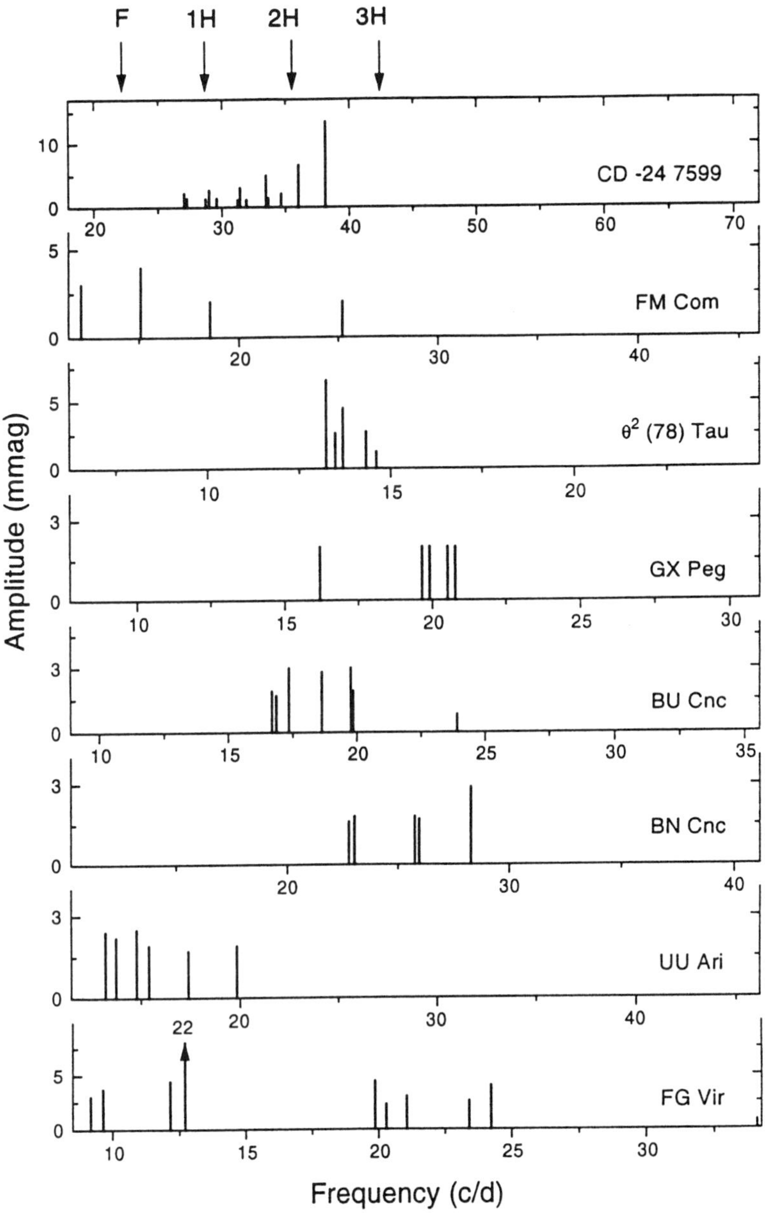


Figure 2. Multiple pulsation modes detected for well-studied $\delta$ Scuti stars. The stars have been ordered in decreasing temperature. For each star, the frequency scale has been normalized by the value of the pulsation constant, Q.

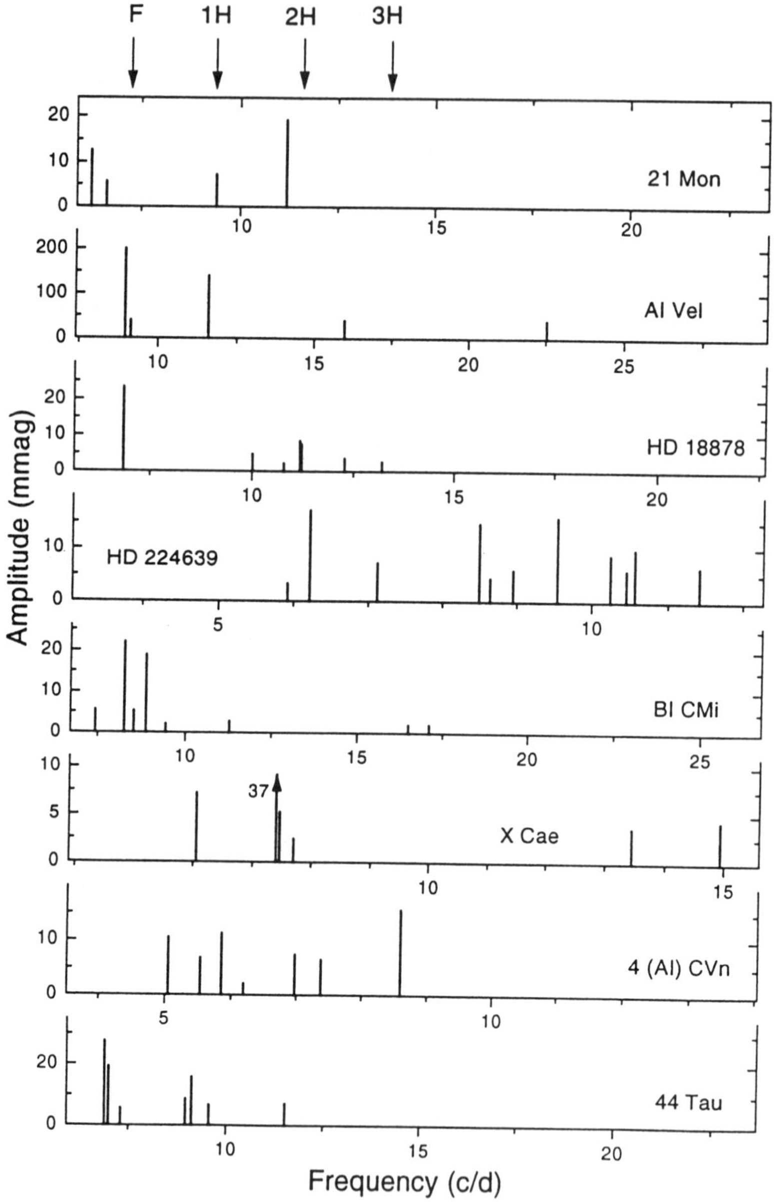


Figure 3. Continuation of Figure 2. In the figure, the theoretically predicted frequencies for low-order radial modes are shown for comparison. Note the lack of systematics of the detected pulsation modes with temperature and luminosity.

available information was summarised during a previous conference of this series (Breger 1990c). A surprising result was the fact that the four Pop. I $\delta$ Scuti giants studied so far with P $\geq$ 0.13 d (DY Her, VZ Cnc, BS Aqr, 4 CVn) all show decreasing periods. Stellar evolution theory, however, predicts increasing periods in that part of the Hertzsprung-Russell Diagram due to increasing radii. An attempt by Guzik & Cox (1991) to explain the period decreases on the basis of helium diffusion was unsuccessful.

One difficulty is that the conclusion of period decreases is based on only four stars. An important observational update is now provided by Rodriguez et al. (1995). They confirm the previous conclusions with new data and also report a period decrease for a fifth $\delta$ Scuti giant, i.e., (1/P) dP/dt = -10.6 $10^{-8}$ for RS Gru.

The disagreement with stellar evolution theory, therefore, still exists. However, one should consider the possibility that observed period changes could be caused by other effects, such as period readjustments caused by discrete mixing events inside the star (Sweigart & Renzini 1979). Such changes would be of either sign. This hypothesis might also explain why period changes could be sudden, rather than continuous. The difficulty with this explanation lies in the fact that the observed period decreases seem to be systematic: positive in the lower part of the HRD and negative in the cool luminous part.

The dilemma might be solved if the available measurements were of insufficient quality and quantity. The situation can be illustrated with one of the better determined stars, DY Her. When the excellent analysis of the times of maxima by Szeidl & Mahdy (1980) is supplemented by more recent results by Yang et al. (1993), the derived period decrease is lowered by about 30%! Furthermore, it is not clear whether the changes are smooth or discontinuous.

Let us look at another $\delta$ Scuti giant, VZ Cnc. Percy, Matthews & Wade (1980) as well as Cao & Jiang (1992) have reported that the period of VZ Cnc is decreasing with a statistical significances of three standard deviations. On the other hand, a different analysis of the new and old times of maximum by Arellano Ferro et al. (1994) led them to interpret the variations in in terms of a binary light-time effect (sinusoidal variations). An inspection of the error bars and the various fits in this paper indicates that other explanations involving a decreasing period, sudden jumps or even a constant period cannot be ruled out for this star and might even be equally probable. The period variations of VZ Cnc need to be regarded as uncertain at this time.

We conclude that four years after the report of the puzzling period decrease of luminous $\delta$ Scuti stars more data have become available, but that the reality of this effect is still tantalising, rather than statistically and astrophysically secure.

## 5. The Delta Scuti Star Newsletter

In order to facilitate rapid communication between researchers in this area, the present author edits a Delta Scuti Star Newsletter, which is issued regularly. Submission of news items and discussions are invited. New items can be found prior to publication on the Internet at http://venus.ast.univie.ac.at, where the contents of previous newsletters can also be accessed.

**Acknowledgments.** Part of this investigation has been supported by the Austrian Fonds zur Förderung der wissenschaftlichen Forschung under Grant number S7304-AST.

## References

Abt, H. A., Bollinger, G., & Burke, E. W., Jr. 1983, ApJ, 272, 196

Arellano Ferro A., Nunez, N. S., & Avila, J. J. 1994, PASP, 106, 696

Balona, L. A., Krisciunas, K., & Cousins, A. W. J. 1994, MNRAS, 270, 905

Bax, J., Wehlau, W. H., & Sharpe, S. 1995, Journ. Roy. Astron. Soc. Can., in press

Belmonte, J. A., Michel, E., Alvarez, M., et al. 1994, A&A, 283, 121

Breger, M., 1990a, Delta Scuti Star Newsletter (Vienna), 2, 13

Breger. M. 1990b, A&A, 240, 308

Breger, M. 1990c, ASP Conf. Ser., 11, 263

Breger, M., Garrido, R., Huang, L., et al. 1989, A&A, 214, 209

Breger, M., Stich, J., Garrido, R., et al. 1993, A&A, 271, 482

Breger, M., Handler, G., Serkowitsch, E., et al. 1995, A&A, in press

Burke, E. W. Jr., Burke, E. W. III, & Lady, S. 1977, IBVS, 1279

Cao, M., & Jiang, S. 1992, Chin. A&A, 16, 148

Cousins, A. W. J. 1966, Roy. Obs. Bull., 122

Dziembowski, W. A., & Goode, P. R. 1992, ApJ, 394, 670

Dziembowski, W. A., & Krolikowska, M. 1990, Acta Astron., 40, 19

Dziembowski, W. A., & Pamyatnikh, A. A. 1991, A&A, 248, L11

Frandsen, S., Viskum, M., Balona, L., & Koen, C. 1994, in Delta Scuti Star Newsletter (Vienna), 7, 5

Goupil, M. J., Michel, E., Lebreton, et al. 1993, A&A, 268, 546

Guzik, J. A., & Cox, A. N. 1991, in Delta Scuti Star Newsletter (Vienna), 3, 6

Handler, G., 1994, in Delta Scuti Star Newsletter (Vienna), 7, 11

Handler, G., 1995, in Delta Scuti Star Newsletter (Vienna), 8, 1

Hildebrandt, G., Schöneich, W., & Lange, D. 1985, Publ. Astrophys. Obs. Potsdam, 112

Kovacs, G. 1981, Acta Astron., 31, 75

Krisciunas, K., & Guinan, E. 1990, IBVS, 3511

Kuschnig, R., Weiss, W. W., Kreidl, T. J., et al. 1990, ASP Conf. Ser., 11, 348

Lehmann, H., Scholz, G., & Hildebrandt, G. 1995, A&A, in press

Mantegazza, L., & Poretti, E. 1992, A&A, 255, 153

Mantegazza, L., & Poretti, E. 1993, A&A, 274, 811

Mantegazza, L., & Poretti, E. 1994, A&A, 281, 66

Mantegazza, L., Poretti, E., & Zerbi, F. M. 1994, A&A, 270, 439

Mantegazza, L., Poretti, E., & Zerbi, F. M. 1995, A&A, in press

Matthews, J. 1990, A&A, 229, 452

Michel, E., Belmonte, J. A., Alvarez, M., et al. 1992, A&A, 255, 139
Ostermann, W. M., Breger, M., Garrido, R., et al. 1991, A&A, 245, 543
Paparo, M., Pena, J., Peniche, R., et al. 1993, A&A, 268, 123
Percy, J. R., Matthews, J. M., & Wade, J. D. 1980, A&A, 82, 172
Poretti, E., Mantegazza, L., & Riboni, E. 1993, A&A, 256, 113
Rodriguez, E., Lopez de Coca, P., & Costa, V. 1995, A&A, in press
Stobie, R. S., Pickup, D. A., & Shobbrook, R. R. 1977, MNRAS, 179, 389
Struve, O. 1955, Sky & Telescope, 14, 461
Sweigart, A. V., & Renzini, A. 1979, A&A, 71, 66
Szeidl, B., & Mahdy, H. A. 1980, Comm. Konkoly Obs., 75, 1
Walraven, Th., Walraven, J., & Balona, L. A. 1992, MNRAS, 254, 59
Yang, D., Jiang, S.-y., Tang, Q.-q., et al. 1993, IBVS, 3831

## Discussion

*Smith*: I find it amusing that you discussed the new group of "slowly" variable F stars, but you did not mention the analogous group of high-degree NRP, rapidly rotating F giants discovered by Gordon Walker and co-workers (Walker et al. 1987, ApJ 320, L139). The co-existence of this latter class with the slow F variables is reminiscent of the controversy surrounding the 53 Per and "SPB" variables among the NRP B stars. This all leads me to wonder whether we are not rushing to judgement in attributing these periods as conventional $g$-modes. Perhaps there is after all room for a rotational excitation mechanism (e.g., the Lee-Saio mechanism).

*Breger*: I did mention these spectroscopically discovered high-degree sectorial modes, but attributed the variability to $p$-modes. In this regard they probably differ from the 53 Per or SPB stars and there seems to be no controversy. At this time the best scenario for these stars might be: (i) photometrically we see low-order, low-degree $p$-modes and possibly some low-order $g$ modes, (ii) line-profile changes show high-degree $p$-modes and some low-degree $p$-modes, (iii) high-order $g$ modes are seen as slow variations.

*Dziembowski*: Mid-F slow pulsators are the likely analogue of SPB stars as Chris Waelkens has already suggested (elsewhere). Just like the metal opacity bump is responsible for $g$ mode pulsations in stars cooler than $\beta$ Cep variables, the HeII ionization bump may be responsible for $g$ mode excitation in stars cooler than $\delta$ Sct variables. Pamyatnykh and I, indeed, have found unstable high-order $g$ modes in models of mid-F stars. Unfortunately, unlike in SPB models, the convective flux is large in the driving zone; therefore we regard our results as not sufficiently reliable and we will not publish them.

*Kovács*: There is a general problem of using linear pulsation models for mode identification in the case of nonlinear, non-radial pulsation. The observed position of a normal mode in the frequency spectrum depends on the following factors: (i) linear non-adiabatic normal frequency; (ii) nonlinear effects; (iii) rotational frequency. For example, if 1:1 resonance is important in the case of a rotationally split set of modes, one can show that in the stationary nonlinear state the observed frequency splitting is equidistant, with a spacing determined by the factors mentioned above (see Goupil & Buchler 1994).

*Breger*: Your important comment underscores the need to do asteroseismology of Delta Scuti stars, i.e., to detect and identify the rotationally split modes from observational data in order to compare with a variety of theoretical predictions. I would speculate that the observations will indicate even more factors than you are listing.

*Frandsen*: In stellar systems (open clusters etc.) the constraints on the stellar parameters are stronger. Therefore, mode identifications become much easier than for isolated stars. Also, phase studies can provide a determination of $\ell$-values that restrict the choice of frequencies to match.

*Breger*: You are correct. For stars in clusters such as Praesepe the distance is quite well known so that the radial order of a mode is more accurately and uniquely determined. Also, in order to obtain the important phase information, photometric measurements should be made in two colours, e.g., the $v$ and $y$ filters of the *uvby* system. Simultaneous spectroscopic measurements (if that could be organised) provide strong constraints.

*Burki*: Do you know the unbiased percentages of stable and pulsating stars in the $\delta$ Scuti region of the HR diagram? And I have the same question concerning the Ap stars (to Don Kurtz).

*Breger and Kurtz*: $\delta$ Scuti stars: Variability surveys yield ~40% variability inside the Lower Instability Strip. Re-measurements of some 'constant' stars have found variability at the later time. This is only partially due to the non-repetitive behaviour of $\delta$ Scuti variation. If we furthermore correct for the low incidence of variability among the classical Am stars, for "normal" stars the variability becomes $\gtrsim$75% at the 0.001 mag level of amplitude.

*Nemec*: Can you give a physical explanation as to why the higher-L $\delta$ Sct stars have decreasing $P$s and the lower-L $\delta$ Sct have increasing $P$s?

*Breger*: If the observed $\dot{P}$ values are correct and not due to some irregular $P$ variations, then *either* the computed evolutionary tracks in this limited part of the instability strip are incorrect and the radii are actually decreasing for a while, *or* radial pulsation is more complicated than believed at the moment. In that case, the periods would depend strongly on the internal structure of the star.

*Astrophysical Applications of Stellar Pulsation*
*ASP Conference Series, Vol. 83, 1995*
*R. S. Stobie and P. A. Whitelock, eds.*

# Probing the Extraordinary Ends of Ordinary Stars: White Dwarf Seismology with the Whole Earth Telescope

Steven D. Kawaler
*Department of Physics and Astronomy*
*Iowa State University*
*Ames, IA 50011 USA*

**Abstract.** During the final evolution of most stars, they shed their outer skin and expose their core of the hot ashes of nuclear burning. As these hot and very dense cores cool into white dwarf stars, they go through episodes of multiperiodic, nonradial $g$-mode pulsation. The tools of stellar seismology allow us to use the pulsation spectra as powerful probes into the deep interiors of these stars. Progress in white dwarf seismology has required significant international cooperation, since another consequence of the complex pulsations of these stars is decoding the true pulsation frequencies requires a coordinated global effort involving high-speed photometric observations. Through one such effort, the Whole Earth Telescope project, we have located subsurface composition changes, detected differential rotation and magnetic fields, and measured fundamental quantities such as stellar mass, luminosity, and distance to extraordinary accuracy.

## 1. Introduction

Models of the evolution of young hot white dwarfs are incompletely constrained by standard techniques of spectroscopy and broad band photometry. A new and powerful tool for probing the depths of these stars is "temporal spectroscopy". This observational tool of asteroseismology can be applied to those objects that show periodic photometric variability. Fortunately, nonradial $g$-mode pulsations are observed in white dwarf stars during three phases of their evolution. Through temporal spectroscopy, we have accurately determined of the mass, rotation rate, cooling/contraction rate, compositional stratification, and magnetic field strength of many white dwarfs over a range of ages.

The phases of white dwarf evolution in which some are observed to pulsate span the time from their birth to middle age. During the hot phases of white dwarf evolution ($T_{\rm eff} > 80,000$ K), some hydrogen–deficient white dwarfs that show carbon and oxygen enrichment are observed to pulsate with periods of 300–1000 seconds. Along with these "DOV" stars, several planetary nebula nuclei with similar C and O enrichment show variations, with periods of 1000–2000 seconds (Bond, Ciardullo, & Kawaler 1993). At cooler effective temperatures (near about 25,000 K) lie the DBV stars, pulsating white dwarfs with helium–dominated spectra. Finally, in a narrow strip near $T_{\rm eff} \approx 12,000$ K we find the DAV (or ZZ Ceti) stars, which are pulsating white dwarfs with hydrogen dominated spectra. Most, if not all white dwarfs in the DAV instability strip are variables (Winget 1988, Bergeron et al. 1995).

Several recent reviews of white dwarf pulsation (i.e., Winget 1988, Kawaler & Hansen 1989, Kawaler 1991) discuss white dwarf seismology as "showing significant promise" as a technique for constraining stellar models. With the success of the Whole Earth Telescope (Nather et al. 1990) in recent years, this promise has become realized. Temporal spectroscopy of white dwarfs addresses many important questions about white dwarf evolution. By measuring the surface compositional stratification, seismological observations directly address the formation of planetary nebula nuclei, the evolutionary status of the PG 1159 and DB stars, and the chemical evolution of white dwarfs of all spectral types. In addition, by measuring the thickness of the surface helium and hydrogen layers, temporal spectroscopy of white dwarf stars has a direct impact on use of the white dwarf luminosity function as a probe of the star formation and age of the galactic disk (Wood 1992). This paper reviews examples of the results of Whole Earth Telescope analyses of DO and DB white dwarfs, and their impact on current models of white dwarf evolution.

## 2. Nonradial pulsation tools

The equations describing nonradial pulsations of a star are fairly complex. However, their asymptotic limits compactly describe the observed pulsation periods of white dwarf stars. Here I present these relations in the form of seismological "tools" that we use in analysis of the pulsation spectra of white dwarfs. We talk of pulsation *modes* as having frequencies $\nu_{n\ell m}$. A given mode has $\ell$ nodal lines on its surface, of which $m$ nodal lines pass through the axis of symmetry; $n$ is the number of nodes in the radial direction. The asymptotic analysis provides expressions for the frequency $\nu_{n\ell}$, or period $\Pi_{n\ell}$, in terms of $n$, $\ell$, and integral properties of the equilibrium model. The periods of nonradial modes do not depend on $m$ unless rotation (or another physical process) disturbs the spherical symmetry of the star. In that case, mode frequencies split into $2\ell+1$ components — assuming all $m$ components are indeed present in the star. In this section we outline how to use period spacings and frequency spacings as probes of the innards of white dwarfs.

### 2.1. *g*-mode period spacings

The asymptotic limit at low frequency (smaller than the radial fundamental pulsation frequency) corresponds to $g$-modes, where buoyancy is the dominant restoring force. For $g$-modes, the pulsation modes are more-or-less equally spaced in period. The periods $\Pi_{n\ell}$ (in seconds) are given by

$$\Pi_{n\ell} = \frac{\Pi_o}{\sqrt{\ell(\ell+1)}}(n+\epsilon), \tag{1}$$

where $\Pi_o$ is a constant that depends only on the the equilibrium structure of the star. Therefore $g$-modes with consecutive values of $n$ are equally spaced in period for modes with the same value of $\ell$. The period spacing for $\ell=1$ modes should be a factor of $\sqrt{3}$ times larger than for $\ell=2$ modes in the same star.

This gives us the first of our seismological diagnostics. Identification of uniform period spacing in a pulsating star gives us $\Pi_o$ and $\ell$, placing immediate and important constraints on the stellar properties. Kawaler (1988) and Kawaler and Bradley (1994) show that, for DO white dwarfs, $\Pi_o$ depends primarily on the mass of the star. For PG 1159 models between $0.4M_\odot$ and $0.8M_\odot$ and $L \geq 10L_\odot$,

$$\Pi_o = 15.5 \left(\frac{M}{M_\odot}\right)^{-1.3} \left(\frac{L}{100L_\odot}\right)^{0.035} \tag{2}$$

with very weak dependence on the thickness of the surface helium layer.

In cooler white dwarfs, where the nondegenerate surface layers determine the pulsation periods, $\Pi_o$ does depend on $T_{\rm eff}$ (i.e., Kawaler 1987, Tassoul, Fontaine, & Winget 1990); however, the instability strips within which the pulsating stars are found are very narrow in temperature, and thus the period spacing is more-or-less dependent again only on mass (Bradley & Winget 1994).

## 2.2. Departures from uniform spacing: mode trapping

Equation (1) holds for idealized stellar models with homogeneous composition. Real white dwarfs are compositionally stratified. The steep subsurface composition gradients produce steep gradients in the density that can act as internal reflecting boundaries for oscillations. This leads to the phenomenon of "mode trapping": modes that have nodes within the composition transition zones can have the amplitude of their eigenfunctions greatly reduced in the interior. The eigenfunctions are, in effect, pinched off below the surface at the composition transition zones. Because trapped modes involve oscillation of a small fraction of the total mass of a star, they require less energy to excite to observable amplitude than untrapped modes. Winget, Van Horn, & Hansen (1981) showed that this property of trapped modes provides a natural mechanism for the preferential selection of trapped modes in ZZ Ceti stars.

More recent work demonstrates how identification of trapped modes allows us to determine the thickness of the surface layers of white dwarfs. The method follows from an analytic analysis by Carl Hansen (1987, private communication). Numerical work confirmed the basic results of Hansen's analysis (see Kawaler & Weiss 1990 and Brassard et al. 1992a). The key element is that trapped modes show departures from uniform period spacing. The period difference between adjacent (in $n$) modes is the principal observational diagnostic of mode trapping. Generally, trapped modes are closer in period to the next lowest–order mode than the mean spacing. Once identified, the periods of the trapped modes allow determination of the depth of composition transition zones; thin surface layers show trapped modes that are much farther apart in period than thick suface layers. An additional result is that the periods of trapped modes of the same $\ell$ should show fixed ratios that are largely independent of the stellar parameters.

Thus the second tool is the period spacing diagram: a plot of $\Delta\Pi$ versus $\Pi$. Minima in $\Delta\Pi$ usually correspond directly to trapped modes. Once trapped modes are identified, the Hansen relation, along with more detailed models, determines the position below the surface at which a mode sees a reflective boundary. For many stars that have been observed with the Whole Earth Telescope, trapped modes have been identified either from minima in the period spacing diagram or from comparison with the trapping period ratios. For more details about the phenomenon of mode trapping and applications to DA, DB, and DOV models see Kawaler & Weiss (1990), Kawaler (1991), Bradley & Winget (1991), Brassard et al. (1992a,b), Bradley, Winget & Wood (1993), Kawaler & Bradley (1994), and Bradley & Winget (1994).

### 2.3. Rotational splitting of $g$-modes

In nonrotating, axially symmetric stars, the eigenfrequency does not depend on the azimuthal quantum number $m$. Hence the frequencies of modes with the same $\ell$ and $n$, but different values of $m$ are identical in such stars. Departures from spherical symmetry will break this $m$ degeneracy, and lead to modes with different $m$ having different frequencies. For slow ($P_{rot} \gg \Pi_{n\ell}$) uniform rotation,

$$\nu_{n\ell m} = \nu_{n\ell} - \frac{m}{P_{rot}}(1 - C_{n\ell}) \tag{3}$$

where $\nu_{n\ell}$ is the frequency of a mode for the nonrotating case, $\nu_{n\ell m}$ is the observed frequency, and $C_{n\ell}$ is a number that is a function of the perturbation eigenfunctions. For white dwarfs, where the amplitude of oscillatory motion in the horizontal direction is much larger than in the vertical, $C_{n\ell} \simeq [\ell(\ell+1)]^{-1}$ (Brickhill 1975).

Slow uniform rotation causes splitting of pulsation frequencies into $m$ components (assuming modes with all available values of $m$ are excited) that are equally spaced in frequency. Since for a given $\ell$ there are $2\ell+1$ possible values of $m$, this means that $\ell = 1$ modes appear as triplets in a power spectrum, and $\ell = 2$ modes appear as quintuplets. The splitting of these multiplets will be different for different values of $\ell$, with $\ell = 1$ triplet splittings being 0.6 times the splitting of $\ell = 2$ quintuplets.

### 2.4. The Whole Earth Telescope

The above effects result in very complex light curve for a star that pulsates in only a few modes. If more than one mode is present in a star, then those modes will beat against one another. The light curve will show variations on time scales of the individual modes and on time scales corresponding to the frequency *differences* between the modes. The resulting light curves are extremely difficult, if not impossible, to fully decode from a single observing site. This is true of most nonradial pulsators: the large number of pulsation modes leads to beating effects with frequencies on the order of 1 cycle per day. Because a single observing site cannot observe its target when clouds, trees, or the Earth gets in the way, diurnal aliases necessarily appear in the Fourier power spectra of the data, and can hide true features of the star's power spectrum.

To solve this problem requires observations that are free from the 1 cycle per day aliases, and the only way to do that is to observe them 24 hours per day. This necessitates coordinated observations at observatories at widely distributed longitudes. Such observations are obtained for pulsating white dwarfs using an extremely remarkable instrument, the Whole Earth Telescope ("WET"). As described in detail by Nather et al. (1990), the WET is a global network of telescopes designed for highly coordinated interactive photometric observations of variable stars. The WET has demonstrated itself to be an extremely powerful instrument; one that has obtained data of extremely high quality on several pulsating white dwarfs. WET data allow us to use the tools of stellar seismology at nearly their full potential.

## 3. Selected results for hot white dwarfs

The PG 1159 stars are hot and luminous pre–white dwarfs. The designation corresponds to the spectral class characterized by CIV and OVI spectral lines with emission cores; see the paper by Werner in these proceedings, and references therein, for more details. Central stars of some planetary nebulae show spectra very similar to the "naked" PG 1159 stars. As the precursors of more ordinary white dwarfs, the PG 1159 stars are the transition objects between stars with active nuclear burning (AGB stars) and stars that are destined to simply cool and fade. Probing their interior structure directly constrains the transition from AGB to white dwarf, and the properties of both classes of stars.

PG 1159-035 was discovered to be a variable star by McGraw et al. (1979), and is the prototype of the DOV class of pulsating white dwarfs. McGraw et al. (1979) immediately recognized the astrophysical importance of these objects: they are rapidly evolving from the planetary nebula phase to the white dwarf cooling track. Thus they are among the hottest, and most rapidly evolving, stars in the Galaxy. NLTE model atmosphere analysis by Werner, Heber, & Hunger (1991) shows that its effective temperature is close to 140,000$K$; the effective temperatures of the PG 1159 spectral class ranges from 80,000$K$ to 160,000$K$.

### 3.1. PG 1159-035

Through temporal spectroscopy with the WET, the prototype of the pulsating hot white dwarfs, PG 1159, has yielded a wealth of information about what one of these stars look like on the inside. The reports of these observations and their analysis (Winget et al. 1991, Kawaler & Bradley 1994) show what such stars can teach us. PG 1159 is wonderful in that it shows all that we expect to see (and more) in these stars. An unbroken sequence of dozens of periodicities match very nicely with the expectations from stellar evolution and pulsation theory.

The noise level in the region of greatest power is less than $7 \times 10^{-4}$ relative amplitude. This sensitivity allows identification of much lower amplitude peaks in the power spectrum that had been possible with single–site data; Winget et al. (1991) completely resolve all of the peaks (or, more precisely, bands of power) present in earlier data into groups of individual peaks. Some of these groups are triplets of individual peaks separated equally in frequency. Other groups show 4 or 5 component peaks, also equally spaced in frequency. The frequency spacing between components in the triplets is 4.2 $\mu$Hz, while the spacing in the quintuplets is 7.0 $\mu$Hz. The expected ratio of frequency splitting for $\ell = 1$ modes to $\ell = 2$ modes is 0.60, which is precisely what is observed! Hence PG 1159 pulsates in $\ell = 1$ and $\ell = 2$ modes; its rotation period is 1.38 days.

The data show a complete string of 20 $\ell = 1$ modes between 430 and 840 seconds. The mean period spacing of the $\ell = 1$ modes is 21.5 seconds, and 12.5 seconds for the $\ell = 2$ modes. The observed ratio of the period spacings is 1.72; the theoretical value is 1.73. Again there is excellent agreement between theory and observation. From these period spacings equation (2) gives the mass of PG 1159 as $0.590 \pm 0.01 M_{\odot}$. Deviations from uniform spacing of the $m = 0$ periods are clearly apparent, and strongly suggestive of mode trapping by a subsurface composition transition. Kawaler & Bradley (1994) use the period spacing diagram to determine the thickness of the surface helium layer in PG 1159 as $3 \times 10^{-3} M_*$. The model that best fits the the mean period spacing and the observed periods had a luminosity of $220 L_\odot$, an effective temperature of 136,000$K$, $\log g = 7.4$, and a surface helium abundance of 30%. From traditional spectroscopy, Werner, Heber, & Hunger (1991) find an effective temperature of 140,000 K and $\log g = 7.0 \pm 0.5$, in excellent agreement with the results from temporal spectroscopy.

### 3.2. PG 2131+066

Another PG 1159 star examined using the WET was PG 2131+066 (hereafter PG 2131). PG 2131 was little–observed (photometrically) following its discovery as a variable star by Bond et al. (1984). Data obtained at McDonald Observatory in 1984 indicated at least 6 periodicities, but the limitations of single–site observing rendered positive period determinations impossible. The Kiel group has analyzed the spectrum of PG 2131, and report an effective temperature of about $80,000 \pm 10,000$ K, and $\log g = 7.5 \pm 0.5$ (Dreizler et al. 1995): they indicate that PG 2131 might be more helium–rich than PG 1159.

PG 2131 was the target of a WET run in September 1992. The star is significantly fainter than PG 1159; therefore the data obtained have a lower S/N level. Despite this, the coverage of PG 2131 was sufficient for Kawaler et al. (1995) to decompose the observed pulsation spectrum of PG 2131. PG 2131 pulsates in $\ell = 1$ modes with several consecutive overtones, and a period spacing of 21.7 seconds. The power spectrum shows these modes are (roughly) equally spaced triplets, giving a rotation period of 5.07 hours. PG 2131 has a mass of $0.61 \pm 0.02 M_\odot$, and there is a hint of a discontinuity in the composition near $6 \times 10^{-3} M_*$ below the surface. The best-fit $0.61 M_\odot$ models have $T_{\rm eff} = 84,000 \pm 2,000 K$ and $L = 10 \pm 4 L_\odot$.

PG 2131 shows some similarities to PG 1159, but also some important differences. The largest amplitude periods are at higher frequency than PG 1159. There are fewer modes present, however, and there is no evidence for $\ell = 2$ modes. The amplitude of the pulsations is slightly larger in PG 2131 than in PG 1159. The differences in the pulsation behaviour are probably correlated to the difference in temperature between PG 1159 and PG 2131.

### 3.3. Other PG 1159 stars

Additional WET campaigns of other hot white dwarfs are currently being analyzed. PG 1707 is another naked PG 1159 star for which WET data exist and are being analyzed, though that star had a gap in coverage (corresponding to central Asia) and therefore has a significant alias problem. WET data have also been obtained for RX J2117+34, the pulsating central star of an enormous old planetary nebula. This star may prove to be extremely important, as it shows periods that are shorter than any other central star, but longer than for the naked PG 1159 stars (Vauclair et al. 1993). It may provide a direct link between the planetary nebula nuclei and hot DO white dwarfs (Appleton, Kawaler & Eitter 1993). Other pulsating PN central stars have been actively investigated by Howard Bond and his collaborators. They have obtained globally coordinated

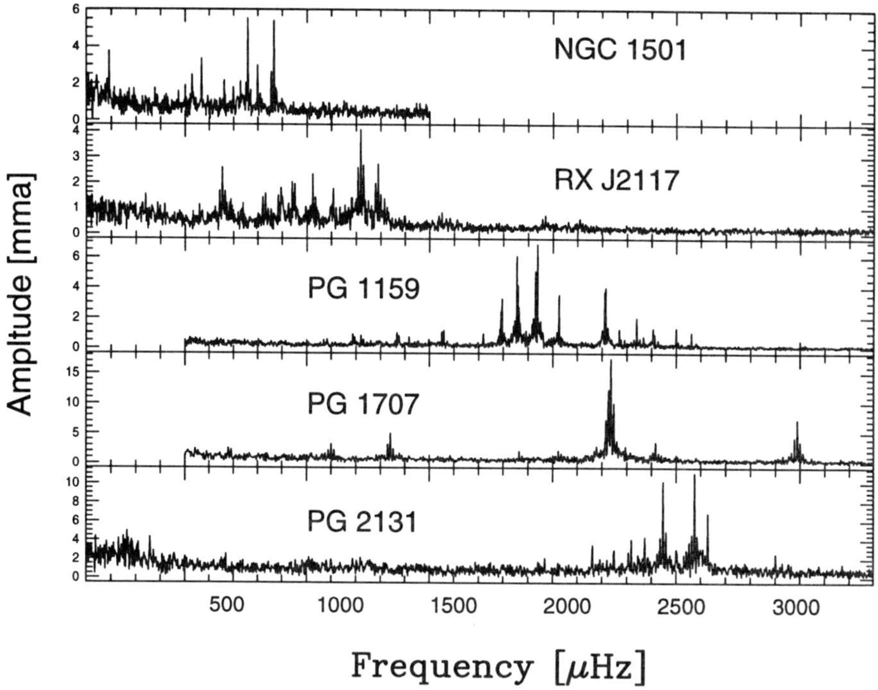


Figure 1. Power spectra of hot pulsating degenerate stars. The data for PG 1159 are from Winget et al. (1991) and for PG 2131 from Kawaler et al. (1995). The other data are from works in preparation: data for NGC 1501 were obtained by Bond et al. (1995), for RX J2117 by Vauclair et al. (1995). PG 1707 data are from an early WET run.

CCD fast photometry of the central star of the planetary nebula NGC 1501 and Sand 3 (see Bond et al. 1993).

Figure 1 shows the power spectra of several hot white dwarfs in order of decreasing dominant period. Note the striking similarities (and differences) between these objects in terms of dominant period and distribution of peak amplitudes. It is left to the reader to derive conclusions based on the temperatures of these objects. The values of $T_{\rm eff}$, in units of $10^3$K, are $96 \pm 3$ for NGC 1501 (Shaw & Kaler 1985), $150 \pm 15$ for RX J2117 (Motch, Werner & Pakull 1993), $140 \pm 10$ for PG 1159, $100 \pm 10$ for PG 1707 (Werner, Heber & Hunger 1991), and $84 \pm 2$ for PG 2131 (Kawaler et al. 1995).

## 4. The DB pulsator GD 358

Another class of pulsating white dwarfs are the pulsating DB (helium–dominated surface) white dwarfs. The prototype of this class is GD 358, which was the first member of the class to be discovered (Winget et al. 1982) following the theoretical prediction by Winget et al. (1981) that DB white dwarfs should show analogous nonradial pulsations to the ZZ Ceti stars. GD 358 has an effective temperature of about 25,000K (Thejl, Vennes, & Shipman 1991). It has an extremely complex light curve that made it an obvious target for study with the WET.

Winget et al. (1994) report on WET observations of GD 358, which have an extremely high duty cycle and clean power spectrum. They found over 180 separate pulsation frequencies in this star, including many triplets. The triplet spacing is measurably non–uniform, with the triplets at long period being systematically wider than at short period. Winget et al. (1994) interpret this as evidence of differential rotation, with the outer layers of GD 358 rotating faster than interior layers. The rotation period near the surface is 0.88 days, and 1.6 days at the deeper levels. In addition, Winget et al. (1994) find a systematic shift of the $m = 0$ peak to lower frequencies, which is the signature of magnetic fields (Jones et al. 1989). Measurement of this shift allows Winget et al. (1994) to determine the magnetic field strength of $1300 \pm 300$G. Other peaks in the power spectrum lie precisely at the sums and/or differences of the dominant "normal" peaks, demonstrating that important nonlinear effects are operating in this star.

Model analysis of GD 358 by Bradley & Winget (1994) successfully reproduces in detail the pulsation frequencies. They determine a mass of GD 358 of $0.61 \pm 0.03 M_{\odot}$, $T_{\rm eff} = 24 \pm 1 \times 10^3 K$, and a luminosity of $0.050 \pm 0.012 L_{\odot}$. The resulting distance to GD 358 is $42 \pm 3$ pc, meaning that the pulsations of GD 358 (with a dose of theory) are a distance indicator. In principle, the same distance determination technique can be applied to the PG 1159 stars, but in those hotter stars the bolometric correction introduces a much larger uncertainty.

Perhaps the most striking result of the modeling by Bradley & Winget (1994) is that the observed periods are best matched with a surface layer of pure helium of approximately $1.2 \times 10^{-6} M_{\odot}$. This is three order of magnitude smaller than PG 1159. This difference in helium layer mass suggests that establishing an evolutionary connection between the PG 1159 stars and DB white dwarfs requires finely–tuned mass loss beyond the PG 1159 stage. Alternatively, these results challenge the notion of there being a direct evolutionary relationship between these objects.

Exploration of this dilemma is another example of how temporal spectroscopy of white dwarf stars drives our growing understanding of their origins and chemical evolution. Dehner (1995) and Dehner & Kawaler (1995) show that the two classes of stars *do* appear to have a direct evolutionary connection. They computed the evolution of an initial model based on the PG 1159 pulsational data, including time–dependent diffusion. As expected, diffusion causes a separation of the elements; a thickening surface layer of nearly pure helium overlays a deepening transition zone where the composition changes to the surface composition of the original model. In the temperature range inhabited by GD 358 and the pulsating DB white dwarfs, this pure helium surface layer is $\sim 10^{-5.5} M_*$ thick. The pulsation periods of the resulting model show a good fit to the WET observations. The model is very similar to the model used by Bradley and Winget (1994) to match the pulsation observations of GD 358. These results demonstrate the plausibility of a direct evolutionary path from PG 1159 stars to the much cooler DB white dwarfs.

## 5. Conclusion

The extremely sensitive data gathering and analysis techniques demonstrated by the WET collaboration have already produced data that push the limits of current theoretical tools. The link between the PG 1159 stars and DB white dwarfs is a tantalizing example of how the powerful techniques of stellar seismology probe the interiors of stars. Because of lack of space, this review has been limited to the DO and DB white dwarfs observed with the WET. Numerous studies of other types of variable stars have been made with this instrument. In these proceedings, Paul Bradley outlines some of the modeling possibilities for DA white dwarfs. J.C. Clemens (1994, 1995) has investigated the temporal spectra of a number of ZZ Ceti stars with the WET. Clemens and others find strong evidence of a fundamental similarity between the ZZ Ceti stars; they appear to have hydrogen layer masses that are entirely consistent with their origin as hydrogen–burning PN central stars that have not lost a significant amount of mass on their descent to become white dwarfs.

Clearly, temporal spectroscopy of white dwarfs, with the WET as a principal instrument, shows much more than promise. The field is extremely exciting, and becoming more exciting than ever as we probe the innards of white dwarfs and reveal the last stages of evolution of stars that were once very much like our own Sun.

## References

Appleton, P.N., Kawaler, S.D., & Eitter, J.J. 1993, AJ, 106, 1973

Bergeron, P., et al. 1995, ApJS, in press

Bond, H., Ciardullo, R., & Kawaler, S.D. 1993, Acta Astr, 43, 425

Bond, H.E., Grauer, A.D., Green, R.F., & Liebert, J. 1984, ApJ, 279, 751

Bond, H.E. et al. 1995, ApJ, in preparation

Bradley, P.A., & Winget, D.E. 1991, ApJS, 75, 463

Bradley, P.A., Winget, D.E., & Wood, M.A. 1993, ApJ, 406, 661

Bradley, P., & Winget,D. 1994, ApJ, 430, 850

Brassard, P., Fontaine, G., Wesemael, F., & Hansen, C.J. 1992a, ApJS, 80, 369

Brassard, P., Fontaine, G., Wesemael, F., & Tassoul, M. 1992b, ApJS, 81, 747

Brickhill, A.J. 1975, MNRAS, 170, 404

Clemens, J.C. 1994, Ph.D. thesis, University of Texas

Clemens, J.C. 1995, in preparation

Dehner, B.T. 1995, Ph.D. thesis, Iowa State University

Dehner, B.T., & Kawaler, S.D. 1995, ApJ, in press

Dreizler, S., et al. 1995, in 9th European Workshop on White Dwarf Stars, ed. D. Koester & K. Werner (Dordrecht: Kluwer), in press

Jones, P.W., Pesnell, W.D., Hansen, C.J., & Kawaler, S.D. 1989, ApJ, 336, 403

Kawaler, S. D. 1987, in Stellar Pulsation: A Memorial to John P. Cox, eds. A. Cox, W. Sparks, & S. Starrfield, (Berlin: Springer–Verlag), 367

Kawaler, S.D. 1988, in IAU Symposium 123, Advances of Helio- and Asteroseismology, ed. J. Christensen–Dalsgaard & S. Frandsen (Dordrecht: Reidel), 329

Kawaler, S.D., & Hansen, C.J. 1989, in I.A.U. Colloquium #114: White Dwarfs, ed. G. Wegner, (Berlin: Springer–Verlag), 97.

Kawaler, S.D., & Weiss, P. 1990, in Proc. of the Oji International Seminar, Progress of Seismology of the Sun and Stars, ed. Osaki, Y.,& Shibihashi, H., (Berlin: Springer), 431

Kawaler, S.D. 1991, in Confrontation Between Stellar Pulsation and Evolution, ed C. Cacciari & G. Clementini, (Provo: PASP), 494

Kawaler, S.D., & Bradley, P.A. 1994, ApJ, 427, 415

Kawaler, S.D., et al. 1995, ApJ, in press

McGraw, J.T., Starrfield, S.G., Liebert, J., & Green, R.F. 1979, in IAU Colloq. 53, White Dwarfs and Variable Degenerate Stars, ed. H.M. Van Horn & V. Weidemann, (Rochester: Univ. of Rochester), 377

Motch, C., Werner, K, & Pakull, M.W. 1993, A&A, 268, 561

Nather, R.E., Winget, D.E., Clemens, J.C., Hansen, C.J., & Hine, B.P. 1990, ApJ, 361, 309

Shaw, R. & Kaler, J. 1985, ApJ, 295, 537

Tassoul, M., Fontaine, G., & Winget, D.E. 1990, ApJS, 72, 335

Thejl, P., Vennes, S., & Shipman, H.L. 1991, ApJ, 370 355

Vauclair, G., Belmonte, J.A., Pfeiffer, B., Chevreton, M., Dolez, N., & Motch, C. 1993, A&A, 267, L35

Vauclair, G., et al. 1995, ApJ, in preparation

Werner, K., Heber, U., & Hunger K. 1991, A&A, 244, 437

Winget, D.E. 1981, Ph.D. thesis, University of Rochester

Winget, D.E., Van Horn, H.M., & Hansen, C.J. 1981, ApJ, 245, L33

Winget, D.E., Robinson, E.L., Nather, R.E., & Fontaine, G. 1982, ApJ, 262, L11

Winget, D.E., Van Horn, H.M., Tassoul, M., Hansen, C., & Fontaine, G. 1983, ApJ, 268, L33

Winget, D.E. 1988, in IAU Symposium 123, Advances in Helio- and Asteroseismology, ed. J. Christensen–Dalsgaard & S. Frandsen, (Dordrecht: Reidel), 305

Winget, D.E., et al. 1991, ApJ, 378, 326

Winget, D.E., et al. 1994, ApJ, 430, 839.

Wood, M.A. 1992, ApJ, 386, 539

## Discussion

*Maeder*: What is the dependence of the period-spacing vs. mass relation on the assumed C/O ratio in the core?

*Kawaler*: The dependence is very small for the DOV (PG1159) models. In these models, the period spacing is determined in the largely degenerate core. Since degeneracy is degeneracy is degeneracy, the precise mix of C/O is not important, as $\mu_E$ is almost unchanged. For lower luminosity models, the dependence of $\Pi_o$ on composition is a bit greater.

*Bradley*: Interior rise in He abundance is also needed to get a good model. I was able to do this with a defect in my evolutionary model building code.

*Goldreich*: Do trapped modes have larger amplitudes than untrapped modes?

*Kawaler*: There is some correlation between amplitude and degree of mode trapping. The largest amplitude modes are almost certainly trapped modes, though surrounding modes also have significant amplitude (compared with the more outlying regions of the power spectrum). Correcting for the apparent "envelope" of power, the correlation is imperfect but may indeed be there.

*O'Donoghue*: Do have an explanation for the sign of the $dP/dt$ term for the 516-s oscillation in PG1159?

*Kawaler*: Yes! In PG1159 models, trapped modes can (and often do) show negative $dP/dt$ even while most modes show positive $dP/dt$. In our best models of PG1159 (Kawaler & Bradley 1994), most modes show $d\ln P/dt$ of about +4 or $5 \times 10^{-14}$ s$^{-1}$, but the trapped modes have $d\ln P/dt = -2 \times 10^{-14}$ s$^{-1}$, which is very close to the value Winget et al. (1991) report for the 516-s mode. In Kawaler & Bradley (1994), we propose that measuring $dP/dt$ for other modes in PG1159 can either confirm or refute this explanation. I hope someone is working on that!

*Astrophysical Applications of Stellar Pulsation*
*ASP Conference Series, Vol. 83, 1995*
*R. S. Stobie and P. A. Whitelock, eds.*

# Seismological Properties of Evolutionary Delta Scuti Models Using OPAL Opacities

Joyce A. Guzik and Paul A. Bradley

*Los Alamos National Laboratory, Los Alamos, NM 87545*

**Abstract.** We present the first results of our project of calculating the evolutionary and seismological properties of $\delta$ Scuti models using Livermore OPAL opacities. We calculate the radial and low-degree nonradial nonadiabatic pulsation frequencies for a 1.8 and 2.0$M_\odot$ model. Our radial first overtone/fundamental mode period ratios are $\sim 0.772$, which agree well with observed period ratios. We find a dense spectrum of unstable nonradial modes including both low order $g$-type modes and low to intermediate order $p-$type modes, confirming that we need a selection mechanism allowing only some modes to grow to observable amplitudes.

## 1. Introduction

Considering the potential for detecting many simultaneous oscillation frequencies in individual $\delta$ Scuti stars through multi-site observing campaigns and satellite observations, we thought it timely to calculate the evolution and pulsation frequencies for a grid of $\delta$ Scuti models with modern opacity and equation of state (EOS) data for asteroseismology of $\delta$ Scuti stars. Here, we present preliminary evolution and pulsation results for $\delta$ Scuti models using the Livermore OPAL opacities (Rogers & Iglesias 1992). We use our evolution and pulsation codes that give excellent results for solar evolutionary and helioseismic models (Guzik & Cox 1993).

## 2. Evolution and pulsation results

Figure 1 shows the evolution tracks of 1.6, 1.8, and 2.0$M_\odot$ stellar models using the Y$=0.28$, and Z$=0.02$ Livermore OPAL opacities and Alexander & Fergusen (1994) low-temperature opacities. For this work, we use the Eggleton et al. (1973) equation of state with Coulomb corrections. For comparison, we show a 1.8$M_\odot$ evolution track (Fig. 1) for models with the old Los Alamos (Cox & Stewart 1965) opacities. The evolution tracks of models using the OPAL opacities are significantly cooler and less luminous for a given stellar mass than the tracks using the old opacities.

We analyze the linear radial and nonradial nonadiabatic periods and growth rates for a 1.8$M_\odot$ model with $T_{\rm eff}$ = 7640K, and a 2.0$M_\odot$ model with $T_{\rm eff}$ = 6990K (marked by triangles in Fig. 1). The 1.8$M_\odot$ model burns hydrogen in a convective core of mass $\sim 0.2M_\odot$, while the 2.0$M_\odot$ model has exhausted core hydrogen. In each model, the radial fundamental and first seven overtone modes

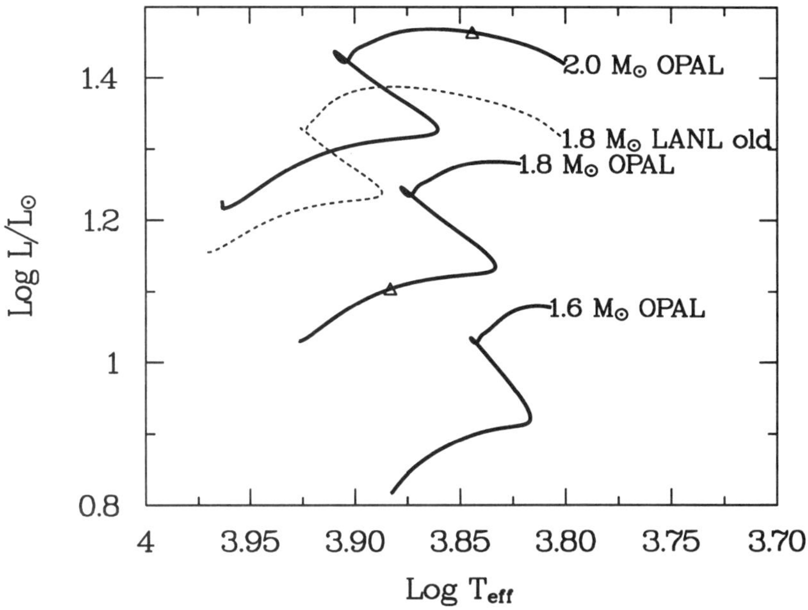


Figure 1. Evolution tracks of stellar models using the OPAL opacities (solid lines), and hotter, more luminous models using old Los Alamos opacities (dashed line).

are unstable, with the 5th and 6th overtones being the most unstable. Our first overtone/fundamental mode period ratios are $\sim 0.772$, in excellent agreement with the observed period ratio of 0.773 (Breger, 1993).

We also find a dense spectrum of unstable low-degree $p$- and $g$-type nonradial modes. Since there are fewer modes observed than predicted to be unstable in $\delta$ Scuti stars, either these modes are present at observationally undetectable amplitudes, or some as yet unexplained mode selection mechanism is operating.

We thank F. Swenson for providing the opacity interpolation routines, and M. Breger, J. Nemec, J. Christensen-Dalsgaard, and A.N. Cox for useful discussions. This research is supported by a NASA Astrophysics Theory Program grant and a LANL LDRD grant.

## References

Alexander, D.R., & Ferguson, J.W. 1994, ApJ, 437, 879

Breger, M., 1993, in IAU Colloq. 139, ed. J. Nemec & J.M. Matthews, (Cambridge Univ. Press), 135

Cox, A.N., & Stewart, J.N. 1965, ApJS, 11, 22

Eggleton, P.P., Faulkner, J., & Flannery, B.P. 1973, A&A, 23, 325

Guzik, J.A., & Cox, A.N. 1993, ApJ, 411, 394

Rogers, F.J., & Iglesias, C.A. 1992, ApJ, 401, 361

*Astrophysical Applications of Stellar Pulsation*
*ASP Conference Series, Vol. 83, 1995*
*R. S. Stobie and P. A. Whitelock, eds.*

# Precision Asteroseismology of DAV (ZZ Ceti) White Dwarf Stars

Paul A. Bradley

*Los Alamos National Laboratory, Los Alamos, NM 87545*

**Abstract.** Recently, Clemens suggested that the similarity in period structure of the hotter DAV (ZZ Ceti) stars could be explained if they pulsated in a common set of modes. Here, I use theoretical pulsation periods from a new grid of evolutionary DA white dwarf models to determine the internal structure of G 117–B15A and R 548. Both stars have hydrogen layer masses of $\sim 1.5 \times 10^{-4} M_\star$ and oxygen-rich cores with a composition gradient starting near $0.75 M_\star$.

## 1. Introduction

In this paper, I derive the structure of G 117-B15A and R 548 by comparing their observed periods to periods predicted by evolutionary white dwarf models. The equilibrium white dwarf models are a subset of those presented by Bradley (1995), so I refer the reader to that paper for additional details about their structure, including the chemical composition profile. Bradley (1995) also presents the general pulsation properties of these models.

I use the effective temperatures and $\log g$ values listed by Koester et al. (1994) and Daou et al. (1990). Both stars have trigonometric parallax data (McCook & Sion 1987) which I use as an independent check on the seismologically predicted distances. G 117–B15A and R 548 have similar pulsation periods (see Table 1), so they should have nearly the same mass and internal structure. A comparison to carbon core models shows the stellar mass is about $0.55 M_\odot$. However, the best fitting carbon core model periods are 215.2 s, 276.5 s, and 341.6 s, so the core cannot be pure carbon. I then try $0.55 M_\odot$ models with various C/O core mass fractions and change the location for starting the switch from a C/O mixture to pure C. These models show close fits when the core is 20:80 C/O, has an inner transition point at $\sim 0.75 M_\star$, and the outer transition point is at $0.90 M_\star$, where the composition is pure carbon. My best fitting model for R 548 has a mass of $0.54 M_\odot$, as opposed to $0.55 M_\odot$ for G 117–B15A. The mass of R 548 is $0.5\sigma$ lower than the spectroscopic mass, while the mass of G 117–B15A is $1.3\sigma$ higher than the most recent spectroscopic mass. Both stars have a hydrogen layer mass of $\sim 1.5 \times 10^{-4} M_\star$ and a helium layer mass of $\sim 1.5 \times 10^{-2} M_\star$. The seismological distance of R 548 is half the parallax distance, indicating a trigonometric parallax value seriously in error or that R 548 is a binary system. We can rule out a binary model for R 548 because we see no spectroscopic evidence for a companion, and the limits on the rate of period change (Tomaney 1987) would correspond to an orbital period of at least $10^5$ yr, far longer than the orbital period of any *hidden* companion.

Table 1. Observed Periods of G 117–B15A and R 548

| Object | $T_{eff}$ (K) | mode indentity ($\ell$, $k$) | | | Trig. Parallax (mas) | Seis. Parallax (mas) |
|---|---|---|---|---|---|---|
| | | (1,2) | (1,3) | (1,4) | | |
| G 117–B15A | 12,350 | 215.2 | 271.0 | 304.4 | $12.1 \pm 5.3$ | 15.9 |
| R 548 | 12,800 | 212.9 | 274.5 | | $14. \pm 2.$ | 29.4 |
| | | 213.1 | 274.8 | | | |

## 2. Summary and conclusions

If the hydrogen layer mass for G 117–B15A and R 548 is typical for DA white dwarfs, then the standard picture of stellar evolution from the AGB to the white dwarf stage (Iben & MacDonald 1986; D'Antona & Mazzitelli 1991) is at least qualitatively correct. The thick hydrogen layer mass will push the mean mass of DA white dwarfs closer to $0.60 M_\odot$ from $0.56 M_\odot$ (Bergeron, et al. 1992). Wood (1995) shows that thick hydrogen layers affect the luminosity function and increases the age estimates for the local Galactic disk by about 0.75 Gyr.

I am grateful to J.C. Clemens, S.D. Kawaler, S.J. Kleinman, R.E. Nather, D.E. Winget, and M.A. Wood for their encouragement and many discussions. This research was supported by a Los Alamos National Laboratory Director's Postdoctoral Fellowship.

## References

Bergeron, P., Saffer, R.A., & Liebert, J. 1992, ApJ, 394, 228

Bradley, P.A. 1995, ApJ, submitted

D'Antona, F., & Mazzitelli, I. 1991, in IAU Symp. 145, ed. G. Michaud & A. Tutukov (Dordrecht: Reidel), 399

Daou, D., Wesemael, F., Bergeron, P., Fontaine, G., & Holberg, J.B. 1990, ApJ, 364, 242

Iben, I.Jr., & MacDonald, J. 1986, ApJ, 301, 164

Koester, D., Allard, N.F., & Vauclair, G. 1994, A&A, 291, L9

McCook, G.P., & Sion, E.M. 1987, ApJS, 65, 603

Tomaney, F. 1987, in IAU Colloq. 95, ed. A.G.D. Philip, D.S. Hayes, & J. Liebert, (Schenectady: Davis), 673

Wood, M.A. 1995, Proc. 9th European Workshop on White Dwarf Stars, eds. D. Koester & K. Werner (Berlin: Springer), in press

*Astrophysical Applications of Stellar Pulsation*
*ASP Conference Series, Vol. 83, 1995*
*R. S. Stobie and P. A. Whitelock, eds.*

# Confining the Edges of the GW Vir Instability Strip

K. Werner and T. Rauch

*Institut für Astronomie und Astrophysik der Universität, Kiel, Germany*

S. Dreizler and U. Heber

*Sternwarte Bamberg, Universität Erlangen-Nürnberg, Germany*

**Abstract.** Ground based optical spectroscopy as well as EUVE and HST observations are used for model atmosphere analyses of PG 1159 stars in order to confine the GW Vir instability strip.

## 1. Introduction

The PG 1159 stars are hydrogen-deficient post-AGB stars within their hottest stage of late stellar evolution. They probably are descendants from the Wolf-Rayet central stars of planetary nebulae and will evolve into helium-rich white dwarfs. The origin of the H-deficiency remains unclear, but we argued that a late thermal pulse causes the removal of the outer stellar layers. The PG 1159 stars span a wide range in $T_{\rm eff}$ and $\log g$ (Fig. 1). Most of the 26 known objects of this spectroscopic class have photospheres which are dominated by helium, carbon, and oxygen and about every other star is associated with a planetary nebula. During the past years we have analysed almost completely the whole ensemble of PG 1159 stars; for a recent review see Dreizler et al. (1995).

## 2. The GW Vir instability strip

Eight PG 1159 stars, including the prototype PG 1159-035 (alias GW Vir), are known as multimode nonradial $g$-mode pulsators. The pulsation driving mechanism is the $\kappa - \gamma$ effect of carbon and oxygen (Starrfield et al. 1984). Large amounts of C and O, a low amount of He as well as the complete absence of H in subphotospheric layers are necessary conditions for driving pulsations. Since these conditions meet the observed photospheric compositions it is widely accepted that C and O opacities are responsible for the instabilities. However, detailed non-adiabatic pulsation modelling (Stanghellini et al. 1991) appears at variance with the observations, because the theoretical instability strip extends too low in $T_{\rm eff}$. It is therefore highly desirable to confine the instability strip by means of spectroscopic analyses of pulsating and non-pulsating members of the PG 1159 group.

Figure 1 shows the results of our analyses, together with the positions of related objects. The grey hatched area comprises the locations of the GW Vir stars (filled symbols). The red and blue edges of the instability region at $\log g = 7$

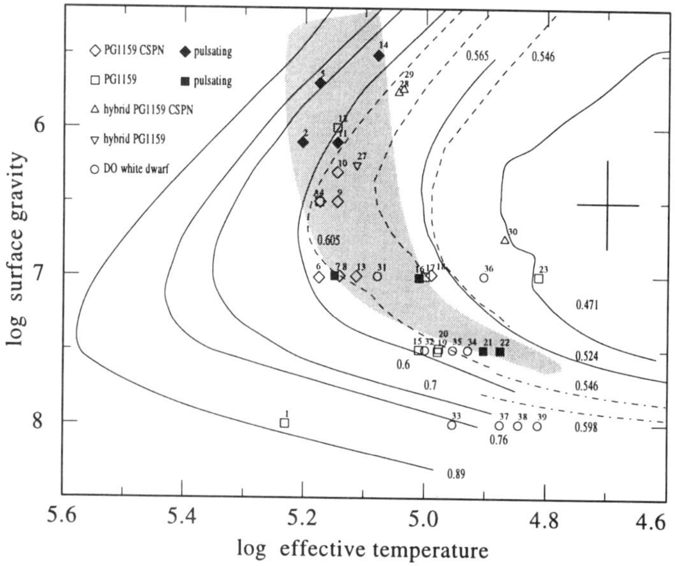


Figure 1. Location of the GW Vir instability strip, comprising the pulsating PG 1159 stars (filled symbols). Evolutionary tracks are labelled with the respective stellar mass. A typical error bar is shown. For details and object identification see Dreizler et al. (1995).

are quite well defined by pulsating/non-pulsating pairs of objects with similar $T_{\rm eff}$. In particular, the blue edge divides PG 1159-035 with $T_{\rm eff}$=140kK from its non-pulsating counterpart PG 1520+525 with $T_{\rm eff}$=150kK. Very precise analyses enabled by HST (UV) and EUVE spectra were required since optical spectroscopy alone is too insensitive to detect such small differences in $T_{\rm eff}$. A word of caution is appropriate. The precise limits of the instability region certainly depend on the total amount of C and O in the driving region as well as on the relative abundance ratio of these species. Several non-pulsating objects are located within the instability strip. This might be a consequence of compositions differing from those in the pulsating stars. Object #27 (HS 2324+3944) for example is a hybrid-PG 1159, i.e., it has detectable traces of H in its atmosphere. If present in the deeper layers, too, hydrogen would poison the pulsations. Photometric observations are still lacking for this recently discovered star.

**Acknowledgments.** This research was supported by grants from the DFG (He 1356/16-2, We 1312/6-1) and the BMFT (50 OR 9409 1).

## References

Dreizler, S., Werner, K., Heber, U. 1995, in White Dwarfs, Lecture Notes in Physics 443, eds. D.Koester and K.Werner, Springer, p. 160

Stanghellini, L., Cox, A.N., Starrfield, S. 1991, ApJ, 383, 766

Starrfield, S., Cox, A.N., Kidman, R.B., Pesnell, W.D. 1984, ApJ, 281, 800

*Astrophysical Applications of Stellar Pulsation*
*ASP Conference Series, Vol. 83, 1995*
*R. S. Stobie and P. A. Whitelock, eds.*

# Helioseismology

Ronald L. Gilliland

*Space Telescope Science Institute*[1], *3700 San Martin Drive, Baltimore, MD 21218 USA*

**Abstract.** Observations of solar $p$-mode oscillations over the last two decades have provided a fundamentally new and unique means of probing the Sun's interior structure. The history of understanding solar oscillations as trapped normal-mode pulsations will be briefly discussed. The large number of excited modes seen in the Sun have been used to constrain the solar core structure, determine the convection zone depth and to fix the variation of angular velocity with depth and latitude. Standard stellar structure and evolution theory does remarkably well in matching the helioseismic results. Further work promises to reveal the solar helium abundance. Upcoming network and space observations should provide the stability and resolution necessary to return detailed information on near-surface structures associated with solar activity.

## 1. Introduction and history

Helioseismology is the study of the solar interior conducted by inferences from properties of global oscillation modes seen at the solar surface. In the early 1960s the first evidence for ubiquitous surface oscillations was reported by Leighton, Noyes and Simon (1962) from two-dimensional Doppler measurements. The now universally adopted explanation of the solar motions as trapped, normal-mode, acoustic-cavity oscillations was first put forward by Ulrich (1970) and Leibacher and Stein (1971). In the 1970s observational data improved dramatically and the theoretical basis for global oscillation modes was solidified, but helioseismology as such did not yet exist. In particular the observational work of Deubner (1975), in conjunction with application of new analysis techniques established the photospheric oscillations as the fundamental modes of standing acoustic waves trapped below the photosphere. A recurring theme of helioseismology has been how well standard theory predicts general oscillation characteristics; this was already seen by Deubner (1975): "we find that the solutions of Ulrich (1970) agree with the observed ridges in all detail to an embarrassing extent." The 1980s saw rapid growth in the observational capabilities (full disk Doppler observations by Claverie, et al. 1979, and Grec, et al. 1980 are of special note), applications of theory to exploit the new data to learn more about the solar

[1]Operated by the Association of Universities for Research in Astronomy, Inc., under contract with the National Aeronautics and Space Administration.

interior and a rapidly burgeoning number of investigators and publications. Today, helioseismology is in its most productive years with fundamental advances, planned now for a decade, in observational capability just months away (GONG and SOHO).

I will provide brief and very incomplete introductions to the theory and observations of solar oscillations in §2. Most of my presentation will examine early successes of helioseismology in constraining global properties of the Sun (§3), and a discussion in §4 of some matters of detail that can be explored. A final section will offer thoughts on what is still to come.

## 2. Background on theory and observations

The physics of pressure wave propagation in an atmosphere has been understood since early in the century (Lamb 1908). In particular an acoustic wave can propagate only at frequencies higher than $c/(2H)$, the acoustic cut-off frequency which varies as the inverse square root of temperature, where $c$ is the sound speed and $H$ is the density scale height. Consider the situation of approaching the outer layers of the Sun from within for a spectrum of acoustic wave frequencies. In the solar atmosphere near the top of the convection zone the temperature drops rapidly near an acoustic cut-off frequency corresponding to about 200 seconds. Waves with periods longer than this cannot propagate beyond this region and are effectively reflected back into the interior. Waves at higher frequencies can escape freely from the photosphere.

Wavelengths for the acoustic waves may be determined from a dispersion relation for the modes. Near the upper reflection point the radial wavelength is large compared to the horizontal wavelength and propagation is nearly vertical. An acoustic wave propagating toward the interior encounters higher sound speeds, a decreasing vertical to horizontal wavelength and is effectively refracted back toward the surface. This lower turning point thus fixes the size of the acoustic cavity for a given mode. Properties of the modes become clear once an appropriate mathematical description, decomposition into spherical harmonics, is adopted and the wave equation for adiabatic oscillations is worked through. Excellent discussions have been provided by Deubner & Gough (1984) and Christensen-Dalsgaard (1988). The normal modes of oscillation are characterised by three wave numbers: the order, $n$, giving the number of nodes in the radial direction, $l$ is the degree which specifies horizontal scales, and with breaking of spherical symmetry (e.g., as with rotation) $m$ corresponds to the number of nodes around the equator. The spatial and temporal wave perturbations are expressed:

$$\xi_{nlm}(r,\theta,\phi,t) = \xi_{nl}(r)Y_l^m(\theta,\phi)e^{-i\omega_{nlm}t}, \tag{1}$$

where $\xi$ is the scalar perturbation (e.g., radial displacement); $r$, $\theta$, $\phi$, and $t$ are the radial coordinate, the colatitude, longitude, and time, respectively, and the $Y_l^m$ are spherical harmonic functions. The amplitudes of solar oscillations are very small and linear perturbation analyses (Unno et al. 1989) on the usual equations of stellar structure suffice to derive theoretical estimates of mode properties.

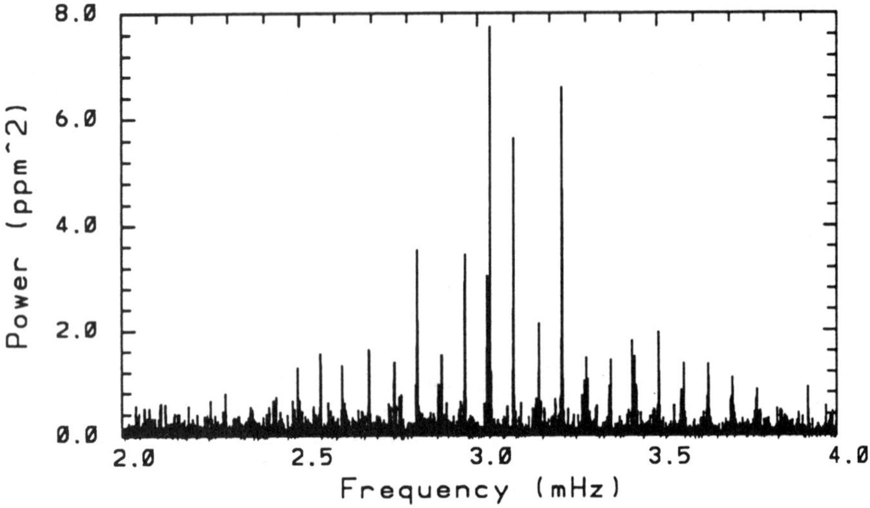


Figure 1. Power spectrum of one month of disk-integrated solar intensity measured with the green channel of the IPHIR experiment (Toutain & Fröhlich 1992).

Solar oscillations have been studied by two primary approaches: 1) Measurements of surface velocity through spectroscopic monitoring of Doppler shifts in atmospheric absorption features. 2) Measurement of intensity variations through photometric monitoring. The solar oscillations consist of literally millions of independent modes, with velocity amplitudes peaking at about 15 cm/s.

A few of the lowest degree oscillations ($l \leq 3$) may be observed with light integrated over the full solar disk. Such observations can be obtained with relatively simple equipment, since no spatial resolution is required. However, with amplitudes of only 15 cm/s, or $\delta L/L$ ~3 ppm (parts per million) the requisite precision is extreme. Figure 1 shows the power spectrum of solar *p*-modes from full-disk data; with spatial resolution the number of observable modes rises dramatically (see, e.g., Libbrecht & Woodard 1990).

As with the theory of solar oscillations, there exist many excellent reviews of observing techniques, and it is unlikely that I could add in either a qualitatively, or even a quantitatively meaningful way to these. The review by Brown (1988) thoroughly discusses the observational challenges, the several methods in use, and includes in an integral way a consideration of mathematical analysis techniques that must be used to extract useful information from the observations.

## 3. Constraining the global solar interior

Only within the past decade have the observations of solar oscillations reached a sufficient level of accuracy to support seismology inferences. In this section I review three areas where helioseismology has provided useful constraints and fundamentally new results. The review by Gough & Toomre (1991) provides a thorough introduction to helioseismic applications.

### 3.1. Core structure

The solar neutrino problem, the observation of a factor of 2-3 fewer than theoretically expected neutrinos from the fusion reactions that power the Sun, has been a significant problem for two decades (see, e.g., Bahcall 1989). Until quite recently most of the solar neutrinos that could be detected from the ground came from a very (exponent of order 15) temperature sensitive, but energetically unimportant, tertiary reaction. Over the past two decades dozens of theories have been put forth to provide modifications to solar models based on best estimates of applicable standard physics, in order to explain the neutrino deficit.

A core temperature decrease of just a few percent below the best estimate solar value would suffice to explain the low neutrino flux. Any modification of the central temperature (e.g., through opacity modifications, or assumed composition differences) results in a change of sound speed. As one might imagine, acoustic wave frequencies are sensitive to integrals of the sound speed over the acoustic cavity. Although mode frequencies primarily reflect conditions throughout the Sun, differences of frequencies for the low $l$-degree modes that penetrate most deeply do provide sensitivity to core conditions. Despite some early indications favouring one class of non-standard solar models (with weakly interacting massive particles to effectively provide a lowered core opacity), helioseismology now substantially supports the standard solar model (Dziembowski et al. 1994), suggesting that an explanation of the neutrino problem is more likely to follow from nuclear physics, than from non-standard physics.

### 3.2. Convection zone depth

Most aspects of solar and stellar evolution modelling are considered to be robust. An acknowledged weak link, however, in computing stellar models is the almost universally used mixing length theory of convection (with a free parameter – the mixing length to pressure scale height ratio). Detailed inference from solar or stellar evolution theory that relies on assumptions about convection zone structure is therefore particularly suspect. The abundances of lithium and beryllium are of cosmic significance in constraining models of the early universe. Lithium and beryllium are easily destroyed at temperatures of a few million degrees. If the bottom of a stellar convection zone extends beyond these critical temperatures these fragile elements are destroyed. A direct calibration of the solar convection zone depth is thus of broad interest.

The bulk of the solar convection zone is nearly adiabatically stratified. Convection occurs when the actual temperature gradient exceeds the adiabatic value. Convective transport at depth is very efficient and regulates the stratification to be just slightly superadiabatic. Although different solar models can have significantly different convection zone depths depending upon assumptions of conditions within the convection zone, they are ubiquitous in showing a rapid transition to strong subadiabatic conditions below the convective boundary. Gough (1986) argued that direct inversions for the sound speed would show the convection zone boundary as a clear break in the gradient of the sound speed. The hard part is of course inverting for the sound speed. For our purposes it is sufficient to point out that the mode frequencies may be expressed as an integral function of the sound speed over the radius. Given many separate oscillation mode frequencies it is in principle possible to invert an integral equation to determine $c(r)$. Christensen-Dalsgaard, Gough & Thompson (1991) have rigourously applied this approach to fix the solar convection zone depth at $0.287 \pm 0.003$ solar radii.

### 3.3. Angular velocity with depth and latitude

The Sun is observed to rotate differentially at the surface as a function of latitude (faster at the equator). The depth dependence of solar rotation is of considerable importance to testing dynamical theories of magnetic field generation. Direct, multi-dimensional numerical models of convection zone dynamics (Glatzmaier 1985; Gilman & Miller 1986) predicted that rotation within the convection zone would have constant angular velocity on cylinders aligned with the rotation.

As with inversions for the sound speed, I will argue that integral equations involving the rotation profile over radius and latitude can be formulated for the amount of splitting in the $m$-dependence of solar oscillation frequencies (this eq. shows the $r$ dependence only):

$$\nu_{nlm} = \nu_{nl0} + \frac{m\beta_{nl}}{2\pi} \int_0^{R_*} \Omega(r) K_{nl}(r) dr, \tag{2}$$

where $\beta_{nl}$ is an order of unity correction factor, $\Omega(r)$ is the solar angular velocity, and $K_{nl}$ is a kernel that may be derived from solar models and is proportional to the local energy density of the mode (Libbrecht & Morrow 1991). A consistent result from such studies (Christensen-Dalsgaard & Schou 1988, Brown et al. 1989; Dziembowski, Goode, & Libbrecht 1989; Thompson 1990; Schou & Brown 1994) has been that the Sun *does not* show constant rotation on cylinders, but rather the latitude dependence remains essentially that directly visible at the surface throughout the convection zone ($r \geq 0.7 R_\odot$). (Actually the results are not considered significant above about 0.85 $R_\odot$.) The radiative interior is found to rotate like a solid body (down to $r \sim 0.5 R_\odot$ where the results are considered reliable) with an angular velocity as seen at mid-latitudes on the solar surface.

The inversions are sensitive to the selection of mode frequencies used as well as the errors. Say that one is interested in the solar rotation rate at a specific radius. Many oscillation modes will have turning points above the radius of interest and therefore have no dependence on the quantity of interest. For modes of lower degree with finite eigenfrequencies in the region of interest, there is some response to conditions at the specified radius, but there is also sensitivity to conditions over a broad range of radii. Such inversions suffer from the same ill-conditioning that plagues any deconvolution process, and small amounts of noise can be amplified to mimic physically interesting, but spurious results.

The rotation rate of the solar core has long been a primary goal of helioseismology. Here it is possible to describe the inherent difficulties with some clarity.

Rotation splits $l$-modes into $m$ = -$l$, . . .,$l$ components, the $l = 0$ (radial) modes do not respond to rotation directly. The size of the splitting is comparable to the angular rotation rate of the Sun, at a period of 27 days, this corresponds to ~440 nHz if the Sun were to rotate as a solid body. The $l = 1$ modes provide the greatest penetration depth and hence sensitivity to core rotation. To resolve a splitting of 440 nHz requires a time series spanning at least one rotation period; to measure the small changes of ~10 nHz that would provide reasonable sensitivity to core rotation requires three years worth of data. At such long baselines two fundamental problems appear: 1) The $p$-modes have finite lifetimes. These generally increase as the frequency (order $n$) decreases. Selection of modes in the low frequency tail of the distribution may provide a long lifetime, but these modes have low amplitudes and hence provide lower S/N. 2) As will be discussed further in §5.1 model frequencies are known to change over the 11 year solar cycle. The amplitude of change for $l = 1$ is comparable to the change of rotational splitting from a rapidly rotating core. Determinations of core rotation ($r < 0.2R_\odot, m < 0.3M_\odot$) have shown large variations, but generally indicate a core that rotates at 2 to 4 times the surface rate (Jiménez, et al. 1994).

## 4. Down to details

Developments over the past decade in helioseismology have shown an increasing range of interesting results. In this section I discuss the surprising extent to which solar oscillation frequencies vary over the sunspot cycle. The physics responsible for solar oscillations may be elucidated through study of waves with frequencies higher than the photospheric acoustic cut-off frequency; such waves can propagate out of the source region and may provide direct information on sources. The local properties of sound waves are perturbed by inhomogeneities in the atmosphere; it may be possible to determine the structure of sunspots at depth through monitoring of such local oscillation changes.

### 4.1. Zeroing in on acoustic sources

I have earlier argued that acoustic waves with frequencies less than ~5.3 mHz (3 minutes) suffer near total reflection in layers near the photosphere; these trapped modes exhibit the discrete frequencies as determined by the eigenmodes of the resonant cavity. The trapped modes can provide information on structural properties of the acoustic cavity, but their source is hidden. Sound waves generated in the interior at frequencies above 5.3 mHz are free to propagate through the photosphere and can therefore exist at a continuum of frequencies. Observations of these oscillations might be expected to provide rather direct inferences to be drawn about excitation sources.

Observations extending well above 5.3 mHz do not show a continuous distribution of frequencies, but rather a decaying sinusoidal envelope of power (~ 150 $\mu$Hz between peaks). Kumar & Lu (1991) and Kumar (1994) have provided an elegant explanation. If the acoustic sources are restricted to a fairly narrow depth, an interference pattern is expected from waves propagating directly toward the observer and those which started toward the interior, but were refracted back toward the surface. The quantitative result is that the sources are 140 ± 60 km (about 1.5 density scale heights) below the photosphere. In a general sense this is consistent with the expectation that $p$-modes result from stochastic excitation by near-surface turbulent convection (see Goldreich et al. 1994 for the latest theory here).

## 4.2. Solar cycle changes

Woodard & Noyes (1985) noted an apparent decrease of oscillation frequencies in the $l$ = 0-2 $p$-modes detected by the Solar Maximum Mission spacecraft as the Sun transitioned from high to low solar activity. Since the frequencies of solar acoustic modes depend upon the interior structure of the Sun, it would be expected that changes of interior conditions associated with the solar cycle would induce changes in the $p$-modes. Over the past decade a consistent picture of frequency changes involving a large range of $l$-modes has begun to emerge; the inferences to be drawn concerning changes to interior properties have not yet become clear.

The best observational evidence for the $l$ = 0-2 modes that can probe the solar core follow from resonant scattering spectrometer studies conducted by the Birmingham group (Elsworth et al. 1994). Data for the oscillation frequencies span 1981 to 1994 and thus cover more than one solar cycle; when the mean frequency of the 24 strongest modes are tracked a remarkably strong correlation with solar activity emerges such that frequencies are lower by $\sim$0.5 $\mu$Hz (out of 3,000 $\mu$Hz) at solar minimum relative to the maximum.

Observation for higher harmonics, $20 \leq l \leq 60$ spanning 1984 to 1990 (minimum to maximum) from Fourier Tachometer observations are reported by Bachmann & Brown (1994). The frequency changes found for these higher degree modes are very similar to those for the low-degree modes. Nearly equal response across a wide range of $l$ suggests that the causal perturbations are in the surface layers of the Sun. A more detailed constraint on this has been provided by Goldreich et al. (1991) who consider frequency shifts as a function of magnetic field stresses in the outer layers of the Sun. Inherent in this analysis is a comparison to the distribution of frequency change with mode order, $n$, which is much smaller at low frequency (Libbrecht & Woodard 1990). Goldreich et al. (1991) find that a variation of the mean magnetic stress at a level of 1% of the gas pressure is sufficient to explain the variations. This theory seems largely confirmed by Woodard et al. (1991) who find tight correlation of solar $p$-mode frequency changes with measured surface magnetic field strength on time scales as short as one month. However, a loose end remains that current theory would predict an order 1% change of solar luminosity (observations show changes 10 times smaller) associated with this.

## 4.3. Probing sunspots

Attempts to determine the structure of individual sunspots at depth through helioseismology is indicative of the increasingly more detailed applications in this field. Just as a local blockage on the surface of a pond can create interesting patterns as surface ripples propagate past it, one would expect the *local* properties of $p$-modes to be altered as they propagate through inhomogeneities in the solar atmosphere. By examining the incoming and outgoing waves associated with an individual sunspot it is possible to set constraints on the sunspot structure.

Observations of $p$-modes on the restricted spatial scales and temporal lifetimes of individual sunspots is quite challenging, nonetheless independent studies (Braun et al. 1992; Bogdan et al. 1993) are now obtaining reasonably consistent results in terms of the observed perturbations to $p$-mode characteristics as a function of frequency and wavelength. The theory of how sunspots are expected to influence the acoustic waves is even more challenging (e.g., Bogdan 1992) and it is fair to say that this promising field has yet to experience a closure between theory and observations.

## 5. Still to come

Helioseismology is a rapidly expanding field of investigation. Both theory and observations have made complementary and fundamental advances in recent years. The near future promises an even more quickening pace thanks to the completion of long planned for experiments.

### 5.1. New experiments: GONG and SOHO

Both the Global Oscillation Network Group (GONG) Project and the Solar and Heliospheric Observatory (SOHO) have been in active planning for over a decade. GONG should have its six longitudinally distributed observing sites deployed in 1995 and SOHO is scheduled for launch in 1995. The two experiments are complementary and each promises fundamental, qualitative advances over previous observations.

GONG (see, e.g., Harvey et al. 1993) will provide nearly continuous coverage of the Sun for 3-11 years through the use of longitude-redundant siting of identical instruments. Each instrument employs a Fourier Tachometer to provide spatially resolved Doppler images at a cadence of one minute. The data resulting from GONG will surpass previous experiments by orders of magnitude in providing the ability to squeeze information out of the multitude of low amplitude solar $p$-modes.

SOHO (see, e.g., Domingo 1988) will be launched by ESA into a continuous Sun-viewing orbit, currently scheduled for October 1995. SOHO will carry several instruments, three of which are for helioseismology. A high resolution spectrometer (GOLF) and solar irradiance monitor (VIRGO) will provide complementary velocity and intensity measurements at low spatial resolution for the study of long period, low degree modes. The solar oscillation imager (SOI) will exploit seeing-free imaging of the full solar disk at high spatial resolution to study oscillation modes at very high degree.

It is expected that both GONG and SOHO will provide impressive new results, particularly for probing those parts of the Sun requiring long-stable measurements (the core), or very high spatial resolution (near surface conditions).

### 5.2. New theoretical frontiers

Making full use of the high quality data about to come on line from GONG and SOHO will require continuing advances, particularly in the area of inversions. Most helioseismology to date has exploited the global, normal-mode characteristic of solar oscillations. Increasingly studies of local characteristics are coming into consideration (e.g., as in probing sunspots). A fundamentally new approach of "time-distance helioseismology" (Duvall et al. 1993; D'Silva & Duvall 1995) may provide a means of probing the near-surface regions through measuring the travel times and distances of solar acoustic waves.

The solar helium abundance is of particular interest because of cosmological and galactic evolution implications. Fixing the helium abundance relies on helioseismic inversions for which the sensitivity of model details of the equation of state used are particularly severe. Thus, while solutions exist for the solar helium abundance (see Antia & Basu 1994 and references therein) I have chosen to list this as a work in progress, where new developments in theory are likely to still be important.

### 5.3. Asteroseismology

Solar-like oscillations have been unambiguously detected on only one star (the Sun) despite many attempts by several groups over the past decade (see Brown & Gilliland 1994 for a thorough review). Although, perhaps the next talk (Bedding 1995) in this conference will provide such a detection. The study of low-degree *p*-modes on other stars, although challenging observationally, will surely become a fundamentally important new area of stellar astrophysics within a decade.

## References

Antia, H.M., & Basu, S. 1994, ApJ, 426, 801

Bachmann, K.T., & Brown, T.M. 1993, ApJ, 411, L45

Bahcall, J.N. 1989, *Neutrino Astrophysics*, (Cambridge University Press: New York).

Bedding, T.R. 1995, this volume

Bogdan, T.J. 1992, in *Sunspots: Theory and Observations*, ed. J.H. Thomas & N.O. Weiss (Dordrecht: Kluwer), 345

Bogdan, T.J., Brown, T.M., Lites, B.W., & Thomas, J.H. 1993, ApJ, 406, 723

Braun, D.C., Duvall, T.L., Jr., LaBonte, B.J., Jefferies, S.M., Harvey, J.W., & Pomerantz, M.A. 1992, ApJ, 391, L113

Brown, T.M. 1988, in *Advances in Helio- and Asteroseismology*, ed. J. Christensen-Dalsgaard & S. Frandsen, (D. Reidel: Dordrecht), 453

Brown, T.M., Christensen-Dalsgaard, J., Dziembowski, W.A., Goode, P., Gough, D.O., & Morrow, C.A. 1989, ApJ, 343, 526

Brown, T.M., & Gilliland, R.L. 1994, ARA&A, 32, 37

Christensen-Dalsgaard, J. 1988, in *Advances in Helio- and Asteroseismology*, ed. J. Christensen-Dalsgaard & S. Frandsen, (D. Reidel: Dordrecht), 3

Christensen-Dalsgaard, J., & Schou, J. 1988, in ESA SP-286, *Seismology of the Sun and Sun-like Stars*, ed. V. Domingo & E.J. Rolfe (Nordwijk: ESTEC), 149

Christensen-Dalsgaard, J., Gough, D.O., & Thompson, M.J. 1991, ApJ, 378, 413

Claverie, A., Isaak, G.R., McLeod, C.P., van der Raay, H.B., & Roca Cortez, T. 1979, Nature, 282, 591

Deubner, F.-L. 1975, AA, 44, 371

Deubner, F.-L., & Gough, D.O. 1984, ARA&A, 22, 593

Domingo, V. 1988, in *Advances in Helio- and Asteroseismology*, ed. J. Christensen-Dalsgaard & S. Frandsen, (D. Reidel: Dordrecht), 545

D'Silva, S., & Duvall, T.L., Jr. 1995, ApJ, 438, 454

Duvall, T.L., Jr., Jefferies, S.M., Harvey, J.W., & Pomerantz, M.A. 1993, Nature, 362, 430

Dziembowski, W.A., Goode, P.R., & Libbrecht, K.G. 1989, ApJ, 337, L53

Dziembowski, W.A., Goode, P.R., Pamyatnykh, A.A., & Sienkiewicz, R. 1994, ApJ, 432, 417

Elsworth, Y., Howe, R., Isaak, G.R., McLeod, C.P., Miller, B.A., New, R., Speake, C.C., & Wheeler, S.J. 1994, ApJ, 434, 801

Gilman, P.A., & Miller, J. 1986, ApJS, 61, 585

Glatzmaier, G. 1985, ApJ, 291, 300

Goldreich, P., Murray, N., Willette, G., & Kumar, P. 1991, ApJ, 370, 752

Goldreich, P., Murray, N., & Kumar, P. 1994, ApJ, 424, 466

Gough, D.O. 1986, in *Seismology of the Sun and the Distant Stars*, ed. D.O. Gough (Dordrecht: Reidel), 125

Gough, D., & Toomre, J. 1991, ARA&A, 29, 627

Grec, G., Fossat, E., & Pomerantz, M. 1980, Nature, 288, 541

Harvey, J., Hill, F., Kennedy, J., & Leibacher, J. 1993, in *GONG 1992: Seismic Investigation of the Sun and Stars*, ed. T.M. Brown, ASP Conf. Ser., 42, 397

Jiménez, A., Herández, F.P., Claret, A., Pallé, P.L., Régulo, C., & Roca Cortés, T. 1994, ApJ, 435, 874

Kumar, P., & Lu, E. 1991, ApJ, 375, L35

Kumar, P. 1994, ApJ, 428, 827

Lamb, H. 1908, Proc. London, Math. Soc.(2), VII, 122

Leibacher, J., & Stein, R.F. 1971, Astrophys. L., 7, 191

Leighton, R.B., Noyes, R.W., & Simon, G.W. 1962, ApJ, 135, 474

Libbrecht, K.G., Woodard, M. 1990, Nature, 345, 779

Libbrecht, K.G., & Morrow, C.A. 1991, in *Solar Interior and Atmosphere*, ed. A.N. Cox, W.C. Livingston, & M.S. Matthews, (Tucson: Univ. Ariz. Press), 479.

Schou, J., & Brown, T.M. 1994, ApJ, 434, 378

Thompson, M.J. 1990, Solar Phys., 125, 1

Toutain, T., & Fröhlich, C. 1992, AA, 257, 287

Ulrich, R.K. 1970, ApJ, 162, 993

Unno, W., Osaki, Y., Ando, H., Saio, H., & Shibahashi, H. 1989. *Nonradial Oscillations of Stars*, (Tokyo: Univ. Tokyo Press), 2nd ed.

Woodard, M.F., & Noyes, R.W. 1985, Nature, 318, 449

Woodard, M.F., Kuhn, J.R., Murray, N., & Libbrecht, K.G. 1991, ApJ, 373, L81

## Discussion

*Christensen-Dalsgaard*: a) A comment on the core rotation. Measurement of low-$\ell$ splitting is very difficult, as the peaks often partially overlap. However, recent data from the BISON group shows clearly separated peaks for $\ell = 1$ and 2. This splitting is consistent with constant rotation in the interior.
b) It is really a big surprise that the standard models are so close to the observations. However, to many of us it was a surprise that He settling is so important in getting the models to agree with the observations. This was regarded as a very non-standard effect until a few years ago, despite pioneering work at Los Alamos.
c) On observational and theoretical accuracy: In much of the Sun, the difference between observation and theory exceeds the observational errors by more than an order of magnitude. Hence, theory has a long way to go. On the other hand, better data are definitely needed, particularly with regards to the core and near-surface structure.

*Goode*: Recent observational results indicate that solar oscillations are not excited by convective drumming from below, but rather by the catastrophic cooling and subsequent collapse of the surface.

*Christensen-Dalsgaard*: It is very interesting that observations now indicate that excitation is dominated by the strong downdrafts. However, as a clarification it should be mentioned that this is really a refinement of the basic convective excitation picture, rather than a drastically new approach.

*Astrophysical Applications of Stellar Pulsation*
*ASP Conference Series, Vol. 83, 1995*
*R. S. Stobie and P. A. Whitelock, eds.*

# More on Solar-like Oscillations in $\eta$ Boo

Timothy R. Bedding

*School of Physics, University of Sydney 2006, Australia*

Hans Kjeldsen

*Institute of Physics and Astronomy, Aarhus University, DK-8000 Aarhus C, Denmark*

**Abstract.** We recently reported strong evidence for solar-like oscillations in the G0 IV star $\eta$ Boo (Kjeldsen et al. 1995, AJ **109,** 1313). We measured small temperature fluctuations produced by oscillations through their effect on the equivalent widths of the Balmer lines. Here we address several issues that were raised at this conference.

**Removing the high-pass filter** To improve the estimates of equivalent-width measurements, the time series was decorrelated against external parameters. This is a well-established procedure (Brown et al. 1991) in which one computes a correction to each data point. The result for $\eta$ Boo was a reduction in the rms scatter of the time series by a factor of 1.9. However, in order to calculate the decorrelation corrections, it was necessary to apply a high-pass filter to remove variations on long timescales. The power spectrum in Kjeldsen et al. (1995) clearly shows the effects of this filter.

It would clearly be desirable to generate a power spectrum that has not been high-pass filtered. This is possible, since the filtering is only necessary while the decorrelation is being performed. Once the decorrelation corrections have been calculated, they can then be applied to the original *unfiltered* time series. The resulting power spectrum for our $\eta$ Boo data is shown in the figure. Comparison with Fig. 1 in Kjeldsen at al. (1995) shows an increase in noise at low frequencies, as expected from a $1/f$ noise source. Apart from that, the two

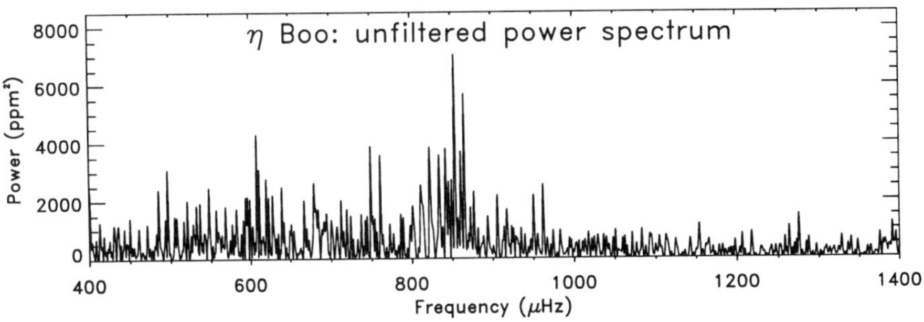

are very similar and the oscillation signature is still clearly visible as a hump of excess power in the frequency range 750–950 $\mu$Hz.

**Solar oscillations** During the daylight hours we obtained spectra of the Sun by observing scattered light from the sky. As described in Kjeldsen et al. (1995), we were able to detect the solar oscillations in a folded power spectrum. At the time, we were puzzled by the detection of the $l = 3$ modes in the Sun, given that the observations did not resolve the solar disk. We now understand this. Model atmosphere calculations show that the Balmer lines are much weaker towards the solar limb, so our measurements of equivalent width were weighted to the central portions of the disk (Bedding et al., in preparation).

**Future observations of $\alpha$ Cen A** We will use the equivalent-width method to search for oscillations in $\alpha$ Cen A with the AAT 3.9 m and ESO 3.6 m over six nights in April 1995. The figure shows simulations of the power spectrum expected from this type of observation. Labels indicate modes with $l = 0$ and 1; modes with $l = 2$ are shown by dashed lines. The insets show the window functions. Input frequencies were from Edmonds et al. (1992). The input amplitudes had a solar-like envelope, randomized about an average peak value of 8 ppm (which is 1.3 times that of the Sun; see Kjeldsen & Bedding 1995). We assumed photon-noise limited observations of the H$\alpha$ line, with a detected photon flux of $4\times10^6\,\mathrm{s}^{-1}\,\mathrm{\AA}^{-1}$ for 9 hours per night for each telescope.

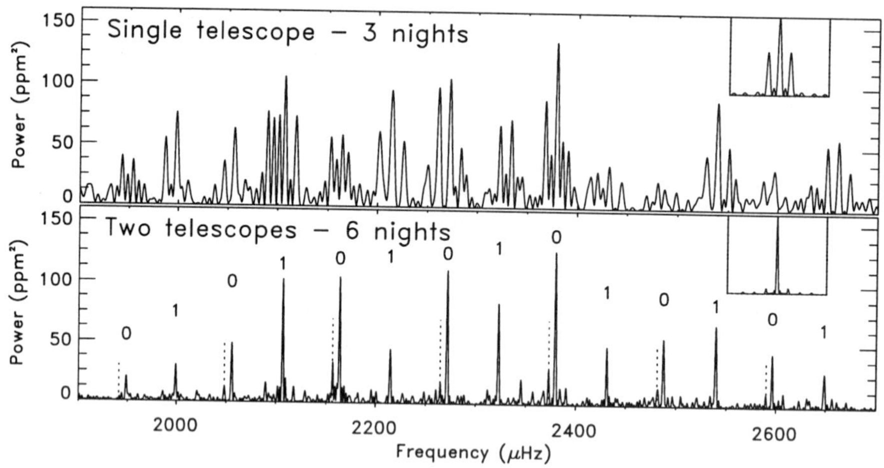


## References

Brown, T.M., Gilliland, R.L., Noyes, R.W., & Ramsey, L.W., 1991, ApJ, 368, 599

Edmonds, P.D., Cram, L.E., Demarque, P., Guenther, D.B., & Pinsonneault, M., 1992, ApJ, 394, 313

Kjeldsen, H. & Bedding, T.R., 1995, A&A, 293, 87

Kjeldsen, H., Bedding, T. R., Viskum, M. & Frandsen, S. 1995, AJ, 109, 1313

*Astrophysical Applications of Stellar Pulsation*
*ASP Conference Series, Vol. 83, 1995*
*R. S. Stobie and P. A. Whitelock, eds.*

# Simulated Power Spectra of Solar-Type Oscillations Driven by an Elementary Excitation Source

I. W. Roxburgh and S. V. Vorontsov[1]

*Astronomy Unit, Queen Mary and Westfield College, Mile End Road, London E1 4NS, UK*

**Abstract.** We consider the synthetic power spectra of multi-mode solar-type acoustic oscillations in stars, computed within a general framework of the linear response of the dynamical system to an arbitrary excitation force which is decomposed in spherical harmonics in space and harmonic functions in time. We discuss possible signatures in the observational data which might be used as diagnostics of the excitation source.

With recent progress in solar seismology, there is an increasing amount of signatures in the observational data which contain valuable information about the excitation source—the high-frequency "pseudo-modes" (Libbrecht 1988; see Kumar 1994 for a recent theoretical discussion) and line asymmetries at intermediate frequencies in the observational power spectra (Duvall et al. 1993; for theoretical discussion, see also Gabriel 1992). The proper interpretation of these observations will hopefully lead to an improvement in our general understanding of the turbulent convection which is responsible for the excitation of solar oscillations (see Goldreich et al. 1994 for a recent discussion). In this contribution, we consider similar observational information which might become available from the seismology of distant stars.

Our analysis is based on a simple asymptotic theory of the high-frequency acoustic oscillations, extended to include an elementary excitation source (Roxburgh & Vorontsov 1995). We use the adiabatic approximation and simulate the energy sink in the dynamical system by allowing free outgoing wave as an upper boundary condition imposed at some level in the upper atmosphere.

The resonant frequencies in the power spectra are described by the equation

$$\int_{r_1}^{R} \left[\frac{1}{c^2} - \frac{(\ell + 1/2)^2}{\omega^2 r^2}\right]^{1/2} dr \simeq \pi \frac{n + \hat{\alpha}(\omega)}{\omega}, \tag{1}$$

where the surface phase shift $\hat{\alpha}(\omega)$ is determined by the atmospheric structure and source location. This frequency equation extends the standard asymptotic eigenfrequency equation to include the high-frequency "pseudo-modes" and the effects of the simple excitation source. The "large frequency separation", the

[1]Institute of Physics of the Earth, Moscow 123810, Russia (permanent address)

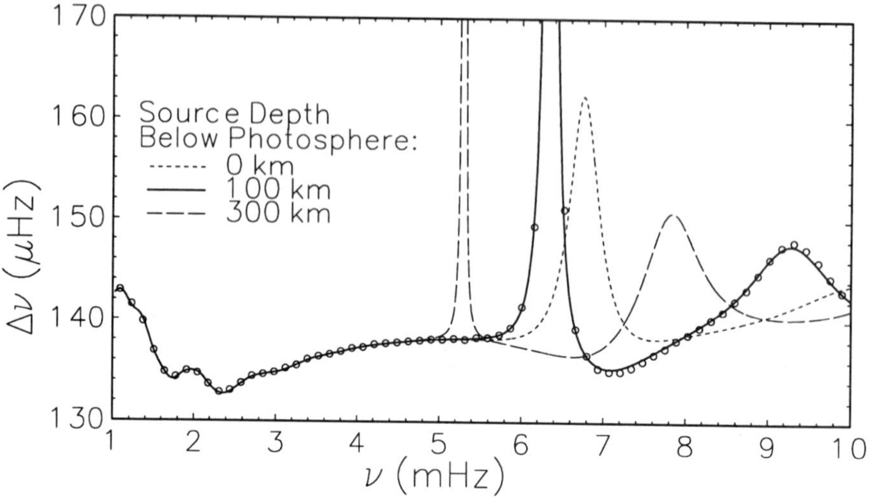


Figure 1. Large frequency separations computed with a dipole source.

measurable quantity in stellar seismology, is

$$\Delta\nu = \nu_{n+1,\ell} - \nu_{n,\ell} \simeq \frac{1}{2\pi}\left(\frac{\partial\omega}{\partial n}\right)_\ell \simeq \left[2\int_0^R \frac{dr}{c} - 2\frac{d(\pi\hat{\alpha})}{d\omega}\right]^{-1} \tag{2}$$

The "large frequency separations" computed for a solar model are shown in Fig.1. A sharp maximum in $\Delta\nu(\nu)$ is clearly seen at intermediate frequencies between the trapped modes and the high-frequency "pseudo-modes"; the position of this maximum is sensitive to the location of the excitation source.

The line profiles in the simulated power spectra also show a quite high sensitivity to the source position and to the relative strength of dipole and monopole components of the excitation. Observational data of exceptionally high quality are needed, however, for measuring the signatures in the power spectra which might be used for the diagnostics of the excitation source; more theoretical work is also needed to employ the diagnostic capability of these measurements, if they become available.

## References

Duvall, T. L., Jr., Jefferies, S. M., Harvey, J. W., Osaki, Y., & Pomerantz, M. A. 1993, ApJ, 410, 829

Gabriel, M. 1992, A&A, 265, 771

Goldreich, P., Murray, N., & Kumar, P. 1994, ApJ, 424, 466

Kumar, P. 1994, ApJ, 428, 827

Libbrecht, K. G. 1988, ApJ, 334, 510

Roxburgh, I. W., & Vorontsov, S. V. 1995, MNRAS, 272, 850

*Astrophysical Applications of Stellar Pulsation*
*ASP Conference Series, Vol. 83, 1995*
*R. S. Stobie and P. A. Whitelock, eds.*

# Radio Observations of Mira Variables, OH/IR Stars and M-Supergiants

Jessica M. Chapman

*Anglo-Australian Observatory, PO Box 296, Epping, NSW 2121, Australia*

H. J. Habing

*Sterrewacht Leiden, PO Box 9513, 2300 RA Leiden, the Netherlands*

N. E. B. Killeen

*Australia Telescope National Facility, PO Box 76, Epping, NSW 2121, Australia*

**Abstract.** We review radio maser observations of the circumstellar envelopes of oxygen-rich Mira variables, OH/IR stars and M-type supergiants.

## 1. Introduction

Towards the end of their lifetimes, most stars lose a considerable fraction of their mass through circumstellar winds. For low and intermediate mass-stars, the largest mass-loss rates of up to $10^{-4}$ $M_{\odot}$ $yr^{-1}$ occur from asymptotic giant branch (AGB) stars. These are luminous, long-period variable stars which include the Mira variables with stellar periods between 100 and 800 days, and the so-called "OH/IR" stars with stellar periods between 300 and 3000 days. Mass-loss rates above $10^{-5}$ $M_{\odot}$ $yr^{-1}$ are also known for a small number of massive M-type supergiant stars with stellar periods above 600 days.

AGB stars and M-type supergiants are evolved stars which have already been through several stages of nuclear burning. The stars have compact cores and extended stellar atmospheres. Strong radial pulsations in the stellar atmospheres are directly linked to the stellar mass-loss in the following way: As the star pulsates, gas in the outer atmosphere is lifted to a height of several stellar radii by pulsation-driven shock waves. At this height the temperature is sufficiently low for the condensation of solid grains. The grains are accelerated outwards by radiation pressure and momentum is transfered to the gas by grain-gas collisions. The outflowing dust and gas forms an expanding circumstellar envelope in which material is transferred from the stellar surface to the interstellar medium at a terminal velocity of typically 15 km $s^{-1}$ (Goldreich & Scoville 1976).

The AGB stars with mass-loss rates above $10^{-6}$ $M_{\odot}$ $yr^{-1}$ are generally invisible at optical wavelengths as the central stars become completely enshrouded by their circumstellar envelopes. However their envelopes can be studied at infrared and radio wavelengths. The infrared emission occurs from circumstellar grains which absorb and re-emit the stellar luminosity, whilst radio emission occurs from molecules which form in the circumstellar envelopes. In particular, for stars with oxygen-rich envelopes, radio maser emission may be detected from SiO, $H_2O$ or OH molecules. Each of these lines requires different physical conditions for its excitations and so originates from a different region of the circumstellar envelope (Chapman & Cohen 1986).

In this paper we review recent radio maser observations of the circumstellar envelopes of oxygen-rich Mira variables, OH/IR stars and M-type supergiants. Table 1 summarizes some properties for the three groups. The parameter $R$ is the far-infrared colour defined as log ($F_{25}/F_{12}$), where $F_{25}$ and $F_{12}$ are the IRAS flux densities at 25 and 12 $\mu$m. This parameter is a useful indicator of the dust optical depth in a circumstellar envelope (van der Veen 1989).

Table 1. Properties of Miras, OH/IR Stars and M-type Supergiants

| | Miras | OH/IR Stars | M-Supergiants |
|---|---|---|---|
| Period (days) | $100 \rightarrow 800$ | $300 \rightarrow 3000$ | $600 \rightarrow 1200$ |
| $L_{star}$ ($L_\odot$) | $10^3 \rightarrow 10^4$ | $10^3 \rightarrow 5 \times 10^4$ | $> 10^5$ |
| $\dot{M}$ ($M_\odot$ yr$^{-1}$) | $10^{-8} \rightarrow 10^{-6}$ | $10^{-7} \rightarrow 10^{-4}$ | $10^{-5} \rightarrow 3 \times 10^{-4}$ |
| $V_{exp}$ (km s$^{-1}$) | $5 \rightarrow 10$ | $5 \rightarrow 30$ | $15 \rightarrow 35$ |
| R | $-0.4 \rightarrow 0.0$ | $-0.4 \rightarrow 0.5$ | $-0.4 \rightarrow 0.0$ |

The Mira variables are traditionally classified as being optically visible, long-period variable stars with large amplitude pulsations ($\Delta M_V > 2.5$). The OH/IR stars are closely related to the Miras but extend their range of properties to thicker circumstellar envelopes, higher mass-loss rates and expansion velocities and longer stellar periods. The OH/IR stars are detected in 18 cm OH maser emission (from one or more of the ground state transitions) and have infrared colours and luminosities which are consistent with their classification as AGB stars.[1] The stars with $R > 0$ are mostly not optically visible.

At present, approximately 1100 OH/IR stars are known (e.g., te Lintel Hekkert et al. 1991; Chengalur et al. 1993). A recent sensitive radio survey of the plane and bulge of our Galaxy, is expected to give a further 2000 new identifications of OH/IR stars (Sevenster et al., these proceedings). Only a small number ($\sim$ 8) of M-type supergiant stars with masers are known. These stars are estimated to have initial masses above 10 $M_\odot$ and are characterized by their high mass-loss rates and expansion velocities but relatively thin circumstellar envelopes.

[1] OH maser emission is also detected from a small number of hotter stars which are likely to be evolving away from the AGB. These stars are not discussed in this paper.

## 2. OH 1612 MHz masers

### 2.1. The location of the OH 1612 MHz masers

OH maser emission from evolved stars is detected at a wavelength of 18 cm from ground state, hyperfine transitions at 1612, 1665 and 1667 MHz. The masers are radiatively pumped by far-infrared emission from warm circumstellar grains. For details of the maser pumping see the review by Cohen 1989.

Aperture synthesis observations of approximately 40 sources have shown that the OH 1612 MHz masers are located in the cool outer regions of the circumstellar envelopes, at radii of 100 stellar radii ($R_\star$) or more (e.g., Booth et al. 1981; Diamond et al. 1985; Chapman & Cohen 1986; Welty et al. 1987). At these radii, the gas temperatures are $\sim$ 100 K and the envelope kinematics are dominated by radial expansion at, or close to, terminal velocity. The strongest maser amplification occurs along the radial directions through the front and back of the circumstellar envelope and the OH 1612 MHz spectral profiles are twin-peaked with the blue- and red-shifted peaks separated by twice the envelope expansion velocity. In most sources, the OH 1612 MHz masers are located within a thin spherical shell. The maser shells however generally show bright knots of emission, indicating some density variations or clumpiness in the mass outflow (Diamond et al. 1985; Welty et al. 1987).

Observationally, there is a good correlation between the OH shell radii and the stellar mass-loss rates (Cohen 1993). This agrees well with theoretical models in which the OH molecules are formed by the photodestruction of $H_2O$ by external UV photons. Stars with higher mass-loss rates and larger envelope densities provide better shielding of the $H_2O$ molecules; and so form OH at larger radii (Huggins & Glassgold 1982).

### 2.2. OH 1612 MHz variability

OH 1612 MHz light curves have been obtained for over 200 Miras and OH/IR stars since Harvey et al. first showed that the OH 1612 MHz masers are strongly saturated and radiatively coupled to the stellar pulsation cycle (Harvey et al. 1974; Schultz et al. 1978; Herman & Habing 1985; van Langevelde et al. 1990, 1993; West et al. 1992; Chapman et al. 1993).

We have monitored the OH 1612 MHz maser emission from a sample of 95 southern OH/IR stars in the bulge and disk of the Galaxy, using observations taken with the Parkes 64-m radio telescope between 1988 and 1993 (Chapman et al. 1993, and in preparation). As examples, Fig. 1 shows the OH 1612 MHz light curves for four stars in the sample. The OH light curves are generally asymmetric with steeper increases towards light maxima and more gradual decays towards minima. Monitoring over several light cycles has shown that the amplitudes of the light curves can vary significantly from cycle-to-cycle, (see also Gaylard & West, these proceedings). These characteristics are also observed in the optical and infrared light curves of Mira variables providing a clear link between the two groups.

88 of the 95 stars in the Parkes sample show periodic variations with maximum/minimum flux density ratios between 1.1 and 3.8. The remaining 7 sources either have irregular light curves or show no significant variations. Of the 88 sources with measurable periods, nearly one third (28 per cent) have periods in

excess of 800 days. Such long periods are not observed for optically visible Mira variables. The OH/IR stars with periods above $\sim$ 800 days appear to have well defined properties with large envelope expansion velocities ($15 \rightarrow 30$ km s$^{-1}$); thick circumstellar envelopes ($R = 0.0 \rightarrow 0.5$), and large shell radii ($> 5 \times 10^{16}$ cm). From their kinematic distances, they have an average scale height in the Galaxy of only $| z | \sim 50$ pc. From these properties, we identify the longest period OH/IR stars as AGB stars with high mass-loss rates, which belong to a young disk population with progenitor masses of at least 4 $M_{\odot}$.

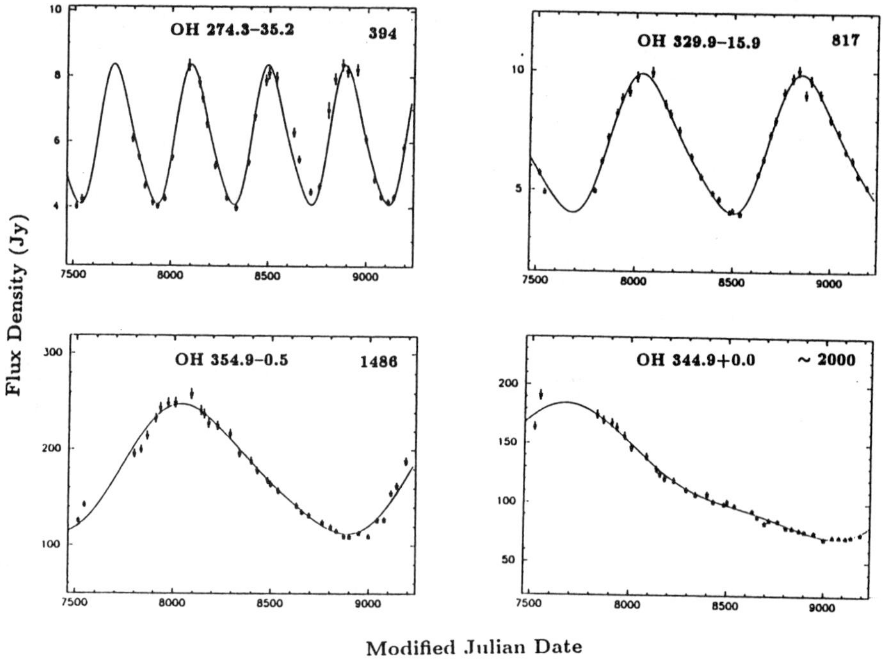


Figure 1. OH light curves for four OH/IR stars in the Parkes montoring programme. Best-fit stellar periods are given in the top right corners.

## 2.3. OH 1612 MHz masers and stellar distances

The geometric distance to a Mira variable or OH/IR star can be obtained by combining the angular and linear diameters of the circumstellar envelope. The angular diameter is obtained from aperture synthesis imaging of the OH maser emission, whilst the linear diameter is determined by measuring the time-delay, or "phase-lag" between the OH light curves from the front and back of the circumstellar envelope. For a spherically symmetric envelope, the linear OH shell diameter is given by the simple relation $D_{16} = \Phi/3.86$, where $D_{16}$ is the shell diameter in units of $10^{16}$ cm, and $\Phi$ is the front-back phase-lag measured in days. Typically the front-back phase lags are only a few per cent of the stellar period, so accurate long term monitoring is needed to determine linear diameters using this method.

The phase-lag technique was first demonstrated by Schultz et al. (1978) who obtained positive phase lags for nine sources. In a much larger programme, Herman & Habing (1985) used the Dwingeloo 25-m radio telescope to monitor the OH 1612 MHz emission from 60 stars over a period of five years. Angular diameters were obtained for 11 of these sources. From this subset, Herman determined the Galactic Centre distance to be $9.2 \pm 1.2$ kpc. More recently van Langevelde et al. (1990) have re-analysed the Dwingeloo data These authors give phase lag diameters for 43 stars and geometric distances for 24 stars. In the best cases, the distances are accurate to $\sim$ 30 per cent. Phase-lag linear diameters have also been determined by van Langevelde et al. (1993) for a few sources near the Galactic centre and by West et al. for sources in a long term monitoring programme of 14 stars at Hartebeesthoek in South Africa (1992, and in preparation).

We have used the Parkes data to determine phase-lags for 75 southern stars. The phase-lags determined using a Fourier series analysis, are between 5 and 120 days, corresponding to linear OH diameters between $1.3 \times 10^{16}$ and $3 \times 10^{17}$ cm. In the best cases, the phase lag diameters are accurate to better than 10 per cent. The phase lag diameters show a clear trend for an increase of OH shell diameter with stellar period. This result can be understood as showing that more massive stars, with longer stellar periods, also have larger mass-loss rates and OH shell radii.

To determine stellar distances we (Chapman, Killeen, van Langevelde) are currently obtaining the angular diameters of the sources in the Parkes programme. As an example, Fig. 2 shows the OH 1612 MHz spectrum and images obtained for the source OH 21.5+0.5 from observations taken with the MERLIN array (U.K.) in November 1993. Images are shown for a number of velocity cuts across the spectrum. For each image, the velocity with respect to the local standard of rest, is given in the top left corner. The first and last images show the emission at the extreme red- and blue-shifted velocities respectively.

The images of OH 21.5+0.5, as for other OH/IR stars, provide strong confirmation for the spherical thin-shell model of Reid et al. (1977). In this model, the OH masers are located in a thin spherical shell of constant radius. The strongest blue- and red-shifted emission, occurs from unresolved regions at the front and back of the circumstellar envelope. At intermediate velocities, weaker emission is detected from rings, centred on the line-of-sight through the stellar position. For well resolved sources, such as OH 21.5+0.5, the angular diameters of the circumstellar envelopes can be accurately determined by fitting the variation in projected ring radius with line-of-sight velocity. For OH 21.5+0.5, we obtain an angular diameter of $1.05 \pm 0.08$ arcsec.

OH 21.5+0.5, is a good example of a high mass, long period OH/IR star. The stellar period is $\sim$ 1800 days and the linear OH diameter, determined by us from the front-back phase-lag, is $(3.1 \pm 0.24) \times 10^{17}$ cm, making it one of the largest OH shells known. Combining the linear and angular diameters yields a distance of $10.0 \pm 1.0$ kpc with a height above the galactic plane of only 10 pc, consistent with a young disk population.

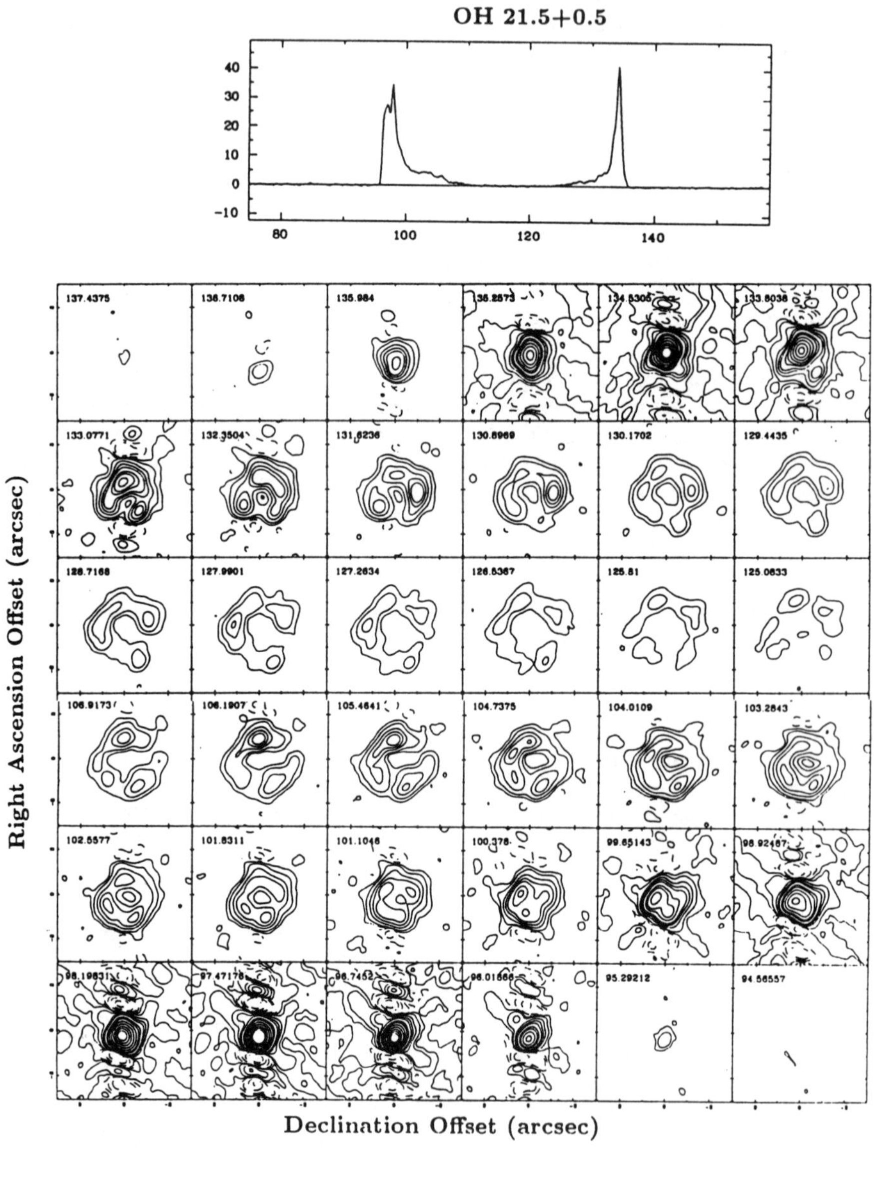


Figure 2. OH 1612 MHz spectrum and MERLIN images obtained in November 1993 for the source OH 21.5+0.5

## 3. OH mainline masers

The OH mainline spectra at 1665 and 1667 MHz are excited closer to the central stars at gas temperatures between $\sim$ 100 and 400 K. Type I OH mainline emission, with stronger maser emission at 1665 and 1667 MHz than at 1612 MHz, is detected from Mira variables with relatively warm circumstellar dust shells. Strong mainline emission is also detected from the M-type supergiant stars.

The OH mainline masers of type I Mira variables and M-type supergiants have complex characteristics. Their spectral profiles are strongly circularly polarized and may show numerous spectral features, including emission at the stellar velocity. Sivagnanam (1989) has shown that the integrated OH mainline emission in Miras follows the stellar pulsation cycles. However, individual features may show strongly erratic variations, or flare-type activity, on timescales of months (Jewell et al. 1980; Le Squeren & Sivagnanam 1985; Chapman et al. 1991). Such variations show that the mainline masers are partially unsaturated.

The circular polarization of the OH mainline masers arises from Zeeman splitting of the OH lines in the stellar magnetic field. Estimates of magnetic field strengths have been obtained for several sources with typical values of a few mG at 20–50 stellar radii (e.g., Claussen & Fix 1982; Chapman & Cohen 1986). For a $1/r^2$ dependence of the magnetic fields, the magnetic field strengths at the stellar photospheres would be around 1–10 G.

The spatial distributions of OH mainline masers are not straightforward to interpret. Bowers et al. (1989) found evidence for axi-symmetric distributions in several Mira variables. For the Mira variable U Orionis, Chapman et al. (1985; 1991) found the OH masers to be located within clumpy structures with a typical clump size of $\sim$ 25 AU. The strongest mainline emission from this source occurs from a ring of radius 60 AU which appears to be part of an expanding, tilted torus. Proper motions were measured for the OH masers on the ring with a mean value of 5.4 $\pm$ 1.4 mas $yr^{-1}$. Non-uniform OH mainline maser distributions have also been observed for the Mira variables R Cas and U Her (Chapman et al. 1994). For these sources, the strongest masers are located within partially filled thick-shells at radii between $\sim$ 50 and 150 AU ($\sim$ 25–75 $R_\star$). The masers of R Cas are highly clumped and show a clear separation of blue- and red-shifted emission indicating that the circumstellar mass-loss may be axi-symmetric.

An issue which remains to be resolved is whether radial acceleration occurs to large radii in the envelopes of some sources. For R Cas, a comparison of the OH and CO velocity ranges indicates that strong radial acceleration may be present. The thermal CO emission from R Cas is located in the outermost parts of the envelope at $\sim$ 1000 $R_\star$, and has a velocity range of 28 km $s^{-1}$, almost twice that of the OH mainline maser emission (Bujurrabal & Alcolea 1991; Chapman et al. 1994). An alternative hypothesis, which needs to be tested, is that the difference between the CO and OH velocity ranges are due to a strongly aspherical envelope geometry in which the different molecules trace regions of considerably different outflow velocities.

The velocity field of R Cas appears to be similar to that of the supergiant star VX Sgr which also has broader CO than OH maser emission. Chapman & Cohen (1986) interpreted the OH mainline and $H_2O$ masers in the circumstellar envelope of VX Sgr, as being located in an accelerating thick-shell with a logarithmic velocity gradient in the shell of $\epsilon \sim 0.5$. Here, we note that the OH mainline maser emission from VX Sgr is exceptionally variable (Chapman et al. 1995 in preparation). OH mainline spectra of VX Sgr taken in 1993 show that the maser emission to be considerably stronger and to extend over a broader velocity range than in the earlier data of Chapman & Cohen (1986). Similar changes have also occured in the $H_2O$ spectral profiles (Bowers et al. 1992). Such changes suggest that intrinsic changes may have occurred in the velocity field of VX Sgr.

## 4. $H_2O$ masers

The $H_2O$ maser emission at 22 GHz occurs from an inversion between two collisionally excited rotational states ($6_{16} \rightarrow 5_{23}$), with radiative decay. These masers are likely to be collisionally excited at temperatures between $\sim 300$ and 1000 K and gas densities of $n_{H2} = 10^{13}$–$10^{15}$ m$^{-3}$ (Yates & Cohen 1994). For collisional excitation, the $H_2O$ masers are expected to lie within a shell with the inner and outer shell boundaries determined by the collision rates. At the inner boundary, the maser inversion is quenched by gas thermalization at high collision rates, whilst at the outer bounday the collision rates are too low for a strong population inversion (Cooke & Elitzur 1985; Menten & Melnick 1991).

The $H_2O$ spectra of evolved stars mostly show two types; a double-peaked spectrum similar to the OH 1612 MHz maser spectra or a strong single peak at the stellar velocity. The masers are likely to be unsaturated (Reid & Menten 1990) and are strongly variable with erratic cycle-cycle variations. In some cases the spectral profiles have changed from double-peaked to single-peaked or vice versa.

The $H_2O$ masers are located in the inner circumstellar envelopes and probe the region where the gas is accelerated to terminal velocity. Images obtained for $H_2O$ masers in $\sim 20$ Mira variables and M-type supergiants have shown that these masers are located within 3 and 30 $R_\star$, corresponding to $10^{14} \rightarrow 10^{15}$ cm for the Mira variables and $10^{15} \rightarrow 10^{16}$ cm for the supergiant stars (Chapman & Cohen 1986; Lane et al. 1987; Reid & Menten 1990; Bowers et al. 1993; Yates & Cohen 1994; Richards et al. 1995). Observationally, the radii of the inner $H_2O$ shell boundaries increase with the stellar mass-loss rates as predicted by the collisional excitation models (Lane 1987; Yates & Cohen 1994).

Strong evidence is seen for acceleration in the $H_2O$ shells of some sources. Fig. 3 shows the integrated $H_2O$ emission from S Per, taken from Richards et al. (1995). The strongest $H_2O$ emission clearly lies on a ring-like structure at a mean radius of $\sim 7$ $R_\star$, with little emission from the front and back of the circumstellar envelope. The simplest explanation for such structures, is that the $H_2O$ masers are tangentially amplified, at radii where strong radial acceleration occurs. For a uniform, spherical envelope with velocity gradient $\epsilon > 1$, the strongest emission occurs at the stellar velocity from a face-on ring centred on the stellar position (Chapman & Cohen 1985). From an analysis of the individual $H_2O$ masers, Richards et al. (1995 and in preparation) model the masers as belonging to an accelerating thick-shell with inner and outer boundaries of 4 and 10 $R_\star$, and corresponding expansion velocities of 10 and 16 km s$^{-1}$. Within this shell the masers show striking filamentary structures, which appear to extend over considerable depths in the envelope and over several km $s^{-1}$.

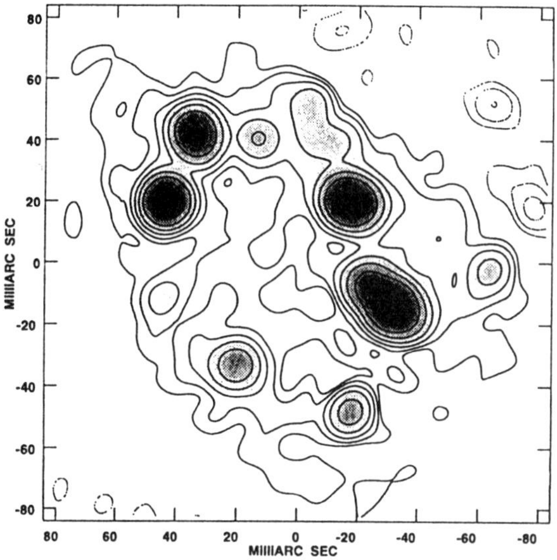


Figure 3. The integrated $H_2O$ maser emission from S Per, taken from Richard et al. 1995

Similar $H_2O$ ring structures are known for the semi-regular variable W Hya (Reid & Menten 1990) and the Mira variable IK Tau (Lane et al. 1987; Yates & Cohen 1994). In each case, the ring of $H_2O$ masers is located within 15 $R_\star$ and shows no systematic velocity gradients. The detection, in several sources, of such rings strongly supports the hypothesis that the $H_2O$ masers are located in accelerating regions rather than in aspherical or disk-like regions which should show more elliptical distributions with systematic velocity gradients.

An interesting possibility discussed by Cohen (1989) is that the most rapid changes observed in the $H_2O$ masers, which occur on timescales of weeks, may be caused by pulsation-driven shock waves, propagating through the inner envelopes at velocities of typically 30 km $s^{-1}$. Such shock waves could provide an additional acceleration mechanism in the inner envelope. If this is so, then time-dependent changes related to the stellar pulsation cycle might be evident in the location and velocities of the $H_2O$ masers.

## 5. SiO masers

The SiO masers are located closest to the stars in the region where the onset of mass-loss occurs. The SiO molecules may act as an efficient coolant in the extended stellar atmospheres (Muchmore et al. 1987). If this is the case, then "run-away" cooling may lead to thermal instabilities giving enhanced grain formation in a two-phase medium with cooler clumps embedded within a less-dense, warmer gas.

The SiO maser emission is observed from several rotational transitions in excited vibrational levels, requiring temperatures above $\sim$ 1000 K. The SiO spectra are strongly variable and linearly polarized with typical linear polarizations detected for Miras and supergiants of tens of per cent. For several sources, weak circular polarization, of a few per cent, has been also detected (Barvainis et al. 1987; McIntosh et al. 1993). The Zeeman splitting effect is very weak for SiO molecules (which are non-paramagnetic) and no Zeeman pairs have been identified. If the circular polarization is due to the Zeeman effect, then strong magnetic fields of tens of Gauss at the stellar photosphere are indicated. Such strong fields appear to be consistent with the weaker fields, measured at larger radii from OH, and would have a dominant influence on the gas motions in the inner envelopes (Chapman & Cohen 1986; Barvainis et al. 1987). However an alternative hypothesis, which needs to be further investigated, is that the circular polarization may occur in regions of much lower magnetic fields due to linearly polarized emission passing through an anisotropic medium (Nedoluha & Watson 1993).

Early VLBI observations of SiO masers in the supergiant stars VY Cma and VX Sgr, and the Mira variable R Cas, showed the masers to be located within several stellar radii (Moran et al. 1979; Lane 1984; McIntosh et al. 1989). For R Cas, McIntosh et al. found the SiO maser emission to be located within a small number ($\sim$ 5) of clumps with each clump having well defined polarization characteristics.

Recent high angular resolution SiO images have been obtained for several sources. The SiO masers show a range of spot sizes from $\sim 10^{12}$ to $10^{14}$ cm (Colomer et al. 1992; Greenhill et al. 1995), and generally have no clear velocity patterns, as expected if strong turbulent motions, rather than a radial outwards expansion, dominate the kinematics close to the star. Ring-like structures centred on the stellar position have been observed for the Mira variables TX Cam and U Her (Diamond et al. 1994) and for the semi-regular variable W Hya (Miyoshi et al. 1994) again indicating strong tangential amplification. However, the acceleration of the SiO masers could presumably arise from both infall or outflow motions.

Greenhill et al. (1995) have studied in detail the SiO masers of the supergiant star VX Sgr and have obtained near-simultaneous measurements of the SiO maser emission and the mid-infrared dust emission. They found that near minimum light, the brightest SiO masers were located on a ring of radius 1.3 $R_\star$ with an inner dust shell radius of 4.6 $R_\star$. These observations, for the first time, clearly locate the SiO masers well within the inner radius of the dust shell, in the extended stellar atmosphere.

## 6. Conclusions

Radio observations of SiO, $H_2O$ and OH masers can be used to study the detailed structures of the circumstellar envelopes of evolved stars, including the Mira variables, OH/IR stars and M-type supergiants. The masers are located in a sequence of layers at radii between one and 100 stellar radii, as expected from considerations of their excitation processes. The OH/IR stars have the most uniform circumstellar envelopes. Observations of the OH 1612 MHz masers in these stars, show that the mass-loss is essentially spherically symmetric, with the masers located in thin-shells expanding at terminal velocity.

The circumstellar envelopes of Mira variables and M-supergiants are far less uniform. Observations of SiO, $H_2O$ and OH mainline masers, show that the mass-loss from these stars is likely to be intrinsically clumpy or irregular. The clumpiness may be enhanced by efficient cooling processes occuring within the extended stellar atmospheres.

Several issues remain unclear, including the role of the stellar magnetic fields and of shock waves in the envelopes, and the cause of clumping and of asymmetries in Miras and supergiant stars.

**Acknowledgments.** Jessica Chapman thanks the scientific organizing committee for providing financial support to attend the meeting, Peter Goldreich for discussions and Jim Moran, Jim Cohen and Anita Richards for providing results prior to publication.

## References

Barvainis, R., McIntosh, G. & Read Predmore, C. 1987, Nature, 329, 613

Booth, R.S., Kus, A.J., Norris, R.P., Porter, N.D. 1981, Nature, 290, 382

Bowers, P.F., Claussen, M.J. & Johnston, K.J. 1993, AJ 105, 284

Bowers, P.F., Johnston, K.J. & de Vegt, C. 1989, ApJ, 340, 479

Bujarrabal, V. & Alcolea, J. 1991, A&A, 251, 536

Chapman, J.M. & Cohen, R.J. 1985, MNRAS, 212, 375

Chapman, J.M. & Cohen, R.J. 1986, MNRAS, 220, 513

Chapman, J.M., Cohen, R.J. & Saikia, D.J. 1991, MNRAS, 249, 227

Chapman, J.M., te Lintel Hekkert, P., Killeen, N.E.B., Caswell, J.L. & Harnett, J.I. 1993, in Lecture Notes in Physics, Astrophysical Masers, eds A.W. Clegg & G.E. Nedoluha, Springer Verlag, 412, p345

Chapman, J.M., Sivagnanam, P., Cohen, R.J. & Le Squeren, A.M. 1994, MNRAS, 268, 475

Chengalur, J.N., Lewis, B.M., Eder, J. & Terzian, Y. 1993, ApJS, 89, 189

Claussen, M.J. & Fix, J.D., 1982, ApJ, 263, 153

Cohen, R.J. 1989, Rep. Prog. Phys., 52, 881

Cohen, R.J. 1993, in Lecture Notes in Physics, Astrophysical Masers, eds A.W. Clegg & G.E. Nedoluha, Springer Verlag, 412, p357

Colomer, F., Graham, D.A., Krichbaum, T.P. et al. 1992, A&A, 254, L17

Cooke, B. & Elitzur, M., 1985, ApJ, 295, 175

Diamond, P.J., Norris, R.P., Rowland, P.R., Booth, R.S., Nyman, L.-A. 1985, MNRAS, 212, 1

Diamond, P.J., Kemball, A.J., Junor, W., Zensus, A., Benson, J. & Dhawanm V. 1994, ApJ, 430, L61

Goldreich, P. & Scoville, N. 1976, ApJ, 205, 144

Greenhill, L.J., Colomer, F., Moran, J.M., Backer, D.C., Danchi, W.C. & Bester, M. 1995, ApJ, in press

Harvey, P.J., Bechis, K.P., Wilson, W.J. & Ball, J.A. 1974, ApJS, 27, 331

Herman, J. & Habing, H.J. 1985, A&AS, 59, 523

Huggins, P.J. & Glassgold, A.E. 1982, AJ, 87, 1828

Jewell, P.R., Webber, J.C. & Snyder, L.E. 1980, ApJ, 249,118

Lane, A.P. 1984 IAU Symp 110, p329 eds R. Fanti, K. Kellerman & G. Setti, Reidel, Dordrecht, Holland

Lane, A.P., Johnston, K.J., Bowers, P.F., Spencer, J.H. & Diamond, P.J. 1987, ApJ, 323, 756

Le Squeren, A.M. & Sivagnanam, P. 1985, A&A, 152, 85

McIntosh, G.C., Predmore, C.R., Moran, J.M., Greenhill, L.J., Rogers, A.E.E. & Barvainis, R. 1989, ApJ, 337, 934

McIntosh G.C., Read Predmore, C. & Patel, N.A. 1993, ApJ, 404, L71

Menten, K.M. & Melnick, G.J. 1991, ApJ, 377, 647

Miyoshi, M., Matsumoto, K., Kameno, S., Takaba, H. & Iwata, T. 1994, Nature, 371, 395

Moran, J.M., Ball, J.A., Predmore, C.R., Lane, A.P., Huguenin, G.R., Reid, M.J. & Hansen, S.S., ApJ, 231, L67

Muchmore, D., Nuth, J. & Stencel, R. 1987, ApJ, 315, L141

Nedoluha, G.E. & Watson, W.D. 1993, in Lecture Notes in Physics, Astrophysical Masers, eds A.W. Clegg & G.E. Nedoluha, Springer Verlag, 412, p47

Reid, M.J. & Menten, K.M., 1990, ApJ, 360, L51

Reid, M.J., Muhleman, D.O., Moran, J.M., Johnston, K.J. & Schwarz, P.R. 1977, ApJ, 214, 60

Richards, A.M.S., Cohen, R.J. & Yates, J.A., 1995, in Circumsteller Matter, Proceedings of ROE Centenary Symposium, Edinburgh, ed. G.D. Watt, Kluwer, Dordrecht, in press

Schultz, G.V., Sherwood, W.A. & Winnberg, A. 1978, A&A, 63, L5

Sivagnanam, P. 1989, Ph.D. thesis, University of Paris-Meudon

te Lintel Hekkert, P., Caswell, J.L., Habing, H.J., Haynes, R.F. & Norris, R.P. 1991, A&AS, 90, 327

van der Veen, W. 1989, A&A, 210, 127

van Langevelde, H.J., van der Heiden, R. & van Schooneveld, C. 1990, A&A, 239, 193

van Langevelde, H.J., Janssens, A.M., Goss, W.M., Habing, H.J. & Winnberg, A. 1993, A&AS, 101, 109

Welty, A., Fix, J.D. & Mutel, R.L., 1987, ApJ, 318, 852

West, M.E., Gaylard, M.J., Combrinck, W.L., Cohen R.J. & Shepherd, M.C. 1992, in Variable Stars and Galaxies, ed B. Warner, ASP Conf. Ser., 30, p277

Yates, J. & Cohen, R.J. 1994, MNRAS, 270, 958

## Discussion

*Kerschbaum*: Concerning the clumps in the OH-maser emission: What is the situation regarding variability in flux and position?

*Chapman*: OH-maser clumps show strong, erratic variations in flux density, but are stable in position over time scales of several years at least. For one Mira variable, $\nu$ Orionis, we (Chapman, Sivagnanam & Cohen) detected proper motions in the OH clumps thus showing their "ballistic" nature.

*Bedding*: The OH rings have beautiful spherical symmetry. Planetary nebulae are often asymmetric, and there is evidence that Miras themselves are asymmetric (Tuthill et al. 1994). Is it possible that the observed sphericity of OH masers does not trace the mass distribution, but rather reflects the symmetric radiation field from the central star?

*Chapman*: I think that the higher mass-loss OH/IR stars are genuinely close to being spherically symmetric, but with incompletely filled and knotty envelopes. Why post-AGB stars have bipolar morphologies is very much an open question. The answer may be related to binary systems, or to rapid evolution of the envelope structure in the post-AGB evolutionary stage.

*Zijlstra*: (i) The $F_{25}/F_{12}$ ratio gives a poor correlation with CO mass loss rates which may be due to the 12-micron flux being affected by variation in the mass loss rate. (ii) The clumpiness seen in 1612 MHz maps is a bit deceptive: you are looking at the tangential maser which is strongly dependent on the degree of saturation of the radial maser, which in turn depends on clumps in the inner part. It seems to me that the evidence for clumpiness in the inner shell is much stronger than for clumpiness in the outer shell.

*Chapman*: (i) Essentially I agree! $F_{25}/F_{12}$ is only a rough indicator of mass-loss rate, but one that is easy to use given IRAS measurements. A better indicator might be the colour K[2.2$\mu$m] - 12. (ii) OH 1612 MHz masers are strongly saturated. For this reason I consider the observed knots of emission at 1612 MHz to reflect genuine density variations.

*Goldreich*: (i) Do you see the amplified stellar image? (ii) What information do you have about the magnetic field in the circumstellar envelope?

*Chapman*: (i) VLBI observations have shown maser spots, for the extreme blue-shifted emission, which do appear to amplify the stellar image. Two examples where this has been seen are $\nu$ Her (a Mira) and OH 127.8 (an OH/IR star). (ii) Magnetic fields can and have been determined from Zeeman splitting of the OH main lines at 1665 and 1667 MHz. Typical field strengths are $\sim$2mG at 50 stellar radii.

*Mattei*: (i) You may be interested in the 30-year light curve of VX Sgr using the AAVSO International Database, that I show in my poster. (ii) I was interested to see the short term variations you found in your radio observations of VX Sgr. Several AAVSO observers have reported variations by as much as 1.0 to 1.5 magnitudes within hours on some Mira variables. Have you found short term variability in other stars, particularly Miras? Could the visual oscillations be related to what you found in VX Sgr?

*Chapman*: Firstly let me say how wonderful the AAVSO data is, allowing us to easily see long term variability trends. VX Sgr is, so far, the only evolved star known to show such rapid OH variations (t $\sim$ 1 hr). I think we may find this effect in some Miras. I am fascinated to hear of corresponding optical variations in Miras. Are the radio and optical variations related? I don't know!

*Busso*: Is there a "canonical" explanation why some AGB stars with periods longer than 300d develop OH/IR maser features and others do not? Is there any influence from the environment, or does the reason lie only in the evolution of the central star?

*Chapman*: Maser emission is more easily detected from stars with thicker envelopes and stronger IR flux densities. However, we (Chapman & Habing 1995, in preparation) have found that some stars with thick envelopes have very little (but not zero) OH. This may be related to the clumpiness of the OH.

*Feast*: Can you distinguish, in individual cases, between supergiants and other OH sources on the basis of OH data alone?

*Chapman*: No, not on OH data alone. Supergiants have high expansion velocities and blue infrared colours - these two *together* should separate supergiants from Miras (low $V_{exp}$, blue colours) and OH/IR stars (redder colours, higher $V_{exp}$).

*Astrophysical Applications of Stellar Pulsation*
*ASP Conference Series, Vol. 83, 1995*
*R. S. Stobie and P. A. Whitelock, eds.*

# Mira Variables: Theory versus Observation

P.R. Wood

*Mount Stromlo & Siding Spring Observatories, Private Bag, Weston Creek PO, Canberra ACT 2611, Australia*

**Abstract.** The identification of the mode of pulsation of Mira variables remains a major problem. Recent angular diameter measurements of solar neighbourhood Miras suggest large radii consistent with first overtone pulsation. On the other hand, the pulsation velocity amplitudes of Mira variables point to fundamental mode pulsation. There is some marginal evidence from multiple modes in Miras that the prime pulsation mode is the first overtone, while distinct (K,logP) sequences in the Magellanic Clouds point to the Miras being fundamental mode pulsators. The above confusing evidence is presented and discussed. Finally, the (K,logP) relations in the SMC and LMC are compared in order to determine if there is any metallicity dependence.

## 1. Introduction

Reliable identification of the pulsation mode of the Mira variables is proving to be very difficult (Willson 1982; Wood 1990b). The basic reason for this difficulty is that Miras do not have sharp edges like our Sun, they just fade gradually as angular diameter increases. For example, in a theoretical model for a Mira atmosphere, the radius changes by a factor of $\sim$2 as Rosseland optical depth changes from 0.01 to 1.0 (Bessell et al. 1989). The definition of radius is not simple in this situation, but modellers usually adopt a radius defined at Rosseland optical depth 1.0 (or $\frac{2}{3}$) (this radius will be referred to herein as the photospheric radius). In spite of the fact that the ratio of fundamental mode to first overtone period in a Mira is typically $\sim$2.5-3.5 (Fox & Wood 1982), observational estimates of Mira radii are so uncertain that a reliable identification of either fundamental mode or first overtone mode as the true pulsation mode has not yet been made.

In the last few years, some new observational and theoretical results relevant to the determination of the Mira pulsation mode have become available. These results are discussed in this review.

## 2. Angular diameter measurements of Mira variables

The angular diameter of a star such as a Mira variable varies greatly depending on the wavelength of observation (Bessell et al. 1989; Scholz & Takeda 1987). This is because of the large variation of opacity with wavelength caused by molecular species such as TiO. In deep bands, where the opacity is high, the diameter (defined at monochromatic optical depth 1.0) can be at least twice as large as the stellar photospheric radius (Rosseland optical depth unity) (Bessell et al. 1989). However, in regions of the spectral continuum longward of 1 $\mu$m (and possibly shortward of 0.4 $\mu$m), the radius at monochromatic optical depth unity appears to coincide well with the photospheric radius. Angular diameter measurements at these wavelengths should provide the most reliable estimates of Mira photospheric radii as there will not be a large correction factor for conversion of the monochromatic observed radius into the photospheric radius. Direct angular diameter measurements for Mira variables have been accumulating for about 20 years using speckle and lunar occultation techniques. Most recently, Tuthill et al. (1994) have used a non-redundant masking technique to measure the angular diameter of R Leo at wavelengths shortward of 1 $\mu$m, and Haniff & Tuthill (1995) have applied the technique to a number of other Miras (see Haniff 1995, this volume). Because of the wavelength of these observations, the observed monochromatic radii need some quite uncertain correction factors for estimation of the photospheric radius.

Some angular diameter measures collected from the literature are given in Table 1. The table gives the monochromatic, uniform disk angular diameter $\phi$ at the wavelength of observation ($\lambda$). Adopting a limb-darkened rather than a uniform disk leads to similar angular diameters (Di Giacomo et al. 1991; Ridgway et al. 1992). No correction to photospheric angular diameter has been made, but since many of the observations were made in the H and K bands or around 0.4 $\mu$m, these corrections should be small in such cases (i.e., the smaller of the measured angular diameters should be approximately equal to the photospheric angular diameter). The distance estimates required to derive radii were obtained by assuming the Miras lay on the (K,logP) relation for LMC Miras derived by Feast et al. (1989) and by using mean K magnitudes from the photometry in Catchpole et al. (1979). Finally, the period-mass-radius (PMR) relations for fundamental mode and first overtone pulsation (Fox & Wood 1982; Wood 1990a) were used to derive radii ($R_0$ and $R_1$, respectively, in $R_\odot$) the stars would have in each mode for a mass of 1 $M_\odot$.

Table 1. Observed Mira diameters

| Star | P(d) | d(pc) | $\lambda(\mu m)$ | $\phi$(mas) | $R_{obs}$ | $R_0$ | $R_1$ | Reference |
|---|---|---|---|---|---|---|---|---|
| R Leo | 310 | 113 | 2.16 | 30-35 | 352-411 | 224 | 405 | Di Giacomo et al. 1991 |
| | | | 0.83,0.90 | 44 | 517 | | | Tuthill et al. 1994 |
| | | | 0.54-0.74 | 30-54 | 352-634 | | | Labeyrie et al. 1977 |
| *o* Ceti | 332 | 113 | <1 | 30-120 | 352-1410 | 232 | 424 | Bonneau et al. 1982 |
| | | | 2.2 | 36.1 | 424 | | | Ridgway et al. 1992 |
| U Ari | 371 | 743 | 1.62 | 6.11 | 472 | 246 | 456 | Ridgway et al. 1979 |
| S Psc | 405 | 1240 | 2.23 | 3.84 | 495 | 257 | 484 | Ridgway et al. 1979 |

It is clear from this table that the large observed radii of Mira variables provide strong evidence that these stars are first overtone pulsators. A similar conclusion was reached by Tuthill et al. (1994) and Haniff & Tuthill (1995). If the Miras are actually pulsating in the fundamental mode, as some of the evidence presented below suggests, then the radii need to be smaller by almost a factor of 2. In this case, some combination of the following factors must apply. (1) The distances to the solar neighbourhood Miras have been overestimated.

Wood (1990b) presented arguments that the (K,logP) relation for Miras should be metal dependent, with the result that the distances derived above should be reduced by 10-20%. By itself, this effect is much smaller than required to reduce the radii to fundamental mode sizes.
(2) The radii were systematically measured at maximum expansion of the star. However, many of the observations were made near maximum light when the star is expanding through mean radius (Bessell, Scholz & Wood 1995). The total variation in photospheric radius (maximum to minimum) in model Miras is 30-50% in the fundamental mode, much less in the first overtone, so that even if all observations were made at maximum expansion, the error in the mean radius estimate would be <25%.
(3) The PMR relations used, which were derived from hydrostatic models, do not apply to Miras pulsating at large amplitude. Very little theoretical work has been done on this problem but some preliminary results presented in Sections 3.1. and 3.2. suggest that the PMR relation for large amplitude pulsators is not greatly different from that derived from linear theory.
(4) The model atmospheres (Bessell et al. 1989; Bessell et al. 1995) which are used to relate the observed monochromatic radii to the photospheric radii are inadequate. There is a large amount of uncertainty in these models. Firstly, the density structure of the models was made up in a semi-empirical manner (Bessell et al. 1989) or derived from piston driven pulsating atmospheres (Bessell et al. 1995). Since schemes to produce Miras pulsating in the fundamental mode have recently been found (see Section 3.1., and Yaari & Tuchman 1995), better model structures will be possible in future. The models of Bessell et al. (1989) are much more extended than those of Bessell et al. (1995) due to the different density structures. But probably the largest uncertainty in the models lies in calculating the thermodynamic equilibrium and opacity in a molecular atmosphere with a velocity gradient throughout.

## 3. Nonlinear pulsation models

Producing nonlinear pulsation models for Mira variables presents significant difficulties. The fundamental mode in Mira models generally is very unstable and the amplitude grows enormously, leading to a series of thermal relaxation oscillations and mass loss unlike the variations seen in real Miras (Wood 1974; Tuchman, Sack & Barkat 1979). Attempts to produce first overtone pulsators with low masses and periods greater than ~300 days have also been generally unsuccessful as the fundamental mode tends to overwhelm the first overtone pulsation.

Yaari & Tuchman (1995) (see also Tuchman 1995, this volume) have reported recently some LMC Mira models that were pulsating stably in the fundamental mode. Two factors seem to have helped stabilise these models: firstly, the models are somewhat less luminous than expected for the LMC ($M_{bol}$,logP) relation, and, secondly, the models were run for a long time starting from numerical noise so that a large envelope structural adjustment occurred before full amplitude was reached. Yaari & Tuchman found that their models contracted and ended up with a smaller period than they had initially - similar results were also found by Wood (1974) and Keeley (1970). In contrast, an example of a fundamental mode pulsator which *expands* and lengthens its period is described below. This model had an abundance and luminosity appropriate for an LMC Mira. For comparison, a first overtone pulsator is also described.

### 3.1. Fundamental mode pulsation

The model discussed here is one of several for which stable fundamental mode pulsation has been produced. The nonlinear pulsations were performed with the code of Wood (1974), updated to include the new OPAL opacities (Iglesias & Rogers 1993) and molecular opacities as described in Chiosi, Wood & Capitanio(1993). When run with standard parameters, the models all showed the typical growth of the fundamental mode pulsation to amplitudes well beyond observed values, mass loss occurred from the surface and ultimately convergence failed (Wood 1974; Tuchman et al. 1979). Since Yaari & Tuchman (1995) had found that Mira models underwent a significant amount of envelope relaxation over hundreds of years of pulsation, a method of stabilising the violent initial pulsation was sought. Basically, more pulsation damping was needed, and this was provided by the artificial viscosity used to treat shock waves (Richtmyer & Morton 1967). The artificial viscosity parameter $a^2$ was increased to the large value of 15 for ~50 years while the pulsating envelope relaxed and was then reduced to a more normal value of 2.5 (together with the addition of Stellingwerf's 1975 velocity cutoff parameter $\alpha$=0.1).

Results for an LMC model with M=M$_\odot$, L=4000L$_\odot$, Y=0.3 and Z=0.008 are shown in Figures 1 and 2. In Figure 1, the large artificial viscosity was used up until time $t \sim 2.6 \times 10^4$ days and was then removed causing the amplitude of surface pulsation to increase enormously. Most envelope readjustment occurred in the stage of initial amplitude growth for $t < 6 \times 10^3$ days, and again when the excess artificial viscosity was removed around $t \sim 2.6 \times 10^4$ days.

Several points should be noted about the models. There is a very large extension of the surface layers caused by pulsation, and in the surface layers, motions with quasi-periods of ~3-12 times the fundamental mode period occur (this is the cause of the variation with time of the mean radius of the outermost zone for $t > 4{\times}10^4$ days). Oscillations at 2 P$_0$ are very common, these oscillations occurring near and above the photosphere (as can be in Figure 2). The total velocity amplitude near the photosphere is ~40 km s$^{-1}$, while the surface layer velocities are small and quite irregular, as observed in optical Mira absorption lines (Joy 1954).

In order to see if the PMR relation derived from linear pulsation analysis of hydrostatic models gives a reasonable approximation to the PMR of pulsating Miras, the mean photospheric radius for the relaxed pulsating model was compared with the value predicted from the linear PMR relation. According to linear theory, for fundamental mode pulsation, $P$ varies approximately as $R^{1.94}M^{-0.9}$ (Wood 1990a) so that $R = kP^{0.52}M^{0.46}$, where $k$ is a constant and R is the photospheric radius in the hydrostatic model. For the present model, $k$ was evaluated using the static radius ($191R_\odot$), period (209 days) and mass ($M_\odot$). Then the relaxed nonlinear period of 243 days used in this (linear) formula predicted a model photospheric radius of 206 $R_\odot$ whereas the mean photospheric radius of the nonlinear model was actually 218 $R_\odot$. This difference between predicted and actual radius is small, at least when it is seen that the linear theory

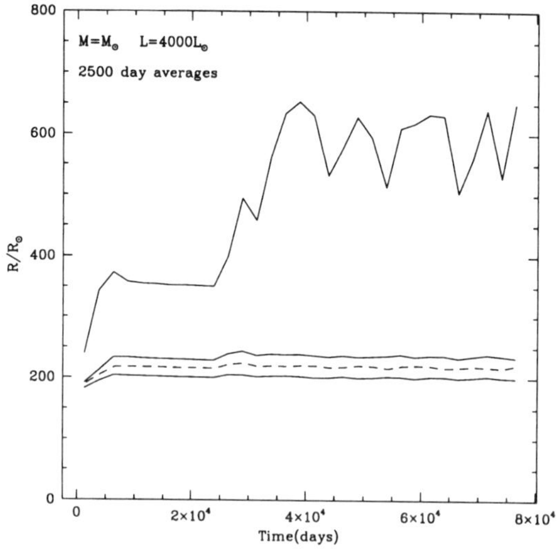


Figure 1. Radii of 3 mass points averaged over 2500 days (~10 pulsation cycles) plotted against time (solid lines). The point with the largest radius is the surface mass zone. The dashed line is the time-averaged photospheric radius.

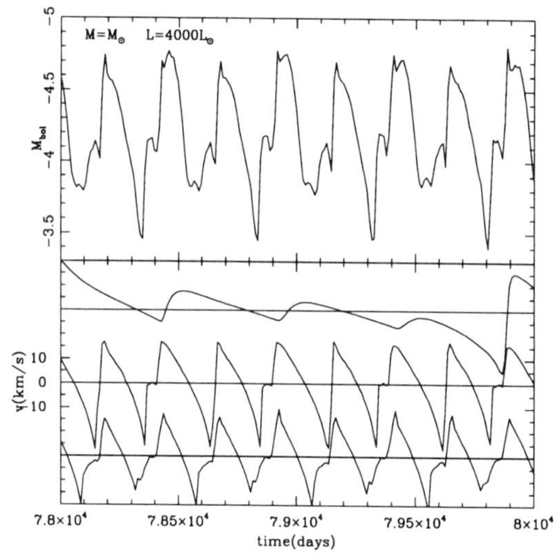


Figure 2. Bolometric luminosity (top panel) and velocity (bottom panel) of the 3 mass zones shown in Figure 1 plotted against time after the pulsation has reached a relaxed state.

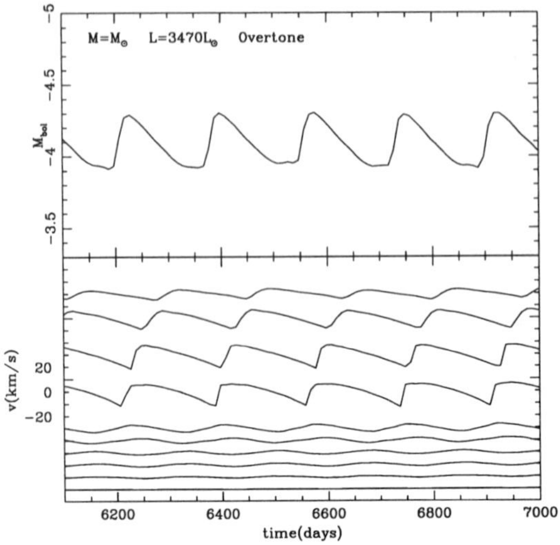


Figure 3. Bolometric luminosity (top panel) and velocity of 10 mass zones (bottom panel) plotted against time after the pulsation has reached a relaxed state. Rosseland optical depth unity lies near the mass zone whose velocity is given by the fourth line down in the bottom panel.

predicts a radius of 400 $R_\odot$ when the nonlinear fundamental mode period of 243 days is used in the PRM for first overtone pulsation. The results indicate that structural adjustment of the envelope due to nonlinear pulsation does not radically alter the PMR relation. However, this statement needs verifying with a larger range of models, particularly models with longer periods.

### 3.2. Overtone pulsation

Production of first overtone pulsation models with masses of $M \sim M_\odot$ and periods $\gtrsim$ 300 days is very difficult because, for these parameters, the fundamental mode always overwhelms the first overtone pulsation. Some overtone pulsators have been reported in the literature (Wood 1974,1975; Tuchman 1991) but always with large mass ($\gtrsim$ 1.5 $M_\odot$) and/or low luminosity and short periods. An example with $M = M_\odot$ and $L = 3470L_\odot$ is shown in Figure 3. The nonlinear first overtone pulsation period of this model showed only small differences between the linear and nonlinear periods and static and mean pulsation photospheric radii of <10%. Another first overtone model run for 160 years showed similar small changes in period and radius (this model had a stable fundamental mode). The PMR relation derived from linear pulsation theory should therefore be quite accurate when applied to full amplitude first overtone pulsators.

The important point to note in comparing Figures 2 and 3 is that the velocity amplitude of fundamental mode pulsation is about 40 km s$^{-1}$ whereas it is only ~25 km s$^{-1}$ for first overtone pulsation. The directly observed Mira pulsation velocity amplitudes are ~24-28 km s$^{-1}$ (Hinkle, Scharlach & Hall 1984; Wood 1987). However, these velocities have not been corrected for geometric and limb darkening effects which, for warmer stars, generally require a factor of 1.3 to 1.4 increase in observed velocity to get the true pulsation velocity. An estimate of the size of this correction factor for Miras can be obtained from the line profiles given for the Mira model atmospheres of Bessell et al. (1995). The line splitting shown for the fundamental mode pulsators in Bessell et al. is ~20 km s$^{-1}$ near maximum light, slightly smaller than the observed line splitting (~24-28 km s$^{-1}$) observed in Miras (Hinkle 1978; Hinkle et al. 1984). The pulsation velocity amplitude in the fundamental mode models is ~36 km s$^{-1}$, so that real Mira pulsation velocities of at least 36 km s$^{-1}$ are predicted by the model atmosphere velocity correction factor. It should also be noted that the first overtone pulsator in Bessell et al. (1995) gave a line splitting of only ~8 km s$^{-1}$ (for a true pulsation velocity amplitude of ~17 km s$^{-1}$).

As noted before in the literature (Willson & Hill 1979; Bowen 1988; Wood 1990b; Bessell et al. 1995), the inability of low mass first overtone pulsators to give pulsation velocity amplitudes $\gtrsim 25$ km s$^{-1}$, and the agreement of theoretical fundamental mode pulsation velocities with observations provides one of the most powerful arguments for fundamental mode pulsation in the Miras. Some disagreement with this point of view has been expressed by Tuchman (1991).

## 4. Evidence for more than one pulsation mode

Possibly the most definitive method of determining the mode of pulsation of the Mira variables is the identification of multiple modes. Since $P_0/P_1 \sim 3$ while $P_1/P_2 \sim 1.2$ (Fox & Wood 1982), the period ratio of the dominant Mira mode to that of secondary modes should allow identification of the primary pulsation mode (under the assumption that the fundamental, first and second overtones are the modes most likely to be present).

### 4.1. Individual LPVs with multiple modes

Barthès & Tuchman (1994) have Fourier analysed AAVSO light curves of the two Miras S CMi and $\chi$ Cyg in order to look for secondary periods. They did find some formally significant secondary periods, compared them with the pulsation periods of models for Mira variables, and concluded that the best fit between observed and theoretical periods could be obtained assuming that the dominant pulsation mode was the first overtone (largely because of the existence of some observed secondary periods >1600 days which could only be explained as due to the fundamental mode). However, the secondary periods found are only marginally significant and some secondary periods have no explanation in terms of radial pulsation modes. Better data is required to confirm these results.

One other characteristic feature of the light curves of Mira variables that is quite common is the existence of an oscillation at a period twice that of the primary period (Wood 1975). This can be explained not as a separate oscillation but as an oscillation induced in the outer layers with period twice that of the interior pulsation, as shown in Figure 2.

### 4.2. LMC sequences corresponding to different modes

In general, pulsation models of AGB stars show that as the star evolves to higher luminosities, lower order modes become unstable. In that case, one might

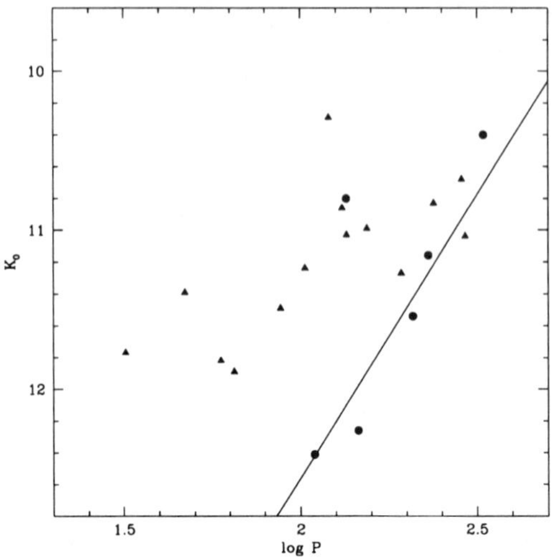


Figure 4. Small and large amplitude red variables in the (K,logP) plane. Periods are given in Sebo & Wood (1995a) for LPVs around NGC 1850 (circles) and Sebo & Wood(1995b) for variables around NGC 2058 and 2065 (triangles). The line is the sequence occupied by large amplitude (Mira) variables in the LMC (Feast et al. 1989; Hughes & Wood 1990).

expect to have groups of AGB stars pulsating in the fundamental mode, the first overtone, and so on. If these stars are in the LMC, then they will occupy separate period-luminosity sequences that will be readily seen in *apparent* magnitude since all the stars are at essentially the same distance. Sebo & Wood (1994, 1995a,1995b) have recently found samples of red variables with both large and small amplitude ($\delta V \gtrsim 0.2$ mag.) in the LMC. A (K,logP) plot for all stars in the sample (Figure 4) shows that some stars lie on the well-known Mira sequence in the LMC but that there appears to be a second sequence (or sequences) at periods shorter by a factor of ~2.2, with no stars at intermediate periods. This situation is just what would be expected if the Miras were fundamental mode pulsators while the second, shorter period sequence corresponded to first overtone pulsation.

## 5. Metal dependence of the Mira PL relation

As noted in section 2., there are good theoretical reasons to expect the Mira (K,logP) relation to depend on metal abundance (Wood 1990a). One test of the metallicity dependence of the Mira PL relation is to compare the (K,logP) relation for LMC and SMC Miras. This is done in Figure 5, from Wood, Moore & Bessell (1995), where the SMC LPVs are plotted along with the LMC relation of Feast et al. (1989) shifted by 0.4 magnitude to account for the different distance moduli of the SMC and LMC according to Cepheids (Caldwell & Laney 1991). It is clear that both the LPVs and Cepheids give the same distance modulus offset between the LMC and SMC. The theoretical Cepheid calculations of Chiosi et al. (1993) predict very little dependence of the Cepheid PL relation on metallicity, at least for a change from SMC to LMC metallicity, so it seems that the SMC and LMC (K,logP) relations for LPVs are similar. Note that if there was a factor of 2 difference in metal abundance between LMC and SMC LPVs (as for young stars in the Clouds - Russell & Bessell 1989), then the formula in Wood (1990a) would predict that the LMC stars were fainter by 0.13 magnitudes in K.

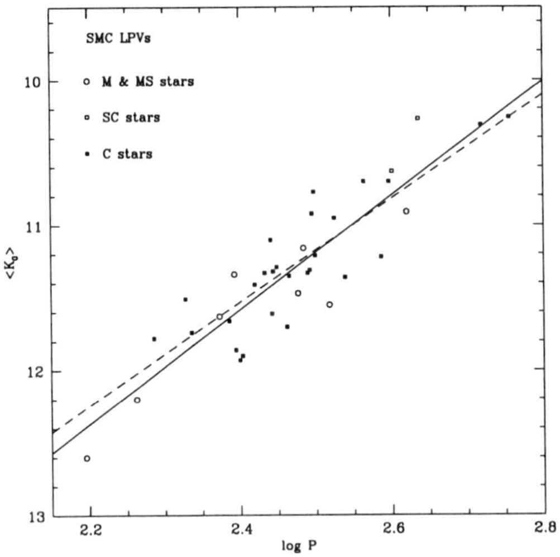


Figure 5. SMC LPVs in the (K,logP) plane. The solid line is a least squares fit to the SMC data while the dashed line is the LMC relation of Feast et al. (1989) shifted by 0.4 magnitude to account for the different distance moduli of the SMC and LMC.

## References

Barthès, & Tuchman, Y. 1994, A&A, 289, 429

Bessell, M.S., Brett, J.M., Scholz, M., & Wood, P.R. 1989, A&A, 213, 209

Bessell, M.S., Scholz, M., & Wood, P.R. 1995, A&A, submitted

Bonneau, D., Foy, R., Blazit, A., & Labeyrie, A. 1982, A&A, 106, 235

Bowen, G.H. 1988, ApJ, 329, 299

Caldwell, J.A.R., & Laney, C.D. 1991, in IAU Symposium 148, *The Magellanic Clouds*, eds. R. Haynes & D. Milne (Kluwer:Dordrecht), p.249

Catchpole, R.M., Robertson, B.S.C., Lloyd Evans, T., Feast, M.W., Glass, I.S., & Carter, B.S. 1979, SAAO Circulars, 1, 61

Chiosi, C., Wood, P.R., & Capitaneo, N. 1993, ApJS, 86, 541

Di Giacomo, A., Richichi, A., Lisi, F., & Calamai, G. 1991, A&A, 249, 397

Feast, M.W., Glass, I.S., Whitelock, P.A., & Catchpole, R.M. 1989, MNRAS, 241, 375

Fox, M.W., & Wood, P.R. 1982, ApJ, 259, 198
Haniff, C.A., & Tuthill, P.G. 1995, MNRAS, in press
Hinkle, K.H. 1978, ApJ, 220, 210
Hinkle, K.H., Scharlach, W.W.G., & Hall, D.N.B. 1984, ApJS, 56, 1
Hughes, S.M.G., & Wood, P.R. 1990, AJ, 99, 784
Iglesias, C.A., & Rogers, F.F. 1993, ApJ, 412, 752
Joy, A.H. 1954, ApJS, 1, 39
Keeley, D.A. 1970, ApJ, 161, 657
Labeyrie, A., Koechlin, L., Bonneau, D., Blazit, A., & Foy, R. 1977,ApJ, 218, L75
Richtmyer, R.D., & Morton, K.W. 1967, Difference Methods for Initial Value Problems (New York: Wiley).
Ridgway, S.T., Benson, J.A., Dyck, H.M., Townsley, L.K., & Hermann, R.A. 1992, AJ, 104, 2224
Ridgway, S.T., Wells, D.C., Joyce, R.R., & Allen, R.G. 1979, AJ, 84, 247
Russell, S.C., & Bessell, M.S. 1989, ApJS, 70, 865
Scholz, M., & Takeda, Y. 1987, A&A, 186, 200
Sebo, K.M., & Wood, P.R. 1994, AJ, 108, 932
Sebo, K.M., & Wood, P.R. 1995a, ApJ, in press
Sebo, K.M., & Wood, P.R. 1995b, in preparation
Stellingwerf, R.F. 1975, ApJ, 195, 441
Tuchman, Y. 1991, ApJ, 383, 779
Tuchman, Y., Sack, N., & Barkat, Z. 1979, ApJ, 234, 217
Tuthill, P.G., Haniff, C.A., Baldwin, J.E., & Feast, M.W. 1994, MNRAS, 266, 745
Willson, L.A. 1982, in *Pulsations in Classical & Cataclysmic Variable Stars*, eds. J.P. Cox & C.J. Hansen (Boulder: Joint Institute for Laboratory Astrophysics), p.269.
Willson, L.A., & Hill, S.J. 1979, ApJ, 228, 854
Wood, P.R. 1974, ApJ, 190, 609
Wood, P.R. 1975, IAU Colloquium 29, *Multiple Periodic Variable Stars*, ed. W.S. Fitch (Dordrecht: Reidel), p.205
Wood, P.R. 1987, in Lecture Notes in Physics, Vol. 274, *Stellar Pulsation*, eds. A.N. Cox, W.M. Sparkes & G.S. Starrfield (Berlin: Springer), p.250
Wood, P.R. 1990a, in *From Miras to Planetary Nebulae*, eds. M.O. Mennessier & A. Omont, Editions Frontières, p.67
Wood, P.R. 1990b, ASP Conference Series, 11, 355
Wood, P.R., Moore, G., & Bessell, M.S. 1995, in preparation
Yaari, A., & Tuchman, Y. 1995, preprint.

## Discussion

*Kawaler*: How does mass loss change the pulsation periods relative to the periods you compute for the models which aren't losing mass?

*Wood*: The effect of mass loss on pulsation has not been explicitly included in these calculations. At the surface, $P_{gas} = 0$ is assumed. For normal Miras, the mass loss rate is too low to be likely to affect the pulsation period.

*Feast*: (i) There is an additional piece of information regarding the mode of pulsation: The $2\mu$m diameter of R Leo from an occultation agrees with overtone pulsation for Miras and all models agree that one gets to the photosphere at $2\mu$m. (ii) Your interesting results for low amplitude variables in the Magellanic Clouds are perhaps a parallel to the semi-regular variables in galactic globular clusters discussed by Patricia Whitelock.

*Wood*: (i) The Di Giacomo diameter at $2.16\mu$m is certainly a strong piece of evidence for a large radius for R Leo. Additional measurements around $2\mu$m should soon be forthcoming and will allow a check on this result. (ii) I agree.

*Kovács*: As far as the theoretical P-L relation is concerned:(i) should not one use the non-linear period rather than the linear one (see Y. Tuchman's talk)? (ii) is not the P-L relation for Miras rather a P-L-C (or some even more involved relation)?

*Wood*: (i) We don't have a theoretical P-L relation for Miras, only a P-M-R relation. The non-linear calculations I presented show that both P and R increase as a result of non-linear effects, but the P-M-R relation is not greatly altered. (ii) Theoretically, yes. Observationally, there is quite a tight P-L relation which indicates some additional constraint on P, M and R, a combination of stability and lifetime considerations, which we cannot currently compute.

*Welch*: Did I understand that the low-amplitude variables found near NGC 1850 and other clusters were being interpreted as having masses near $1M_{\odot}$? The turnoff mass of these clusters is closer to $5M_{\odot}$.

*Wood*: The LPVs are field variables, not showing any spatial connection with the clusters.

*Haniff*: Can observed radial velocities be used to infer the corresponding changes in "photospheric" radius as a function of pulsation phase? And if so, are these radius changes consistent with your theoretical modelling?

*Wood*: The observed velocities in the visible range always correspond to infall onto the star (absorption lines). Clearly, different mass zones are seen at different pulsation phases so that we can't integrate the velocity with time to derive motion of a mass point, or the photosphere. Similar remarks also apply to infrared velocities.

*Maeder*: Would you please comment on the driving mechanism of Miras' pulsations; in particular, is there any contribution from the molecules?

*Wood*: Driving is produced largely in the hydrogen ionization zone, and the helium ionization zones as well. There is essentially no driving produced in the outer, molecular layers. Details can be found in Fox & Wood (1982).

*Kanbur*: In order to get overtone models you might try giving it a 'high kick' so that it reaches its limit cycle from above. This works in the case of building ~10d Cepheid models.

*Astrophysical Applications of Stellar Pulsation*
*ASP Conference Series, Vol. 83, 1995*
*R. S. Stobie and P. A. Whitelock, eds.*

# Long Term Non Linear Thermal Effects in the Pulsation of Mira Variables

A. Yaari and Y. Tuchman
*Racah Institute, Hebrew University of Jerusalem, Jerusalem 91904, Israel*

The main conclusions that can be drawn from most of previous non-linear simulations of LPV (Keeley 1970, Rose & Smith 1972, Wood 1974, Tuchman et al. 1978, 1979) are the following:

In all cases where the fundamental mode is predicted, by the linear analysis, to be the dominant, the pulsations are seen to diverge and lead to a rather violent mass-loss process (Wood 1974, Tuchman, Sack & Barkat 1979).

A steady pulsation with (roughly) a limit cycle does set in only in those cases where the linear analysis predicts that the dominant mode (with the largest value of the growth rate) is the first overtone (or higher). In all these cases the actual limit cycle is found to be in a good agreement with the linear predictions, with regard to the pulsation mode and period (Wood 1974, Tuchman, Sack & Barkat 1979). However, it is important to note that periods, for these overtone pulsators, are always found to be smaller than 300 days (Wood 1974, Tuchman, Sack & Barkat 1979). Increasing the luminosity of these models, in order to increase their periods, results in a switch to the fundamental mode, followed by the violent relaxation oscillations, mentioned above.

The problem is that observed LPVs (Miras) exhibit a relatively clear periodic pulsation where most have periods between 300 and 400 days (Wood & Cahn 1977). "Simple" modifications of the theoretical models, such as varying the mixing length, altering surface conditions, etc., are not found to help resolve the difficulty.

An additional problem of these first overtone pulsators is associated with their velocity amplitude which is found to be small compared to the observed values Hill & Willson (1979).

In all previous works, calculating nonlinear dynamics of Miras, *including our own*, there was no clear identification that the star has indeed reached its limit cycle. To our surprise, it turned out, after continuing earlier calculations much farther (from about 50 years to more than 600 years!, Figure 1), that the non-linearity in the pulsation of a LPV causes a significant change of its internal energy structure (i.e., the entropy profile, see Figure 2). This change is accomplished in few hundred years ($\sim$ 300 years which is about the thermal time scale of the envelope). As a new stable modified entropy profile is reached, a "true" limit cycle is seen to set in and is rather different from the earlier (obtained for $\sim$ 50 years) "false" limit cycle (Figure 1). Pulsation has shifted from first overtone ($\sim$ 175 days) to a *new* fundamental mode ($\sim$ 330 days, no nodes) whose period is now between the first overtone and fundamental mode ($\sim$ 508 days) predicted by the linear analysis of the initial equilibrium model.

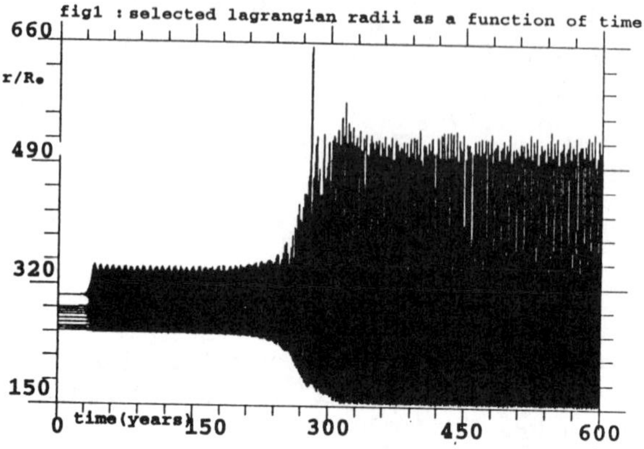


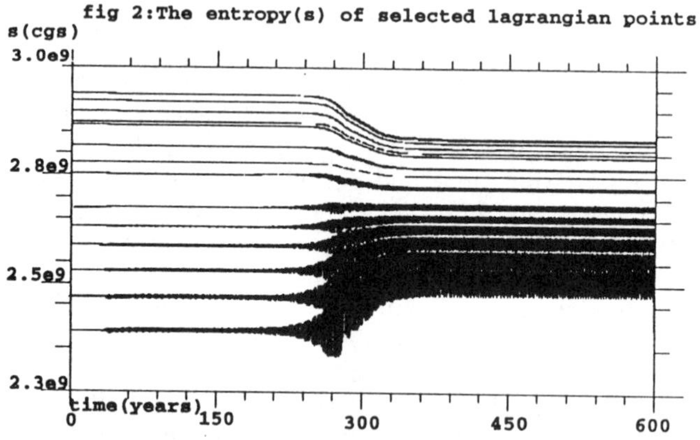


The pulsation features – periods, luminosity and velocity amplitudes – of these new nonlinear oscillating models are actually found to solve the two difficulties described above and to fit nicely the observational data.

The most important conclusion of these new results is that it is not possible to use the simple linear analysis for static models of Miras or Long Period Variables in order to determine their oscillatory behaviour, even their periods, as was and still is usually done.

## References

Hill, S.A., & Willson, L.A., 1979, ApJ, 229, 1029

Keeley, D.A., 1970, ApJ, 161, 656

Rose, W.K., & Smith, R.L., 1970, ApJ, 159, 903

Tuchman, Y., Sack, N., & Barkat, Z., 1978, ApJ, 219, 183

Tuchman, Y., Sack, N., & Barkat, Z., 1979, ApJ, 234, 217

Wood, P.R., 1974, ApJ, 190, 609

Wood, P.R., & Cahn, J.H., 1977, ApJ, 211, 499

*Astrophysical Applications of Stellar Pulsation*
*ASP Conference Series, Vol. 83, 1995*
*R. S. Stobie and P. A. Whitelock, eds.*

# The Mass Loss Rate – Period Relation in Carbon Miras

Martin Groenewegen

*Institut d'Astrophysique de Paris, CNRS, 98 bis Boulevard Arago, F-75014 Paris, France*

**Abstract.** The relation between mass loss rate and pulsation period in carbon Miras is discussed. The dust mass loss rate is very low (about 2 $10^{-10}$ $M_\odot$/yr) up to about $P$ = 380 days, where there is a sudden increase. For $P > 400$ days there is a linear relation between $\log \dot{M}$ and $P$. The change in the mass loss rate near 380 days may be related to radiation pressure on dust becoming effective in driving the outflow.

## 1. Introduction

The mass loss rate is the determining factor in the evolution of Asymptotic Giant Branch (AGB) stars since the time scale on which mass loss takes place is usually much shorter than the time scale of the nuclear processes.

As on the other hand it is well known that most AGB stars are variable and that it is believed that mass loss and pulsation are connected it is useful to study the relation between mass loss and pulsation period. The $\dot{M} - P$ relation has been studied in quite some detail for oxygen-rich Miras (Schild 1989, Wood 1990, Whitelock 1990, Whitelock et al. 1994) but never for carbon-rich Miras. In this contribution such a relation is presented (also see Groenewegen 1995, where this relation is compared to the ones derived for oxygen-rich Miras).

## 2. Results

Thirty-two carbon Miras were selected from the General Catalog of Variable Stars, Jones et al. (1990) and Le Bertre (1992). The spectral energy distributions and IRAS LRS spectra were fitted using the dust radiative transfer model of Groenewegen (1993).

The relation between the dust mass loss rate and pulsation period is shown in Fig. 1 (top panel). Two features stand out: there is a tight correlation above 400 days, and there appears to be a significant increase in the dust mass loss rate near $P$ = 380 days.

A different tracer of the dust formation process is the dust-to-gas (DTG) ratio. In the bottom panel the DTG ratio is plotted versus $P$. The gas mass loss rates as derived from CO measurements were taken from Loup et al. (1993) and scaled to the adopted distances. The DTG ratio increases with period for $P \gtrsim 400$ days, suggesting a more efficient dust production.

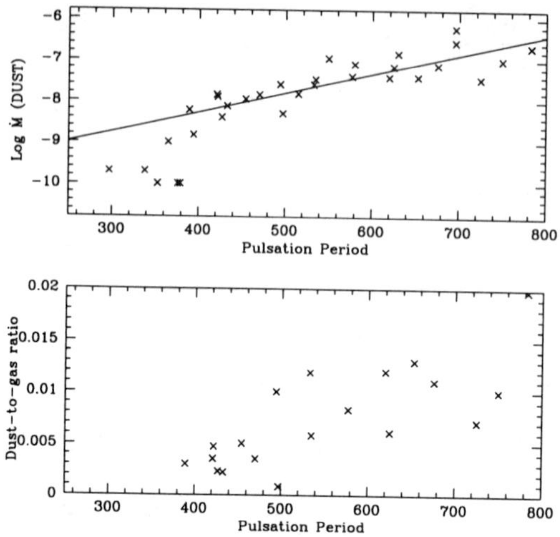


Figure 1. The dust mass loss rate and dust-to-gas ratio in the sample of carbon Miras. The solid line is a least-squares fit for $P > 400$ days.

As is evident from Fig. 1 there is very little data available for shorter periods. I am engaged to obtain near-infrared photometry, mid-infrared spectra and CO data for more short period carbon Miras.

Recently, Netzer & Elitzur (1993) and Habing et al. (1994) predicted a lower limit of about $10^{-7}$ $M_{\odot}$/yr to the (total) mass loss rate to sustain an outflow driven by radiation pressure on dust. With $\Psi = 0.003$ near $P = 380$ days, a total mass loss rate of 8 $10^{-8}$ $M_{\odot}$/yr is derived, close to this lower limit. Thus, the increase in the dust mass loss rate near 380 days may be related to radiation pressure on dust becoming effective and hence provide observational evidence for the theoretical prediction of Netzer & Elitzur and Habing et al.

## References

Groenewegen, M.A.T. 1993, Ph.D. thesis, University of Amsterdam

Groenewegen, M.A.T. 1995, in: Circumstellar Matter, ed. G. Watt, in press

Habing, H.J., Tignon, J., & Tielens, A.G.G.M. 1994, A&A, 286, 523

Jones, T.J., et al., 1990, ApJS, 74, 785

Le Bertre, T. 1992, A&AS, 94, 377

Loup, C., Forveille, T., Omont, A., & Paul, J.F. 1993, A&AS, 99, 231

Netzer, N., & Elitzur M. 1993, ApJ, 410, 701

Schild, H. 1989, MNRAS, 240, 63

Whitelock, P. 1990, ASP Conf. Ser., 11, 365

Whitelock, P., et al. 1994, MNRAS, 267, 711

Wood, P.R. 1990, in: From Miras to Planetary Nebulae, Editions Frontieres: Gif-sur-Yvette, p. 67

*Astrophysical Applications of Stellar Pulsation*
*ASP Conference Series, Vol. 83, 1995*
*R. S. Stobie and P. A. Whitelock, eds.*

# Yellow Supergiants and R Coronae Borealis Stars

P.L. Cottrell
*Mount John University Observatory, Dept of Physics & Astronomy, University of Canterbury, Christchurch, New Zealand*

**Abstract.** The study of luminous variable stars has important consequences both with respect to the scale of the Universe and for our understanding of the later stages of stellar evolution.

This review will look at a number of groups of these stars, in particular the RV Tauri stars, the hydrogen-deficient Carbon (HdC) stars, including the R Coronae Borealis (RCB) stars and some aspects of Cepheid variables.

Both photometric and spectroscopic observations will be presented to show the amount of progress that has been made in this area over the last several years. Links will be made to various models which are being used to understand the observational data.

The most important conclusion is that there is still enormous scope for the acquisition and detailed analysis of data on the relatively bright members of these groups of stars. These can be used to provide further understanding of the more distant objects of each group and of the wider problems in astrophysics.

## 1. Introduction

This paper will investigate a subset of objects presented in John Percy's 1993 paper, "Supergiant variables: recent observational results". I will confine myself to the yellow supergiant variables of low mass, as well as some aspects of Cepheid variables.

The talk will reinforce one of Percy's comments. *The study of non-linear processes in supergiant variables requires long series of observations of good and consistent quality.* In addition, it is also appropriate to add Percy's next sentence. *Long-term projects are often not encouraged or supported in astronomy.* They are sometimes regarded as being simply 'stamp-collecting', a term Ernest Rutherford used to describe all aspects of science apart from physics.

To gain recognition for this type of research it is imperative that the focus must be to provide observational tests of existing models or to prompt specific developments in this modelling. Although not exclusively, this has generally been confined to stellar envelope models, but with the acquisition of high quality spectral data, there is a need to develop appropriate dynamical model atmospheres to interpret the observational data.

The late stages of stellar evolution provide objects which can be used for a multitude of astrophysical purposes. Several examples of these pieces of work

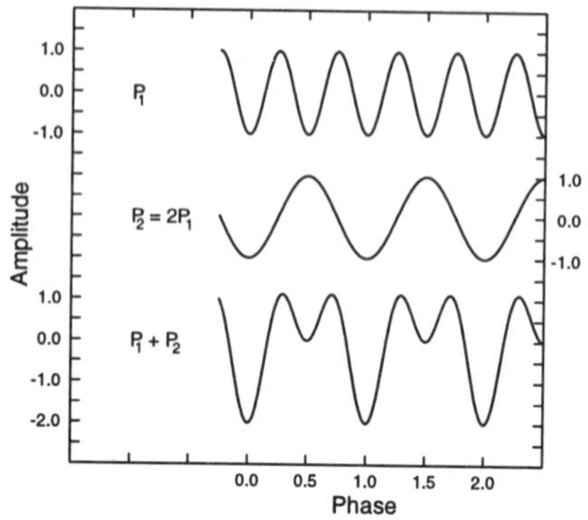


Figure 1. A schematic representation of the alternating behaviour of RV Tauri stars (bottom), using two precisely phased sinusoids of periods $P_1$ and $P_2$.

are: studies of stellar pulsation; tests of stellar evolutionary theory; investigation of circumstellar shells; determination of stellar masses; evolutionary links between high luminosity objects (e.g., W Vir to RV Tau to SRd and RCB to planetary nebula to white dwarf); determination of the distance scale; understanding the precision of this scale.

There is also scope for investigating other types of connections, in addition to evolutionary ones, as they can provide insights into the physical processes which are occurring. Two examples of this type of analogy are *Shock waves in stellar atmospheres and breaking waves on an ocean beach* by Wallerstein & Elgar (1992) and *Of stars & trumpets* by Sasselov (1993).

## 2. Observational program

As an example of the type of long-term observational program that is required to address some of the issues mentioned above, I will discuss the program that we have developed at Canterbury. I will combine it with other researchers' programs to complement the discussion.

The Mount John University Observatory (MJUO) program has involved a symbiotic relation between the facilities available and the types of astrophysical programs which we wanted to undertake. It follows some of the ideas of Warner (1986) by using well-equipped and well-calibrated instruments on small telescopes, gaining regular access to telescope time, employing resident observers and doing appropriate research. This has lead to a number of long-term photometric and spectroscopic studies of regular and semi-regular variable stars.

The specific types of objects that I will consider are RV Tauri stars, HdC stars, with particular attention to the RCB stars, and Cepheid variables. In most of these studies we combine the photometric and spectroscopic data, the latter in the form of line profile variations, stellar radial velocities and abundance

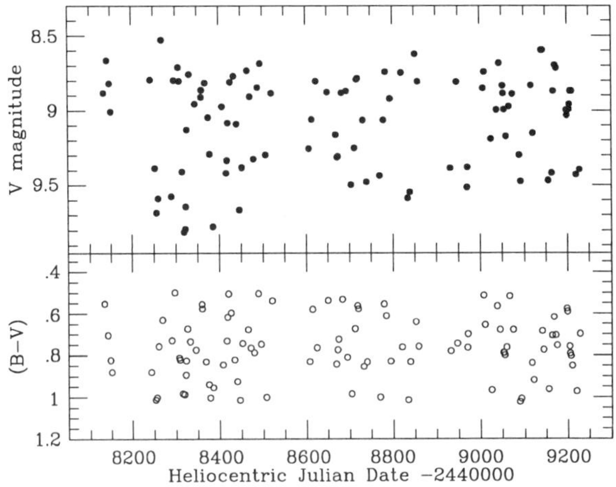


Figure 2. MJUO *V and (B–V) photometry of the RVa star, RU Cen.*

determinations, to understand the evolutionary and dynamical properties of these systems.

## 3. RV Tauri stars

Much of the new work described in this section is the PhD thesis work of Karen Pollard. It exemplifies the discussion in the previous section and presents some significant new work.

### 3.1. Previous work

It seems appropriate to begin my overview using Wahlgren (1992) as the starting point. This paper discussed the class of stars in significant detail, in particular the alternating photometric behaviour (which is the principle defining characteristic of these stars, see Figure 1), their infrared excesses and the general spectroscopic properties of the systems (see also Preston et al. 1963).

There is a general consensus that the RV Tauri stars are low mass objects in transition from AGB to white dwarf phase. The infrared excess is indicative of a relatively recent phase of mass loss which has contributed towards an extensive circumstellar envelope. The more recent spectroscopic work (Lèbre & Gillet 1991a,b; Gillet 1992) has included higher resolution than the earlier work and has enabled more precise radial velocities to be obtained and investigation of atmospheric stratification effects and the propagation of shock waves through the photospheric layers.

Before discussing the new data it is appropriate to define the major divisions in this class of star. Photometrically, they have been divided into two sub-classes as a consequence of their light amplitude. The RVa sub-class are stars which, although showing the alternating behaviour, tend to have a maximum brightness which is constant, whereas the RVb sub-class have a maximum

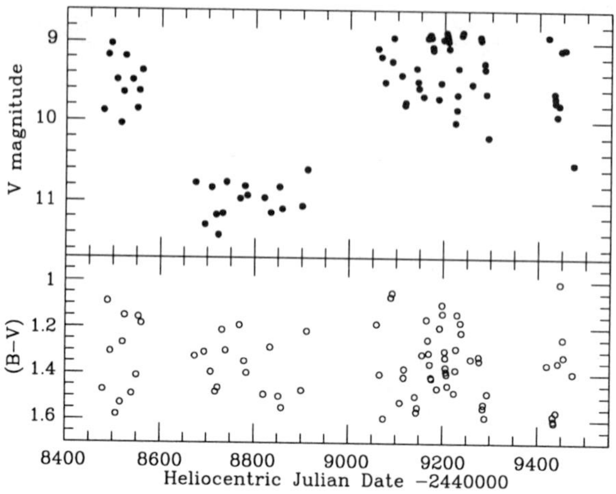


Figure 3. MJUO *V and (B–V) photometry of the RVb star, AI Sco.*

brightness which changes over timescales of 500 to 2000 days. Spectroscopically, Preston et al. (1963) defined a number of classes as well, using 'A', 'B' and 'C' as the designators, respectively corresponding to G-K type variables and Fp type with enhanced and weak molecular carbon features.

One important paper which relates well to the type of study which I outlined in the general introduction is the work by Shenton et al. (1992), which used contemporaneous IUE, optical and infrared data to model a substantial fraction of the electromagnetic spectrum of AC Her.

There have been a number of spectroscopic studies which have determined photospheric compositions for one or more of these stars (see e.g., Luck & Bond 1984, 1989; Giridhar et al. 1994). Most studies have shown that these stars have varying degrees of metal deficiency and with some evidence of carbon enhancement in type 'B' objects.

### 3.2. MJUO program

A photometric *BVRI* program has been undertaken on 11 stars and has extended over about 4 years. Figures 2 and 3 show examples of the photometry for RU Cen (RVa) and AI Sco (RVb). It has been supported by contemporaneous spectroscopy at MJUO (and supplemented by some spectra obtained on the Mount Stromlo Observatory 1.9-m telescope). These spectra enabled the investigation of the phase dependent behaviour of H$\alpha$ and some of the metallic lines and the general systematic trend in velocity as a function of phase to be determined.

To enable a detailed comparison of the class of stars as a whole, all these photometric data have been phased (using the spacing between successive deep minima as the defining period). (For the RVb sub-class the long-term trends in the data were removed prior to phasing.) This has shown that there are significant differences between individual stars and also between different cycles of the same star.

### 3.3. Modelling

There are two main models which have been developed to describe the alternating behaviour of the RV Tauri stars. These are the resonance hypothesis (see Takeuti & Petersen 1983; Tuchman et al. 1993; Fokin 1994) and a period doubling processes (Buchler & Kovács 1987; Aikawa 1993; Buchler et al. 1995), which these latter authors use to describe the RV Tauri stars as being indicative of a transition from regular to irregular (chaotic) pulsations. Certainly the semi-regular nature of the photometric and spectroscopic observations is going to require some individual attention in describing each of these stars. Buchler et al. (1995) also reinforce the idea that continuous (or almost continuous) observations of irregular variable stars could provide testbeds for chaos theory.

Table 1. Orbital parameters for $\upsilon$ Sgr and LSS 4300

| | $\upsilon$ Sgr (1929) | $\upsilon$ Sgr (1990) | $\upsilon$ Sgr (1992) | LSS 4300 |
|---|---|---|---|---|
| $P$ (d) | 137.95 | 137.943±0.004 | 137.89±0.17 | 52.09±0.02 |
| $\gamma$ ($kms^{-1}$) | 13.3 | 7.4 (fixed) | 13.0±1.3 | -5.1±4.2 |
| K ($kms^{-1}$) | 49.1±0.8 | 47.4±0.8 | 46.6±1.7 | 56.9±3.4 |
| e | 0.060±0.016 | 0 | 0.065±0.039 | 0.38±0.06 |
| $\omega$ (°) | 16.8±9.5 | – | 23±24 | 47.1±7.3 |
| $T^a$ | -4.7±0.1 | 7916.2±0.4 | 4889±16 | 5025.6±2.0 |
| asin$i$ (Gm) | 90.3 | – | 88.21±3.3 | 37.7±2.5 |
| F(m) | 1.677 | 1.52±0.08 | 1.44±0.16 | 0.79±0.15 |

[a]zero epoch is JD 2440000

## 4. HdC stars

In this area I will describe two types of semi-regular variables for which our understanding has been significantly enhanced by the type of approach that we have taken. One type is the hydrogen-deficient binaries (HdB) and the other is the RCB stars.

### 4.1. HdB

Some of the new work that I will describe was undertaken by David Frame, a master of science student at Canterbury.

*Previous work* There has been limited photometry and spectroscopy on this group of stars, particularly as three ($\upsilon$ Sgr, LSS4300 and CPD-58°2721) of the four confirmed members are southern objects. Greenstein, between 1940 and 1950, undertook an extensive spectroscopic study of $\upsilon$ Sgr, in which he deduced the hydrogen deficiency of the atmosphere, confirmed the orbital variations of some features and the non-orbital variation of others, particularly H$\alpha$. Malcolm & Bell (1986) and Morrison (1988) obtained limited photometric data on a

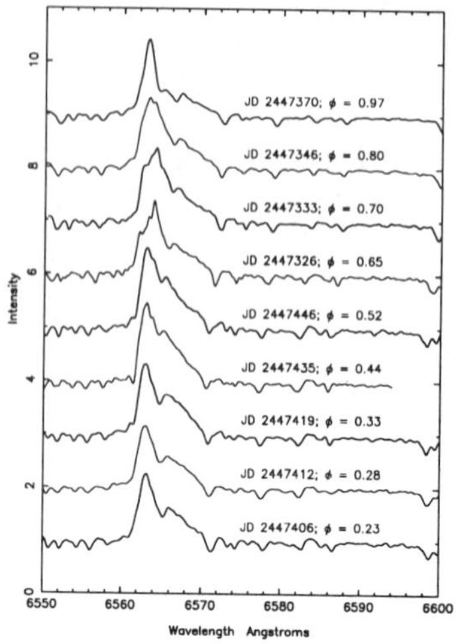


Figure 4. Sample of $\upsilon$ Sgr spectra as a function of orbital phase.

number of these stars, but were only able to deduce that there were some short-term (~10–20 d) variations.

More recently Dudley & Jeffrey (1990, 1993) have combined various data, including IUE spectra, to recalculate the binary star orbit and through a model atmosphere analysis derived temperatures and luminosities for the primary star (the hydrogen-deficient member) of the system.

*MJUO program and modelling* We have undertaken long-term photometry of the three southern HdB stars. In addition, spectroscopy was obtained over an extended period for $\upsilon$ Sgr and LSS4300 and orbits determined (see Table 1). The $\upsilon$ Sgr orbit (labelled 1992) is similar to that obtained by Dudley & Jeffrey (1990), labelled 1990, and Seydel (1929), labelled 1929. The orbit determined for LSS4300 must be regarded as somewhat provisional at this stage, because it is based on relatively few points (7 from our work and 8 from Jeffery et al. 1987).

The spectra were also used to examine the variability, and extent (with velocities ranging from -400 to +550 km s$^{-1}$), of the H$\alpha$ profile, both as a function of orbital phase and over the time span of the study (~4 years). Figure 4 shows a sample of these spectra, stacked as a function of orbital phase. There is no dependence of the emission feature with phase, but the strength of the emission varies greatly over the period of the observational program. This was also noted by Greenstein in his series of papers.

There are still not sufficient data to enable a definitive model to be developed for this very unusual system, but there seems to be more than one mass loss process occurring in this binary system. A model developed by Nariai (1967) made some predictions about an anomalous high velocity H$\alpha$ absorption line and some aspects of the emission, but the non-orbital related high velocity emission

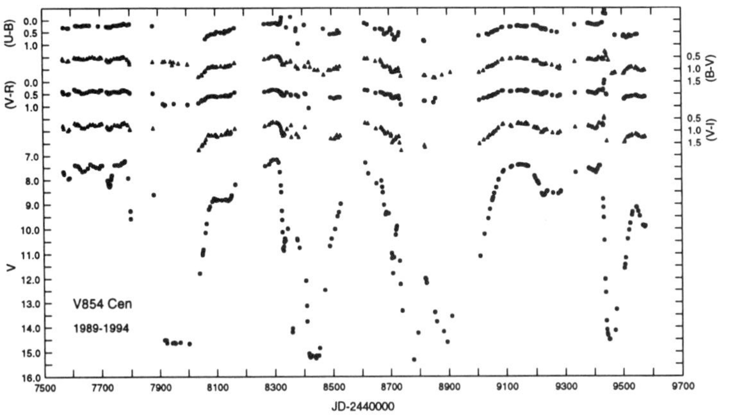


Figure 5. MJUO *V and colour photometry of V854 Cen.*

must originate in a separate stream of material. A paper on these observations is in press (Frame et al. 1995).

## 4.2. RCB stars

Over the last 15 years there has been increased attention paid to this class of peculiar star. Their peculiar abundances (see Lambert & Rao 1994) and anomalous variability (see Cottrell et al. 1990; Lawson et al. 1990), where they undergo large scale declines in brightness, have interested a wide range of photometrists, spectroscopists and modellers. There are also implications for stellar chromosphere studies, dust formation and evolution, and the later stages of stellar evolution.

*MJUO photometry program* At MJUO we have built up a database consisting of ~9 years of *(U)BVRI* photometric data on up to 15 RCB stars. (A similar long-term program, on R CrB itself, has been carried out by Fernie & Seager (1994) using APT techniques.) These data have enabled detailed studies to be made of the pulsational characteristics of the stars (Lawson et al. 1990), with the more recent data containing photometry of many declines. Examples of the more recent data for the 'ultimate galactic roller coaster' star, V854 Cen is shown in Figure 5. Analysis of the extensive data set is in progress (Cottrell et al. 1995).

There have been a number of papers which have shown or endeavoured to make predictions about the time of decline (Dick & Walker 1992) and whether there is any correlation between the decline onset time and the pulsations (Pugach 1977; Lawson et al. 1992). Although there are some indications, there is not a general trend.

As indicated in the introduction, the need to have magnitude and colour information available on these types of objects has been more than justified by the requests that we have had from other researchers. Generally they have obtained infrequent spectroscopic observations during the decline phase, and wish to know what the photometric state of the star was at that time. It cannot be stressed too strongly how vital that photometric information has become.

*Spectroscopic program* As indicated in the previous sections, there have been numerous studies of RCB stars both for abundance work in our Galaxy (see Lambert & Rao 1994) and extragalactic systems (see Pollard et al. 1994), as well as to investigate the spectroscopic variations (at ultraviolet and optical wavelengths) during the pulsations at maximum light (see Clayton et al. 1994) and during various phases of the declines. Many more studies, particularly of the decline phase, are required to enable a consistent model of this phenomenon to be developed.

*Other work and modelling* There have been a number of new observational developments and some progress in the modelling of the decline events.

Recent discoveries of nebulae around RCB, hot RCB stars or related objects (see Pollacco et al. 1991; Pollacco & Hill 1994) and 2 new RCB stars (Benson et al. 1994; Jurscik 1994) have rekindled the search for the remnants of mass loss or mass ejection processes from these stars and increased the possibility that with systematic search strategies there is some hope for an increase in the number of RCB or related stars. This can only help with various statistical studies about decline characteristics and provide further input for evolutionary modellers in terms of the expected lifetime for these stars.

There have been a number of groups working on the dust problem around RCB stars. It is now generally thought that the dust forms close to the photosphere (Clayton et al. 1992) and this group has worked on empirical models to show the general characteristics of this. Pugach and his co-workers (see Pugach & Kovalchuk 1994) have developed a geometrical interpretation to the decline light curve, sometimes having a number of clouds which are simultaneously causing dimming of the RCB star. Some dust formation specialists (e.g., Goeres & Sedlmayr 1992) have also been involved in some detailed modelling projects to form dust in the proximity of RCB stars. However, there are still serious problems in getting the dust to form sufficiently close (within 2 $R_*$) of the stellar surface, a requirement of the more recent observational data.

## 5. Cepheids

The final area that I would like to address in this review of yellow supergiants is in Cepheid star research, far removed from the semi-regular or irregular stars already discussed. In particular, I wish to consider the spectroscopic characteristics of a number of these stars. This work began with the visit of George Wallerstein to Canterbury in 1988 and has continued since then with Michael Albrow and others.

### 5.1. Previous work

Wallerstein, Jacobsen and their various co-workers, as well as the CORAVEL consortium have provided much of the radial velocity work on Cepheids. (Apologies to other researchers who have been involved, but space precludes listing everbody!) Most of this work has centred on getting well-defined velocity curves to combine with various photometric data in order to use the Baade-Wesselink method to deduce radii and then distances, as well as tests for stellar evolutionary theories.

### 5.2. MJUO and other work

We were also involved in the quest for radial velocity curves (Wallerstein et al. 1992), but got 'side-tracked' by our series of observations which showed that there were small, but significant, asymmetries in the metallic spectral lines of all the Cepheids that we were investigating using MJUO facilities.

The analysis of these data involved the use of the line bisector method to determine the wavelength (and hence velocity) of a spectral line at a range of depths in the profile (Albrow & Cottrell 1994). Different techniques have been adopted by others (Sasselov & Lester 1990; Butler 1993) to make the appropriate measurements to quantify the line formation depth dependence of a spectral line.

### 5.3. Modelling

This detailed line profile analysis and line asymmetry has lead to further refinements in hydrodynamical models of Cepheids. In particular, there have been a number of studies which are seeking to provide full dynamical envelope and atmosphere models which will provide an opportunity to really understand the details of line formation processes in dynamic atmospheres.

Albrow & Cottrell (1995a,b) have constructed hydrodynamic envelope models from which 'snap shot' dynamic atmospheres have been produced and synthetic line profiles have been calculated for comparison with the observational data. As a consequence of this, new projection factors have been calculated for use with Cepheid analyses and a new approach to the understanding of the formation of spectral lines in moving atmospheres has been developed using flux contribution functions.

Gillet and various co-workers (see Breitfellner & Gillet 1993) have also been considering ways in which line profiles are formed. These studies have confirmed the need for a better understanding of the line formation processes in dynamic stellar atmospheres.

## 6. Summary

This paper has highlighted the importance of and need for continued long-term, contemporaneous photometric and spectroscopic work on a variety of variable yellow supergiant stars for a variety of astrophysical reasons.

Observers need to incorporate observations from a variety of other wavelengths in order to more completely define the parameters of a given system, e.g., binaries, circumstellar shells, chromospheres and dust formation.

There is also a need for more detailed hydrodynamical envelope and model atmosphere work to complement the available spectroscopic and photometric observations. This could lead to more complete models of a number of systems, in particular the semi-regular (e.g., the RV Tauri and RCB stars) and regular (e.g., Cepheids) variables.

**Acknowledgments.** The invitation to present this paper would not have been possible without the involvement of my collaborators at Canterbury, who have helped me in this work over the last 10 years: Michael Albrow, Mike Clark, David Frame, Alan Gilmore, Pam Kilmartin, Warrick Lawson, Karen Pollard and the technical support available at the Observatory and University. I would also like to thank Karen Pollard and Michael Albrow for comments on the manuscript and Karen for generating Figures 3 and 4.

## References

Aikawa, T. 1993, MNRAS, 262, 893
Albrow, M.D., & Cottrell, P.L. 1994, MNRAS, 267, 548
Albrow, M.D., & Cottrell, P.L. 1995a, these proceedings
Albrow, M.D., & Cottrell, P.L. 1995b, MNRAS, in press
Benson, P.J., et al. 1994, AJ, 108, 247
Breitfellner, M.G., & Gillet, D. 1993, A&A, 277, 524
Buchler, J.R., & Kovács, G. 1987, ApJ, 320, L57
Buchler, J.R., et al. 1995, Phys.Rev.Lett, 74, 842
Bulter, R.P. 1993, ApJ, 415, 323
Clayton, G.C., et al. 1992, ApJ, 397, 652
Clayton, G.C., et al. 1994, ApJ, 432, 785
Cottrell, P.L., Lawson, W.A., & Buchhorn, M. 1990, MNRAS, 244, 149
Cottrell, P.L., et al. 1995, in preparation
Dick, J.S.B., & Walker, H.J. 1992, A&A, 252, 701
Dudley, R.E., & Jeffrey, C.S. 1990, MNRAS, 247, 400
Dudley, R.E., & Jeffrey, C.S. 1993, MNRAS, 262, 945
Fernie, J.D., & Seager, S. 1994, PASP, 106, 1138
Fokin, A.B. 1994, A&A, 292, 133
Frame, D.J., et al. 1995, in press
Gillet, D. 1992, A&A, 259, 215
Giridhar, S., Rao, N.K., & Lambert, D.L. 1994, preprint
Goeres, A., & Sedlmayr, E. 1992, A&A, 265, 216
Greenstein, J.L. 1950, ApJ, 111, 20
Jeffery, C.S., Drilling, J.S., & Hill, P.W. 1987, MNRAS, 225, 1005
Jurscik, J. 1994, Acta Astron., 43, 353
Lambert, D.L. & Rao, N.K. 1994, JA&A, 15, 47
Lawson, W.A. et al. 1990, MNRAS, 247, 91
Lawson, W.A. et al. 1992, MNRAS, 256, 339
Lèbre, A., & Gillet, D. 1991a, A&A, 246, 490
Lèbre, A., & Gillet, D. 1991b, A&A, 251, 549
Luck, R.E., & Bond, H.E. 1984, ApJ, 279, 729
Luck, R.E., & Bond, H.E. 1989, ApJ, 342, 476

Malcolm, G.J., & Bell, S.A. 1986, MNRAS, 222, 543
Morrison, K. 1988, MNRAS, 233, 621
Nariai, K. 1967, PASJ, 19, 564
Percy, J.R. 1993, Ap&SS, 210, 123
Pollacco, D.L. & Hill, P.W. 1994, MNRAS, 267, 692
Pollacco, D.L., et al. 1991, MNRAS, 248, P1
Pollard, K.R., Cottrell, P.L., & Lawson, W.A. 1994, MNRAS, 268, 544
Preston, G.W., et al. 1963, ApJ, 137, 401
Pugach, A.F. 1977, IBVS, 1277
Pugach, A.F., & Kovalchuk, G.U. 1994, Astr Zh, 71, 250
Sasselov, D. 1993, unpublished preprint
Sasselov, D., & Lester, J.B. 1990, ApJ, 362, 333
Seydel, F.L. 1929, PAAS, 6, 278
Shenton, M., et al. 1992, A&A, 262, 138
Takeuti, M., & Petersen, J.O. 1983, A&A, 117, 352
Tuchman, Y., et al. 1993, A&A, 271, 501
Wahlgren, G.M. 1992, ASP Conf. Series, 45, 270
Wallerstein, G., et al. 1992, MNRAS, 259, 474
Wallerstein, G., & Elgar, S. 1992, Science, 256, 1531
Warner, B. 1986, in IAU Symp 118, 3

## Discussion

*Habing*: This is a political statement in line with your opening remarks. There is a permanent danger that small telescopes will be shut down because large telescopes are so costly to build or to optimise. I think that a constant pressure should be exercised against this danger. Conferences like these and commissions of the IAU are the bodies from where we can exercise counterpressure.

*Cottrell*: I agree that we must exert pressure, but we need to ensure that excellent projects are proposed for these small telescopes and that specific objectives are set to answer specific astrophysical problems.

*Percy*: What fraction of the yellow supergiants are RV Tauri stars and what fraction are SRds?

*Cottrell*: I would estimate that a small fraction of yellow supergiants are RV Tauri stars and that a somewhat larger fraction are classified as of SRd type. However, the boundaries between the different types may be quite blurred and hence subdivision in this way may not be a perfectly satisfactory way of classifying semi-regular variables.

*Percy*: Could you comment on the rather frequent occurrence of RVb stars, with their interesting and puzzling obscurations?

*Cottrell*: The long period obscuration episodes in the RVb subclass of RV Tauri stars has been described by a number of models, although our new photometric and spectroscopic data indicate that the orbit is such that some type of active interaction phase occurs when the stars are at their closest approach and the RV Tauri star is closest to the observer (see Pollard et al., these proceedings).

*Mattei*: You may be interested in my poster in which I show 30-year light curve of U Mon using the observations from the AAVSO International Database.

*Cottrell*: The record of the AAVSO offers key data for long-term study of many variables. However, there is also a need for photoelectric colours to enable other information to be inferred about these systems.

*Mattei*: There are two southern long period variables, R Cen and R Nor, that show RV Tauri (RVa) type variability, with deep and shallow alternating minima. Would you please comment if and how they may be related to the RV Tauri stars?

*Cottrell*: I would expect that these variables are related to the RV Tauri stars and that the alternating deep-shallow effect is some form of double mode resonance that may be ascribed to a much larger group of stars. With more extensive photometric and spectroscopic coverage we may be able to combine some of the wide variety of star classes into a more compact set.

*Welch*: Could you explain why the colour changes are different in nature from decline to decline in the RCB stars?

*Cottrell*: The different colour changes for different declines in the RCB stars are probably a combination of a number of effects, including the height of formation in the stellar atmosphere of the obscuring material, its growth rate and the level of excitation of the upper atmospheric (chromospheric) layers. This combination can consequently give very 'blue' declines when much of the photosphere but little of the chromosphere is initially obscured, to 'red' declines where either the chromosphere is obscured at the same rate as the photosphere or the chromosphere is in a quiescent state. The latter two situations would then not exhibit large numbers of emission lines in the spectra of the declines and hence would not have excessively blue colours.

*Waelkens*: Can you comment on the correlation between declines and pulsations in RCB stars? The absence of enhanced infrared emission during declines suggests to me that a decline is more a line-of-sight effect and that outbursts occur much more often on all sides of the star. It seems then hard to understand how the observed start of a decline correlates well with a pulsation phase, the more so that during most pulsation cycles no decline is observed.

*Cottrell*: There are at least two stars (V854 Cen and RY Sgr) in which there is evidence for a correlation between declines and pulsations (Lawson et al. 1992; Pugach 1977) and there is further evidence that the obscuring material must be forming close to the star (Clayton et al. 1992). I agree that the declines are a line-of-sight effect, where the obscuring cloud forms and blocks out successively more and more of the photosphere (and chromosphere). The onset of the decline could be considered to be a pulsation in which the atmospheric conditions (e.g., the temperature and pressure in the outer atmosphere) are such that grain formation can commence. The characteristics of the decline will then depend upon the geometry of the cloud with respect to the star and the observer and the rate of grain formation.

*Astrophysical Applications of Stellar Pulsation*
*ASP Conference Series, Vol. 83, 1995*
*R. S. Stobie and P. A. Whitelock, eds.*

# Classical Cepheids and Galactic Structure

J.D. Fernie

*David Dunlap Observatory, University of Toronto, Toronto ON, M5S 1A7, Canada*

**Abstract.** In this review I consider recent results on the space distribution of classical Cepheids, first dealing with their Z-distribution normal to the galactic plane, and second considering their X,Y components in the plane. In doing so, use is made of a new database of 505 stars known to be or likely to be classical Cepheids which has recently been compiled by Fernie et al. (1995) and which is available electronically on the World Wide Web at the URL http://ddo.astro.utoronto.ca/cepheids.html, and by anonymous ftp at perseus.astro.utoronto.ca (128.100.77.18) in the directory pub/cepheids.

Following the discussion of space distribution, a review of recent work on Cepheid kinematics and the resulting determination of the distance to the galactic centre will be given.

## 1. The Z-distribution

Perhaps the most straightforward and least controversial data relating to Cepheids and galactic structure are their galactic latitudes and longitudes. Fig. 1 shows a plot of latitude b versus longitude $\ell$ for Cepheids contained in the above database and binned into 30° intervals of longitude. We see that not only do the majority lie at negative latitudes, but that there is a sinusoidal variation present. This must mean that the Cepheids lie below the galactic plane (and/or the Sun above it), and that the Cepheid plane is tilted with respect to the galactic plane. I have used galactic latitude here to show that the effect is independent of any distance scale, but obviously it is the actual Z-coordinate that is the physical quantity to be used.

Replacing latitude with Z produces a diagram much like Fig. 1. Using individual values rather than binned values, and fitting a sinusoid by least-squares gives

$$\begin{array}{ccccccc} Z & = & -36 & + & 28 sin(\ell & - & 27^\circ)\ pc \\ & & \pm 8 & & 11 & & 23 \end{array} \quad (1)$$

The 27° offset is evidently not significant, meaning that the tilt of the Cepheid plane is cross-wise to the radial direction from the galactic centre through the Sun, and that it is upwards in the direction of galactic rotation. Reducing the equation to two unknowns leads to the solution

$$\begin{array}{ccccc} Z & = & -35 & + & 24 sin\ell\ pc \\ & & \pm 8 & & 10 \end{array} \quad (2)$$

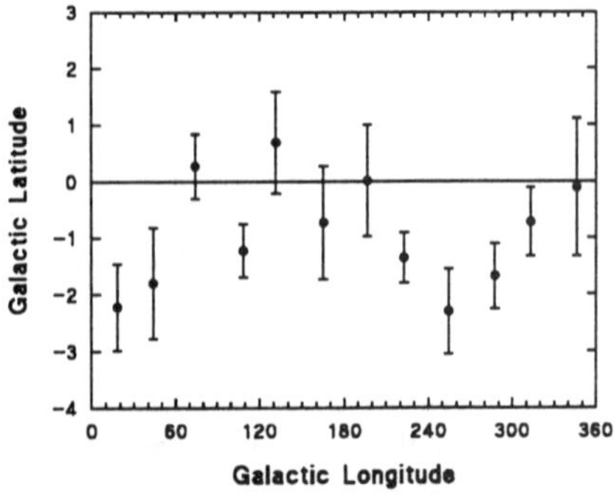


Figure 1. The average latitude of classical Cepheids as a function of longitude. Points are averages in 30° longitude bins.

Interestingly, Conti & Vacca (1990) find a similar result for the distribution of Wolf-Rayet stars:

$$Z = -33 + 54 sin(\ell + 33^\circ)\ pc \tag{3}$$

They do not quote the uncertainties in these numbers, but the relations are probably not statistically different, especially since there are far fewer Wolf-Rayet stars involved than Cepheids and the uncertainties will be correspondingly larger. The fact that quite different populations give comparable results strengthens the case for the physical reality of the phenomenon.

Why should it exist? The explanation that comes most readily to mind is that it is an age effect. The Sun, after all, is 100 times older than the typical Cepheid, and with each having a Z-component of space velocity, there is no reason why they should coincide in Z value. The existence of the tilt, however, is not so easily explained, unless there is some precession or non-axisymmetric effect involved in galactic rotation.

If age is a factor then short-period Cepheids should show a different behaviour from long-period ones, since the former can be up to 10 times the age of the latter. Fig. 2 shows the |Z| distance from the Cepheid plane plotted against log P with a scale of ages shown along the top. I interpret the absence of points in the upper right part of the diagram as being due to the longer-period, younger stars not having had as much time to diffuse away from the plane as have their older brethren. The dashed line shows the expected envelope relation based on a period-age relation and arbitrary average Z-velocity; it seems a reasonable fit. Suppose we now take two groups of stars representing different ages; on the basis of Fig. 2 I adopt Cepheids with $P < 6$ days as being "old", and those with $P > 10$ days as being "young". Repeating the earlier analysis we now get

Old Cepheids:

$$\begin{array}{ccccccc} Z & = & -46 & + & 29 sin(\ell & - & 64^\circ)\ pc \\ & & \pm 13 & & 21 & & 33 \end{array} \tag{4}$$

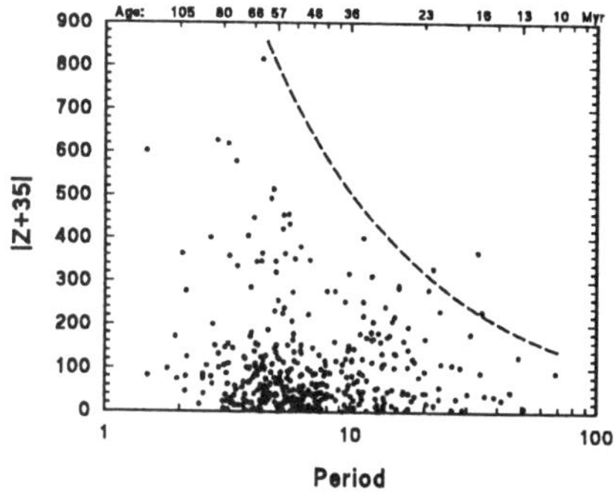


Figure 2. Z-distribution as a function of period (i.e., age). The curve is based on the assumption that the envelope of drift from the plane is proportional to age.

which suggests that the longitude term is not significant and should be dropped. In this case we arrive at a simple

$$Z = -49 \pm 13\ pc \tag{5}$$

Young Cepheids:

$$Z = \underset{\pm 13}{-3} + \underset{16}{51} sin(\ell + \underset{23}{15°})\ pc \tag{6}$$

which suggests that the expression can be reduced to

$$Z = \underset{\pm 16}{50}\ sin\ell\ pc \tag{7}$$

There is thus the rather startling but apparently significant result that the Z-distribution of Cepheids depends quite strongly on period (i.e., age), in the sense that older Cepheids lie well below the plane but in an untilted sheet, while younger ones lie close to the plane but in a tilted sheet.

This implies, of course, that there is a continuum between these extremes, which complicates any discussion of such things as the Cepheid scale height, since the plane from which it should be measured is a function of period both in position and tilt. Even our sample of some 400 Cepheids is not really sufficient to do this in a statistically meaningful way, so I have simply determined the average scale height from a plane 35 pc below the galactic plane. The result is 101 ± 10 pc, which may be compared to 55 pc found by Janes & Phelps (1994) for young open clusters and to 45 pc found by Conti & Vacca (1990) for Wolf-Rayet stars. This is probably reasonable, since the average Cepheid at about 50 million years is older than those other populations.

A useful plot is one of |Z| against distance from the sun in the X,Y plane (= dist x cos b). This has the potential for uncovering distance scale errors where the error is itself dependent on distance, e.g., a reddening error which increases with increasing reddening. This produces a plot in which the upper envelope

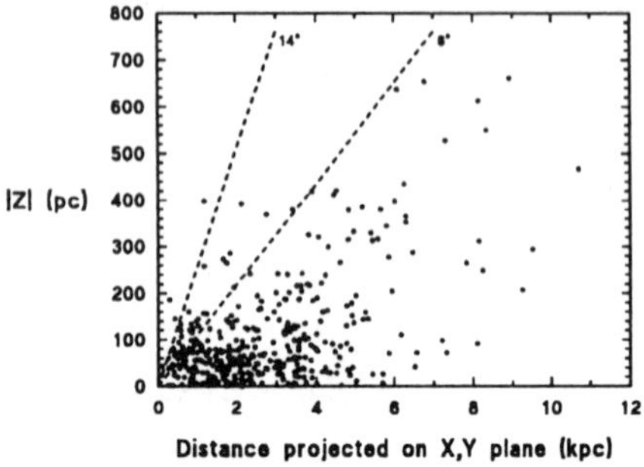


Figure 3. The apparent Z-distribution vs. X-Y distance out to 12 kpc from the sun. The dashed lines indicate latitudes of 14° and 6°. Strong selection effects are apparent.

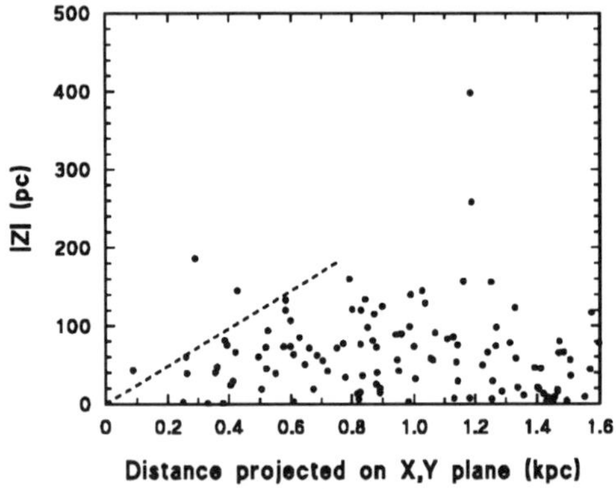


Figure 4. The Z-distribution out to only 1.6 kpc. Note the seeming absence of Cepheids close to the Sun. The dashed line corresponds to a latitude of 14°.

of points appears to vary with distance from the Sun. Fig. 3 shows such a plot out to 12 kpc, and indeed suggests an alarming increase in the upper envelope with distance. Before dealing with this, however, let us examine the left-hand portion of the plot out to only 1.6 kpc, as shown in Fig. 4.

Here there is a satisfactorily flat distribution, aside from two isolated points. At first sight it would seem that our neighbourhood out to about 200 pc is lacking in Cepheids, and there may be two reasons for this. The dashed line is placed merely to divide the relatively packed area to its right from the emptier region to its left; it corresponds to b = 14°, and it may be that the search for Cepheids at higher latitudes should be expanded. However, such searches have been made (e.g., Fernie & Hube 1971) without major success. For the most part, the gap near the Sun likely just reflects the fact that at greater distances we sample larger areas of the plane. For example, an annulus on the plane between 600 and 700 pc from the Sun has an area 13 times the area of a circle of radius 100 pc, and therefore presumably should contain 13 times the number of Cepheids.

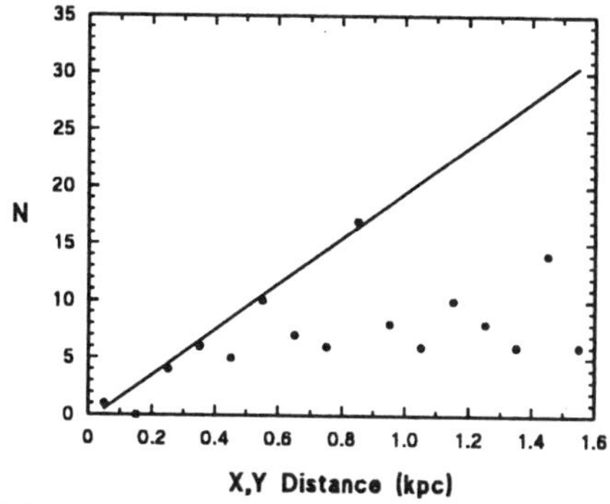


Figure 5. The observed numbers of Cepheids in 100 pc-wide annuli as a function of X,Y distance (points). The line shows the expected count based on the growing size of the annuli with distance. Significant incompleteness sets in and grows beyond about 500 pc.

In this case, the surprise is rather the relative constancy in the density of points beyond 0.4 kpc in Fig. 4. If we count the number of Cepheids in successive annuli of width 100 pc and centred on distances of 0.5, 1.5, 2.5, etc. kpc, we get the points shown in Fig. 5. The line shows the growth in area of these annuli, and thus number, N, of expected Cepheids, normalised to one star within 100 pc of the Sun. Incompleteness sets in with some severity from about 0.5 or 0.6 kpc, and implies we are missing about two-thirds of the Cepheids near 1 kpc, with the situation rapidly worsening at larger distances.

Returning to Fig. 3, we see again the $b = 14°$ line, but now there is still a paucity of stars to its right. I take this to mean that as we go to fainter stars, searches are made ever nearer the galactic equator. Thus we see in Fig. 3 the second line at $b = 6°$ plays the role of the 14° line in Fig. 4 when we consider stars beyond about 3 kpc. Logically, this trend should continue as one goes to greater distances, but eventually the increasing interstellar extinction makes lower-$|Z|$ Cepheids fainter than the magnitude limit of the search and they are lost. Thus there appear to be no Cepheids within 100 pc of the plane beyond 8 kpc — a classic case of Malmquist bias! Feast (1994), among others, has pointed out that most such reddening problems could be obviated by working at K magnitudes, where extinction is less than 10% that for V magnitudes.

I conclude, then, that the overall trend in Fig. 3 tells us little or nothing about the distance scale, but rather is the result of very considerable selection effects.

## 2. The X,Y distribution

Cepheids are quite often invoked as being useful indicators of spiral structure. To this end, Fig. 6 is a plot of about 400 classical Cepheids as projected on the galactic plane, and while the informed eye will readily discern the Puppis-Vela and Carina-Crux arms towards roughly 245° and 290° respectively, there is not much else to enthuse about. Very similar plots are shown by Caldwell

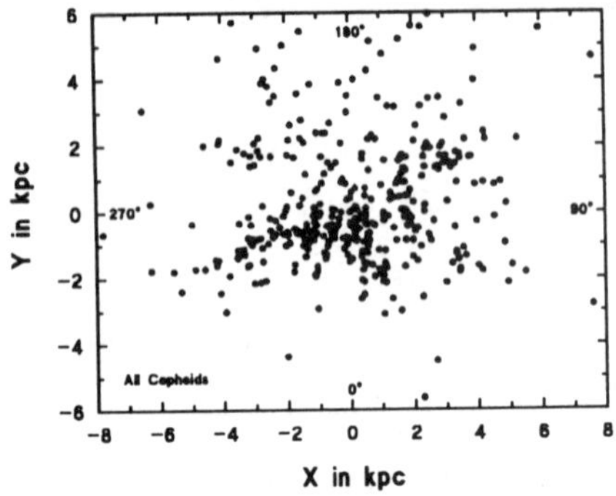


Figure 6. The distribution of about 400 classical Cepheids projected on the galactic plane.

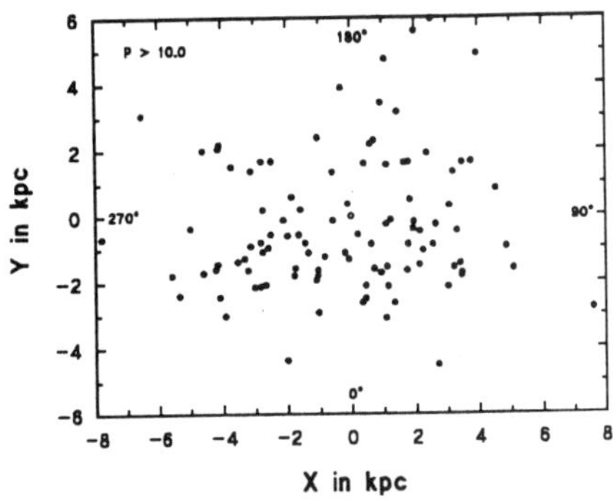


Figure 7. The X,Y distribution for 105 classical Cepheids having periods greater than 10 days. Although these are the younger stars with ages below about 30 Myr, they still do not delineate the spiral arms satisfactorily.

& Coulson (1987), Opolski (1988), Berdnikov & Efremov (1993), and Pont et al. (1994). But of course the majority of these stars will be the shorter-period and thus older ones, while one would expect that their longer-period, younger counterparts would better delineate spiral features. Fig. 7 shows the same plot limited to Cepheids with $P > 10$ days, but although these younger stars do lie closer to the known spiral arms, they are too few in number to delineate the arms convincingly. In any case, with ages greater than 10 Myr they are less concentrated to the arms than are the younger open clusters, HII regions, and the like, for which see Elmegreen (1985). The corresponding diagram of Conti & Vecca (1990) for Wolf-Rayet stars also shows the arms more clearly. I conclude that most Cepheids are too old, and the youngest ones too few, to contribute much to the delineation of overall spiral structure.

Efremov (1983), in discussing Cepheids and spiral structure, has made an interesting comparison between the Galaxy and M31, finding that in M31 the spiral structure at around 10 kpc from the centre is much more obvious than it

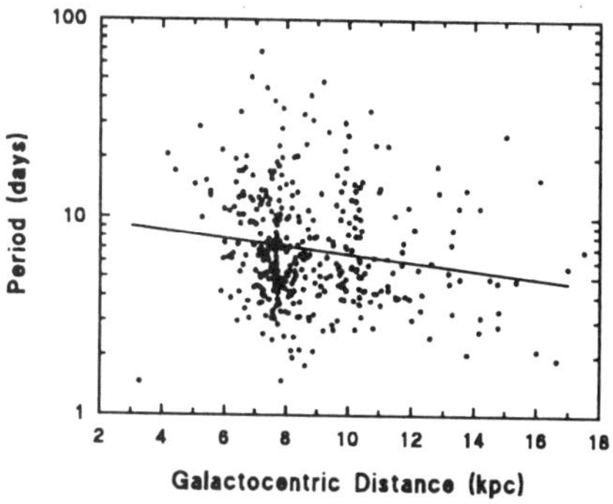


Figure 8. Period as a function of distance from the Galactic centre. The least-squares line has a slope of $-0.0203 \pm 0.0066$ dex $kpc^{-1}$ and thus is likely significant.

is in the solar neighbourhood of the Galaxy. He suggests that the prominence of spiral structure is determined by the component of the difference between the galactic rotation velocity and the spiral pattern speed that is perpendicular to the spiral arm, and accordingly by the distance from the corotation radius. He proposes that in the Galaxy this corotation radius is near the Sun's galactocentric distance, and the spiral pattern thus weak, but that it lies much further out in M31, and the pattern then correspondingly stronger at 10 kpc.

Nevertheless, Fig. 6 does contain some interesting features. The Carina-Crux arm, stretching towards about 290° shows considerable structure within a few kpc of the Sun. In particular, there seem to be lanes of Cepheids lying across the arm, and since none of them point directly at the Sun this is presumably not a 'finger of God' artifact caused by errors in distance. Berdnikov & Efremov (1993) see these as enormous complexes of stars at varying Z-values, related to the corrugations in the galactic plane topography studied by Alfaro et al. (1992), and suggest that Parker instabilities in ordered magnetic fields play an important role in their development. Pavlovskaya & Filippova (1989), as well as Opolski (1988), have similarly considered the possible groupings of Cepheids.

It has long been known that there is a (weak) correlation of average Cepheid period on galactocentric distance, and Fig. 8 illustrates this for about 400 stars in the new database. The least-squares line has a slope of $-0.020 \pm 0.007$ dex/kpc, so appears significant, although to what extent it may be dependent on selection effects is unknown. This is usually taken to be due to a radial metallicity gradient, for which an excellent discussion is given by Caldwell & Coulson (1987).

Finally, if the average interstellar extinction rate (Av/kpc) within 2 kpc of the Sun is examined as a function of longitude one arrives at Fig. 9, showing a sinusoidal variation. Many years ago (Fernie 1962) it was found that other quantities, such as the surface brightness and colour of the Milky Way, show similar variations. Fig. 9 is consistent with these earlier findings, which suggested that the Sun is located near the southern edge of a local interstellar dust cloud of characteristic dimensions about 500 pc. A similar study has been made by Vardanyan et al. (1993).

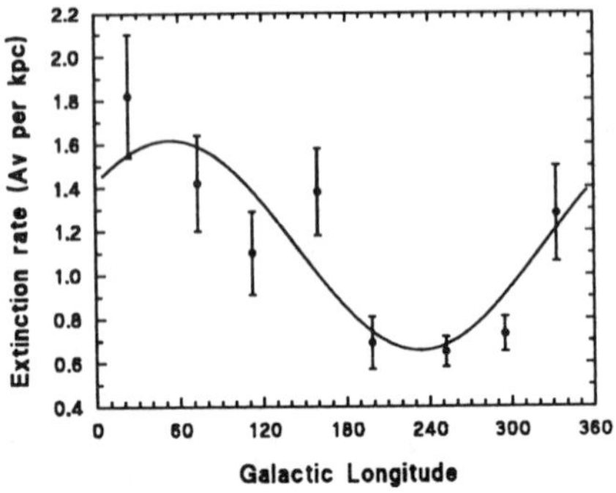


Figure 9. The average extinction rate within 2 kpc of the Sun as a function of longitude. The amplitude of $0.55 \pm 0.12$ mag kpc$^{-1}$ is significant and correlates with other features of the Milky Way.

## 3. Cepheid kinematics

The study of galactic rotation has become much more sophisticated in recent years, compared to earlier studies that were limited by the available data to not much more than determinations of the local Oort constants and solar motion. Current work includes considerations of non-axisymmetric, non-circular motions. See, for example, Kuijken & Tremaine (1994) for a general discussion, and Caldwell et al. (1992) for a discussion specific to Cepheids. Of particular importance have been specific searches by the Caldwell, Metzger, Schechter group for distant Cepheids in longitudes near 300° and 60°, where they carry high weight in determining the distance to the galactic centre.

A study now rather old, but nevertheless worth mentioning, was that by Karimova & Pavlovskaya (1981). They used new proper motions on the FK4 system for 97 Cepheids, and considered a variety of kinematic models. They also determined box orbits, finding that about 70% of Cepheids have $e < 0.1$.

Two recent major studies of Cepheid kinematics are those of Caldwell & Coulson (1987) and Pont et al. (1994). Both are very extensive and are highly recommended reading for anyone working in this area. Terms of higher order than the usual first ones were included either explicitly or implicitly, and a variety of effects explored. Two quantities of particular interest, $R_0$ and $2AR_0$, were found to be $7.8 \pm 0.7$ kpc, $228 \pm 19$ km s$^{-1}$ (Caldwell & Coulson), and $8.1 \pm 0.3$ kpc, $257 \pm 7$ km s$^{-1}$ (Pont et al.). Pont et al. conclude that these differences (although hardly significant) are mostly accounted for by an increased sample size and better quality radial velocities in the later study, rather than by differences in distance scales.

Pont et al., in an interesting account of the interdependence among quantities, point out that the good agreement of $R_0$ with other independent methods (Reid 1989) suggests the Cepheid distance scale is already good to 0.1 mag or better in modulus, and that if new independent methods yield $R_0$ with much improved precision the problem could be inverted and $R_0$ and A used to improve the Cepheid distances.

Finally, Pont et al. have addressed the long-standing K-term problem (a systematic difference of a few km $s^{-1}$ between predicted and observed radial velocities) by comparing the radial velocities of cluster Cepheids with those of other cluster members and finding that on average they agree to about 0.05 km $s^{-1}$, suggesting that the problem is not peculiar to Cepheids.

## References


Alfaro, E.J., Cabrera-Cano, J., & Delgado, ApJ, 399, 576

Berdnikov, L.N., & Efremov, Yu.N., 1993, AstLett, 19, 389

Caldwell, J.A.R., Avruch, I.M., Metzger, M.R., Schechter, P.L., & Keane, M.J. 1992, in ASP Conf. Series 30, Variable Stars and Galaxies , ed. B. Warner, p111

Caldwell, J.A.R., & Coulson, I.M., 1987, AJ, 93, 1090

Conti, P.S., & Vacca, W.D., 1990, AJ, 100, 431

Efremov, Yu.N. 1983, Sov.Ast.Lett. 9, 51

Elmegreen, D.M. 1985, in IAU Symp. 106, The Milky Way Galaxy , ed. by H. van Woerden, R.J. Allen, and W.B. Burton, p255, Reidel, Dordrecht.

Feast, M.W., 1994, Ap&SS, 217, 169

Fernie, J.D. 1962, AJ, 67, 224

Fernie, J.D., & Hube, J.O. 1971, ApJ, 168, 437

Fernie, J.D., Beattie, B., Evans, N.R., & Seager, S. 1995, IBVS No. 4148

Janes, K.A., & Phelps, R.L. 1994, AJ, 108, 1773

Karimova, D.K., & Pavlovskaya, E.D., 1981, Sov.Ast., 25, 561

Kuijken, K., & Tremaine, S. 1994, ApJ, 421, 178

Opolski, A. 1988, Acta Astron., 38, 375

Pavlovskaya, E.D., & Filippova, A.A., Sov.Ast., 33, 602

Pont, F., Mayor, M., & Burki, G. 1994, A&A, 285, 415

Reid, M.J., 1989, in IAU Symp. 136, The Center of the Galaxy , ed. M. Morris, p37. Kluwer, Dordrecht.

Vardanyan,R.A., Ambaryan, V.V., & Pogosyan, A.V. 1993, Astrophysics. 36, 122


## Discussion

*Welch*: The distribution of Cepheids on the sky is strongly affected by selection biases introduced by absorption and survey differences as a function of galactic latitude and longitude. Could you comment on how this affect your conclusions about the distribution of Cepheids in Z and R?

*Fernie*: I fully agree there are major selection effects at work here, and that they must affect details of the results. However, I think the broad conclusions are probably OK.

*Feast*: I wonder whether your results on the latitude distribution of younger Cepheids refers to a larger mean distance than that for older Cepheids and is, therefore, perhaps more representative of the overall distribution.

*Fernie*: A good point. I have not looked at that, but I would be surprised if the overall results changed dramatically.

*Habing*: In one or two of your XY diagrams I seemed to notice that there are significantly more Cepheids in the hemisphere containing the Galactic Centre than in the other. Could this be an effect of decreasing density of the galactic disk - thus a large scale effect?

*Fernie*: Probably yes. Although, one must keep in mind that other factors can intrude, witness the fact that WR stars are very deficient in the Galactic anti-centre direction.

*Burki*: Which kind of criteria did you use to select classical Cepheids from Pop. II Cepheids.

*Fernie*: Basically, we adopted the GCVS listings, supplemented by Hugh Harris' lists of Pop. II Cepheids.

*Astrophysical Applications of Stellar Pulsation*
*ASP Conference Series, Vol. 83, 1995*
*R. S. Stobie and P. A. Whitelock, eds.*

# Pulsating Cool Stars and Galactic Structure

Patricia Whitelock
*SAAO, P O Box 9, Observatory, 7935, Western Cape, South Africa.*
*(e-mail: paw@saao.ac.za)*

**Abstract.** The review concerns the use of Miras and OH/IR stars as probes of Galactic structure and concentrates on groups of stars with both kinematic and luminosity data. The reliability of distance determinations for these stars is covered in some depth. Kinematic and scale height data for local Miras suggest that those with pulsation periods of around 200 day are members of the thick disk or inner halo. Longer-period stars are kinematically cooler and have smaller scale heights. Stars with moderate dust shells, which may be more evolved, seem to belong to the same population as thin-shelled sources with the same pulsation period. Our current knowledge of the structure of the Galactic Bulge is reviewed, as is our understanding of its rotation as derived from observations of SiO Masers from Mira variables. Infrared and radio surveys, in progress or planned, offer us the potential to extend the study of low-mass metal-rich stars well beyond the solar neighbourhood.

## 1. Introduction

This paper concentrates on the use of Miras and their long-period analogues, the OH/IR stars, (henceforth collectively referred to as Miras) as probes of Galactic structure. Other aspects of these stars, their evolutionary status, their pulsation mode(s), etc., are dealt with in other papers in this volume, in particular those by Chapman, Wood and Tuchman.

I want to start by looking at why we might be interested in using Miras as probes of Galactic structure, and at the advantages they have over more traditional probes. First, we know from observations of Miras in globular clusters and from the narrowness of the period-luminosity (PL) relation, that stars of a particular age or initial mass will become Miras with a particular period and that there can be rather little evolution of that period with time. Therefore, by choosing Miras within a particular period range, we can isolate a particular population. Secondly, the Mira evolutionary phase is the most luminous one that any given star ever reaches. For low-mass stars a Mira will be several thousand times brighter than its main sequence progenitor. Thus, in principle, the star can be observed at far greater distances than it can at other stages of its evolution. Thirdly, the Mira phase is short lived so every Mira represents at least $10^4 M_\odot$ of stars, with the same age, metallicity, similar mass etc. Fourthly, their circumstellar environments produce Maser emission which can be detected over very large distances unhindered by interstellar extinction. These Masers have enabled kinematic studies to be made of regions largely inaccessible by other techniques, in particular near the Galactic Centre.

There are of course a few difficulties associated with the use of these AGB stars as Galactic structure probes. Some of these are fundamental and others merely a function of our current ignorance. In the latter category, the most serious is our rather poor understanding of the pulsation and evolution of these stars; there are no satisfactory models for their atmospheres nor unanimity on their pulsation mode; neither do we understand how to integrate mass-loss, the dominant evolutionary process, into the evolutionary theory. This lack of understanding has led to obvious questions about the applicability of the PL relation, particularly for stars with periods in excess of about 420 day (see §2). The short lifetimes of the stars are a disadvantage as well as an advantage: we are required to examine large volumes in order to find statistically significant samples. Perhaps the biggest disadvantage is the time-consuming nature of the observations: typically a dozen measurements spaced over several years are required to establish the approximate period of one of these stars. The bolometric amplitudes are large, over a magnitude for long-period stars, so a single measurement is not necessarily a good approximation to the mean magnitude. Other useful measurements, such as phase-lag distances are extremely time consuming.

## 2. Measuring distances

The following five methods are the most commonly used to determine distances of Miras: (1) Assume they all have the same bolometric or infrared luminosity; (2) A colour-dependent bolometric correction to an adopted luminosity; (3) Kinematic Distance; (4) PL Relation; (5) Phase-lag method for OH/IR stars. Neither the assumption that all AGB variables have the same luminosity nor the use of a colour dependent bolometric corrections works very well unless the sample is carefully selected to have a very narrow range of properties. Kinematic distances are derived assuming the stars follow circular orbits within a model of Galactic rotation. It is clear from the pioneering kinematic studies by Baud et al. (1981) that only the OH/IR stars with large OH-expansion velocities follow circular orbits. These stars with large expansion velocities are the most luminous and presumably have the most massive progenitors. Habing (e.g., 1988) has made this particular point at various times but even in the recent literature you will find attempts to disprove the validity of the PL relation via kinematic distances using OH/IR stars with only moderate expansion velocities.

The period-luminosity relation is an important method and, judging from the recent literature, a source of considerable confusion. The PL relation for Miras with periods less than 420 day (which constitute about 70% of the Miras in the solar neighbourhood) was discussed by Whitelock et al. (1994). They showed that a single PL relation is obeyed by Miras in the LMC, the Galactic globular clusters and the solar neighbourhood. Feast (1988) and Peter Wood (1995 private communication) have shown that this same relation is also obeyed by Miras in the SMC. It is clear that there is no justification for making theoretical metallicity corrections to the *observed* PL relation. Relations for both bolometric and K magnitude have been derived. Of the two, the bolometric one is to be prefered because the K relation breaks down for sources with thick dust shells when circumstellar extinction becomes important at 2.2$\mu$m. Nevertheless the K relation does have the advantage that the same relation applies to both oxygen- and carbon-rich Miras.

The LMC contains Miras with periods in excess of 420 day which clearly lie above an extrapolation of the PL relation for shorter period stars (e.g., Feast et al. 1989). At very much longer periods (P $\gtrsim$ 1000 day) OH/IR stars are found in the LMC which again lie on the extrapolation of the PL relation derived from short period Miras (Wood et al. 1992). One might anticipate that stars exist with 420 $<$ P $<$ 1000 day which also obey the PL relation but they will have shells sufficiently thick to make them optically invisible and of too low a luminosity to be detected by IRAS. We expect to find out about them from surveys using the ISO satellite.

Our knowledge of Galactic stars with periods over 420 day is still very limited. Lindqvist et al. (1992) discovered 134 OH/IR stars within a degree of the Galactic Centre, which we might reasonably assume to be at the distance of the Galactic Centre. Van Langevelde (1992) determined periods (mostly 450 $<$ P $<$ 1000 day) and luminosities for several of these OH/IR stars and finds that on average they fall distinctly below an extrapolation of the PL relation *if the distance to the Galactic Centre is 8 kpc.* Recently Jones et al. (1994) redetermined periods and luminosities for some of these sources, finding somewhat different results and on average a higher luminosity at a given period. My impression is that the measurements are too uncertain to draw definite conclusions on the PL relation, but further observations of these stars are very important and the whole problem is well worth pursuing. It should be made clear that these are very difficult measurements to make; the crowding is horrendous and the corrections for interstellar extinction are large and uncertain.

Ultimately we might hope to get the best distances to OH/IR stars from the phase-lag technique (Chapman 1995) but the published measurements to date are not sufficiently accurate to critically test the PL relation (Whitelock 1990). This situation should change in the near future as results from the Australia Telescope and Hartebeesthoek become available.

To summarise the status of Mira luminosities: the PL relation is well established for Miras with periods less than about 420 day. The absolute luminosities of longer-period stars remains very uncertain. There is also a PL relation for C Miras in the LMC but we have no real idea how well it applies to Galactic stars. Thus, from the point of view of using Miras as Galactic structure probes among populations of low mass stars, the situation is quite good. Most Miras in the solar neighbourhood have periods less than 420 day. We only know about the very long-period OH/IR sources as a result of Galactic Plane OH surveys to very large distances.

## 3. The solar neighbourhood

Before discussing recent work on Miras in the solar neighbourhood I want to remind you of the main result from Feast's (1963) kinematic analysis. This work demonstrated rather clearly that the asymmetric drift and velocity dispersion of the Miras are functions of their pulsation period and that different period Miras belong to different populations. It is this conclusion from Feast's work which has, more than anything else, been responsible for the approach of those of us working from South Africa, and for the importance we have placed on measuring the periods of the Miras we have studied.

Note that Galactic Structure studies are limited by the completeness of surveys for Mira variables. In recent years the compilation of the GCVS has been supplemented by new variables discovered largely via the two-micron sky survey, the AFGL survey, the IRAS survey and various OH surveys of the Galactic Plane. None of these covered the entire sky, although IRAS was not far off. Follow-up observations are necessary to determine the period of a Mira once its nature has been recognised. The follow-up work, even on the surveys finished some years ago, is incomplete, so we do not yet have a complete sample of Miras with known periods even within one kpc of the Sun. Although it is probably very close to being complete for Miras with periods of over 300 day within one kpc. If the DENIS and 2MASS surveys are followed by systematic efforts to determine periods, etc. the situation will improve dramatically.

I want to concentrate on a few recent studies, in particular those by Jura & Kleinmann (1992, henceforth JK92), Jura et al. (1993, henceforth JYK93), Jura (1994) and Whitelock et al. (1994). The approach taken by Jura and collaborators in looking at the space distribution of the Miras is a particularly useful one; they divided the Miras into groups according to their periods and used infrared data, rather than the less reliable optical data, to derive distances. They use a modified version of the K PL relation to derive distances. As shown by Whitelock et al. (1994) this modified PL relation is incorrect and in the following discussion all distances are corrected to the PL relation used by Whitelock et al. In addition to discussing separately short-, intermediate- and long-period Miras Jura & Kleinmann (1989) also isolate a separate group of *dust enshrouded AGB stars* and examine their properties, in this case without considering their periods. It is possible to trace periods and other data on a large fraction of these stars which shows them to comprise: 25 oxygen- and 16 carbon-rich Miras with $270 < P < 1600$ day, 7 oxygen- and 11 carbon-rich Miras with unknown periods and 4 non-Miras. Although this constitutes an interesting group of stars worth further investigation, it is too inhomogeneous a group to be useful for Galactic structure studies.

The stars examined by Whitelock et al. (1994) were colour-selected IRAS sources from the South Galactic Cap. There were 61 stars, 35 of which were already known to be Miras by the GCVS, although not all had known periods. Periods were determined for all of them. The kinematics of this sample were also examined, using velocities from the literature together with new ones determined by John Menzies. Figure 1 shows the radial velocity and the height above the plane as functions of period. The distances were calculated from the PL relation. The open circles are C stars, all of the others are oxygen-rich Miras. These diagrams qualitatively, at least, reinforce what we saw from Feast's work. The 200 day Miras have a large velocity dispersion and occur at large distances from the plane. Longer period stars belong to kinematically cooler populations and presumably originate from more massive progenitors.

Table 1 lists some standard numbers for different populations for comparison with the Miras: the scale height ($Z_0$) above the Galactic plane, the velocity dispersion ($\sigma_Z$) in the same direction and the asymmetric drift ($V_A$). These

Table 1. Local Populations (Freeman 1987; Carney 1993)

| | $Z_0$ (pc) | $\sigma_z$ (kms$^{-1}$) | $V_A$ |
|---|---|---|---|
| Halo | - | 75 | ~ 220 |
| "Thick Disk" (IP II) | 1000 | 40 | 30 to 40 |
| Old Disk | 350 | 20 | 15 |

values are typical of those quoted by many people over the last few years. Table 2 lists the same quantities for Miras in the various period groups. The kinematic information is from Feast et al. (1972) and the scale heights are from JK92/JYK93. Although the samples are not drawn from the same volume, there is overlap between the stars from Feast and those from Jura. The scale heights were obtained via the median height above the plane. As incompleteness will be most serious among distant stars the scale heights are treated as lower limits. The shortest period group ($P < 145$ day) do not fit the pattern of the other stars; they are something of an enigma and there is no room to discuss them here. The 150-250 day Miras have kinematics which are intermediate between the halo and the so called 'thick disk' (or intermediate population II). This is an interesting observation in itself as current ideas for the formation of the thick disk tend to favour a discontinuity between the disk and halo populations. JK92 derived a single scale height for all stars with periods less than 300 day; however, their sample is dominated by the 200 to 300 day group. For the 145 to 200 day group alone the scale height calculated from the data in JK92 is about 700 pc. On the basis of the kinematics of this period group, particularly the asymmetric drift, we expect these stars to be transition objects between the thick disk and the halo. The scale height is more suggestive of the thick disk. The longer-period stars have much smaller scale heights, as the kinematics suggest they should, and look more like old-disk stars.

Jura (1994) published a list of 14 stars (his table 4) of which he suggested most (about 10) were members of the thick disk ("E/T disk"), the kind of object he suggests the short-period (P<300) Miras would evolve into towards the end of their Mira phase when they have relatively thick dust shells. These stars were selected from $|b| \geq 30^o$, with $F_\nu(12\mu m) \geq F_\nu(2.2\mu m)$ and $|Z| \geq 600$ pc. Jura calculated the distances assuming a bolometric luminosity of $10^4 L_\odot$. Of the 14 stars listed, one is a semi-regular variable (03287–1535: see Whitelock et al. 1995), one is a carbon star (02152+2822) while the others are oxygen-rich Miras with periods (available for 11 out of 12 of the stars) in the range 425 to 559 day. Their kinematics (see Table 3, from velocities in the literature) suggest an old-disk population, with a scale height between 300 and 400 pc. The velocities and distances from the plane of these stars are compared in Fig. 1 with those of the IRAS selected group from Whitelock et al. (1994), also listed in Table 3 (there are 7 stars in common). There are no grounds for associating these stars with the kinematically hotter, shorter-period group. Jura (1994) also suggested that the circumstellar gas outflow velocities for these stars were lower than for typical dusty stars. A comparison with Fig. 1 from Sivagnanam et al. (1989) indicates that the expansion velocities are within the normal range found for stars of the same period. More broadly speaking we see from the scale heights,

Table 2. Miras (Feast et al. 1972; JK92; JYK93)

| P (d) | Feast no. | $\sigma_z$ (kms$^{-1}$) | $V_A$ | JK92/JYK93 no. | $Z_0$ (pc) |
|---|---|---|---|---|---|
| < 145 | 22 | $38 \pm 28$ | $20 \pm 13$ | 14 | |
| 145-200 | 46 | $60 \pm 67$ | $98 \pm 22$ | 33 | > 700 |
| 200-250 | 71 | $62 \pm 15$ | $48 \pm 9$ | 72 | |
| 250-300 | 77 | $27 \pm 23$ | $20 \pm 10$ | 71 | |
| < 300 | | | | 190 | > 560 |
| 300-350 | 83 | $36 \pm 10$ | $19 \pm 6$ | | |
| 350-400 | 54 | $26 \pm 10$ | $10 \pm 8$ | | |
| 300-400 | | | | 106 | $\gtrsim 270$ |
| > 400 | 35 | - | $2 \pm 8$ | 68 | $\gtrsim 270$ |

Table 3. IRAS selected stars from $|b| > 30^{\circ}$

| P (d) | no. | $\sigma_{RV}$ (kms$^{-1}$) | $V_A$ | $Z_0$ (pc) |
|---|---|---|---|---|
| Candidate very-dusty thick-disk stars from Jura (1994) | | | | |
| 425-559 | 13 | $20 \pm 4$ | $9 \pm 9$ | ($Z > 600$) |
| Dusty Miras (Whitelock et al. 1994) | | | | |
| 250-350 | 26 | $38 \pm 6$ | $99 \pm 17$ | > 710 |
| 350-450 | 16 | - | $19 \pm 14$ | $\gtrsim 460$ |
| > 450 | 13 | - | $\sim 0$ | $\gtrsim 280$ |

velocity dispersions and asymmetric drifts in Table 2 that the 200 day Miras are plausibly an extreme part of the thick disk and that there is a gradual change of properties as we go to longer periods, until we get to the Miras with periods of around 400 day with properties which suggest an association with the old disk. Presumably the long-period OH/IR stars belong to a still younger population. The one outstanding inconsistency between this picture and the data presented here is the asymmetric-drift of the 250 to 350 day Miras from Table 3; this is similar to the large drift of the much shorter period Miras from Table 2. The velocity dispersion of this group is what we would expect for their periods, but the uncertainties are too large to enable us to draw a firm conclusion from this. There seem to be two possible explanations for this asymmetric drift. Majewski (1993) has summarised evidence suggesting that the asymmetric drift of the thick disk increases with distance from the plane. More recent work (e.g., Ojha et al. 1994) suggests that the gradient in asymmetric drift is small and possibly zero. If there is a gradient in the asymmetric drift then it might explain the difference between the drift of the Miras from the solar neighbourhood (Table 2) and those from the South Galactic Cap (Table 3) as the latter sample contains a large fraction at considerable distances from the plane. This point is obviously critical to the interpretation of these kinematics. An alternative explanation is that these dusty 250-350 day Miras have evolved from the shorter period stars without dust shells. If this were the case we would have overestimated their distances with the PL relation and the scale height would be smaller. There are two crucial measurements which would clarify this point. One is good distances to a number of the dusty Miras, and the second is a much more complete survey

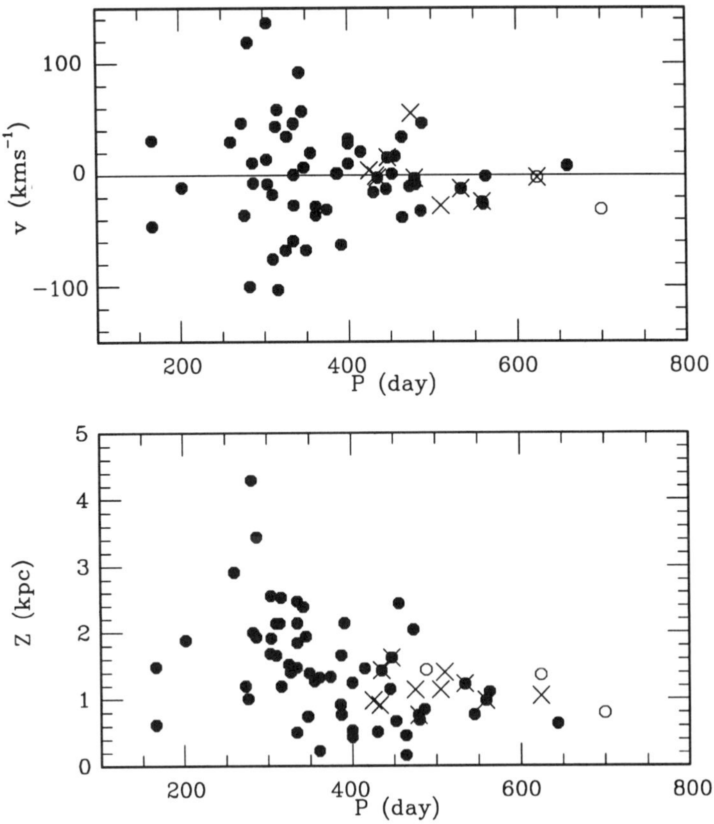


Figure 1. A comparison of the height above the Galactic Plane (Z) and the velocity (v) as a function of pulsation period (P) for the stars from Whitelock et al. (1994) (closed circles: M stars; open circles: C-stars) and the stars from Jura (1994) (crosses).

of the Galactic Caps for Miras of both the dusty and ordinary variety, followed by detailed analysis of their kinematics, periods and luminosities. It is worth noting that work on Miras in the Sgr I window of the Bulge by Glass et al. (1995) does not support the notion that those with thicker dust shells are less luminous than thin shelled sources of the same period. Possibly we must await follow-up studies after the DENIS and 2MASS surveys for a solution.

Note that the kinematic results on the Miras do not suggest a discrete thick and old disk, but that a continuum exists from the old disk to the thick disk and maybe beyond.

## 4. The Galactic Centre

The Galactic Centre is a very interesting region for many reasons. In the current context it is notable that it contains many OH/IR stars to which we have an independent, if uncertain, distance estimate. Lindqvist et al. (1992) discovered 134 OH/IR sources within 100pc of the centre and van Langevelde (1992) and Jones et al. (1994) have measured luminosities for a few of them (see §2). They seem to be rather different from the Miras in the Bulge in that they include stars with periods up to 1000 day. This is presumably indicative of rather more massive stars than are found in the Bulge. This may be because the population is younger or it could be the result of numerous binary mergers in the dense environment around the Centre. Lindqvist et al. showed that the OH sources are in rapid rotation around the centre. For this reason it has been suggested that these stars may be more closely related to the disk than to the Bulge. However the existence of at least some stars on non-circular orbits is more consistent with a halo than a disk population. I understand that more work is being done on these stars by Habing and associates and that we can look forward to seeing proper motions in the future.

## 5. The Galactic Bulge

What have we learned about the structure of the Bulge from Mira-like variables? The distribution of periods of Bulge Miras is of some interest as it we have seen above that the kinematics of local Miras are a function of their period. There are in fact still rather few regions in the Bulge for which there is anything approaching a complete sample of all periods, although a large number of Miras are catalogued. Most of what we know about the period distribution comes from the Baade windows surveyed by Lloyd Evans (1976) with some additional long-period stars discovered as IRAS sources (Glass 1986). IRAS did not detect the short-period Miras at the distance of the Bulge. The period distributions were illustrated in Fig. 3 of Whitelock (1993). Again the long period tail may be the result of binary mergers. There is a somewhat larger fraction of long-period stars in the inner Bulge than in the outer Bulge. There seem to be significant numbers of the stars which locally we would associate with the thick disk and slightly smaller numbers of those we would associate with the old disk, and everything in between. There is however no evidence, as yet, of a kinematic dependence on period among the Bulge Miras.

Miras have played a key role in the recent discovery of the bar structure associated with the Bulge. Nakada et al. (1991) looked at the distribution of colour-selected IRAS sources (most of which we have good reason to believe will be Miras) and found a longitudinal asymmetry which they interpreted as a tilt in the Bulge. Whitelock & Catchpole (1992) looked at IRAS Miras for which we had periods and luminosities and found a tilt in the Bulge. This has been confirmed in a recent analysis of the data from the DIRBE experiment on the COBE satellite (Dwek et al. 1995).

Whitelock & Catchpole (1992) examined the distances to IRAS Miras in two strips across the Bulge. The distances were determined from the PL relation. They found that the stars at positive longitudes are closer than those at negative longitudes, and were able to model this with an ellipsoidal Bulge tilted to the line of site by about 45°. This approach, using distances to individual stars, is time consuming but enormously powerful in determining detailed structure compared to examining the overall light distribution at any given wavelength. The interpretation of the latter requires detailed models of the numbers and types of stars in the Bulge.

Menzies (1990) obtained optical spectra of the brightest of these Miras and demonstrated that at these latitudes the Bulge is rotating at 70 $kms^{-1}kpc^{-1}$. A few of the Bulge IRAS sources were also detected in OH as part of the Parkes survey of colour-selected IRAS sources by te Lintel Hekkert (1990) although most of the Bulge sources do not have sufficiently thick dust shells to make them detectable OH emitters at the distance of the Bulge. A far larger number of Bulge sources have been detected as SiO Masers from the Nobeyama radio observatory by Izumiura et al. (1994). Initially this group examined the IRAS sources in the $7^o < |b| < 8^o$ strip that Whitelock et al. (1991) had demonstrated were Miras. This was so successful that they decided to extend the survey over a much larger area of the Bulge, viz. $3^o < |b| < 15^o$ and $l = \pm 15$ (Izumiura et al. 1995). They measure a Bulge rotation of 75 $kms^{-1}kpc^{-1}$, with no evidence of departure from cylindrical rotation, although the velocity dispersion is high (76 $kms^{-1}$) and the number of sources in a given latitude strip not large. (The OH sources near the Galactic centre are rotating much more rapidly at 1200 $kms^{-1}kpc^{-1}$). There are some tantalising asymmetries in the rotation curve and in particular an offset from zero of $18 \pm 8$ $kms^{-1}$. This may indicate that the local standard of rest is moving towards the Galactic Centre or that there are streaming motions of the stars in the bar.

## 6. Future prospects

Finally, a few remarks about the future: We have the potential to investigate the detailed disposition and motions of old stars over a large fraction of the Galaxy; we are no longer limited to a tiny volume around the sun. Using radio observations of molecular Masers we have the ability to measure the velocities and proper motions of variables at least as far as the Galactic Centre, independent of intervening extinction. Using the results from HIPPARCOS for optical Miras and phase-lag distances to OH/IR stars we can test, calibrate and refine the distance scale. ISO observations of stars in the Magellanic Clouds will aid our understanding of the long-period variables. The DENIS and 2MASS surveys will reveal red variables throughout a large fraction of the galaxy. To make proper use of these data we will have to determine pulsation periods, and this will be time consuming. Accurate periods are a by-product of the searches for gravitational lenses in the Bulge and in the LMC. Photographic work in progress with the Anglo-Australian Schmidt will also provide periods. In the very crowded and obscured regions multiple observations with infra-red arrays will be required, and such measurements are becoming easier to make. The observations which form the basis of our understanding of the Galactic disposition of old populations are largely of stars from the extended solar neighbourhood. It will be fascinating to see how they stand up to a test on the grand scale, and this we should be able to do in the not too distant future.

**Acknowledgments.** I would like to thank Michael Feast and John Menzies for helpful discussions.

## References

Baud, B. et al. 1981, A&A, 95, 156

Carney, B. W. 1993, in: Majewski S. R. (ed), Galaxy Evolution: The Milky Way Perspective, ASP Conf. Ser. 49, p. 83

Chapman, J. 1995, these proceedings

Dwek, E. et al. 1995, preprint

Feast, M. W. 1988, in: van den Bergh, S. & Pritchet, C. J. (eds) The Extragalactic Distance Scale, ASP Conf. Ser. 4, p. 9

Feast, M. W., Woolley, R., & Yilmaz, N. 1972, MNRAS, 158, 23

Feast, M. W. et al. 1989, MNRAS, 241, 375

Freeman, K. C., 1987, ARA&A, 25, 603

Glass, I. S. 1986, MNRAS, 221, 879

Glass, I. S. et al. 1995, MNRAS, in press

Habing, H. J. 1988, A&A, 200, 40

Izumiura, S. et al. 1994, ApJ, 437, 419

Izumiura, S. et al. 1995, submitted to ApJS

Jura, M. 1994, ApJ, 422, 102

Jura, M., & Kleinmann, S. G. 1989, ApJ, 341, 359

Jura, M., & Kleinmann, S. G. 1992, ApJS, 79, 105

Jura, M., Yamamoto, A., & Kleinmann, S. G. 1993, ApJ, 413, 298

Jones, T. J. et al. 1994, AJ, 107, 1111

Lindqvist, M., et al. 1992, A&AS, 92, 43

Lloyd Evans, T. 1976, MNRAS, 174, 169

Majewski, S. R. 1993, ARA&A, 31, 575

Menzies, J. W. 1990, in: Jarvis B. J. & Terndrup D. M. (eds), Bulges of Galaxies, ESO/CTIO Workshop, p. 115

Nakada, Y. et al. 1991, Nature, 353, 140

Ojha, D. K. et al. 1994, A&A, 290, 771

Sivagnanam, P. et al. 1989, A&A, 211, 341

te Lintel Hekkert, P. 1990, thesis, Leiden

van Langevelde, H. J. 1992, thesis, Leiden

Whitelock, P. A. 1990, in: Cacciari, C. & Clementini, G. (eds), Confrontation between Stellar Pulsation and Evolution, ASP Conf. Ser., 11, p. 365

Whitelock, P. A. 1993, in: Dejonghe, H. & Habing, H. J. (eds), Galactic Bulges, IAU Sym 153, Kluwer, p. 39

Whitelock, P. A., & Catchpole, R. M. 1992, in: Blitz L. (ed), The Center, Bulge and Disk of the Milky Way, Kluwer, p. 103

Whitelock, P. A. et al. 1994, MNRAS, 267, 711

Whitelock, P. A. et al. 1995, submitted to MNRAS

Wood, P. R. et al. 1992, ApJ, 397, 552

## Discussion

*Stift*: You find no metallicity dependence of the PL relation. How does the assumed LMC distance modulus affect this conclusion?

*Whitelock*: We find the distance of the Miras in the LMC and the globular clusters with respect to the RR Lyrae stars. So if you change the distance to the Clouds you will change the distance to the clusters in the same way. The magnitudes of the Miras with spectroscopic and trigonometric parallaxes are not tied in the same way to the LMC.

*Bedding*: For how many Miras will HIPPARCOS give reasonable distances?

*Whitelock*: Probably about half a dozen or so. It depends on the final accuracy of the HIPPARCOS positions (Catherine Turon is nodding!).

*Chapman*: From present observations of stellar periods against other quantities, it looks very unlikely that the P-L relationships will be found to be valid for P $\gtrsim$ 800 days. I interpret this in the sense that the most massive stars ($\gtrsim 4M_\odot$) *do* show strong evolution in P - so that a star of given mass and luminosity could have a large range of periods between, say, 800 and 2000 days.

*Habing*: With respect to the question how much the period of a given star can increase during these late phases of evolution: the Feast (1963) paper showed clearly that there is a relation between period and kinematics - how much freedom does this paper leave one to speculate about period increases?

*Whitelock*: I think all we can say is that there is no observational evidence for changes in $\log(P)$ of more than about 0.05. But we obviously cannot rule out the possibility that much larger changes occur very rapidly, so that we do not actually observe stars in that state.

*Shibahashi*: I was very much impressed by using the beautiful P-L relationship that you showed us. However, I also got complaint by such a simple relation between the periods and the luminosities of Mira variables. The periods of a star are mainly governed by two parameters, its mass and radius. Could you say some words to explain the unique P-L relation of Mira variables by which the period is related with only one parameter, L?

*Whitelock*: It is a very curious situation. What we seem to be finding is that there is a particular mass/age/metallicity relationship which is the same in various parts of the Galaxy and in the Magellanic Clouds.

*Astrophysical Applications of Stellar Pulsation*
*ASP Conference Series, Vol. 83, 1995*
*R. S. Stobie and P. A. Whitelock, eds.*

# Observations and Interpretation of Luminous Blue Variables

Henny J.G.L.M. Lamers

*Astronomical Institute and SRON Laboratory for Space Research, Princetonplein 5, NL-3584 CA Utrecht, The Netherlands, hennyl@sron.ruu.nl*

**Abstract.** The different types of variations of LBVs are discussed. The "typical LBV variations" have amplitudes of $\Delta V \simeq 0.5$ to 2.0 magnitudes and irregular time-scales of months to years. This is due to changes in the stellar radius and the effective temperature. Modelling of this variability for one star, S Dor, shows that the radius of the star varies between 100 and 380 $R_\odot$, the effective temperature between 20,000 and 9,000 K, and the luminosity between log $L_* = 6.10$ to 5.9. The variation of the radius is not an apparent variation of the effective radius of the wind due to a variable mass loss rate (which has often been assumed) but it is a true variation of the radius of the star itself. The changes in $L_*$ suggest that about $10^{-3}$ to $10^{-2}$ $M_*$ takes part in the expansion of the star. The irregular microvariations with amplitudes of about $\Delta V \simeq 0.2^m$ on time-scales of weeks are probably due to non-adiabatic pulsations with mode-interaction. We argue that LBVs are close to *their* effective Eddington Limit and discuss a qualitative scenario to explain their location in the HR-diagram.

## 1. Introduction

Luminous Blue Variables (LBVs) form a small class of very luminous irregular photometric variables, originally called "P Cygni type stars" in our Galaxy, "S Doradus variables" in the LMC, and "Hubble Sandage variables" in other galaxies. The variability occurs on all time-scales from weeks to decades, with amplitudes of $\Delta V \simeq 0.2^m$ to $2^m$. Some LBVs have also suffered "giant eruptions" when they brightened by $3^m$ or more and ejected a considerable amount of mass on the order of 1 $M_\odot$ in a few decades. The most famous examples of these are the eruptions of P Cygni around 1600 and of $\eta$ Car between 1830 and 1860. From small-number statistics together with evolutionary considerations and from the observations of ring nebulae around some LBVs it is estimated that most, if not all, LBVs suffer giant eruptions at intervals on the order of $10^3$ years.

Fig. 1 shows the light curve of the LBV AG Car over more than a decade. Notice the irregularity of the brightness variations without any periodicity. At best there is a "characteristic time-scale" (sometimes called "semi-period") of about 2 years and a "characteristic amplitude" of $\Delta V \simeq 1.5^m$ to $2.0^m$. These variations on a time-scale of years with amplitudes of $0.5^m$ to $2.0^m$ have been called the "typical LBV variations" or "S Dor variations" or "moderate LBV

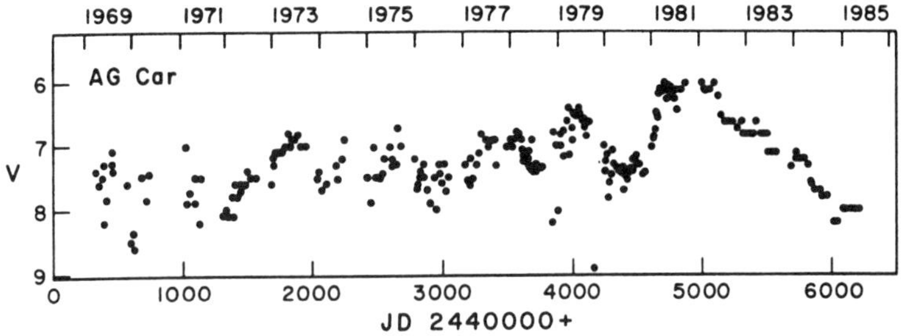


Figure 1. The visual lightcurve of the LBV AG Car between 1969 and 1985. Notice the variability on all time-scales from weeks to years. (from Lamers, 1989)

variations" (although they are not moderate at all!). I will refer to them as the typical LBV variations. Fig. 1 also shows the presence of both faster and slower variations.

LBVs represent a short post main-sequence phase of about $3.10^4$ years in the lives of stars with an initial mass $M \leq 50\ M_{\odot}$. However, there is also a group of lower luminosity LBVs with initial masses between about 35 and 50 $M_{\odot}$. These stars might be in a post Red Supergiant phase.

In this paper I will discuss some aspects of the variability of LBVs. I will not discuss the giant eruptions although these are fascinating and crucial for the evolution of the LBVs into Wolf-Rayet stars. Instead I will concentrate on the "typical LBV variations" and on the "micro-variations" because these are possibly due to instabilities which are of relevance for stellar pulsations, the topic of this conference. I will discuss the observations and their interpretation, but I will leave the theory of the instability mechanism to be discussed by Cox et al. (these proceedings).

An excellent review about the LBVs has been published by Humphreys & Davidson (1994). The reader is referred to that paper and to the book "Physics of Luminous Blue Variables" (Davidson et al. 1989) for topics and aspects that are not discussed here.

## 2. Typical LBV variations

### 2.1. The observations

Typical photometric LBV variations are shown in Fig. 2. The amplitude of the variations in V ranges from about $2.0^m$ for AG Car and R127 to $0.5^m$ for HD160529.

Wolf (1989) has pointed out that there is a relation between the maximum amplitude of the variations and the luminosity of the star. This relation can be used to derive extragalactic distances.

The typical LBV variations in V are associated with changes in the colours: the stars are bluest when they are at visual minimum and reddest at visual

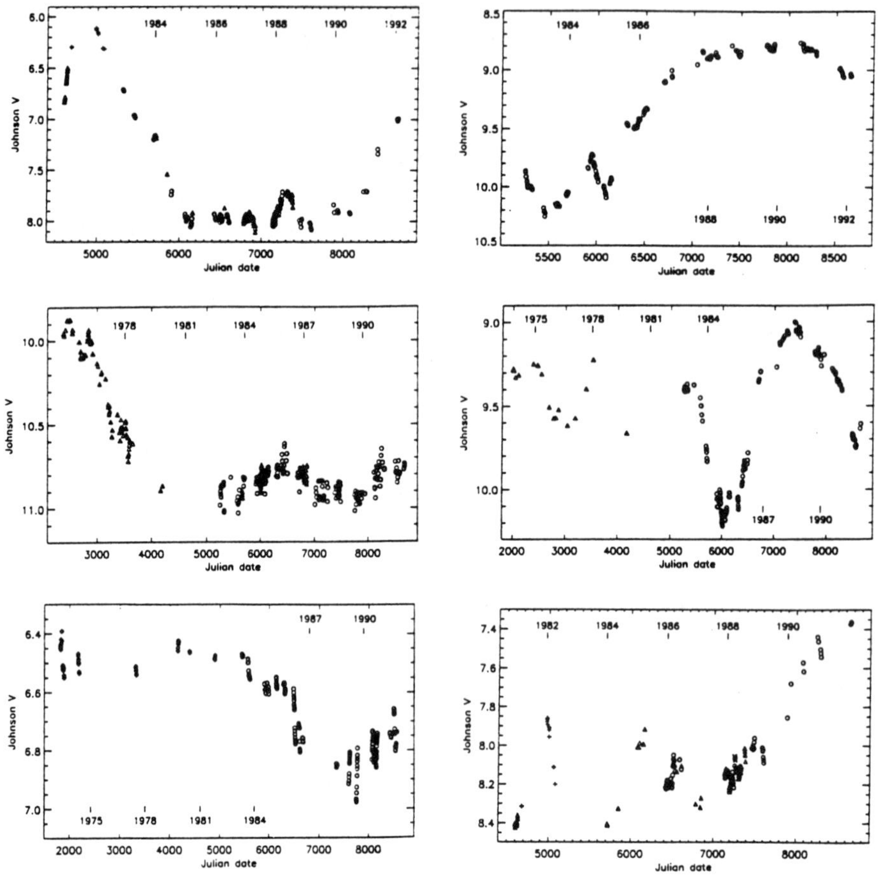


Figure 2. The visual lightcurves of six LBVs measured with the LTPV project for monitoring photometric variations of long-term variables in Strömgren-photometry. Left from top to bottom: AG Car, R71, HD160529. Right from top to bottom: R127, S Dor, HR Car. Notice the irregular nature of the "typical LBV-variations" and the fast "microvariations" during visual minimum (from Spoon et al. 1994)

maximum. This is shown in Fig. 3 for the star R127. Between 1984 and 1988 the visual brightness increased from $10.3^m$ to $8.8^m$. The colours followed the change in V. These colour changes indicate that the star is hottest when it is faintest in V and coolest when it is brightest in V. This implies that the changes in V are largely due to changes in temperature and in bolometric correction (BC).

## 2.2. Modelling the LBV variations

We have started a program in Utrecht to model the variations of LBVs in detail. The photometric variations of LBVs are modelled with the ISA wind code for modelling stars with expanding atmospheres in non-LTE (de Koter et al. 1993). This is probably the fastest non-LTE code of its kind and it can be used very efficiently to model observations of LBVs. The input of the models is: the stellar mass $M_*$, the radius $R_{in}$ of the star at optical depth $\tau_{Ross} \simeq 10$ to $10^2$, the stellar luminosity, the mass loss rate and the velocity law of the wind. In the first studies we assumed that the atmospheres consist only of H and He. (We are now in the process of including the metals in the models). With these input parameters we calculated the model atmospheres and the resulting energy distributions, Strömgren-colours and V-magnitude. These are compared with the observed Strömgren-photometry during the variability of LBVs.

The results for the star S Dor are shown in Fig. 4. We adopted a mass of 45 $M_\odot$, derived from the luminosity and the models of Schaerer et al. (1993). This stellar mass is kept constant in all the models of S Dor. For mass loss rates of $\dot{M} \simeq 10^{-6}$ to 3 $10^{-4}$ $M_\odot yr^{-1}$ the energy distribution is insensitive to the mass loss rate and the velocity of the wind. This is because in this mass loss range the wind is optically thin in the visual and UV continuum (de Koter et al. 1995). Since the mass loss rates of LBVs are in the above mentioned range (e.g., Lamers, 1987) the modelling of the energy distribution and colours of LBVs is almost independent of the assumed values of $\dot{M}$ and $v_\infty$. The energy distribution is most sensitive to changes in the luminosity and the stellar radius, in fact to the flux $L/4\pi R_{in}^2$ and to the gravity at the inner radius of the atmosphere, i.e., to $M_*/R_{in}^2$. A change in the gravity results in a change in the scale height of the atmosphere. A change in $L$ and $g$ results in a change in the temperature structure of the atmosphere which affects the Strömgren-colours and V directly. Therefore the Strömgren-colours and V are very sensitive to changes in $L$ and $R_{in}$ for a given mass.

The top part of Fig. 4 shows the variations in V and $u - b$. The lower panels show the derived values of $R_{in}$ and $L_*$ during the LBV variations and the resulting values of $T_{eff}$ and $g_{eff}$. The effective temperature is defined by $L = 4\pi R_{2/3}^2 T_{eff}^4$, where $R_{2/3}$ is the radius at an optical depth $\tau = 2/3$ at 5500 Å. The effective gravity is $g_{eff} = GM_*(1 - \Gamma)/R_{in}^2$, where $\Gamma = g_{rad}/g_N$ is the ratio between the radiation pressure force due to electron scattering and the gravity. Fig. 4 shows that the radius of the star S Dor changes from 100 $R_\odot$ at visual minimum to 380 $R_\odot$ at visual maximum. These are changes in the radius of the star itself and not only changes in the effective radius $R(\tau = 2/3)$ of the wind. A similar change in radius was found for the LBV R71 by Leitherer et al. (1989). The effective temperature changes from 20 000 K to 9 000 K and the luminosity changes from log $L = 6.10$ to 5.90 between visual minimum and

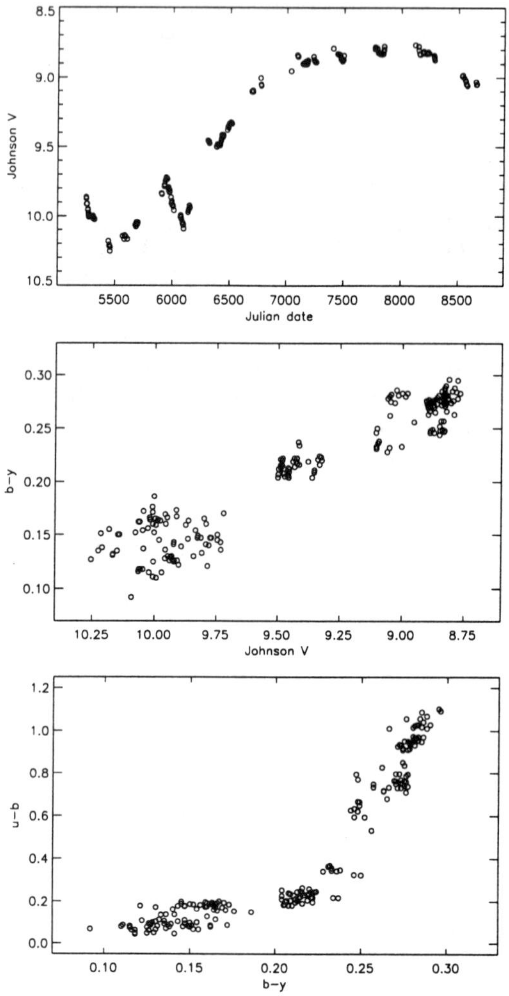


Figure 3. This figure shows the variations of the LBV R127 in the LMC in Strömgrenphotometry. Notice that there is a colour-magnitude relation and a colour-colour relation. The star is hotter when it visually faint and cooler when it is visually bright. (From de Koter & Lamers, 1993)

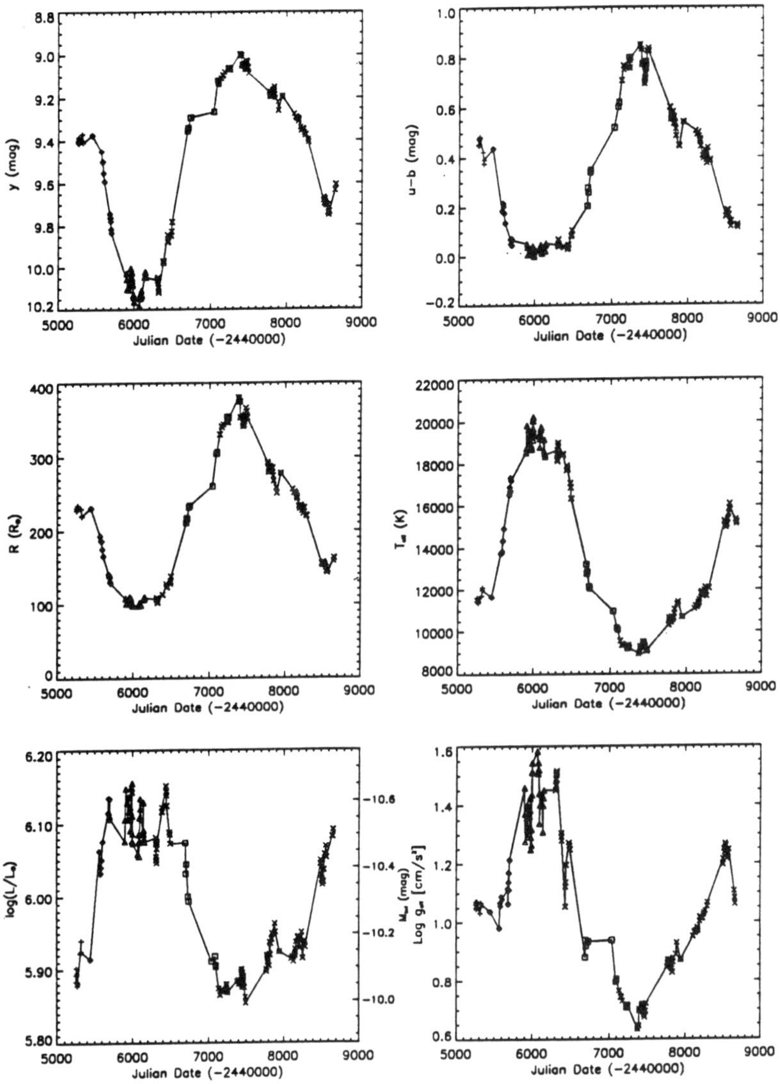


Figure 4. Results of the modelling of the photometric variations of S Dor. The top part shows the observed variations in magnitude and colour. The lower panels show the derived variations of the stellar parameters: $R_*$, $T_{\rm eff}$, log L and log $g_{\rm eff}$. Notice the large variations in $R_*$ and $T_{\rm eff}$ and the variation in $L_*$. During visual maximum the star is largest, coolest and has the lowest luminosity. (From Vennix et al. 1995)

maximum. The bolometric magnitude decreased by $\Delta m_{bol} = 0.5^m$ and the V magnitude increased by $\Delta\ V = 1.2^m$. So the bolometric correction must have changed by $1.7^m$ between visual minimum and maximum.

The variation in L is particularly interesting because it is contrary to the common assumption that the luminosity of LBVs remains essentially constant during typical LBV variations. This assumption is based on observational studies of the visual and UV energy distributions of a few LBVs observed with the IUE satellite. It was found by Wolf et al. (1981) for the star R71, by Leitherer et al. (1985) for S Dor, by Stahl & Wolf (1986) for R127, and by Lamers et al. (1989) for AG Car. However, the uncertainty of about 10 percent in the absolute calibration of IUE and the uncertainty in the extinction correction implies that the previous conclusions about the constancy of $L$ is no more accurate than about 0.2 dex at best. Our results of the detailed modelling show that for the star S Dor $M_{bol}$ does vary by about 0.5 mag. Obviously we have to check this result for more LBVs (we are presently doing this) and to improve the accuracy of the models by using fully blanketed NLTE expanding model atmospheres in the very near future (de Koter, in preparation).

If the derived variation in L of the LBVs turns out to be a common characteristic of LBVs it helps to understand the origin of the variations in the underlying star. This is discussed in the next section.

*We conclude that the typical photometric variations of LBVs are due to changes in the radius of the underlying star and that the luminosity of the LBV S Dor is not constant but varies by about 0.2 dex.*

## 2.3. The origin of the typical LBV variations?

We can use the changes in the luminosity of the LBV S Dor to obtain a first order estimate of the amount of mass involved in the radial expansion of the star by a factor four. This estimate is based on simple energy considerations and is probably correct within a factor three or so.

The mean luminosity of S Dor in the period between JD +5000 to +9000 is $L \simeq 1.0\ 10^6\ L_\odot$ but it dropped to about $8\ 10^5\ L_\odot$ during the the visual maximum between JD +6800 to +8500 (see Fig. 4). This implies a decrease in luminosity by $\Delta L \simeq 2\ 10^5\ L_\odot$ during about 1700 days which corresponds to a missing radiative energy of $\Delta E_{rad} = \Delta L \cdot \tau \simeq 1\ 10^{47}$ erg over this time interval.

Suppose that this missing energy was used to expand the outer layers of the star from the observed radius of 100 $R_\odot$ to 380 $R_\odot$. Let $M_{exp}$ be the amount of mass that takes part in the expansion. Since the *outer* radius of the expanding layer increased its size from 100 to 380 $R_\odot$, but the *inner* radius did not expand at all, we assume a mean increase in radius of the expanding layer from about 75 to 200 $R_\odot$. If all the missing energy $\Delta E_{rad}$ was used to expand a layer of mass $M_{exp}$ from 75 to 200 $R_\odot$ we can write

$$\Delta E_{rad} = \frac{G\ M_{eff}\ M_{exp}}{R_\odot}\{\frac{1}{75} - \frac{1}{200}\} \tag{1}$$

where $M_{eff} = M_*\ (1 - \Gamma)$ is the effective mass of the star, i.e., corrected for the outward force due to radiation pressure, with

$$\Gamma = g_{rad}/g_N = 2.66\ 10^{-5}(L_*/L_\odot)/(M_*/M_\odot) \tag{2}$$

For S Dor we adopt $M_* = 45\ M_\odot$, $L_* = 1.0\ 10^6\ L_\odot$ which results in $M_{eff} = 18\ M_\odot$. Substitution of these values into eq. (1) gives an estimate of the amount of mass involved in the expansion

$$M_{exp} \simeq 0.17\ M_\odot \simeq 0.003\ M_*. \tag{3}$$

In this estimate we have neglected the thermal energy. If the star remains in thermal equilibrium, the cooling of the expanding layers also releases energy. This can increase the estimate of $M_{exp}$ by about a factor 2.

We conclude that about 3 to 6 $10^{-3}\ M_*$ takes part in the expansion of S Dor during visual maximum. This is only a small fraction of mass of the star. However, because of the very steep density profile of the early type supergiants, the expanding layer may extend to much deeper than half the stellar radius at visual minimum and deeper than 90 percent of the stellar radius at visual maximum.

*This suggests that the typical LBV variations, which are in fact due to drastic changes in the stellar radius, are due to some instability in the star which is located at about* $10^{-3}$ *to* $10^{-2}\ M_*$ *below the photosphere.*

## 3. The microvariations

LBVs show microvariations of $\Delta V \simeq 0.10^m$ to $0.30^m$ on time-scales of 10 to 50 days. These variations are more pronounced at visual minimum than at visual maximum. The microvariations have been studied extensively by van Genderen (see, e.g., van Genderen 1991 and van Genderen et al. 1992). Normal supergiants in the same area of the HR diagram as the LBVs also show these types of variations but with smaller amplitudes. They are usually referred to as $\alpha$ Cygni variations. Fig. 5 shows a typical example of the microvariations of R71.

The time-scale of the microvariations is changing continuously in an irregular manner. For instance, the microvariations of R71 shown in Fig. 5 had a characteristic ("peak-to-peak") time-scale of 24 days in 1983-1985. However it changed to 14 days in 1986-1987. A recent Fourier analysis by Spoon et al. (1995) of the microvariations of LBVs has shown that the Fourier spectrum of every observing run (typically six months) is different.

Van Genderen et al. (1992) has studied the colour variations of the LBVs in the Walraven photometric system. He found that when the star is brighter in V it is bluer and hotter. This is opposite to the colour variations during the typical LBV variations! Van Genderen pointed out that temperature variations alone cannot explain both the magnitude and the colour variations and concluded that microvariations are due to changes in both temperature and radius. This suggests some kind of pulsational instability. It should be noted however that there is no phase lag between the colour variations and the brightness variations as in the Cepheids. So the variations are not simply due to irregular radial pulsations.

Gautschy (1992) has argued that the photometric variations in the supergiant $\alpha$ Cygni are due to non-adiabatic pulsations where mode-interaction is responsible for the irregular behaviour. The similarity between the photometric

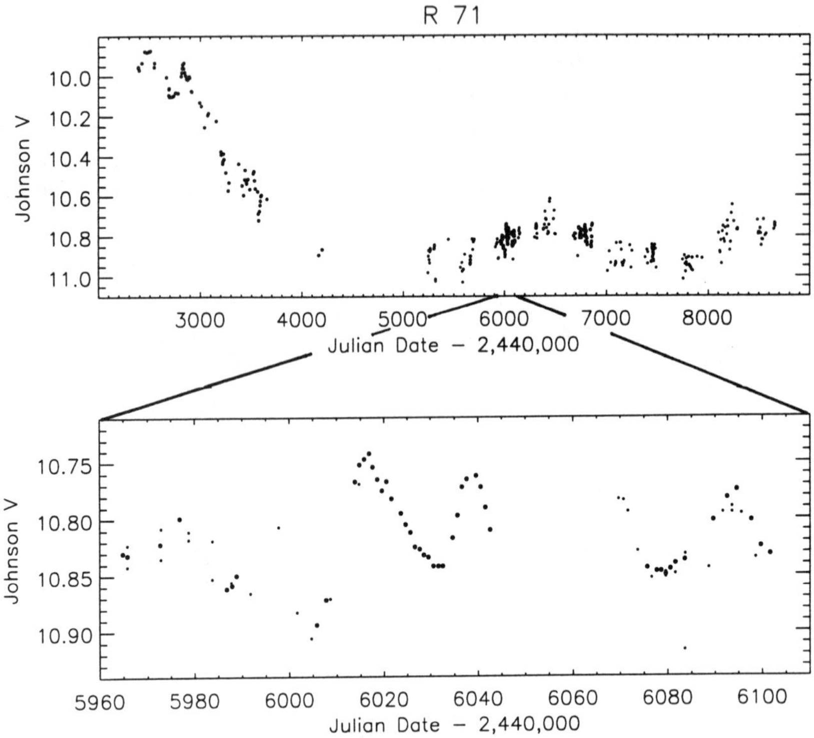


Figure 5. The lightcurve of R71 in Strömgrenphotometry measured in the LTPV-project. The upper panel shows the lightcurve over an interval of 7000 days. The error bars are on the order of 0.005 mag. The upper figure already shows the presence of short-term variability. It is shown most clearly in the expanded plot of the lower figure which covers an interval of 150 days. The "microvariations" have an amplitude of about $0.15^m$ and a characteristic "peak-to-peak"-time of 24 days. (From Spoon et al. 1994)

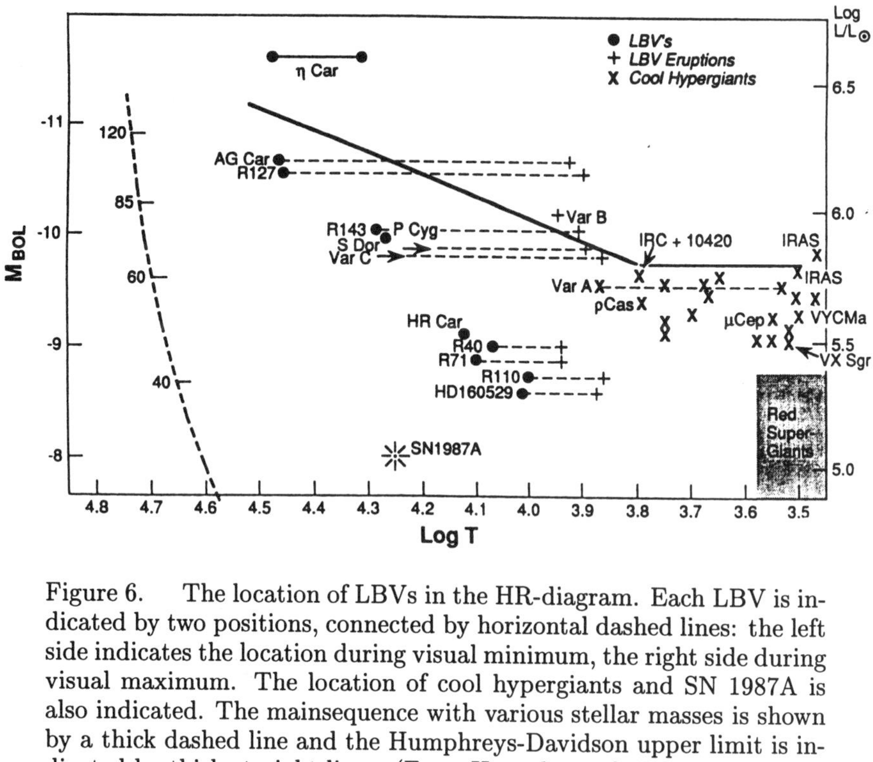


Figure 6. The location of LBVs in the HR-diagram. Each LBV is indicated by two positions, connected by horizontal dashed lines: the left side indicates the location during visual minimum, the right side during visual maximum. The location of cool hypergiants and SN 1987A is also indicated. The mainsequence with various stellar masses is shown by a thick dashed line and the Humphreys-Davidson upper limit is indicated by thick straight lines. (From Humphreys & Davidson, 1994)

variations of $\alpha$ Cyg and the LBVs suggests that the same mechanism might also explain the microvariations of the LBVs.

## 4. LBVs and the Eddington limit

Fig. 6 shows the location of the LBVs in the HR-diagram [1]. The figure suggests that the luminosity is constant. However I have shown in section 2.2 that this may not be the case.

---

[1]The concept of "effective temperature" looses its meaning for stars with extended atmospheres when the radius of the photosphere, e.g., $R(\tau_\lambda = 2/3)$ is a function of wavelength. In that case $T_{\rm eff}$ is no longer defined by $T_{\rm eff} = \{L/4\pi R_*^2\}^{1/4}$. The value of $T_{\rm eff}$ plotted in Fig. 6 is that of a plane parallel model atmosphere that produces about the observed energy distribution

At visual maximum the LBVs all reach about the same effective temperature of $T_{\rm eff} \simeq 8000$ K. At this temperature the bolometric correction is about $0^m$ so V at visual maximum is a good indicator of $m_{bol}$.

The LBVs at visual minimum are located in a narrow strip which runs from about (log L, log $T_{\rm eff}$) = (6.2, 4.47) to (5.3, 4.00). The process that is responsible for the LBVs must operate on stars in an evolutionary phase which corresponds to this instability strip. There are also other, i.e., more normal, supergiants in the LBV-strip. This implies that the instability of LBVs is not only related to the location in the HR-diagram but that there must be something special in the structure or the evolutionary phase that makes a star an LBV.

Humphreys & Davidson (1984), Lamers (1986) and Appenzeller (1986) have suggested that LBVs are unstable because they are close to their effective Eddington limit. The Eddington limit is defined by $g_{rad} = g_N$ where the radiation pressure force equals the force of gravity. [2]

This can be written in terms of an L/M ratio at the Eddington limit of

$$\frac{L_E}{M_E} = \frac{4\pi cG}{\kappa_F} \tag{4}$$

The opacity in the photosphere increases with decreasing $T_{\rm eff}$ from 40 000 K (where most of the opacity is from electron scattering) to 10 000 K (where the multitude of metallic lines in the Balmer continuum results in a maximum flux-mean opacity $\kappa_F$). The opacity decreases again to lower temperatures at $T_{\rm eff} \leq$ 10 000 K.

Lamers & Fitzpatrick (1988) have calculated the Eddington limit using the mass luminosity relation for stars at the end of the H-burning phase. Lamers & Noordhoek (1993) have calculated the location of the Eddington limit in the HR-diagram for galaxies of different metallicities using the new opacities in the model atmospheres and in the new evolutionary tracks from Schaerer et al. (1993). They find that the Eddington limit forms a "trough" in the HR-diagram with a minimum near 10 000 K, because the opacity reaches a maximum near that temperature. The "instability strip" of the LBVs is approximately parallel to the left side of the predicted effective Eddington limit, but it does not coincide. The effective gravity of the atmospheres of LBVs is so low that they are only loosely bound to the star. It is easy to imagine that if the underlying star is unstable even a small variation may have a large effect on the loosely bound atmospheres of LBVs.

The question about the special conditions of the LBVs still has to be answered since the majority of the stars in that same region of the HR-diagram do not show the typical LBV variations. Lamers & Fitzpatrick (1988) have argued that the LBVs have a lower mass than other stars in this region of the HR-diagram. They suggested the following scenario:

When massive stars with $M_* \geq 50\ M_\odot$ evolve off the main sequence they expand and decrease in $T_{\rm eff}$ at almost constant luminosity. They will eventually

[2] In the calculation of the "classical" Eddington limit only the radiation pressure by electron scattering is taken into account. In the calculation of the "effective" Eddington limit the flux mean opacity in the photosphere is taken into account. The photospheric opacity can easily be ten times larger than the electron scattering opacity

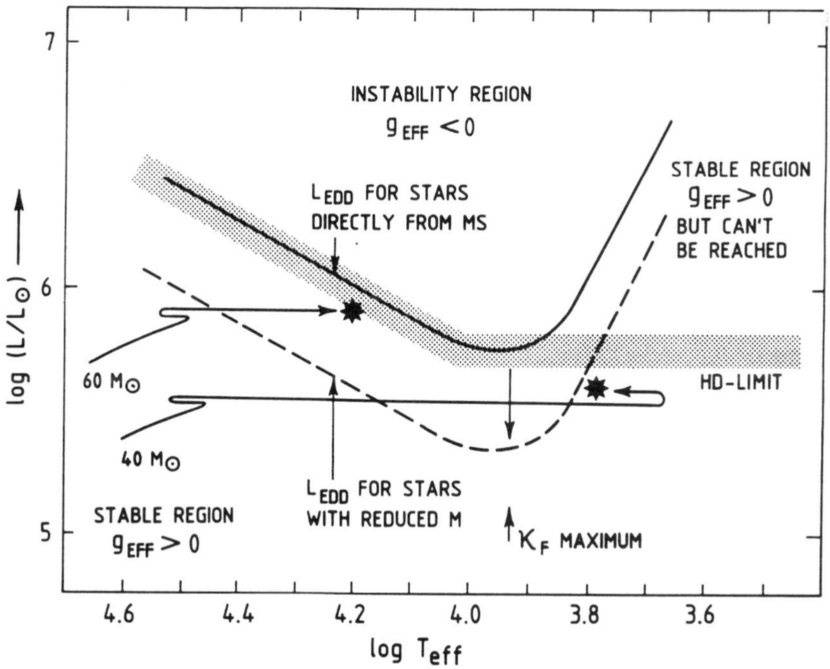


Figure 7. Schematic representation of the effect of the Eddington limit on the upper part of the HR-diagram. The Eddington limit forms a "trough" in the HR-diagram, with a minimum luminosity at $T_{\rm eff} \simeq 8000$ K. When the star looses mass, its Eddington limit in the HR-diagram is lowered. Since massive stars evolve at almost constant luminosity the Eddington limit of a star in the post main sequence phase moves to the left in the HR-diagram when the star looses mass. However, the evolution tends to move the stars to the right. So massive stars may run frequently into "their" Eddington limit. Stars with initial mass $M_i < 50\ M_\odot$ will not reach their Eddington limit in the post main sequence phase but in the post red giant phase when they have lost a considerable fraction of their mass. (From de Koter, 1993)

approach their Eddington limit. This will result in a very high mass loss rate, possibly even to large eruptions of the type of P Cygni or $\eta$ Car. In such eruptions as much as 1 $M_{\odot}$ may be ejected. After the ejection, the star is found slightly to the left of its previous location in the HR-diagram and will evolve again to the right (Stothers & Chin, 1979). Each phase of mass ejection will result in a *decrease* of the Eddington limit $L_E$. This is sketched in Fig. 7.

This implies that *the Eddington limit for stars that evolve at constant luminosity moves to the left in the HR-diagram after each episode of high mass loss.* So each star has its own Eddington limit which is determined by its present mass and luminosity. Lamers & Fitzpatrick suggested that the LBVs are older than the other stars in the same region of the HR-diagram and that they have lost more mass. So their Eddington limit is to the left of that for normal (younger) stars. This can explain why normal stars can be found in the LBV-instability strip. This is supported by the fact that the LBVs have a higher N-abundance in their atmosphere, which indicates that the layers which contain the products of the CNO-cycle have already reached the surface. It means that LBVs are closer to *their* Eddington limit than the other stars and therefore they are more unstable.

The LBVs of the lowest luminosity, with $L < 10^{5.8}\ L_{\odot}$, are probably post red supergiant stars which are on their way to the left in the HR-diagram. When stars of $L < 10^{5.8}\ L_{\odot}$ evolve off the main sequence they do not reach their Eddington limit because the L/M ratio of these stars is lower than for the more luminous stars. However, these stars lose a considerable fraction of their mass as red supergiants. Therefore, when they evolve back to the left in the HR-diagram their Eddington limit will be lower than in the post main sequence phase and they may still become unstable LBVs. This is supported by the calculations of the Eddington limit for post red giant stars by Lamers & Noordhoek (1995).

**Acknowledgments.** I am grateful to the giraffes, zebras, rhinocerosses, impalas, baboons and louries of the Hope Game Reserve for pleasant distractions during the writing of this paper and to Frieda the ostrich for her continuous (but seldomly appreciated) interest in my work with pens and papers.

## References

Appenzeller, I. 1986 in *Luminous stars and Associations in Galaxies*, eds. C.W.H. de Loore, A.J. Willis, P. Laskarides, Reidel, p 139

Davidson, K., Moffat, A., Lamers, H.J.G.L.M. (eds) 1989 *Physics of Luminous Blue Variables*, Kluwer

de Koter, A. 1993, *Studies of the Variability of LBVs*, thesis Univ. Utrecht

de Koter, A., Lamers, H.J.G.L.M. 1993 in *Studies of the Variability of LBVs*, thesis Univ. Utrecht, p 195

de Koter, A., Lamers, H.J.G.L.M., Schmutz, W. 1995 A&A (in press)

de Koter, A., Schmutz, W., Lamers, H.J.G.L.M. 1993, A&A 277, 561

Gautschy, A. 1992, MNRAS 259, 82

Humphreys, R.M. & Davidson, K. 1994, PASP 106, 1025

Lamers, H.J.G.L.M. 1986 in *Luminous stars and Associations in Galaxies*, eds. C.W.H. de Loore, A.J. Willis, P. Laskarides, Reidel, p 157

Lamers, H.J.G.L.M. 1987 in *Instabilities in Early Type Stars*, eds H.J.G.L.M. Lamers and C. de Loore, Reidel, p 99

Lamers, H.J.G.L.M. 1989 in *Physics of Luminous Blue Variables*, ed. Davidson, Moffat and Lamers, Kluwer, p135

Lamers, H.J.G.L.M., Fitzpatrick, E. 1988, ApJ 324, 279

Lamers, H.J.G.L.M., Hoekzema, N., Trams, N.R., Cassatella, A., Barylak, M. 1989 in *Physics of Luminous Blue Variables*, ed. Davidson, Moffat and Lamers, Kluwer, p271

Lamers, H.J.G.L.M., Noordhoek, R. 1993, in *Massive Stars: their Lives in the Interstellar Medium*, eds J. Cassinelli and E. Churchwell, ASP p517

Lamers, H.J.G.L.M., Noordhoek, R. 1995 (in preparation)

Leitherer, C., Appenzeller, I., Klare, G., Lamers, H.J.G.L.M., Waters, L.B.F.M., Wolf, B. 1985, A&A 153, 168

Leitherer, C., Schmutz, W., Abbott, D.C., Hamann, W.-R., Wesselowski, U. 1989, ApJ 346, 919

Maeder, A., Meynet, G. 1987, A&A 182, 243

Schaerer, D., Charbonnel, C., Meynet, G., Maeder, A., Schaller, G. 1993, A&AS 102, 339

Spoon, H.W.W., de Koter, A., Lamers, H.J.G.L.M., Sterken, C. and Stahl, O. 1994, A&AS 106, 141

Spoon, H.W.W., de Koter, A., Lamers, H.J.G.L.M., Sterken, C. and Stahl, O. 1995, in preparation

Stahl, O., Wolf, B. 1986, A&A 154, 243

Stothers, R., Chin, C.W. 1979, ApJ 233, 267

van Genderen, A.M. 1991 in *Rapid Variability of OB stars*, ed D. Baade, ESO Conf and Workshop Proc. Nr 36, p117

van Genderen, A.M. et al. 1992, A&A 264, 88

Vennix, R., Lamers, H.J.G.L.M., de Koter, A. 1995 (in preparation)

Wolf, B. 1989, A&A 217, 87

Wolf, B., Appenzeller, I., Stahl, O. 1981, A&A 103, 94

## Discussion

*Pfleiderer*: Expansion and contraction over about 2 A.U. within about a year should cause observable radial velocities. Do you have spectroscopic evidence not only for the expansion which might as well be stellar wind but also for the contraction phase?

*Lamers*: This would a very interesting independent check of the radius variations. Assuming a typical increase in the radius of about 300 $R_\odot = 2\ 10^{13}$ cm during a typical time of 2 years = $6\ 10^7$ s, we expect photospheric expansion and contraction velocities of only about 3 km s$^{-1}$. The practical problem is twofold: a) there are very few photospheric lines which are *not* affected by the stellar wind, so one has to study weak photospheric lines, b) the photospheric lines have typical widths of about 50 km $s^{-1}$, so one has to look for velocity variations less than 10% of the line width for weak lines. This is presently difficult but not impossible with new high resolution, high S/N spectra.

*Habing*: Your model requires rapid motion of stellar mass, travelling over large distances in short times. But could it be an apparent effect, in the sense that there is an overall increase in matter in the region between small and large radii, so that the effective stellar radius ($\tau \sim 2/3$) appears to move very quickly outward ('phase velocity' against 'true velocity')?

*Lamers*: We have considered this possibility. An increase in the density between 100 $R_\odot$ and 400 $R_\odot$ can be reached in two ways: (a) either you increase the density in this range by increasing the stellar radius or (b) by increasing the density in the wind so that the layer where $\tau \sim 2/3$ is reached moves outward. However, this second possibility is in disagreement with the line profiles, because this requires $\dot{M} > 3\ 10^{-4} M_\odot$ /yr whereas the emission lines and P Cygni profiles indicate $\dot{M} \lesssim 5\ 10^{-5} M_\odot$ /yr. If the star increases its radius, as in (a), by 300 $R_\odot$ in 2 years the velocity at the surface is not very large and only 3 km $s^{-1}$. This is much smaller than the typical wind velocity of 200 km $s^{-1}$.

*Guzik*: Could you elaborate on the assumptions that went into your energy distribution modelling with the non-LTE atmosphere code? What mass loss rate was used/implied?

*Lamers*: We made spherical expanding non-LTE model atmospheres of different values of L and $R_{in}$ (where $R_{in}$ is the radius of the underlying star at $\tau_R \sim 10$). The mass was kept constant. The mass loss rate was also varied, but we found that this does not affect the energy distribution because the wind remains optically thin for all reasonable values of $10^{-6} \lesssim \dot{M} \lesssim 3\ 10^{-4}\ M_\odot$ /yr. So as a consequence, we could not derive the mass loss rates yet. This will come from a study of the line profiles. I guess that the mass loss rate will turn out to be several times $10^{-5} M_\odot$ /yr at visual maximum.

*Smith*: Bob Wing describes pulsating stars of low gravity as "jellyfish" - you poke one somewhere and only part of the star responds. Your arguments here are predicated on spherical symmetry. For example, you infer changes in stellar radius from changes in spectral type. If, instead, a disturbance distorts a star over a confined area, the temperatures inferred from different spectral regions can be quite varied. If panchromatic observations turn out to show a variety of spectral types at one time, then it will bolster your case that the disturbance is relatively close to the star's surface because it will imply the local jellyfish response.

*Lamers*: You are right. We assumed spherical models so we derived the parameters of spherical LBV's. In the future we can test the spherical assumption when we include the study of the variations of the line profiles.

*Percy*: In the case of $\rho$ Cas, the 'micro-variations' are due to pulsation, but there are occasional large, long cycles which probably result in enhanced mass loss. Could the pulsations in LPV's also trigger 'outbursts'?

*Lamers*: I doubt this. Firstly, because the typical time scale of the LPV outbursts is between 2 and 10 years, whereas the pulsations are on time scales of 10 to 40 days; secondly, the amplitudes of the pulsations are small, i.e., about 0.1 to 0.2 mag in V; thirdly, the pulsations are probably non-radial.

*Maeder*: You have given a nice argument based on energy conservation to show that about 1 or 2% of the LBV mass is lifted during the outbursts. There is another line of arguments which support your result. After a major change of radius, the star re-adjusts thermally during the inter-burst phase. From the observed characteristic times, we may ask what is the thickness of a layer which has just the thermal time scale corresponding to the observed ones. Interestingly enough, this thickness also corresponds to 1 - 2% of the total mass. Thus, both the energy and time scale arguments are consistent.

*Kovács*: It is important to keep in mind that the linear adiabatic normal-mode frequencies have little connection with the observed ones for stars of extended envelopes. Non-adiabatic and nonlinear effects change the adiabatic values considerably (see Y. Tuchman's paper in these proceedings). Therefore, any conclusion about the non-radial nature of the pulsation is premature at this point (assuming that the conclusion is based solely on a period argument).

*Astrophysical Applications of Stellar Pulsation*
*ASP Conference Series, Vol. 83, 1995*
*R. S. Stobie and P. A. Whitelock, eds.*

# Theoretical Pulsations of Luminous Blue Variables

Arthur N. Cox, Joyce A. Guzik, Michael S. Soukup

*Los Alamos Astrophysics*

Siobahn M. Morgan

*University of Northern Iowa*

**Abstract.** Both radial and low degree and order $g$-mode nonradial pulsations are predicted for luminous blue variables that occur in the blue supergiant region of the Hertzsprung-Russell diagram. It is found that the radial strange modes have very large growth rates due to helium ionization in models at surface effective temperatures between 10,000 and 20,000 K.

## 1. Introduction

Luminous blue variables (LBVs) are the most massive stars in galaxies with original masses between 30 and over 100 $M_\odot$. Luminosities are near $10^6$ $L_\odot$. They are observed to have microvariations as well as short and long time scale outbursts, and the former seem to be pulsations with periods typically of 30 days. LBVs exist close the the Humphreys-Davidson line in the Hertzsprung-Russell diagram (HRD) that defines the cool limit and maximum luminosity of observed blue supergiants. Mass loss from all these very luminous stars reduces the mass by 30% or more as they evolve into the LBV region. These stars may not survive the LBV stage to produce blue looping evolution tracks.

## 2. Evolution

Evolution tracks in the HRD have been calculated by the Iben code with an approximate allowance for the opacity changes from the Iben fit to the new OPAL opacities. We discuss here only two models with initial masses of 80 and 50 $M_\odot$. Mass loss rates used produce masses of 47.28 and 31.13 $M_\odot$ at surface effective temperatures of $logT_{eff}$ = 4.26 and 4.10, and luminosities of $logL_\odot$ = 6.10 and 5.77. Both models are in the post-main sequence stage, burning hydrogen in a shell surrounding an almost pure helium core. Mass loss exposes surface helium enhanced to Y= 0.54 and 0.38 for the two models, and the internal composition structures, homogenized in several layers by active or fossil convection zones, are used for models to examine pulsational stability.

## 3. Theoretical pulsations

As reported before (Cox, Morgan, Soukup, & Guzik 1993, BAAS, 25, 1441 and Soukup, Cox, Guzik, & Morgan 1994, BAAS, 26, 907) nonradial g-modes of radial order between 10 and 20 with periods near 10 days for $\ell=1$, are often found when convection is frozen-in. These earlier papers suggested that convection in the 200,000 K temperature region, caused by the large OPAL iron line opacities, has a time scale very close to the nonradial periods. Thus it is likely that, if convection cannot be suppressed, it will destroy the almost pure kappa effect driving, just as it does at the red edge of the Cepheid instability strip. Suppression was sought by a composition gradient set up by diffusion or merely the result of the helium gradient caused by evolution, but high radiation pressure envelopes do not produce a large enough Ledoux term.

At surface temperatures cooler than 25,000 K, it is discovered that ionization of the enhanced helium gives important pulsation driving by the classical kappa, gamma, and radius effects. In these outer very low density layers, convection occurs but it carries usually less than 1% of the luminosity. Thus any time-dependent convection destruction of driving is minimal, as tests show. Since the LBVs are found in the temperature region all the way down to 10,000 K, independent of the varying Z abundances of our Galaxy and the two Magellanic Clouds, it appears that the iron line kappa effect from the OPAL opacities cannot be the main cause of the LBV microvariations.

When a helium abundance of Y=0.32 or greater exists in the envelopes of these blue supergiants, a density inversion occurs centered around 30,000 K. The high helium opacity requires a high temperature gradient to transport the emergent luminosity. But then only an inverted density gradient can produce the needed pressure gradient for envelope hydrostatic balance. The deeper large iron line opacity also tends to create a density inversion, but there convection is more effective in energy transport, and the inversion is smaller or nonexistent.

This density inversion produces the so-called strange modes found in high luminosity-to-mass ratio envelopes. The temperature gradient variation during a pulsation cycle is another driving mechanism, but it is usually small compared to the classical kappa effect. But for strange modes, the radial perturbation increases with depth in this thin density inversion layer, and the resulting temperature gradient damping overcomes the usual driving. Then, deeper than the density inversion region, the temperature gradient perturbation is much steeper than normal, and now considerable new pulsation driving occurs. The strange mode perturbation structure extends deeper than for normal acoustic modes, and its periods are typically 30 rather than 10 or 20 days for the regular modes. With so much driving, growth rates are often near 100% per period as Kiriakidis, Fricke, & Glatzel (1993, MNRAS, 264, 50) and recently Glatzel (1994, MNRAS, 271, 66) discuss.

We propose that microvariations of LBVs are mostly or all radial pulsations that occur when mass loss exposes enough helium to produce strange modes. The red edge of the pulsation region is caused by the lack of stars that can survive the LBV outbursts when the surface helium is sufficiently enhanced. Whether these radial modes cause outbursts is an unanswered question.

*Astrophysical Applications of Stellar Pulsation*
*ASP Conference Series, Vol. 83, 1995*
*R. S. Stobie and P. A. Whitelock, eds.*

# Rotation Curve and Metallicity of the Outer Disc from Cepheids

F. Pont, D. Queloz, M. Mayor and P. Bratschi

*Geneva Observatory, Sauverny, Switzerland*

**Abstract.** A group of faint, remote Cepheids in the outer disc of the Galaxy has been observed in spectroscopy and photometry. These Cepheids are used as tracers to measure the rotation curve of the disc beyond R=11 kpc, and to study the metallicity gradient in the outer disc.

## 1. Rotation curve of the outer disc from Cepheids

Cepheids have been used to estimate the rotation curve of the Galaxy since Joy (1939), and the present sample of Cepheids measured in radial velocity determines the rotation curve for galactocentric radii R between 6 and 11.5 kpc (Caldwell & Coulson 1987, Pont et al. 1994). At larger radii, our knowledge comes mainly from HII regions, which are not very accurate distance indicators and show an ambiguous, ill-defined rise of the rotation curve beyond 11 kpc.

We have been measuring since 1993 radial velocities for a group of faint ($m_v = 11-15$ mag), distant (R>11 kpc) Cepheids of the outer disc, with the aim of obtaining a precise determination of the rotation curve up to R=16 kpc.

Preliminary results were presented in Pont et al. (1995) and outline a rotation curve significantly lower than that given by HII regions. This can mean either a distance scale difference between Cepheids and HII regions, or a real difference in their kinematics.

## 2. Metal deficiency of outer disc Cepheids

The spectra that we used to measure radial velocities allow us to address another question : the metal deficiency of outer disc Cepheids. A good knowledge of the metallicity gradient in the outer disc is necessary to correct Cepheid distances for metallicity effects; it is also an essential constraint on the chemical history of the Milky Way.

We collected for each Cepheid up to 8 high-resolution, low signal-to-noise ratio spectra distributed over the pulsation phase. The spectra were then cross-correlated with a numerical template. The area of the cross–correlated function (CCF) is a measurement of the mean area of the spectral lines. It is therefore a sensitive function of temperature, metallicity, gravity and microturbulence (Queloz 1995).

In order to isolate the metallicity effect, we determine the influences of the other parameters by following Cepheids along their pulsating cycles (see Fig.

1a). We do not yet have an absolute calibration of the CCF area in terms of [Fe/H]. However, the small scatter of the CCF area gradient as a function of R (Fig. 1b), especially for solar neighbourhood calibrators, indicates that this technique is full of promises. We shall complete the calibration soon and thus get good quality *spectroscopic* metallicity estimates for the faint Cepheids of our sample.

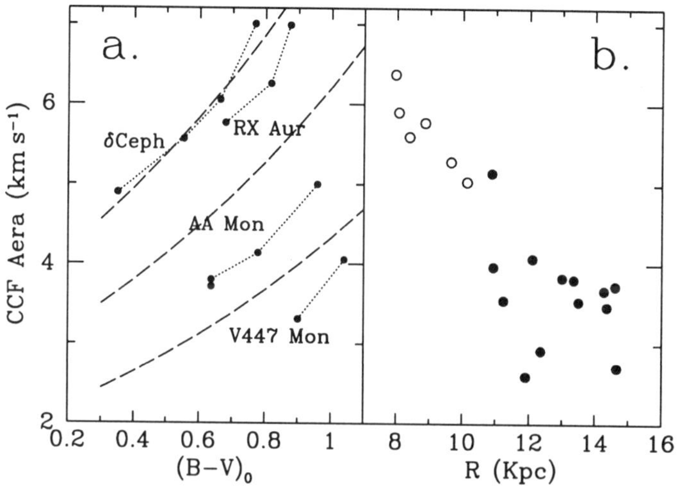


Figure 1. (a) Variation of the area of the CCF during the pulsation cycle for Cepheids of various metallicities, as a function of dereddened colour (δ Cep, RX Aur: solar neighbourhood calibrators; V447 Mon, AA Mon: deficient outer disc Cepheid). The dashed lines represent loci of constant metallicity obtained by an analytical fit. Changes induced by microturbulence were removed, to the first order, by ignoring spectra taken when microturbulence is important, namely during the ascending part of the light curve. (b) Average CCF area for each Cepheid of our sample corrected from the temperature effect, as a function of distance form the galactic centre. The solar neighbourhood calibration Cepheids and the outer disc Cepheids are shown as open and filled circles respectively. The metallicity gradient in the disc appears clearly, showing the CCF area to be a very sensitive metallicity indicator.

## References


Caldwell, J.A.R., & Coulson, I. 1987, AJ, 93, 1090

Joy A. H. 1939, ApJ, 89, 356

Pont, F., Mayor, M., & Burki, G. 1994, A&A, 285, 415

Pont, F., et al. 1995, in: IAU Symp. 169, ed. L. Blitz, in press

Queloz, D. 1995, in: IAU Symp. 167, ed. A.G. Davis Philip, in press

*Astrophysical Applications of Stellar Pulsation*
*ASP Conference Series, Vol. 83, 1995*
*R. S. Stobie and P. A. Whitelock, eds.*

# Dust Formation and the Doubly-Periodic Variable Stars

T. Lloyd Evans

*South African Astronomical Observatory, P. O. Box 9, Observatory 7935, South Africa*

**Abstract.** The doubly-periodic semiregular variable V Hydrae became redder and developed resonance emission lines and bands during the 1992-94 deep minimum, which therefore resulted from the production of a circumstellar dust cloud. This removes the need to postulate a secondary pulsation period, ten times as long as the main pulsation period, in the stellar as opposed to circumstellar envelope.

## 1. Introduction

RV Tauri variables of the RVb subgroup and many red semiregular variables of type SRb show long secondary periods, which are typically ten times longer than the pulsation period for the semiregular variables and twenty times longer in the case of the RVb stars (Payne-Gaposchkin 1954). Such large ratios imply the presence of a high overtone as well as the fundamental if both periods result from pulsation (Wood 1975). The observation that the doubly-periodic RVb stars have larger near-infrared excesses than the singly-periodic RVa stars suggested that the occurrence of double-periodicity is related to the presence of dust near the star (Lloyd Evans 1985). The less-certain finding that the doubly-periodic semiregular variables of spectral class M have more pronounced silicate emission than other oxygen-rich variables (Lloyd Evans 1987) provided some support for this idea.

V Hydrae, a doubly-periodic (530, 6670 days) variable which is also a carbon star, is the most spectacular of all these variables with minima which last for several years and are 3.5 mag deep, while the pulsations have an amplitude of 1.2 mag. Spectroscopic observations started in 1986 and have been continued to date.

## 2. Observations

V Hydrae started to fade early in 1992. Spectra taken then show a unique appearance: the normally steep spectral gradient in the blue became much flatter and the absorption bands of $C_2$ strengthened while the emission of $C_2$ at 4737 Å, a common feature in carbon-rich Miras and especially marked in V Hydrae in earlier years, disappeared. The blue region became difficult to observe as the carbon star faded because of the greater relative prominence of the blue continuum (Lloyd Evans 1991). Red spectra showed the development of emission lines of the alkali metals, Na I, K I and Rb I. Na I, for which observations are most extensive, showed two emission maxima separated by an interval of several months during which the D lines were in net absorption although emission was still present. Strong emission of the 4737 Å and 5165 Å bands of $C_2$ appeared and reached maximum strength at the same time as the emission lines. Continuation of the 530 day variation in the H-alpha emission strength suggests that the pulsation continued as normal. $JHKL$ photometry shows that the maximum fading was 2 mag at $J$ and 1 mag at $K$, implying considerable reddening.

The Mira variable R Lep, also a carbon star, underwent a faint cycle in 1994 and showed reddening and spectroscopic phenomena similar to those of V Hya. R Lep was also observed repeatedly starting in 1986 and this was the first such event, although the spectrum was much more variable from cycle to cycle than in the case of V Hya.

## 3. Discussion

These observations show that the slow variations of V Hydrae, and by implication of other doubly-periodic cool variables, result from absorption by circumstellar matter which appears to have been ejected contemporaneously by the star. The same explanation evidently applies to the occasional faint cycles of Mira variables.

Winters et al. (1994) have made hydrodynamical models of circumstellar dust shells and find that the dust shell may impose a longer-term modulation on the variations of a long period variable. One of their models gives a good representation of the light-curve of R For which had deep minima in 1983 and 1990 (Le Bertre 1992). A new dust layer is formed at each of the deep minima. Fleischer et al. (1992) found that the ejection of a dust shell occurred in every pulsation period if the C/O ratio is large but became less frequent if C/O is reduced towards one. This is at variance with the systematics of the doubly-periodic semiregular variables for which $P_2/P_1$ is 9.4 for M stars but 12.2 for carbon stars (Payne-Gaposchkin 1954). The absence of such well-defined relationships for the Mira variables, which differ from semiregulars primarily in their larger amplitudes, suggests that we do not yet have the complete picture.

## References

Fleischer, A. J., Gauger, A. & Sedlmayr, E. 1992, A&A, 266, 321

Le Bertre, T. 1992, A&AS, 94, 377

Lloyd Evans, T. 1985, MNRAS, 217, 493

Lloyd Evans, T. 1987, in: Circumstellar Matter, A. Appenzeller & C. Jordan, Dordrecht: Reidel, 541

Lloyd Evans, T. 1991, MNRAS, 248, 479

Payne-Gaposchkin, C. H. 1954, Harvard Annals, 113, 191

Winters, J. M., Fleischer, A. J., Gauger, A. & Sedlmayr, E. 1994, A&A, 290, 623

Wood, P. R. 1975, in: Multiple Periodic Variable Stars, W. S. Fitch, Dordrecht: Reidel, 69

*Astrophysical Applications of Stellar Pulsation*
*ASP Conference Series, Vol. 83, 1995*
*R. S. Stobie and P. A. Whitelock, eds.*

# Pulsating Stars and the Distance Scale - I

Alistair R. Walker

*Cerro Tololo Inter-American Observatory*[1]*, Casilla 603, La Serena, Chile*

**Abstract.** A summary is given of recent measurements, by various methods, of the absolute magnitudes for RR Lyrae stars. Calibrations within our galaxy, using the Baade-Wesselink method and by Statistical Parallaxes, give a result that is some 0.2-0.3 mag fainter than is obtained by all other methods considered. The resolution of this conundrum will be likely in the near future when new and improved datasets become available.

## 1. Introduction

RR Lyrae stars are representative of the oldest recognisable stellar population. They occur in the general field, and in globular clusters. For the latter, ages are constrained by use of theoretical isochrones to fit the main-sequence turn-off, if distances are known. Since globular clusters are thought to have formed very shortly after the formation of our galaxy, their ages define a lower limit to the age of the galaxy. The distribution of field RR Lyraes in our galaxy allows us to compare population models with the observations, and to determine the size of the galaxy. In local group galaxies, the RR Lyrae distance scale can be compared with those determined from other distance indicators, such as Cepheids and Miras. The Large Magellanic Cloud (LMC) is crucial in this respect.

Recently, it has become apparent that the RR Lyrae distance scale zeropoint differs significantly depending on whether it is calibrated via the Cepheid distance scale in the LMC, or by distances determined to galactic halo RR Lyraes, using Baade-Wesselink analyses or the Statistical Parallax method. Chaboyer (1995) in a very useful discussion of all the theoretical and observational parameters affecting the ages of galactic globular clusters, points out that the uncertainty in the distances to the clusters is the single largest effect, some 20 percent. Also uncertain is the slope of the RR Lyrae absolute magnitude-metallicity relation. Some pulsation analyses (Sandage 1993, Cox 1994) indicate a high slope of 0.30-0.40, whereas most other methods give a low slope, 0.15-0.22. Higher values imply that the galactic globular clusters are coeval, or nearly so, whereas smaller values indicate that the globular cluster system formed over a period of some Gyr.

---

[1]Cerro Tololo Inter-American Observatory, National Optical Astronomy Observatories, is operated under cooperative agreement with the National Science Foundation.

The distance scale for the RR Lyraes, via the ages that result, is thus of crucial importance in the discussion of theories of formation of our galaxy, and is the best observational lower limit to the age of the Universe. In the sections that follow I will summarise the various methods recently used to determine RR Lyrae absolute magnitudes.

## 2. Methods

### 2.1. Comparison with Cepheids in the LMC and other local group galaxies

RR Lyraes occur in seven LMC clusters and for all of them mean magnitudes are available from CCD photometry, as summarised by Walker (1992). These clusters have a rather small range in metallicity and to a good approximation all can be assumed to have $[Fe/H] = -1.9$. If the LMC is assumed to have a distance modulus of 18.5 mag then the RR Lyraes have $M_v = 0.47 \pm 0.03$ The precise value depends on whether the clusters are forced to conform to some LMC geometry, such as the HI disk, but since they are well-distributed over the face of the LMC the mean absolute magnitude changes by very little. The Cepheid modulus of the LMC is often taken as $18.47 \pm 0.15$ (Feast & Walker 1987), but more recent work (see e.g., M. Feast, this conference) would increase this value by perhaps 0.1 mag (and with smaller error), partially due to a slightly larger modulus adopted for the Pleiades, and partially due to more recent work on the calibrating clusters (eg Turner 1993, Gieren & Fouqué 1993). This of course would make the RR Lyraes even brighter.

There have been a handful of surveys for field RR Lyraes in the MC, most recently those by Reid & Freedman (1992) and Hazen & Nemec (1992) deserve mention. Far more extensive is the approximately 8000 RR Lyraes measured via the MACHO survey (K. Cook, this conference). Kinman et al. (1991) summarise the earlier observations, to find that (referred to an LMC modulus of 18.5) the field RR Lyraes have $M_v = 0.52$. Suntzeff et al. (1992) has made a detailed comparison between RR Lyraes in the galactic halo and in the LMC, to find that the two populations appear indistinguishable when scaled by galaxian mass, and a reasonable mean metallicity for the LMC field RR Lyraes is $[Fe/H] = -1.6$. The result for the field RR Lyraes is consistent with that for the clusters, allowing for the expected increase in brightness at the metallicity decreases. The actual value for the slope of the magnitude-metallicity relation is not well-constrained due to the size of the errors in combination with the rather small baseline in metallicity.

An interesting test, suggested by D. Welch (private communication), is to measure field RR Lyraes in the vicinity of LMC clusters that are rich in Cepheids, thus minimising any effects due to interstellar absorption when making the Cepheid-RR Lyrae comparison. Such a test has been made for NGC 1866 (Welch & Stetson 1993, Walker 1995). The result (for four RR Lyraes only) is consistent with the above measurements, thus the evidence from the LMC is that if the Cepheid distance scale is adopted then metal-poor RR Lyraes are relatively bright, with $M_v \approx 0.45$.

A convenient summary of the Cepheid-RR Lyrae comparison for other local group galaxies has been given by van den Bergh (1995). The data are much less extensive than for the MC, the RR Lyraes are faint, and in some cases there are unresolved questions relating to differential reddening between the fields containing the RR Lyraes and the Cepheids. The best comparison is possibly that for IC1613, where Saha et al. (1992) find the RR Lyraes bright relative to the Cepheid measurements by Madore & Freedman (1991), by about the same $\approx 0.3$ mag as found in the equivalent LMC comparison. For M31 and M33 the evidence is less clear, and van den Bergh (1995) shows that there appears to be better agreement for these more metal-rich galaxies (M31, M33, and our own galaxy by definition). He goes on to suggest that this might be due to the Cepheid PL relation having a metallicity-dependent zeropoint. HST measurements of Cepheids in environments of different metallicities at fixed distance (e.g., in M31 and M33), together with comparison with other distance indicators, should soon definitively be able to test this hypothesis.

### 2.2. Comparison with the distance to the Galactic Centre

It has been long realized that the RR Lyraes can be used to find the distance to the Galactic centre by looking for the maximum in space density of RR Lyraes along a line-of-sight passing close to the galactic centre. Two such analyses are those by Oort & Plaut (1975) and Walker & Mack (1986). Because of high interstellar absorption, the lines-of-sight possible are limited to the very few low absorption windows, such as Baade's Window, and even for these the absorption is a major problem. Working in the IR alleviates this, two recent such investigations are those of Fernley et al. (1993) in Plaut's field 3 low, and Fulbright et al. (1995), working in Baade's window. The RR Lyrae galactic centre distance can be compared with those obtained by other methods. Kerr & Lynden-Bell (1986) adopted the simple approach of taking the mean of estimates for $R_0$ published between 1974 and 1986. They found $R_0 = 8.5$ kpc. Reid (1993) pursued the approach of accounting for systematic and statistical errors, and took account of covariance where necessary. He obtained $R_0 = 8.0 \pm 0.5$ kpc. Both these estimates include RR Lyraes themselves as part of the mix of distance indicators used to arrive at $R_0$. The analysis by Fulbright et al. (1995) for all the IR measurements (128 stars in two fields) is somewhat inconclusive, in that if the RR Lyraes are assumed to have absolute magnitudes as suggested by the LMC results above, i.e.,

$$M_K = -2.33 logP - 1.18,$$

then $R_0 = 8.8$ kpc, while if a 0.3 mag fainter zeropoint is chosen (see below) then the distance falls to $R_0 = 7.7$ kpc. It thus appears that the distance to the galactic centre is not well known enough to unambiguously constrain RR Lyrae absolute magnitudes. This situation should change in the near future with new survey material, for OH/IR masers from the Leiden-Sydney survey (J. Chapman, this conference), and for Cepheids expected to be found in the 2MASS and DENIS surveys (M. Feast, this conference).

### 2.3. Statistical parallax analysis

The latest RR Lyrae statistical parallax analysis (Layden et al. 1994) uses many new proper motions (80 percent from the Northern PM survey) plus a homogeneous set of radial velocities and abundances by Layden (1994). There are now 214 stars in the sample, sufficient to allow separation into disk and halo populations. This is done by a combination of metallicity and kinematic indicators, in a number of ways. The results do not appear to be sensitive to the precise way in which the separation is made, nor on the initial assumptions needed to convert observed motions into space velocity. For the halo sample, which has mean $[Fe/H] = -1.6$, $M_v = 0.74 \pm 0.12$, while for the disk sample with mean $[Fe/H] = -0.65$, $M_v = 1.1 \pm 0.35$. The disk solution unfortunately has error large enough so that the slope of the magnitude-metallicity relation is not well-determined. The result for the halo stars is almost identical to previous determinations (Hawley et al. 1986, Strugnell et al. 1986) that used a smaller number of stars. A major criticism of the earlier work was that it was not possible to calculate independent solutions for the halo stars and the metal-rich stars. HIPPARCOS proper motions, available soon for at least 150 RR Lyraes, should allow a more definitive solution for both disk and halo populations.

### 2.4. Astrometric distances to globular clusters

This is a very straightforward method which can give distances directly. In principle, all that is required is to equate dispersions of proper motions and radial velocities, after taking account of possible rotation (Cudworth & Peterson 1988). Due to the large distances involved, a long baseline in time is needed to measure the proper motions to reasonable accuracy. Even with a baseline of several decades the errors are still large, two of the more accurate results are those for M2 by Cudworth & Rauscher (1987), $M_v = 0.65(+0.37, -0.31)$, $[Fe/H] = -1.62$, and for M92 by Rees (1992), $M_v = 0.44(+0.47, -0.38)$, $[Fe/H] = -2.15$.

### 2.5. Baade-Wesselink method

The Baade-Wesselink method, as applied to RR Lyraes, has received extensive discussion in the recent literature (e.g., Liu & Janes 1990a, Cacciari et al. 1992, Jones et al. 1992). All this work has used infrared K-band photometry and high-quality radial velocity measurements. The phase interval considered has been restricted to that for which the atmosphere is expected to be "quiet". There are now a number of measurements for cluster RR Lyraes (e.g., Liu & Janes 1990b, Storm et al. 1994, and references therein) as well as for the brighter field stars. No systematic differences in luminosity are found between field and cluster stars, for a given metallicity. The several groups engaged in this work obtain very consistent results, despite detail differences in the methods employed. The formal error bars are such that the magnitude-metallicity relation is defined quite accurately, typically

$$M_v = 0.16[Fe/H] + 1.02.$$

Critics of the method note that it is difficult to estimate the systematic errors; changes to the temperature scale and to the conversion factor ($p$) relating radial velocity to pulsation velocity could significantly alter the zeropoint of the magnitude-metallicity relation.

### 2.6. Horizontal branch evolutionary models

In the past five years several different groups have presented evolutionary tracks for stars evolving off the ZAHB (e.g., Lee & Demarque 1990, Catelan 1993,

Dorman 1993). They vary in their use of Los Alamos or OPAL opacities, in the helium abundance adopted, whether they use scaled solar abundances or enhance oxygen or the $\alpha$ elements, etc. However the luminosities calculated are in relatively good agreement. All predict a rather small value for the slope of the magnitude-metallicity relation. Before comparing with observations the ZAHBs should be brightened using determinations of the vertical extent of the instability strip such as those made by Sandage (1990). If this is done then a typical representation for the absolute magnitudes of RR Lyraes is

$$M_v = 0.19[Fe/H] + 0.81.$$

### 2.7. Pulsation analyses

The pulsation equation (van Albaba & Baker 1971) can be written as

$$logL = 1.19logP + 0.81logM + 4.193logT - 13.687.$$

Thus, given $(P, M, T)$ luminosities can be derived. This can be done in a number of ways. The procedure used by Sandage (1993) is as follows. From plots of $logP_{ab}$ vs. $[Fe/H]$ for cluster and field stars he derives the period shift at the Blue Fundamental Edge as

$$\Delta logP_{ab} = (-0.12 \pm 0.02)[Fe/H] - (0.50 \pm 0.01).$$

A large value such as adopted for the coefficient of the metallicity term is supported by recent cluster analyses (e.g., Walker 1994 for M68). Masses are found from horizontal branch models, in this case those of Dorman (1993), by

$$logM = -0.059[Fe/H] - 0.288.$$

Due to the large coefficient on the temperature term in the pulsation equation, correct representation of the temperature is critical. Using $(B - V)_{mag}$, and forcing

$$dlogP/dlogT = -3.7$$

then

$$logT_e = 0.012[Fe/H] + 3.865,$$

and substituting in the pulsation equation,

$$M_{bol}(RR)_{bfe} = (0.36 \pm 0.12)[Fe/H] + (1.00 \pm 0.10).$$

Updating using the Cox (1994) formulation of the pulsation equation, and slightly higher masses which are now to be preferred, makes very little change, and with

$$BC = 0.06[Fe/H] + 0.06$$

then

$$M_v(RR) = 0.34[Fe/H] + 1.00.$$

Thus a rather steep slope for the magnitude-metallicity equation is obtained. This has the property of agreeing with the LMC calibration for the metal poor stars, and also makes the age spread amongst the great majority of globular clusters small or non-existent.

A similar analysis by Fernley (1993) uses a smaller value for the period-shift coefficient ($\approx -0.07$), and a mass relation from Kovacs et al. (1991) that implies a very small mass difference between OoI and OoII clusters. The pulsation equation is parameterised as a function of metallicity, and the BC equation originates with Kurucz (1979), and in any case is very similar to that given above. Temperatures are found via the convenient tabulation of B-V and V-K for temperatures, metallicities and gravities relevant to RR Lyraes derived by Lui & Janes (1990a) from Kurucz (1979) models. Then,

$$M_v = 0.19[Fe/H] + 0.84.$$

A third application of the pulsation equation is that by Smith & Silbermann (1995) in their analysis of their new photometry of the M15 RR Lyrae variables. Adopting $0.75M_\odot$ for the RRab stars, and using their V-R photometry with a Kurucz conversion to $T_e$, they find for these metal poor stars, which have $[Fe/H] = -2.15$, that $M_v = 0.47 \pm 0.12$ where the error estimate accounts for only errors in their photometry and the reddening. They note that the Longmore et al. (1990) temperature scale is some 300K cooler, which if adopted would make the RR lyraes have $M_v = 0.63$.

The final method is that by Cox (1994), who avoids the temperature calibration in the following manner. Firstly, RRd masses are calculated using OPAL opacities, which results in masses of $0.70M_\odot$ for OoI clusters such as M3 and M5, and $0.76M_\odot$ for OoII clusters such as M15 and M68. He then draws the instability strip and mode switch boundaries from Bono & Stellingwerf (1994) on the HR diagram, and adds the lines of constant period from the pulsation equation (updated version). The observational data (periods, relative luminosities) from the four clusters mentioned above are then fitted to these HR diagrams. The colours are used only to calculate bolometric corrections to V, and to evaluate the reddening. He finds that the OoI clusters have $M_v(RR) = 0.60$ and the OoII clusters $M_v(RR) = 0.37$, thus the metal poor RR Lyraes are relatively bright and a steep slope for the magnitude-metallicity relation is implied.

## 2.8. Fourier fitting

This method involves matching RRc light curves to hydrodynamic pulsation models (Simon & Clement 1993). If the light curves are represented by Fourier series,

$$magnitude = A_0 + \sum A_j(cos j\omega t + \phi_j)$$

and defining

$$R_{j1} = A_j/A_1,$$

$$\phi_{j1} = \phi_j - j\phi_1$$

then it is found that the mass, luminosity, and temperature can be parametrised in terms of the Fourier parameter $\phi_{31}$, e.g.,

$$logL = 1.04logP - 0.058\phi_{31} + 2.41.$$

Because RRc light curves are nearly sinusoidal, light curves of excellent quality, with very well-sampled phase cover, or preferably both, are needed in order to derive accurate values of $\phi_{31}$. Only a few of the more recent CCD studies (eg Clement et al. 1992, Walker 1994, Smith & Silbermann 1995) have produced data of the required quality. These results (for eight clusters) indicate that the RR Lyraes are relatively bright, with a small slope for the magnitude metallicity relation.

## 3. Summary

Recent work on the absolute magnitudes of RR Lyraes by a variety of methods have not yet reached convergence. Observationally, results from the HIPPARCOS mission will have a major impact, as they will provide improved statistical parallaxes for many RR Lyraes, and also provide parallaxes for subdwarfs. The latter, along with HST colour-magnitude diagrams for galactic globular clusters, will allow the measurement of relatively accurate distances to globular clusters via main-sequence fitting. The results from the present review are summarised in Table 1. In view of the domination of distance scale research by unexpected or under-estimated systematic errors, the non-inclusion of error estimates in Table 1 is deliberate.

Table 1. RR Lyrae Absolute Magnitude Estimates

| Method | [Fe/H] | | | | |
|---|---|---|---|---|---|
| | -2.15 | -1.9 | -1.6 | -1.0 | -0.65 |
| Stat. Parallax | | | 0.74 | | 1.1 |
| Baade-Wesselink | 0.68 | 0.72 | 0.76 | 0.86 | 0.92 |
| ZAHB models | 0.40 | 0.45 | 0.51 | 0.62 | 0.69 |
| Puls. (Sandage) | 0.27 | 0.35 | 0.44 | 0.66 | 0.78 |
| Puls. (Fernley) | 0.43 | 0.48 | 0.54 | 0.65 | 0.72 |
| Puls. (Smith) | 0.47 | | | | |
| Puls. (Cox) | 0.37 | | 0.60 | | |
| Fourier Fits | 0.35 | 0.42 | 0.47 | 0.62 | |
| Astrometry | 0.44 | | 0.65 | | |
| LMC Cepheids | | 0.45 | | | |

## References


Bono, G., & Stellingwerf, R.F. 1994, ApJS, 93, 233

Cacciari, C., Clemintini, G., Fernley, J. 1992, ApJ, 396, 219

Catelan, M. 1993, A&AS, 98, 547

Chaboyer, B. 1995, ApJ, in press

Clement, C.M., Jakulac, M., & Simon, N.R. 1992, ApJ, 395, 192

Cox, A.N. 1994, preprint

Cudworth, K.M., & Peterson, R. 1988, In The Extra-galactic Distance Scale, S. van den Bergh & C. Pritchet, ASP Conf. Ser no. 4, p. 172

Cudworth, K.M., & Rauscher, 1987, AJ, 93, 856

Dorman, B. 1993, in The Globular Cluster - Galaxy connection, J.P. Brodie & G.H. Smith, 11th Santa Cruz workshop

Feast, M.W., & Walker, A.R. 1987, ARAA, 25, 345

Fernley, J. 1993, A&A, 268, 591

Fullbright, J., Carney, B.W., Walker, A.R., & Terndrup, D. 1995, in prep

Gieren, W.P., & Fouqué 1993, AJ, 106, 734

Hazen, M.L., & Nemec, J.M. 1992, AJ, 104, 111

Jones, R.V., Carney, B.W., Storm, J., & Latham, D.W. 1992, ApJ, 386, 646

Kerr, F.J., & Lynden-Bell, D. 1986, MNRAS, 221, 1022

Kinman, T.D., Stryker, L.L., Hesser, J.E., Graham, J.A., Walker, A.R., Hazen, M.L., & Nemec, J.M. 1991, PASP, 103, 1279

Kovacs, G., Buchler, J.R., & Marom, A. 1991, A&A, 252, L27

Kurucz, R.L. 1979, ApJS, 40, 1

Layden, A.L. 1994, AJ, 108, 1016

Layden, A., Hanson, R., & Hawley, S. 1994, Poster paper presented at the 184th meeting of the AAS

Lee, Y-W., & Demarque, P. 1990, ApJS, 73, 709

Longmore, A.J., Dixon, R., Skillen, I., Jameson, R.F., & Fernley, J.A. 1990, MNRAS, 247, 684

Liu, T., & Janes, K.A. 1990a, ApJ, 354, 273

Liu, T., & Janes, K.A. 1990b, ApJ, 360, 561

Madore, B.F., & Freedman, W.L. 1991, PASP, 103, 933

Oort, J.H., & Plaut, L. 1975, A&A, 41, 71

Rees, R.F. 1992, AJ, 103, 1573

Reid, M.J. 1993, ARA&A, 31

Reid, N., & Freedman, W. 1994, MNRAS, 267, 821

Saha, A., Freedman, W.L., Hoessel, J.G., & Mossman, A.E. 1992, AJ, 106, 763

Sandage, A. 1990, ApJ, 350, 603

Sandage, A. 1993, AJ, 106, 703

Simon, N.R., & Clement, C.M. 1993, ApJ, 410, 526

Smith, H.A., & Silbermann, N.A. 1995, AJ, in press

Storm, J., Carney, B.W., & Latham, D.W. 1994, A&A, 290, 443

Suntzeff, N.B., Schommer, R.A., Olszewski, E.W., & Walker, A.R. 1992, AJ, 104, 1743

Turner, D.G. 1993, In New Perspectives on Stellar Pulsation and Pulsating Variable Stars, J.M. Nemec & J.M. Matthews, Cambridge University Press, p. 90

van Albaba, T.S., & Baker, N. 1971, ApJ, 169, 311

Walker, A.R. 1992, ApJ, 390, L81

Walker, A.R. 1994, AJ, 108, 555

Walker, A.R. 1995, AJ, submitted

Walker, A.R. & Mack, P.M. 1986, MNRAS

Welch, D.L., & Stetson, P.B. 1993, AJ, 105, 1813

## Discussion

*Turner*: In the distance ladder established using Cepheids, it seems that some of the steps may be unnecessary. With the recent Hubble Telescope results for Virgo Cluster galaxies, galactic cluster Cepheids can be used to establish distances beyond the Local Group without the intermediary of the LMC and Local Group galaxies. Interestingly enough, they give different distances than those based on the LMC calibrations. Any comment?

*Walker*: This is certainly true. However, since many of the new Cepheid distances to galaxies beyond the Local Group are referenced to the LMC, it seems likely that the LMC will continue to be used as a link in the distance scale chain.

*Barnes*: 1. Given the uncertainty in RR Lyrae statistical parallax this result can *almost* be made consistent with the Cepheid distance scale. A new HIPPARCOS based statistical parallax study of RR Lyraes (Daffyd Evans, John Fernley, Suzanne Hawley & myself) should improve the uncertainty.
2. Also I believe there are likely enough systematic uncertainties in the Baade-Wesselink analyses for RR Lyraes to give me pause in believing believing that distance scale (e.g., $v_{puls}/v_{rad} = f(phase)$).
With 1 and 2 I conclude that the faint RR Lyrae scale is less credible than the bright RR Lyrae scale and the Cepheid scale.

*Laney*: The B-W radii (based on JHK photometry) of Cloud and Galactic Cepheids are consistent with the usual distance scale.

*Feast*: With regard to your RR Lyrae distance to the centre it is interesting that Glass et al. (MNRAS in press), have found 8.7 kpc from Miras (also based on a Cepheid distance to the LMC).

*Astrophysical Applications of Stellar Pulsation*
*ASP Conference Series, Vol. 83, 1995*
*R. S. Stobie and P. A. Whitelock, eds.*

# Properties of the SX Phœnecis Stars in the Metal-Poor Globular Cluster NGC 5053, and Pulsation Modes of Three of the Newly-Discovered SX Phe Stars in 47 Tuc[1]

James M. Nemec

*Department of Physics & Astronomy, York University, ON Canada*

**Abstract.** The properties of variable blue stragglers (i.e., SX Phe stars) in the extremely metal-poor globular cluster NGC 5053 are reviewed. Pulsation modes are derived for three of the SX Phe stars recently discovered in the central regions of 47 Tuc by Gilliland et al. (1995a,b).

## 1. Properties of the SX Phœnecis stars in NGC 5053

Nemec et al. (1995b) have recently monitored for variability 16 of the 28 presently-known blue stragglers (BSs) in the very metal-poor globular cluster NGC 5053, and identified five of the 16 stars as SX Phe stars. As a group the five stars have the shortest periods ($49 < P < 57$ min), lowest luminosities ($3.0 \leq M_V \leq 3.3$ mag) and smallest amplitudes ($0.08 \leq A_V \leq 0.25$ mag) of the SX Phe stars that are presently known. According to their colours and locations in both $B$ and $V$ period-luminosity diagrams, four of the five SX Phe stars (Nos. 11, 13, 14 and 15 in the Nemec & Cohen 1989 catalog) appear to be fundamental-mode pulsators, while the fifth and bluest star, No. 7, appears to be a first-overtone pulsator. One of the stars, No. 11, exhibits cycle-to-cycle light variations that are suggestive of multi-mode pulsations (unfortunately, not enough photometry is available to derive the secondary pulsation period). Estimated masses for the stars, derived both directly from the pulsation equation and indirectly by comparison with the RR Lyr stars, range from 0.53 $M_\odot$ for the candidate first-overtone star NC7, to 1.8 $M_\odot$ for NC13. While the latter is comparable with twice the main sequence turnoff mass, the very low mass for NC7 is surprising unless the derived colour is too blue (additional colour observations are needed to rule out this possibility). Because of their short periods the NGC 5053 SX Phe stars are important for establishing period-luminosity-metallicity relations. Based on the new photometry of the SX Phe stars, and contemporaneous photometry of the RR Lyrae stars (Nemec, Mateo & Schombert 1995a), the estimated slopes of the $P$-$L$-[Fe/H] relations are $-2.16\pm0.46$ ($B$) and $-2.13\pm0.46$ ($V$), with assumed metallicity coefficients $c_B$=0.35, $c_V$=0.32, and extinction coefficients $R_B$=0.41 and $R_V$=3.1 (see Nemec, Linnell Nemec & Lutz 1994). Assuming $E_{B-V}$=0.06, the estimated true distance modulus of NGC 5053 is $(m{-}M)_0$=16.06.

[1]Dedicated to the memory of Thomas E. Lutz and William H. Wehlau.

In addition to the SX Phe stars, several non-variable NGC 5053 stars were found inside the Pop. II instability strip. These may be the result of rapid rotation which prevents the stars from pulsating. Measurements of the rotation velocities of the NGC 5053 blue stragglers will be required to test this idea.

## 2. Pulsation modes of three SX Phe stars in the core of 47 Tuc

At this conference Gilliland et al. (1995a) announced the exciting discovery of three SX Phe stars in the central region of the metal-rich globular cluster 47 Tuc. These stars, with their high metallicities, should prove useful in the investigation of possible dependencies of the *P-L* relations on [Fe/H]. One of the stars is doubly-periodic with pulsation periods of 2.447 hours and 1.890 hours (see also Gilliland et al. 1995b), the ratio of which, $P_1/P_0$=0.772±0.001, is similar to that of SX Phe itself and is indicative of simultaneous pulsation in the fundamental and first-overtone modes. The other two stars have pulsation periods of 91.07 min and 80.20 min, and mean *V* magnitudes of 15.3 and 15.8 mag, respectively. The mean period of the 47 Tuc SX Phe stars is much longer than that of the NGC 5053 SX Phe stars, suggesting a possible dependence of pulsation period on [Fe/H]. The stars also extend the *P-L* relations to longer periods than previously available. Comparison of the mean *V* magnitudes and periods to the period–adjusted-magnitude relations for SX Phe stars (Nemec et al. 1995b) demonstrates that all three 47 Tuc stars lie on or near those lines, and suggests that the 80-min star is a fundamental-mode pulsator and that the 91-min star is a first-overtone star. A preliminary analysis suggests that with the inclusion of the 47 Tuc stars the slope of the *V* relation increases from $-2.13 \pm 0.46$ (Nemec et al. 1995b) to $-2.28 \pm 0.39$. On the other hand, the derived distance modulus for 47 Tuc appears to be ~0.7 mag smaller than previous estimates, indicating that either the preliminary mean *V* magnitudes are too bright or the assumed metallicity dependence is in error. Final reductions of the 47 Tuc HST data are eagerly awaited.

The author thanks his collaborators, and Michael Breger, Alistair Walker and Doug Welch for questions that helped improve this paper.

## References

Gilliland, R.L., Edmonds, P.D., Petro, L. Saha, A. & Shara, M.M. 1995a, poster paper, these proceedings.

Gilliland, R.L., Edmonds, P.D., Petro, L. Saha, A. & Shara, M.M. 1995b, ApJ, STScI Preprint No. 909.

Nemec, J.M. & Cohen, J.G. 1989, ApJ, 336, 780

Nemec, J.M., Linnell Nemec, A.F. & Lutz, T. 1994, AJ, 108, 222

Nemec, J.M., Mateo, M. & Schombert, J.M. 1995a, AJ, 109, 618

Nemec, J.M., Mateo, M., Burke, M. & Olszewski, E.W. 1995b, AJ, in press

*Astrophysical Applications of Stellar Pulsation*
*ASP Conference Series, Vol. 83, 1995*
*R. S. Stobie and P. A. Whitelock, eds.*

# Pulsating Stars and the Distance Scale - II

Michael Feast
*Department of Astronomy, University of Cape Town, Private Bag, Rondebosch 7700, South Africa. (e-mail: mwf@uctvax.uct.ac.za)*

**Abstract.** The current Cepheid zero-point is equivalent to an LMC distance modulus of $18.57 \pm 0.10$. The zero-point from corrected Baade-Wesselink data is probably not significantly different from this. A re-examination of the Baade-Wesselink data for RR Lyrae variables leads to an LMC modulus of 18.51, an age difference between $\beta$- and $\alpha$-group galactic globular clusters of $+1.46 \pm 0.70$ Gyr, and an $M_v$ - [Fe/H] slope in agreement with theory. Other questions discussed include; Avoiding bias in using the Cepheid PL relation; Metallicity spread amongst Cepheids; Cepheids and $H_o$.

## 1. Introduction

Cepheids remain central to the distance scale problem. This paper is concerned with them and with the RR Lyrae variables.

For well observed Cepheids with multicolour photometry (e.g., BVI) a PLC relation gives relative distances accurate to about 2 percent (0.04 mag). These good relative distances are satisfactory for some problems; for instance, to investigate the 3-D structure of the Magellanic Clouds (Caldwell & Coulson 1986). But for many problems we need accurate absolute distances. One such problem is the determination of the distance to the centre of the Galaxy ($R_o$) and other galactic constants from Cepheid distances and motions as in the continuing work described by Caldwell et al. (1992) or that of Pont et al. (1994) (see also Fernie 1995). Cepheids are likely to remain important for studies of galactic structure and dynamics, partly because follow-up studies of the near-infrared sky surveys, DENIS and 2MASS, should lead to the identification of significant numbers of distant Cepheids, possibly increasing the number of known variables of this type by a factor of 5 to 10 (Feast 1994a). Accurate Cepheid distances are of course needed to determine the distances of the Magellanic Clouds and more distant galaxies, with the aim of investigating the structure and dynamics of the local group, calibrating other distance indicators, and providing a basis for the estimation of $H_o$.

## 2. Cepheid PL and PLC relations

PL and PLC relations with arbitrary zero points are best set up in the Large Magellanic Cloud (LMC). PL relations at V,J,H and K have recently been derived by Laney & Stobie (1994) who combined LMC and other data. The scatter about their relation at K is small (0.17 mag) but not as small as in a PLC relation (negligible intrinsic scatter). Laney & Stobie do not include a rederivation of the PL relations at R and I and in fact the photometry of Magellanic Cloud Cepheids at these wavelengths is considerably less extensive than one would wish. Much of the existing LMC I data, for instance, was obtained with the limited objective of having a few observations per star so as to derive individual (BVI) reddenings (see, Martin et al. 1979; Caldwell & Coulson 1985). This is somewhat unfortunate since current work on extragalactic Cepheids with the HST is at V and I, and ground-based work (e.g., Pierce et al. 1994) is at R. Reasonably satisfactory PL relations at R and I are available (e.g., Madore & Freedman 1991) but this is an area which could do with further work.

It was established a long while ago from BVI photometry of LMC Cepheids (Martin et al. 1979) that the scatter in the PL relation at V was intrinsic and not due either to observational error or differential reddening as had at one time been suggested. This showed, empirically, the need for a PLC relation, a result confirmed by Laney & Stobie (1986) who demonstrated the existence of a PLC relation in the infrared. A further, useful, confirmation comes from the results of the EROS project (Beaulieu 1995, Beaulieu et al. 1995) which show rather clearly that an instability strip of finite width is necessary to interpret overtone pulsators in the LMC. It has not so far been possible to establish an independent PLC relation using Baade-Wesselink (BW) type distances to galactic Cepheids. Amongst reasons for this is the fact that the instability strip narrows towards shorter periods (see e.g., Caldwell & Coulson 1986) and the galactic BW sample tends to have somewhat shorter periods than the LMC sample on which the derivation of the PLC relation depends. However, as was recognised by Gieren et al. (1993), the main reason for their failure to find a PLC relation is the uncertainty of the individual BW distance estimates.

Determining unbiased values for the coefficients in the PLC relation is more difficult than simply establishing that a PLC relation exists. This is so because the problem is multivariant and the results depend on how one chooses to deal with the problem of observational errors. Combination of the EROS work on the colours of LMC overtone and fundamental Cepheid pulsators with the MACHO data on double-mode Cepheids may perhaps allow an estimate of the PLC coefficients free of some of the uncertainties of bias. This is an important consideration if the aim is to compare empirical coefficients with theoretical predictions. It is much less important if one is using the PLC relation for distance estimates since in a particular solution a change in one PLC coefficient is largely offset by a change in the others. Thus the difference in derived absolute magnitudes between two rather extreme solutions is less than 0.1 mag even at the edges of the instability strip (Feast & Walker 1985).

Whether one employs a PL or a PLC relation must depend on the particular application and the extent and accuracy of the available observations. The sensitivity of the various relations to a variety of effects is discussed by Feast (1991).

## 3. The Cepheid zero point

The zero-point calibration of the PL and PLC relations depends primarily on Cepheids in galactic clusters. The distances to these clusters are based on main-sequence fitting procedures calibrated using field main-sequence stars with relatively well determined trigonometrical parallaxes or adopting distance estimates for the Hyades and/or Pleiades clusters. The available evidence was reviewed recently (Feast 1993). That discussion can now be up-dated by including the following revisions and additions; A revised mean trigonometrical parallax for the Hyades (Gatewood et al. 1992) giving a distance modulus of $3.20 \pm 0.06$; A new moving cluster solution for the Hyades (Zhao & Chen 1994) giving a modulus of $3.29 \pm 0.03$; A preliminary determination of the trigonometrical parallax of the Pleiades from HIPPARCOS data (Penston 1994) giving a modulus of 5.67 (uncertainty unstated). Treating all the data as in Feast (1993) leads to an effective modulus for the Pleiades of $5.62 \pm 0.05$ and an LMC modulus of $18.57 \pm 0.10$. From a slightly different treatment of the cluster and association data, Laney & Stobie (1994) obtained a result which leads to an LMC modulus of 18.60 when referred to the new Pleiades modulus. These results are very close to the modulus derived from the SN1987A ring, $18.50 \pm \geq 0.13$ (Panagia et al. 1991; Dwek & Felten 1992; Plait et al. 1994). In the above discussion I have not included a recent determination of the Pleiades modulus based on surface brightnesses to obtain stellar angular diameters together with star-spot rotational periods and stellar rotational velocities to obtain linear diameters (O'Dell et al. 1994). This interesting new method gives a Pleiades modulus of $5.60 \pm 0.16$ and so is not yet competitive with other methods. A trigonometrical parallax of $\delta$ Cephei (Gatewood et al. 1993) which leads to a distance modulus of $7.61 \pm 0.67$ is not inconsistent with a PLC (V,B–V) value based on an LMC distance of 18.57 (=7.19). Weidemann's (1993) suggestion that the Hyades modulus is 3.35, based on white dwarf masses and surface gravities, is compatible with the value adopted here.

Obviously one hopes for improved Hyades and Pleiades moduli from HIPPARCOS. A useful main sequence may also be derived from HIPPARCOS parallaxes of nearby stars provided it can be decided which stars are on the zero-age main sequence and what their metallicities are. In addition in the future we may hope to find more galactic clusters and associations containing Cepheids. Catalogues of galactic clusters are at present quite incomplete beyond about 3 kpc. The infrared sky surveys, DENIS and 2MASS may be expected to find new OB star clusters and associations in a volume near the galactic plane which is about a factor five larger than that currently surveyed (see Feast 1994a). A few of these will, one hopes, contain Cepheids.

Several groups have put a very large effort into the determination of Baade-Wesselink type radii and distances for galactic Cepheids. Gieren et al. (1993) obtained a PL relation based on BW analyses of V,R photometry and radial velocities for 100 galactic Cepheids. When this is applied to the $V_0$ data for LMC Cepheids as tabulated by Laney & Stobie (1994) one finds an LMC modulus of 18.80 or, following the discussion of Laney & Stobie and applying a small metallicity correction, 18.83 (±0.04 internal). However a recent detailed analysis, also by Laney & Stobie (1995), establishes that their BW work based on infrared photometry is free of certain problems which affect radii and distances determined from optical photometry. A comparison of the Laney-Stobie radii obtained from K,J–K or V,V–K, for 29 stars in common with Gieren et al. and reddening for 24 in common between Gieren et al. and Laney & Stobie (1993), shows that the Laney-Stobie radii are a factor 0.94 smaller and their $E_{(B-V)}$ values 0.022 mag less. Other things being equal this will make their distance scale 0.20 mag smaller. A discussion of infrared and optical BW data for the single Cepheid U Sgr (Welch 1994) agrees, within the uncertainties, with this difference in distance scales. Thus a preliminary conclusion is that the Laney-Stobie BW scale gives about 18.6 for the distance modulus of the LMC, very close to the value based on the cluster distance scale. The definitive analysis of the Laney-Stobie data is awaited with interest (see also Laney 1995). (Note that the zero-point of the BW scale adopted by Di Benedetto (1994) is based on the Gieren et al. radii)

Various attempts have been made to apply the BW technique directly to Magellanic Cloud Cepheids. Further work in this direction is desirable. The estimates so far available do not compete in accuracy with those just discussed. In view of Laney & Stobie's discussion, the photometric part of this work will best be carried out in the infrared.

## 4. Comparison with the RR Lyrae scale

One is obviously interested in a comparison of results from Cepheid and RR Lyrae zero points. Extensive BW work on RR Lyraes shows good evidence of an $M_V$ - [Fe/H] relation. However the slope of this relation derived by Carney et al. (1992) (= 0.15) was considerably less than that predicted by Sandage (e.g., 1993) from an explanation of the Oosterhoff effect (=0.30). In addition the BW zero point, when applied to RR Lyraes in LMC clusters gave an LMC distance modulus 0.35 mag less than the Cepheid scale adopted above (see Walker 1992). Both these discrepancies have been somewhat alleviated in recent times. Skillen et al. (1993) revised the $M_V$ - [Fe/H] slope to 0.21 using new BW data and their zero point was revised by Fernley (1994) using a new limb-darkening coefficient. Fernley then finds an RR Lyrae distance modulus for the LMC of 18.43. This is less than the Cepheid modulus, though perhaps not seriously so. However both the slope and zero point problems can be made to go away entirely by the following consideration.

The $M_V$ - [Fe/H] relation generally adopted from BW data is obtained from a regression of $M_V$ on [Fe/H]. However use of this relation is subject to bias of the generalised 'Malmquist' type. For objects which are all at the same true distance this bias is avoided by using the inverse regression. (The bias correction factor for the inverse relation (equation (2) of Feast (1994b) becomes unity in this case except for the exponential part which is the (small) correction in converting mean modulus to mean distance and which is not directly relevant here). The data used by Fernley (1994) give the following two regressions:

Direct

$$M_v = 0.97 + 0.21[Fe/H] \quad (1)$$

Inverse

$$M_v = 1.13 + 0.37[Fe/H] \quad (2)$$

The second equation and Walker's (1992) LMC data give an LMC modulus of 18.51 in almost embarrassingly good agreement with the Cepheid modulus (18.57) or the SN1987A modulus (18.50). (Note that this RR Lyrae result may not be entirely free from bias since the method adopted here assumes no selection bias in [Fe/H] for the galactic calibrators and this may not in fact be the case). The slope of the relation required for a comparison with Sandage theory will lie close to midway between the two regressions if the errors in the two variables are similar as is suggested by the work of Skillen et al. (1993, Table 16). Thus theory and observation are evidently in good agreement.

There is an interesting application of the inverse relation in connection with the distances and corresponding ages of galactic globular clusters. Lee et al. (1994) have developed an age criterion based on metallicity and horizontal branch morphology. On this basis metal-poor clusters can be divided into an coeval group (mainly in the inner Galaxy, the $\beta$ group of van den Bergh 1993) and a group (mainly in the outer Galaxy, the $\alpha$ group of van den Bergh) which is about 2 Gyr younger. Using RR Lyrae absolute magnitudes from the inverse relation above, assuming bias effects will be roughly equal for the two groups, and taking other data from Sandage (1993b), one finds an age difference between ten $\beta$-clusters and six $\alpha$-clusters of $+1.46\pm0.70$ Gyr which is marginally significant and close to the theoretical value of Lee et al.

## 5. Overtones, metallicities and reddenings

In applying Cepheid PL and PLC relations to specific problems the effects of overtone pulsators and of metallicities and reddenings must be taken into account.

It has long been known that some low amplitude Cepheids in the Magellanic Clouds stand above the PL and PLC relations and that these are best interpreted as overtone pulsators. The work of Smith et al. (1992) on Cepheids in the SMC showed that at short periods the overtone pulsators were an appreciable fraction of the total number and the EROS and MACHO projects are finding large numbers of overtone pulsators, mostly at shorter periods, in the LMC. It seems likely that the overtone pulsators can be identified with good success from their light curves (see for instance Beaulieu et al. 1995).

The effects of metallicities and reddenings have been discussed previously (Feast 1991). Metallicity effects on the PL relation arise from changes in line blanketing, which are particularly important at shorter wavelengths, and from temperature shifts of the instability strip (see for instance Laney & Stobie 1994). Direct tests of the latter effect at metallicities above solar are particularly desirable. If metallicity dependent temperature shifts of the instability strip are uncertain (e.g., at high metallicities) a PLC relation (with blanketing corrected colours) may be safer to use than a PL relation.

In principle BVI photometry of a group of Cepheids at a common distance allows one to estimate their mean metallicity, reddening and distance (Feast 1988). It is however clearly desirable to have independent spectroscopic evidence for Cepheid metallicities. The data of Luck & Lambert (1985,1992) yield the results in Table 1. The final column lists the mean values for young objects in the Magellanic Clouds adopted by Feast (1989). The large scatter in the Luck and Lambert results for LMC Cepheids may be due to problems in the correct phasing of the spectra with the IR photometry used to estimate temperatures (see Feast 1995).

Table 1. Cepheid metallicities

| System | [Fe/H][a] | s.e.[b] | N[c] | [Fe/H][d] |
|---|---|---|---|---|
| The Galaxy | 0.00 | 0.03 | 15 | (0.0) |
| LMC | −0.30 | 0.16 | 5 | −0.2 |
| SMC | −0.48 | 0.04 | 5 | −0.6 |

[a][Fe/H] from Luck and Lambert (1985,1992)
[b]s.e. = Internal standard error of the mean
[c]N = The number of Cepheids observed
[d][Fe/H] for young objects from Feast (1989)

Recently Andrievsky et al. (1994) have presented spectroscopic results suggesting that galactic double-mode Cepheids have a large range in [Fe/H] (−0.43 to −0.03) and that this correlates with $P_1/P_0$. This result is particularly surprising in view of the MACHO work (Alcock et al. 1995) which shows narrow $P_1/P_0$ versus $P_o$ sequences for the Galaxy and the LMC respectively, displaced from one another in the sense expected for the lower metallicity of the LMC Cepheids. This strongly suggests, at least, that at a given period there is a very small range of metallicities in the samples from either galaxy.

## 6. Magnitude bias

Sandage (1988) was probably the first to discuss in a systematic way the bias in the use of PL relations with a magnitude limited sample of Cepheids all at the same true distance. The basic reason for this bias is the finite range in absolute magnitude at a fixed period. For reasons similar to those given in section 4, this problem can be circumvented in certain cases by using a PL relation derived from a regression of log P on magnitude which is the inverse of the procedure normally adopted. As an illustration the data at B for the galaxy NGC3109 (Sandage & Carlson 1988, and Capaccioli et al. 1992) have been analysed in this way. Table 2 shows the slopes, $\alpha$, of the inverse regression derived for this galaxy and for the LMC, using in the latter case the data listed by Sandage (1988) restricted to approximately the same range in $M_B$ as in the NGC3109 sample. The difference in slope between the LMC and NGC3109 is not significant in the case of the Capaccioli et al. data. The difference in the case of the Sandage & Carlson data suggests a possible scale error in this early photometry. Using the inverse relation, the distance modulus of NGC3109 from the data of Capaccioli et al. is 0.23 mag less than that given by the direct method.

Effects such as these will be less significant at longer wavelengths since the width of the PL relation decreases with increasing wavelength. However they should be taken into account for the highest accuracy.

Table 2. Inverse Regression Slope Test in NGC3109

| Reference | $\alpha$ |
|---|---|
| LMC | -0.310±0.036 |
| Sandage & Carlson 1988 | -0.467±0.044 |
| Capaccioli et al. 1992 | -0.270±0.072 |

## 7. Cepheids and $H_o$

The detection and photometry of three Cepheids in NGC4571, a probable member of the Virgo cluster by a group using the CFHT (Pierce et al. 1994) was a major technical achievement and shows the sort of work that is now becoming possible with ground-based telescopes. This work raises an interesting problem. We are used to expressing the intrinsic scatter in a PL relation by a Gaussian standard deviation, ($\sigma$). However the scatter about Cepheid PL relations is not Gaussian but approximates to a uniform distribution through a strip of the expected width ($\sqrt{12}\sigma$). For instance the frequency distribution of residuals from a PL($V_o$) relation in the LMC (restricted to logP $\geq$ 0.8 since the instability strip narrows at shorter periods) is non-Gaussian at the 1 percent level whilst a uniform distribution is a reasonable fit. The NGC4571 observations were made at R where the PL relation in the LMC has a width characterised by $\sigma \sim 0.25$ (Madore & Freedman 1991) so that the total width ($\sqrt{12}\sigma$) is about 0.87 mag. It is perhaps surprising in view of such a width that the three Cepheids discovered in NGC4571 lie almost exactly along a line of the same slope as the mean LMC PL relation. However given that they do, a uniform distribution of residuals about this relation implies that the observed NGC4571 PL line can lie anywhere in the PL strip over an absolute magnitude range of 0.87 with equal probability. Thus simply adopting the other assumptions that go into the derivation of $H_o$ by Pierce et al. one finds that this constant can lie anywhere in the range 71 to 106 km/s/Mpc with equal probability, and outside this range with some decreasing probability. The value quoted by Pierce et al. (87±7) is only the centre of the possible range. The alternative to the above interpretation would of course be that for some reason the Cepheid instability strip in NGC4571 was very narrow. One might then worry about the appropriate zero point of the PL relation.

We come finally to the HST work on extragalactic Cepheids. Table 3 lists the salient results from the papers known to me. These papers contain a large amount of important material. For instance it is very informative to compare the rather different methods used by the various groups to obtain their basic photometric zero point, on which of course everything else depends. Freedman et al. (1994a) give an extensive and valuable discussion of photometric and other possible sources of error. One might think that their analysis of the errors applied rather generally to HST work of this kind. Freedman et al. (1994b) obtain their best estimate of $H_o$ (= 77 ± 16 km/s/Mpc) by taking M100 as a member of the Virgo cluster and assuming a relative distance for Virgo and Coma. (A value of $H_o$ from Virgo alone is rather sensitive to the (uncertain) cosmological recession velocity of Virgo). Saha et al. (1994, 1995) use their galaxy distances to calibrate

Table 3. Distance Moduli from HST Cepheid Observations

| Reference | Galaxy | Modulus | Number |
|---|---|---|---|
| Freedman et al. (1994a) | M81 | 27.80±0.20 | 30 |
| Freedman et al. (1994b) | M100 | 31.16±0.20 | 20 |
| Saha et al. (1994) | IC4182 | 28.36±0.09[a] | 28 |
| Saha et al. (1995) | NGC5253 | 28.08±0.10[a] | 14 |
| Tanvir et al. (1995)[b] | M96 | 30.32±0.15 | 8 |

[a]Not taking into account errors in the adopted reddening
[b]provisional result

SNIa as distance indicators and deduce $H_o = 54 \pm 8$. If we take these results at their face value, there seem to be at least two possible ways to proceed. We can note that the quoted values each lie within $1\sigma$ of 62 km/s/Mpc. Or, we can allow for $2\sigma$ errors, which after all are not so improbable. In that case the two sets of data confine $H_o$ to the range 45 to 70. Whatever value of $H_o$ one adopts as a working hypothesis one should bear in mind the wise comment of Freedman et al. (1994b);

*"We do not wish to mislead the reader into believing that the problem of determining $H_o$ has been solved. It has not."*

What the recent fine work has shown is that the distance to individual galaxies as far away as Virgo can be obtained with an uncertainty of about 10 percent. The derivation of $H_o$ requires the use of secondary indicators; SNIa, the Tully-Fisher relation, the various methods used to estimate the depth and structure of Virgo and its distance relative to Coma, etc. None of these indicators have as yet the certainty which we attach to the Cepheid scale. The testing and calibration of these secondary indicators depends heavily on the determination of Cepheid distances to many more galaxies.

## 8. Addendum

In his paper at this meeting Alistair Walker mentioned a recent determination of the distance to LMC SN1987A by Gould. This depends on a rather detailed modelling of the SN ring emission. A limit to the LMC modulus ($< 18.37 \pm 0.04$) was derived. The quoted error is based on plausible variations in the adopted model. The difference from the result obtained by Panagia et al. (1991) (= 18.50) emphasises the importance of variations in the modelling procedure as was stressed by Dwek & Felten (1992). The work of Gould still leaves some aspects of the SN ring to be understood and it is therefore perhaps unwise to depend solely on the SN distance in setting up an extragalactic distance scale. The best option is probably to average Gould's upper limit (which is essentially also his best estimate) with the Cepheid and RR Lyrae results discussed above. In that case the LMC modulus would be close to 18.5.

If, alternatively, one chose to adopt the view that the difference of Gould's modulus and the Cepheid result ($> 0.20$ mag) rested entirely with an error in the

Cepheid scale then this could be due either to an error in the absolute calibration or to a difference of unknown origin between Cepheids in different galaxies (in this case the LMC and our Galaxy). This latter possibility would throw doubt on the value of Cepheids as even relative extragalactic-distance indicators. It seems unnecessary to adopt such an extreme position but clearly more work is desirable both on the SN ring modelling and on the Cepheid and RR Lyrae scales.

## 9. Acknowledgements

I am grateful to a number of colleagues for discussions and for preprints and to the University of Cape Town for a research grant.

## References

Alcock, C., et al. 1995, AJ, submitted

Andrievsky, S.M., Kovtyukh, V.V., Usenko, I.A., Klochkova, V.G., & Galazutdinov, G.A. 1994, A&AS, 108, 433

Beaulieu, J.P. 1995, this volume

Beaulieu, J.P., et al. 1995, A&A, submitted

Caldwell, J.A.R., & Coulson, I.M. 1985, MNRAS, 212, 879

Caldwell, J.A.R., & Coulson, I.M. 1986, MNRAS, 218, 223

Caldwell, J.A.R., Avruch, I.M., Metzger, M.R., Schechter, P.I., & Keane, M.J. 1992, in: Variable Stars and Galaxies, B. Warner (ed.), ASP Conf. Ser., 30, 111

Capaccioli, M., Piotto, G., & Bresolin, F. 1991, AJ, 103, 1151

Carney, B.W., Storm, J., & Jones, R.V. 1992, ApJ, 386, 663

Di Benedetto, G.P. 1994, A&A, 285, 819

Dwek, E., & Felten, J.E. 1992, ApJ, 387, 551

Feast, M.W. 1988, in: The Extragalactic Distance Scale, S. van den Bergh & C.J. Pritchet, ASP Conf. Ser., 4, 9

Feast, M.W. 1989, in: The World of Galaxies, H.G. Corwin & L. Bottinelli, Springer-Verlag, 118

Feast, M.W. 1991, in: Observational Tests of Cosmological Inflation, T. Shanks et al. (eds.), Dordrecht;Kluwer, 147

Feast, M.W. 1993, in: New Aspects of Magellanic Cloud Research, B. Baschek, G. Klare & J. Lequeux, (eds.), Springer-Verlag, 239

Feast, M.W. 1994a, in Science with Astronomical Near-Infrared Sky Surveys, N. Epchtein, A. Omont, B. Burton & P. Persi, Dordrecht;Kluwer, 169

Feast, M.W. 1994b, MNRAS, 266, 255

Feast, M.W. 1995, in Stellar Populations, P.C. van der Kruit, Dordrecht;Kluwer, in press

Feast, M.W., & Walker, A.R. 1985, ARA&A, 25, 345

Fernie, J.D. 1995, this volume

Fernley, J. 1994, A&A, 284, L16
Freedman, W.L., et al. 1994a, ApJ, 427, 628
Freedman, W.L., et al. 1994b, Nature, 371, 757
Gatewood, G., Castelaz, M., De Jonge, J.K., Persinger, T., Stein, J., & Stephenson, B. 1992, ApJ, 392, 710
Gatewood, G., de Jonge, J.K., & Stephenson, B. 1993, PASP, 105, 1101
Gieren, W.P., Barnes, T.G., & Moffett, T.J. 1989, ApJ, 342, 467
Gieren, W.P., Barnes, T.G., & Moffett, T.J. 1993, ApJ, 418, 135
Laney, C.D. 1995, this volume
Laney, C.D., & Stobie, R.S. 1986, MNRAS, 222, 449
Laney, C.D., & Stobie, R.S. 1993, MNRAS, 263, 921
Laney, C.D., & Stobie, R.S. 1994, MNRAS, 266, 441
Laney, C.D., & Stobie, R.S. 1995, MNRAS, in press
Lee, Y-W., Demarque, P., & Zinn, R. 1994, ApJ, 423, 248
Luck, R.E., & Lambert, D.L. 1985, ApJ, 298, 782
Luck, R.E., & Lambert, D.L. 1992, ApJS, 79, 303
Madore, B.F., & Freedman, W.L. 1991, PASP, 103, 933
Martin, W.L., Warren, P.R., & Feast, M.W. 1979, MNRAS, 188, 139
O'Dell, M.A., Hendry, M.A., & Collier Cameron, A. 1994, MNRAS, 268, 181
Panagia, N., Gilmozzi, R., Macchetto, F., Adorf, H-M., & Kirshner, R.P. 1991, ApJ, 380, L26
Penston, M.J. 1994, The Observatory, 114, 2
Pierce, M.J., Welch, D.L., McClure, R.D., van den Bergh, S., Racine, R., & Stetson, P.B. 1994, Nature, 371, 385
Plait, P.C., et al. 1994, Harvard-Smithsonian preprint 3831
Pont, F., Mayor, M., & Burki, G. 1994, A&A, 285, 415
Saha, A., et al. 1994, ApJ, 425, 14
Saha, A., et al. 1995, ApJ, 438, 8
Sandage, A.R. 1988, PASP, 100, 935
Sandage, A.R. 1993a, in: New Perspectives on Stellar Pulsation and Pulsating Variable Stars, J.M. Nemec & J.M. Matthews (eds.), CUP, 3
Sandage, A.R. 1993b, AJ, 106, 719
Sandage, A.R., & Carlson, G. 1988, AJ, 96, 1599
Skillen, I., Fernley, J., Stobie, R.S., & Jameson, R.F. 1993, MNRAS, 265, 301
Smith, H.A., Silbermann, N.A., Baird, S.R., & Graham, J.A. 1992, AJ, 104, 1430
Tanvir, N.R., et al. 1995, private communication
van den Bergh, S. 1993, ApJ, 411, 178
Walker, A.R. 1992, ApJ, 390, L81
Welch, D.L. 1994, AJ, 108, 1421
Weidemann, V. 1993, A&A, 275, 158
Zhao, J.L., & Chen, L. 1994, A&A, 287, 68

**Discussion**

*Stift*: (i) You stated that metallicity effects decrease with wavelength. That is at variance with basic astrophysics and I warmly recommend my poster for detailed arguments.
(ii) You use rather wide strip widths (0.8 - 0.9 mag in R). I estimated ~0.5 mag width in V from BV photometry, using a uniform distribution of Cepheids in the instability strip in my numerical simulations.
(iii) The PLC relation is notoriously sensitive to metallicity. Using it for the determination of the 3-D structure of the Magellanic Clouds without metallicity corrections could result in any depth structure being an artefact of metallicity gradients.

*Feast*: (i) Metallicity effects are more important at B than at V (see section 5 para 3 and Feast 1991)
(ii) In discussing the work of Pierce et al. (section 7) I adopted for consistency the value of $\sigma$ which they quote (from Freedman & Madore). I quite agree that the width of the strip at R is not particularly well determined. For the LMC data discussed in section 7, the strip width at V is about 0.74 mag.
c) The 3-D structure of the SMC, though seen most clearly in the available BVI(PLC) data, is confirmed by other Cepheid and non-Cepheid work (see, e.g., Laney & Stobie 1986). For discussions of the PLC and PL relations as well as reddening and metallicity determinations see Feast 1988, 1989, 1991.

*Barnes*: I think you have been unfair to the Baade-Wesselink optical results on Cepheids by comparing $(V - M_V)_{BW}$ *uncorrected* for metallicity with $(V - M_V)_{PLC}$ *corrected* for metallicity. The LMC $(V - M_V)_{BW}$ *corrected* is 18.67, negligibly different from your $(V - M_V)_{PLC}$ of 18.57. If you believe the (preliminary) HIPPARCOS Pleiades result of 5.67, your result would be 18.62. Whatever the relative merits of optical and IR BW methods (and I agree that the IR approach is superior), the optical results agree with the ZAMS method in the Galaxy and in the LMC and in the SMC (Barnes, Moffett & Gieren, 1994, ApJ Letters).

*Feast*: I have not entirely understood your comment since a metallicity correction to a PL relation at V (Galaxy to LMC) is expected to be small and to increase the adopted LMC modulus (e.g., Laney & Stobie 1994 Table 6). However, this correction is not directly relevant here, since we are in effect simply comparing PL relations derived in different ways. The (uncorrected) PL(V) modulus of the LMC based on clusters and associations is 18.50 (Laney & Stobie 1994) or 18.55 when adjusted to the cluster scale of the present paper. The uncorrected LMC modulus based on the same LMC data and the PL(V) relation from Baade-Wesselink work at V,R (Gieren et al. 1993) is distinctly larger (18.80). This difference is a direct result of the difference between the galactic PL(V) relations derived by the two different methods. It is this difference which now seems understandable in the light of the IR work of Laney & Stobie.

*Lamers*: Can you explain why it is better to use the 'inverse' relation for the PL relation, rather than the normal relation?

*Feast*: For Cepheids all at the same distance and selected only by magnitude at the relevant wavelength, the mean logP at a given magnitude is obviously independent of selection effects and the 'inverse' relation, by definition, gives a best estimate of mean logP at any given magnitude. This leads to the use made of the inverse relation in the text. On the other hand, the mean magnitude at a given logP is strongly affected by selection according to magnitude. Complications can arise in principle (e.g., varying interstellar extinction from star to star). Also if one worked at a wavelength at which the evolutionary tracks sloped steeply through the instability strip then a variation in mass function from galaxy to galaxy might have some effect on the inverse relation, but this is likely to be small at least at visual wavelengths. (The mass function effect could be important in using the direct relation.)

*Hughes*: The Malmquist bias in measuring distances to the galaxies using fits to PL relations arises from incompleteness in faint magnitudes, which gives an asymmetric distribution about the PL strip. If this is not corrected, then a regression using either V or logP will still give the wrong fit.

*Feast*: If the only problem in the case you mention is due to incompleteness or faint magnitudes, then it is correctly dealt with by using the 'inverse' relation (see sections 4 and 6, and my reply to Lamers, also Feast 1994 and Schechter AJ 85, 801, 1980).

*Turner*: Let me present a less optimistic view on the potential for finding additional Milky Way cluster Cepheid calibrators with fainter searches for Cepheids. Most Cepheids are old enough that their parent clusters are well on the way towards dynamical dissolution. The result is that most calibrating clusters in the local sample are very poorly populated. Distant clusters of the same type would, therefore, be extremely difficult to recognise.

*Feast*: The importance of the DENIS and 2MASS surveys is that the IR data may well allow clusters and associations of (reddened) early type stars to be picked out rather easily (see Feast 1994a)

*Evans*: I would just like to make a comment of the use of binary Cepheids for determining the PLC zero point. At the moment we have 9 - 10 with very low reddening. This includes several interesting stars such as a double-mode Cepheid plus two singly-excited overtone Cepheids. The external error as determined by comparing with the Feast & Walker PLC is ~0.35 mag (s.d. per star). Future work is to use IUE spectra of the Pleiades which Ron Pitts and I are working on to directly ratio with the comparison spectra, giving us a distance directly related to the HIPPARCOS results.

*Cacciari*: The $M_{bol}$ - [Fe/H] relation for RR Lyraes you quoted from Sandage (1990) has a slope 0.37 which is based entirely on the assumed mass-metallicity relation. Sandage assumed the relation from double-mode pulsators known in 1990 (~0.1 $M_{\odot}$ difference between M3 and M15) but later results using OPAL opacities show a much flatter dependence on metallicity, which would lead to a $M_{bol}$ - [Fe/H] slope around 0.25. We (Fusi Pecci et al.) have very recent HST data of globular clusters in M31, to obtain CM diagrams and the luminosity level of the horizontal branch. From this data we should be able to define independently, at least, the slope of the $M_{bol}$ - [Fe/H] relation very soon.

*Astrophysical Applications of Stellar Pulsation*
*ASP Conference Series, Vol. 83, 1995*
*R. S. Stobie and P. A. Whitelock, eds.*

# Variable Stars in the MACHO Collaboration[1] Database

Kem H. Cook[2]

*Institute of Geophysics and Planetary Physics, Lawrence Livermore National Laboratory, MS L-413, Livermore, CA 94550, U.S.A.*

C. Alcock[2]

*Lawrence Livermore National Laboratory, Livermore, CA 94550, U.S.A.*

R.A. Allsman, T.S. Axelrod, K.C. Freeman, B.A. Peterson, P.J. Quinn, A.W. Rodgers

*Mt. Stromlo and Siding Spring Observatories, Weston, ACT 2611, Australia*

D.P. Bennett[2], J. Reimann

*Center for Particle Astrophysics, Univ. of California, Berkeley, CA, U.S.A.*

K. Griest[2]

*Univ. of California, San Diego, CA 92093, U.S.A.*

S.L. Marshall[2]

*Univ. of California, Santa Barbara, CA 93106, U.S.A.*

M.R. Pratt[2], C.W. Stubbs[2]

*Univ. of Washington, Seattle, WA 98195, U.S.A.*

W. Sutherland[2]

*Univ. of Oxford, Oxford OX1 3RH, U.K.*

and D. Welch

*McMaster Univ., Hamilton, ON L8S 4M1, Canada*

**Abstract.** The MACHO Collaboration's search for baryonic dark matter via its gravitational microlensing signature has generated a massive

[1]The MACHO Collaboration is a collaboration between the Lawrence Livermore National Laboratory (DOE), Mount Stromlo and Siding Spring Observatories (ANU), and the Center for Particle Astrophysics at the University of California, Berkeley (NSF).

[2]Center for Particle Astrophysics, University of California

database of time ordered photometry of millions of stars in the LMC and the bulge of the Milky Way. The search's experimental design and capabilities are reviewed and the dark matter results are briefly noted. Preliminary analysis of the $\sim 39,000$ variable stars discovered in the LMC database is presented and examples of periodic variables are shown. A class of aperiodically variable Be star is described which is the closest background to microlensing which has been found. Plans for future work on variable stars using the MACHO data are described.

## 1. Introduction

The MACHO Collaboration is surveying fields toward the Large Magellanic Cloud (LMC), the Small Magellanic Cloud (SMC) and the bulge of the Milky Way to detect the transient brightenings of stars in those fields due to gravitational microlensing caused by intervening compact, massive objects. The primary goal of this project is to determine the contribution of baryonic dark matter in the form of massive, compact, halo objects (MACHOs) to the total mass of the Milky Way. It is straightforward to calculate that, at any given instant, about one in two million LMC stars will appear at least 34% brighter than normal due to microlensing if the dark halo is only composed of MACHOs (Paczynski 1986, Griest 1991). If the halo extends beyond the distance of the LMC, then the probability that an SMC star will be lensed is higher. There are two reasons to monitor stars toward the centre of the galaxy. First, it provides a proof-of-principle for the experiment because microlensing is expected from the known, faint-end of the stellar luminosity function; second, the probability of seeing microlensing due to MACHOs is not much less than that expected toward the LMC (Griest et al. 1991).

Microlensing occurs when a MACHO passes close to the line of sight to a more distant star. The MACHO's mass creates a gravitational lens resulting in two images of the source star. These images are separated by less than a milliarcsecond for halo object masses less than thousands of solar masses and so the phenomenon is called microlensing. The total flux in the two images is more than in the unlensed situation, and so the source star appears brighter. Because the MACHO, the solar system and the source star are all in relative motion, the brightening is transient; because of the equivalence principle the brightening is achromatic. The brightening is expected to be symmetric because the apparent motions are constant over the time frame of the event. The event duration depends upon the motion of the MACHO relative to the line of sight, the mass of the MACHO and the relative distance of the MACHO and the source (see Paczynski 1986). The amplification of the source star depends only upon how close to the line of sight the MACHO approaches. For a lens at 10 kpc and a typical, relative velocity of 200 km/s, and a source in the LMC, the time that the source appears more than 1.34 times brighter than normal is $\hat{t} \sim 140\sqrt{M/\ M_{\odot}}$ days. Thus, a credible MACHO search requires monitoring millions of stars with reasonable photometric accuracy on time scales from hours to years. It also requires the demonstration that microlensing can be clearly distinguished from all possible sources of stellar variability.

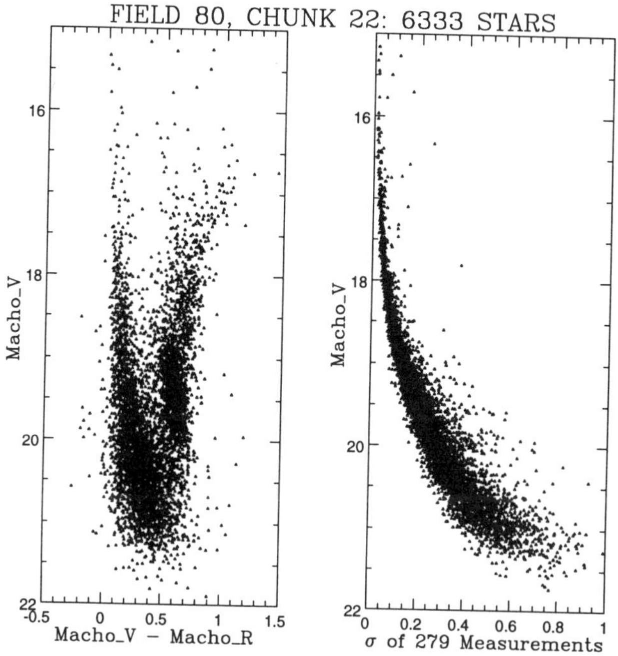


Figure 1. The left panel is a colour magnitude diagram prepared from the median of 279 measurements of the stars in a 3x3 arcmin region of the LMC bar. The right panel plots the standard deviation of the time series used to calculate the median MACHO_V magnitude.

## 2. Data Collection and Analysis

The MACHO Collaboration has the dedicated use of the refurbished, 1.27m, Great Melbourne telescope for searching for microlensing. We have equipped it with a prime focus reimager-corrector with an integral dichroic beam splitter which provides a 1 degree field in two passbands simultaneously. These are each sampled with an 2x2 array of 2048x2048 Loral CCDs which are read out simultaneously via two amplifiers per CCD in about 70 seconds (Stubbs et al. 1992). These arrays produce images which cover 0.5 square degrees with 0.63 arcsec pixels.

Exposures are 300 s for the LMC and 150 s for the bulge. The telescope and data taking are automated (Alcock et al. 1992) although there is an observer present to optimise data taking and field selection during bad weather and to protect the telescope.

The observing strategy for the data taken in 1992-1993 was to observe the LMC if possible; if not, the bulge was observed. The telescope is restricted to hour angles between about ± 9 hours for the LMC because of the south support pier for the polar axle. Significant data on the SMC has only been collected since

1994. For each primary target, a set of field centres is defined so that there would be little, if any, overlap between images of adjacent fields. Each night fields are sampled such that denser fields are sampled first and progressively less dense fields are sampled later. For the LMC, this resulted in a spiral sampling cantered on the bar. After sampling roughly the top 20 fields, the spiral was begun again. Images are taken whenever the weather allows. Images are bias subtracted, flat fielded and crosstalk corrected before photometry and archiving on 8mm tape. One of the best seeing images of each field is reduced to create a template of source positions and magnitudes. Routine reductions of each field are done on 3x3 arcmin chunks using about 20 bright, uncrowded stars from the template to determine the astrometric solution and define an analytic point spread function and a photometric zero point which are used to determine the brightnesses of the chunk's stars relative to these standards. The photometry code is a derivative of DoPhot (Schecter et al. 1994) which we have named Sodophot. Sodophot can photometer about $10^6$ stars per hour using a Sparc 2. The microlensing search only requires accurate relative photometry and as a result the zero points for each field and the transformation from the MACHO bands (4500-5900Å and 5900-7800Å) to standard astronomical bands have not yet been accurately determined although the data has now been collected. Current magnitude estimates are probably good to $\pm$ 0.15 mag. It should be noted that other collaborations are searching for MACHOs and finding many variable stars— in particular, see the paper by Beaulieu describing Cepheid results from the EROS experiment.

Figure 1 shows a colour-magnitude diagram (CMD) for one chunk of data from an LMC bar field and the standard deviation for the light curve of each star in the CMD. This figure shows how well relative photometry can be done as a function of magnitude. The top fields from two LMC seasons and one bulge season have been reduced and analysed for microlensing (Alcock et al. 1993, 1995a, 1995b, 1995c). The LMC data represent data on about 9 million stars with 140 to 350 epochs of data spanning 400 days and the bulge data represent 11 million stars with about 100 epochs of data spanning 150 days. The three microlensing events seen in this LMC data set suggest that we are detecting only about 20% of a standard dark halo in MACHOs of sub-stellar mass. The bulge results suggest that the structure of the inner portion of the Milky Way is not known because standard models predict much less lensing than is observed.

## 3. Periodic Variable Stars in the MACHO Database for the LMC

As part of the analysis of every source's photometry time series, a suite of statistics is calculated. For each light curve a robust $\chi^2$/dof (where the extreme 20% of points are not used in the fit) for a fit to constant brightness is calculated and if this is larger than a (normal) $\chi^2$/dof which can occur by chance 1% of the time, the source is considered to be variable. For sources where the psf fitting did not converge in the template frame, no analysis is done. This cut, coupled with the fact that LMC stars brighter than about V = 14 are saturated in our data, results in the exclusion of the brightest variables in the LMC from our data. It was recognised that this variable criterion selects against very long period eclipsing binaries. So, we have selected a further set of variables on the basis of their having a (non-robust) $\chi^2$/dof $> 3$ for downward excursions

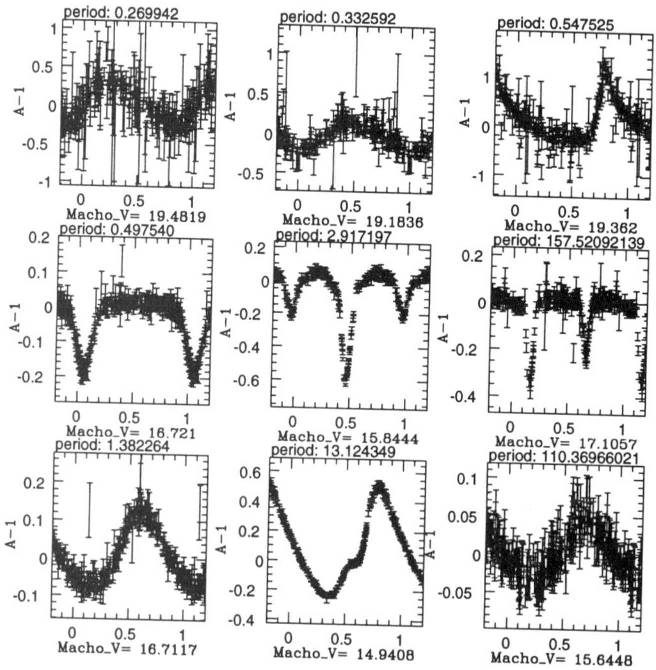


Figure 2. This figure shows phased MACHO_V light curves in flux units relative to the mean flux. The top row shows examples from the three peaks in the RR Lyrae period histogram. The second row shows a range of periods for eclipsing binaries. The third row shows an overtone and a fundamental mode Cepheid and a low amplitude Long Period Variable.

from the mean. The initial criterion yielded 38753 candidate variables out of 9 million light curves. The analysis for eclipsing binaries yielded another 10962 candidates. The candidate variable star light curves were then folded and tested for periodicity in the range of 0.1 day to the full data span which was usually about 400 days. The code used was developed by James Reimann as part of a statistics thesis at UC Berkeley (Reimann 1994) and employs the super-smoother algorithm to model the folded light curve. A measure of the goodness of fit of the model and the data is calculated. This technique basically determines whether the trial period has produced a smooth folded light curve and is fast and robust. The light curves were divided into categories on the basis of their mean magnitudes and colours and sorted on the basis of the goodness of fit parameter of the best period. Our categories include Cepheids, RR Lyrae, eclipsing binaries and red variables. This initial data set is not long enough to determine periods for many of the Long Period Variables and so we have chosen to call all variables which are redder than $V-R = 0.9$ red variables until we have longer time spans to fold.

A preliminary analysis of this data provides the following census:

- ~ 8000 RR Lyrae
- ~ 1500 Cepheids
- ~ 1200 eclipsing binaries from normal variable cuts
- ~ 1000 additional eclipsing binaries from special analysis
- ~ 19000 red variables

Figure 2 shows typical time series and folded light curves for each of these classes. It should be noted that since we observe the LMC throughout the year, and potential observing start times are dependent upon weather as well as the LMC's altitude, there is no strong aliasing and we are able to sample all phases of long and short period variables. In the following sections, I will briefly discuss each of these categories.

### 3.1. Cepheids

A clear indication that the MACHO database has much to offer in the study of variable stars came when a preliminary period-luminosity relation was plotted for periodic variables with the colours and magnitudes appropriate for Cepheid variables. This plot (Figure 3) makes clear the abundance of first overtone pulsators in the LMC—roughly 1/3 of the Cepheids are 1st overtone pulsators. More than 50% of these Cepheids are new identifications. It should be noted that much of the scatter in this plot is due to the fact that the 22 fields from which these light curves were drawn have not yet had a photometric zero point determined. A more complete discussion of work on Cepheids in the MACHO database can be found in this volume in the review by Welch et al. (1995).

### 3.2. RR Lyrae

RR Lyrae are faint enough that the typical measurement error of 10% is due to the mean seeing of 2.2 arcsec in this data set and the crowding in the bar of the LMC. Yet, periods are easily determined, and we have identified almost 8000 in 10.5 square degrees in the centre of the LMC. Examination of the candidate variables from the eclipsing binary search has revealed additional, low amplitude and high average error RR Lyrae light curves, but we estimate that these 8000 represent at least 80% of the underlying population. The central density of RR Lyrae in this sample is about 1100 per square degree. This is significantly higher than estimates made by extrapolating from outer field densities assuming an exponential disk distribution (Kinman et al. 1991). This high central density may signal the presence of a PopII halo in the LMC. An histogram of the periods for this set of RR Lyrae is shown in Figure 4. Three peaks are seen at periods of 0.58, 0.34 and 0.28 days. The first two peaks represent RRab and RRc type stars which are fundamental and first overtone oscillators respectively. The peak at 0.28 days may represent second overtone oscillation. Light curves of these stars have small amplitudes but are more asymmetric than RRc stars. We would designate these RRe stars. The RRab have a broad period distribution and represent ~ 80% of the RR Lyrae. This set RR Lyrae represents a powerful probe of the older, more metal poor population of the LMC as well as the physics of stellar pulsation.

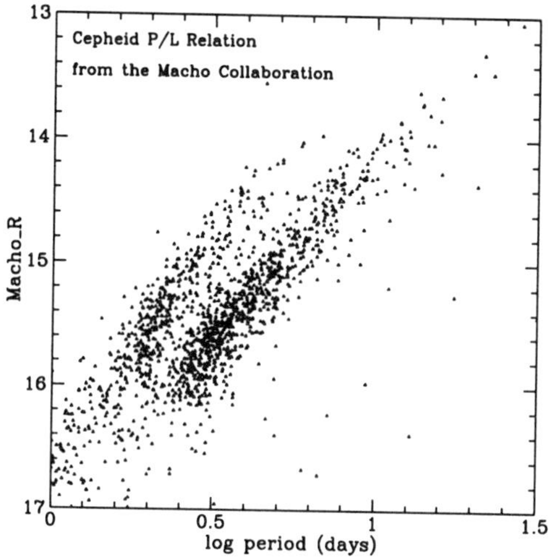


Figure 3. P/L relation for 1540 Cepheids from the first year's analysis of MACHO data.

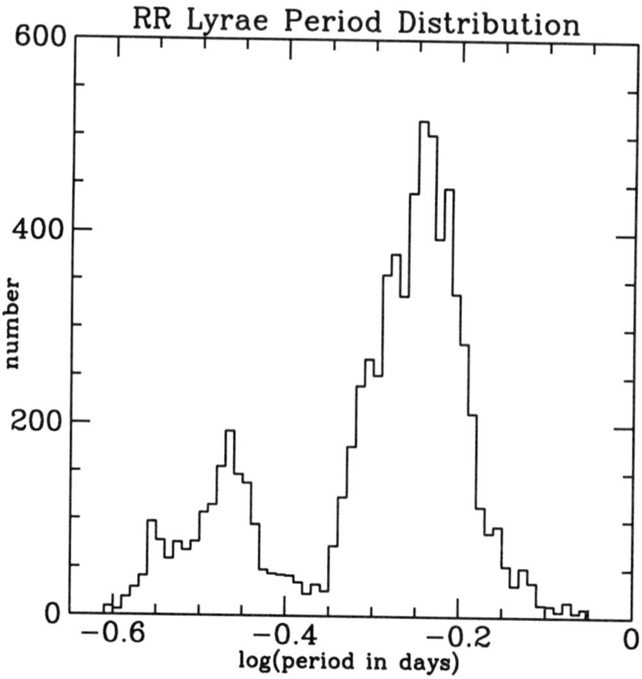


Figure 4. Histogram of periods found for 7652 RR Lyrae in the central 10.5 square degrees of the LMC.

### 3.3. Eclipsing Binaries

A great deal of stellar physics can be probed by study of the large number of eclipsing binaries found in the MACHO database. Much of this work will require that the data be placed upon a consistent photometric system so that model atmosphere calculations can be used to model observed light curves. It will also be necessary to determine our detection efficiency for eclipsing binaries with a large range of characteristics so that we can make statistically meaningful statements about the distribution of separations and eccentricities found during star formation in the LMC.

### 3.4. Red Variables

The bright, large amplitude, long period variables of the LMC have been well studied. The MACHO data on these stars provide unusual time resolution which may allow these variables to be more easily studied for the presence of chaotic behaviour. The MACHO data will also yield direct, statistically meaningful, measures of the variability of such stars as M giants and carbon stars. Low amplitude, red variables are not well studied or understood. Overtone pulsations in the extended atmospheres of late type giants are expected to produce low amplitude light variations. Wood (this volume) points out that there is an unresolved question about the pulsation mode of Mira variables which would be answered by the identification and study of a first overtone Mira sequence. This seems quite likely because many very red, low amplitude, periodic variables have been found in the MACHO data. The last panel in Figure 2 shows the light curve of one such variable.

## 4. Aperiodic and Transient Variables in the LMC Data

Transient brightenings of stars are of particular interest to our collaboration since they may represent a potential source of confusion with microlensing. Our microlensing analysis has, in fact, identified a type of variable star which has a constant luminosity for extended periods of time, but shows occasional outbursts of 10-30% which we have termed 'bumps'. These bumpers are bright, main sequence stars with $V \sim 15 - 17$. These stars most often brighten more in our red passband than our blue passband. For more than half of these stars the episodes are clearly asymmetric and shorter than about 50 days with a more rapid brightening than dimming (Figure 5, top). The rest show more symmetric, and longer episodes (Figure 5, bottom). We have obtained spectra of seven of these stars and two of them showed strong H$\alpha$and H$\beta$ emission while the rest showed evidence for filled cores in the Balmer lines. These stars are probably related to, or may be identical to, galactic Be stars. Percy and co-workers (1988, 1992) have noted possible outburst of a similar nature in galactic Be stars. Essentially all of the bumpers detected in our first year's data have shown further outbursts in our subsequent year's data. The apparent constancy of these stars between bumps argues that they are not a low amplitude pulsator.

Our data system is now reducing and analysing images as they are accumulated, resulting in alerts being generated when a source deviates significantly from its previous behaviour. This will allow us to alert the world to interesting

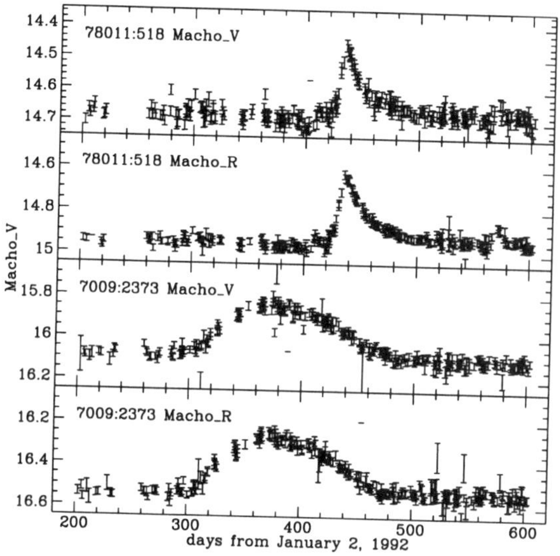


Figure 5. Light curves demonstrating the two general types of outbursts observed in the bumpers.

transient stellar phenomena as well as to the beginning of microlensing events. If this system had been operational in November of 1992, we would have alerted the world to an LMC nova about two days earlier than it was first noticed, because this was our first data point during its rise. This capability will be used to identify bumpers in outburst for further spectroscopic study.

## 5. Bulge Variables

We have identified roughly 50,000 variables in our bulge data. Their light curves have been searched for periodicity in the same manner as the LMC data. The results of this phasing have not been analysed at this time. It is possible to note that there are few, if any, bumpers in the bulge data, since these appear in the analysis for microlensing as well as variable stars.

## 6. Future Work

We are in the process of determining accurate magnitudes and transformations of our photometry to Cousins's V and R. Analysis of the bulge variables is just beginning and more LMC fields and bulge fields are currently being analysed. There are currently more than 30,000 dual-colour, 0.5 square degree images in our database, representing greater than 2 Tbytes of data. The photometry database is larger than 120 Gbytes and rapidly growing. We are currently reducing data in 'real' time allowing us to recognise microlensing events as the begin, and allowing us to detect the beginning of transient stellar phenomena as well. Perhaps the most exciting prospect is that our ability to alert the world to microlensing may allow concerted follow-up efforts to find details in microlensing curves indicating the existence of planets around low mass stars acting as lenses.

**Acknowledgments.** Work performed at LLNL is supported by the DOE under contract W7405-ENG-48. Work performed by the Center for Particle Astrophysics on the UC campuses is supported in part by the Office of Science and Technology Centers of NSF under cooperative agreement AST-8809616. Work performed at MSSSO is supported by the Bilateral Science and Technology Program of the Australian Department of Industry, Technology and Regional Development.

## References

Alcock, C. et al. 1992, in: "Robotic Telescopes in the 1990s", ASP Conf. Ser. 34, (ed.) A.V. Fillipenko, p. 193

Alcock, C. et al. 1993, Nature, 365, 621

Alcock, C. et al. 1995, AJ, 109, in press

Alcock, C. et al. 1995a, Phys.Rev.Lett, 74, 2967

Alcock, C. et al. 1995b, ApJ, in press

Alcock, C. et al. 1995c, ApJ, in preparation

Griest, K. 1991, ApJ, 366, 412

Griest, K., et al. 1991, ApJ, 372, L79

Pacyznski, B. 1986, ApJ, 304, 1

Kinman, T.D., et al. 1991, PASP, 103, 1279

Percy, J.R., et al. 1988, PASP, 100, 1555

Reimann, J. 1994, Ph.D. Thesis, Univ. Cal., Berkeley

Schechter, P.L., Mateo, M., & Saha, A. 1994, PASP, 105, 1342

Stubbs, C.W., et al. 1993, in: "Charge-coupled Devices and Solid State Optical Sensors III", (ed.) M. Blouke, Proc. SPIE, 1900, 192

## Discussion

*Sullivan*: (i) Are you aware that there is another MACHO-style collaboration in progress? It involves astronomers and physicists from NZ and Japan. The Japanese are providing a large format CCD mosaic array, and this will be mounted on a 61cm aperture telescope at Mt John in NZ. There is a poster illustrating progress. (ii) Could you expand on your success in automating the observing and data reduction? Also, a brief comment on how you improved dome seeing would be appreciated.

*Cook*: (i) Yes, I am aware of the NZ/Japan collaboration. Having run short of time, I neglected to mention other microlensing detection experiments. There are 5 other groups in addition to the MACHO collaboration of which I am aware. (ii) The telescope, data reduction and real time analysis are essentially completely automated. We do have two dedicated observers who start and stop observing in response to weather - rain in particular. We have improved our seeing by a combination of tuning the correction optics, insulating and ventilating the dome and ventilating the primary.

*Kovács*: Current evolutionary models do not support the existence of 2nd overtone RR Lyrae stars. Have you checked the position of your 2nd overtone candidates on the colour-period diagram?

*Cook*: We have waited for the final standardisation and zero pointing before looking at details of colour - but, I will check the 2nd overtone candidates colours as soon as I return to the U.S.

*Breger*: Stars with RR Lyrae star colours are periods around 0.2 - 0.3 probably are $\delta$ Scuti stars. Your observed amplitude of 0.4 mag is no problem: quite a few slowly rotating $\delta$ Scuti stars (AI Vel, etc.) have large amplitudes.

*Percy*: Your so called 'bumpers' appear to be standard Be emission episodes (amplitude $\sim$ 0.2 mag, time scale weeks) which are observed in many Be stars. We monitor these regularly at the University of Toronto. (see PASP, 100, 1555 (1988)).

*Bedding*: Can you comment on the suggestion by Sahu (1994) that the LMC lensing events are caused by normal stars in the LMC itself?

*Cook*: Our collaborations' calculations suggest that LMC/LMC lensing is less than 1/10th as likely as lensing by a Milky Way object based upon generally acceptable surface mass density, velocity dispersion and luminosity functions for the LMC.

*Butler*: Given the work of Campbell & Walker which suggests that 'Jupiters' are rare at best, will MACHO be able to detect Saturn mass objects? Will MACHO be able to place any mass limits on planets that might be detected?

*Cook*: Saturn mass objects can produce detectable deviations of lensing light curves. The important measure is the planet's orbit diameter compared to the Einstein Ring of its primary. If information about the distance and mass of the primary can be obtained then system model dependent masses of the planet can be extracted.

*Mattei*: May I ask what you are planning to do with variables that MACHO is discovering? Is it possible to obtain data on some of these variables? And how? What is your data policy?

*Cook*: The variables are being investigated by members of the team and by collaborators with our team called affiliates. At the current time, our data are not in a calibrated and uniform format so that we can freely distribute it. The MACHO collaboration does accept applications to work with us on well defined topics using our data.

*Kawaler*: How much longer will the MACHO collaboration be taking data?

*Cook*: Our collaboration has a Memorandum of Understanding for dedicated use of the 50-inch telescope at Mt Stromlo Observatory which will terminate in 1996. We are hoping to negotiate a renewal.

*Astrophysical Applications of Stellar Pulsation*
*ASP Conference Series, Vol. 83, 1995*
*R. S. Stobie and P. A. Whitelock, eds.*

# Cepheids in the Magellanic Clouds

D.L. Welch[1]

*Dept. of Physics and Astronomy, McMaster Univ., Hamilton, ON, L8S 4M1, Canada*

C. Alcock[2], D.P. Bennett[2], K.H. Cook

*Lawrence Livermore National Laboratory, Livermore, CA 94550*

R.A. Allsman, T.S. Axelrod, K.C. Freeman, B.A. Peterson, P.J. Quinn, A.W. Rodgers

*Mt. Stromlo and Siding Spring Observatories, Australian National Univ., Weston, ACT 2611, Australia*

K. Griest[2]

*Dept. of Physics, Univ. of California, San Diego, CA 92093*

S.L. Marshall

*Dept. of Physics, Univ. of California, Santa Barbara, CA 93106*

M.R. Pratt, C.W. Stubbs[2]

*Dept. of Astronomy, Univ. of Washington, Seattle, WA 98195*

and W. Sutherland (The MACHO Collaboration)

*Dept. of Physics, Univ. of Oxford, Oxford, OX1 3RH, U.K.*

**Abstract.** In the past few years, the Magellanic Clouds have been the targets for several major variable star surveys. The results of these surveys are now becoming available and it is clear that a Renaissance in LMC and SMC variable star research will result. In this review, I will describe the results of such surveys and review the questions that are likely to be answered by further work.

With respect to results, I will concentrate on LMC MACHO Project data, including beat Cepheids, discovery statistics, mode identification,

[1]MACHO Project Affiliate, The MACHO Project is a joint collaboration between the Lawrence Livermore National Laboratories (DOE), Mount Stromlo and Siding Spring Observatories (ANU), and the Center for Particle Astrophysics of the University of California (NSF)

[2]Center for Particle Astrophysics, University of California

Fourier decomposition of lightcurves, and the differences between the LMC and galactic sample.

## 1. Introduction

The Magellanic Clouds continue to play a central role in our understanding of Cepheid variables. Long duration, wide-field surveys have now begun to yield large volumes of high-quality photometric data which will replace and extend the results of photographic surveys published over thirty years ago. Unlike these past surveys, the individual photometric measurements for each star will be available to all researchers. The impact of these surveys on our understanding of pulsating stars is not likely to be surpassed for decades to come.

## 2. Recent work

Let me briefly mention work carried out since the review of Welch, Mateo & Olszewski (1993). The largest and most comprehensive published work is that of Smith et al. (1992) who surveyed 1.3 square degrees in the SMC for new variables using automated photographic photometry techniques. They discovered 78 new variables, many of them new Cepheids. Their mean B period-luminosity relation clearly shows two sequences of points which are interpreted as being due to Cepheids pulsating in the fundamental (F) and first overtone (1H) modes. They also find that 34% of the total sample are 1H pulsators. Sebo & Wood (1995) report on the detection of Cepheids in the rich, young clusters NGC 1850 and NGC 330. Storm et al. (1995) have undertaken Baade-Wesselink solutions for several SMC Cepheids. Poretti, Antonello & Mantegazza (1995) have produced the first results from a study of 21 short-period LMC Cepheids which compares the Fourier sequences in the Galaxy and the LMC in the neighbourhood of the long-period s-Cepheid sequence.

The two major intensive photometric surveys of LMC fields reported at this meeting are those of the EROS project by Beaulieu (1995) and the MACHO Project, described in more detail later in this paper, and by Cook (1995). The EROS project obtained some 15,000 images of a single 0.4 square degree field in both $B_J$ and $R_C$. They found 96 Cepheids in their field, 72 of which are new and found that 1H pulsators were 30% of their sample.

## 3. The MACHO Project

The work I would like to describe in somewhat more detail involves a portion of the MACHO Project photometry database. This survey project has been described by Alcock et al. (1992). Its principal goal is to detect microlensing events in order to estimate the mass spectrum of objects in the disk and halo. Due to the rarity of such events, it is necessary to observe large numbers of stars repeatedly, and in the process large numbers of variable stars are also detected and recorded. The 1.27m telescope at Mount Stromlo, Australia is currently dedicated full-time to this project. A camera built specifically for the project

(Stubbs et al. 1993), has a field-of-view of 0.5 square degrees and images in a 'blue' (450-630nm) and 'red' (630-760nm) bandpass simultaneously.

To date the first 5500 frames, distributed over 22 fields and 400 nights, have been analysed. At present we have identified over 1500 Cepheids in these fields. Typical datasets contain between 100 and 320 two-colour observations.

### 3.1. Beat Cepheids

The passage of time is one of the few things in extragalactic research that can be measured with high precision. Beat Cepheids are stars which pulsate in two radial modes simultaneously. Only about 14 are known in our Galaxy, with CO Aur and EW Sct (Cuypers 1985; Figer et al. 1991) added to the list of Balona (1985). The ratio of the two periods can be determined with high precision and is expected to be dependent on metallicity. Beat Cepheids are difficult to discover, in practice, because of the large number of datapoints required to recognise and characterise them. Existing surveys for Cepheid variables in other galaxies have not detected beat Cepheids because they have lacked either the total number of observations required, the photometric precision, or both.

The MACHO photometric database was searched for stars whose phased lightcurves exhibited a photometric scatter in excess of that expected from photometric errors. A total of 45 beat Cepheids were discovered and are reported in Alcock et al. (1995). The mode identification was based on analysis of CLEAN'ed power spectra and position in the P-L and reddening-free P-L-C relations. MACHO Project lightcurves for a singly-periodic LMC Cepheid and an LMC beat Cepheid are shown in Figure 1. While the photometric calibration to a standard system is not complete, a number of conclusions can be drawn:

- Thirty are F/1H pulsators, fifteen are 1H/2H pulsators.
- The ratio of the shorter period ($P_S$) to the longer period ($P_L$) is near 0.7 and 0.8 for the F/1H and 1H/2H pulsators, respectively.
- The period ratios for LMC beat Cepheids are systematically higher than galactic beat Cepheids.
- The period ratio difference is presumably due to metallicity and is in the sense predicted by theory.
- Beat Cepheids seem to be found in bands across the instability strip (a result which must be confirmed by the photometric calibration).
- About 20% of the stars with fundamental periods shorter than 2.5 days are beat Cepheids.
- Several of the beat Cepheids were found in or near young clusters which will allow their evolutionary state to be determined.
- None of the previously reported beat Cepheid candidates common to our fields were confirmed.

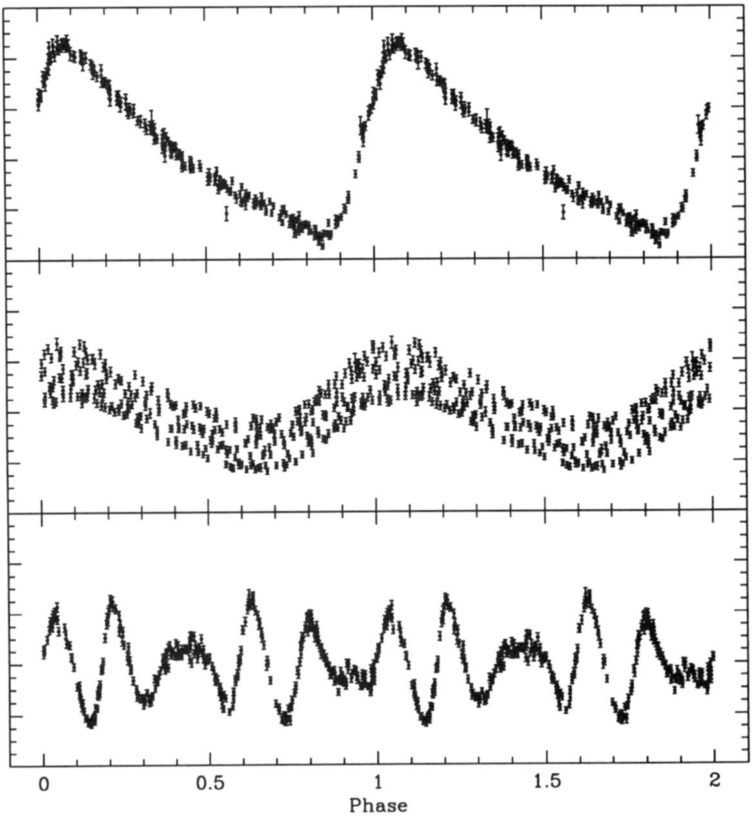


Figure 1. A $V_{\mathrm{MACHO}}$ lightcurve of one of the 1500 Large Magellanic Cloud Cepheids in the MACHO Project photometry database is shown in the top panel. This fundamental pulsator has a period of 4.1 days. The middle panel illustrates the lightcurve of a so-called 'beat' Cepheid which is pulsating with two different periods simultaneously, when it is phased with on of the two periods (4.84 days). The bottom panel shows what the lightcurve would look like if followed continuously. The pattern in the lightcurve of this 'beat' Cepheid repeats every 24.2 days. There are a total of 249 observations for the normal Cepheid and 319 for the 'beat' Cepheid. Each point is plotted twice to reveal continuity. The vertical scale for each lightcurve is 1.0 mag.

Petersen & Christensen-Dalsgaard (1995) have already begun work on the theoretical implications of the LMC beat Cepheid data and hopefully will provide predictions for the expected period ratios for such stars with SMC-like metallicities.

### 3.2. Discovery statistics

Of the 1500 Cepheids identified to date, 950 of these are F pulsators and 550 are 1H pulsators. We have cross-correlated our variable list with the catalogue of Payne-Gaposchkin (1971) and find that new discoveries comprise 57% and 94% of the F and 1H samples, respectively. The lack of 1H pulsators in existing surveys is due to their relatively small photometric amplitudes. This survey is expected to be very nearly complete for Cepheids with amplitudes greater than 0.05 mag throughout the Cepheid period range.

### 3.3. Fourier decomposition

The MACHO photometry database is a near-ideal sample for studying lightcurve behaviour. A systematic description of Cepheid lightcurve shapes in terms of Fourier series was first introduced by Simon & Lee (1981) as a means to compare observational light and radial velocity curves with the predictions of pulsation models. One drawback of the galactic Cepheid data sample is its observational inhomogeneity, despite heroic observing campaigns by numerous observers. Observations of LMC Cepheids allow a direct interpretation on the pulsation mode using the P-L and P-L-C relations - something not possible for their galactic counterparts.

We have undertaken Fourier decomposition of LMC Cepheid lightcurves. An example of the the run of the phase difference $\phi_{21}$ with $\log_{10} P$ is shown in Figure 2. In addition to allowing a comparison with theory, we are producing an algorithm by which the shape of a lightcurve of a given period and amplitude may be predicted. This will aid in the estimation of mean brightness of incompletely-sampled extragalactic Cepheid lightcurves.

### 3.4. Young clusters containing Cepheids

It is easy to get the impression that all young clusters containing Cepheids in the LMC and SMC have been surveyed. However, the work reviewed by Welch, Mateo & Olszewski (1993) covers only a small fraction of known young clusters and concentrates on those far from the bar of the LMC. We have cross-correlated positions of catalogued clusters with the existing set of 1500 Cepheid variables and find many new Cepheid-rich clusters. We plan to publish a list of these objects in the near future, with finder charts.

## 4. Future avenues of research

The mining of the MACHO Project photometric database has only begun. There are many avenues of research which we plan to explore. These include:

- We will search for evidence of singly-periodic 2H pulsators, and, if found, determine their lightcurve properties.

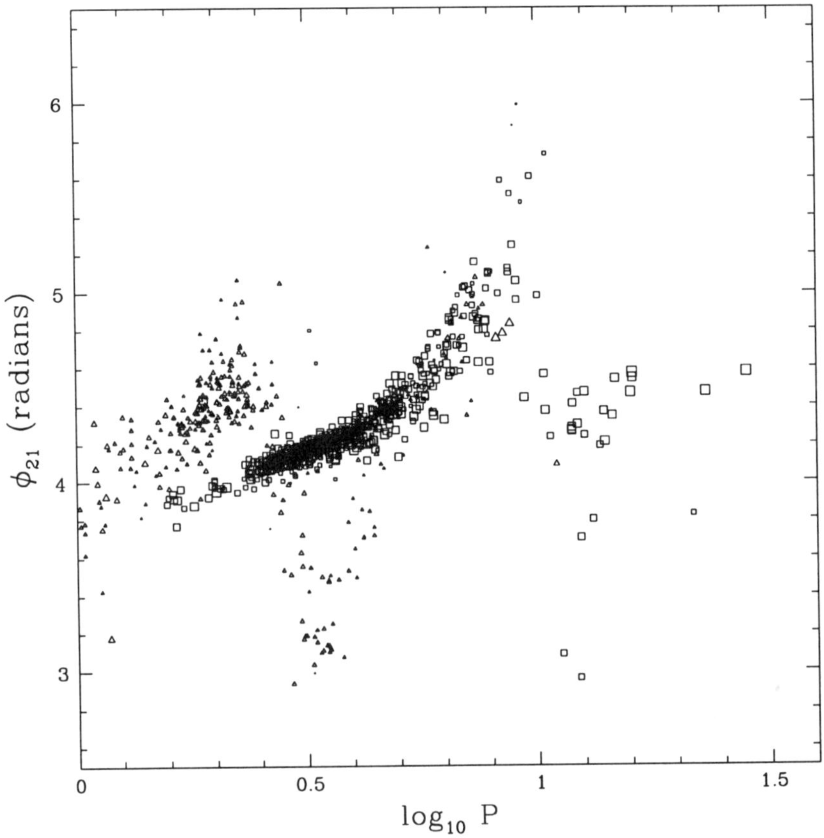


Figure 2. The Fourier parameter $\phi_{21}$ plotted as a function of $\log_{10} P$ for over 1500 Cepheids. This particular sample was selected by requiring the photometry lists to contain over 100 datapoints with fewer than 4 5-$\sigma$ outliers (which were excluded from the fit), a fitted RMS error smaller than 0.025 mag, and an error in $\phi_{21} < 0.15$ radians. The sample has not been carefully culled for contaminating objects such as foreground Type II Cepheids. The triangles and squares indicate an early (and imperfect!) classification of overtone and fundamental pulsation mode. The size of the symbols is proportional to the Fourier amplitude $R_1$. The sequence of long-period s-Cepheids is definitely present and the well-known resonance at 10 days are also clearly visible.

- When reductions for the six SMC fields observed by the MACHO Project become available, they will be searched for beat Cepheids and their period ratios will be compared with the LMC and galactic samples.
- We will make a systematic search for objects which change their lightcurve amplitude on a timescale of several years to determine whether or not objects like $\alpha$ UMi and V473 Cyg exist in the LMC.
- The remainder of the LMC MACHO fields will be searched and more recent data will be included to improve existing period ratios.
- Clusters containing beat and singly-periodic Cepheids will be studied for evolutionary information.
- A catalogue of lightcurve parameters for the current sample will be compiled and released.
- The lightcurve data for all Cepheid variables will be compiled and released.

**Acknowledgments.** Work performed at LLNL is supported by the DOE under contract W7405-ENG-48. Work performed by the Center for Particle Astrophysics on the UC campuses is supported in part by the Office of Science and Technology Centers of NSF under cooperative agreement AST-8809616. Work performed at MSSSO is supported by the Bilateral Science and Technology Program of the Australian Department of Industry, Technology and Regional Development. DLW was a Natural Sciences and Engineering Research Council of Canada (NSERC) University Research Fellow during this work.

## References

Alcock, C. et al. 1995, AJ, 109, in press

Alcock, C. et al. 1992, in: "Robotic Telescopes in the 1990s", ASP Conf. Ser. No. 34, ed. A.V. Fillipenko, p.193

Balona, L.A. 1985, in: "Cepheids: Theory and Observations", IAU Colloquium No. 82, ed. B.F. Madore (Cambridge: Cambridge University Press), p.17

Beaulieu, J.-P. 1995, this volume

Cook, K.H. 1995, this volume

Cuypers, J. 1985, A&A, 145, 283

Figer, A., Poretti, E., Sterken, C., & Walker, N. 1991, MNRAS, 249, 563

Mantegazza, L. 1983, A&A, 118, 321

Payne-Gaposchkin, C.H. 1971, "The Variable Stars of the Large Magellanic Cloud", Smithsonian Contributions to Astrophysics, 13, 1

Petersen, J.O. & Christensen-Dalsgaard, J. 1995, this volume

Poretti, E., Antonello, E., & Mantegazza, L. 1995, this volume

Sebo, K.M. & Wood, P.R. 1995, this volume

Simon, N.R. & Lee, A.S. 1981, ApJ, 248, 291

Smith, H.A., Silbermann, N.A., Baird, S.R., & Graham, J.A. 1992, AJ, 104, 1430

Storm, J et al. 1995, this volume

Stubbs, C.W. et al. 1993, in "Charge-coupled Devices and Solid State Optical Sensors III", ed. M. Blouke, Proc. of the SPIE, 1900, 192

Welch, D.L., Mateo, M., & Olszewski, E.W. 1993, in IAU Colloquium 139, New Perspectives on Stellar Pulsation and Pulsating Variable Stars, J.M. Nemec & J.M. Matthews, Cambridge: Cambridge University Press, p.359

## Discussion

*Stift*: What is the period-number distribution like? It should tell you something about metallicity and mass loss.

*Welch*: The most prominent feature in the period-number distribution seems to be due to burst of star formation about 2 $10^8$ yr ago. We will be making the catalogue of mean properties available soon.

*Sasselov*: On your period-ratio diagram - is there an overlap at P = 1.25d ? If there is, could it be used to determine a scale of mode selection, whether it is an evolutionary time scale or some other scale.?

*Welch*: Yes, there is an overlap - a small one. I haven't checked the percentage relative to the evolutionary time scale.

*Moskalik*: The 1.25d period is not very significant; it is just a coincidence. For one group (F/1H) it is an F-mode period, for another group (1H/2H) it is a 1H-mode period. You cannot compare them directly, you have to use fundamental (or overtone) periods for both groups in order to have a meaningful comparison.

*Poretti*: In our studies of galactic Cepheids we found three stars (DT Cyg, V1334 Cyg, V1726 Cyg) whose light curves are perfectly sinusoidal, i.e., the amplitude of the $2f_1$ term is not significant. Have you found similar light curves in the LMC Cepheids?

*Welch*: Yes, there are many near-sinusoidal light curves.

*Antonello*: I am happy with the results on the first overtone pulsators (Fourier parameters), because they confirm the analogous results on Galactic Cepheids obtained by our group, and definitely confirm the first overtone pulsation of these Galactic Cepheids. Now it is important to improve the models in order that they are able to reproduce satisfactorily the observed features of the Fourier diagrams.

*Matthews*: (i) Is the fraction of first overtone to fundamental pulsators relatively constant from field to field, or does it depend on position within the LMC? (ii) Have you examined the V-R colour curves for the Cepheids for temperature and phase lag information?

*Welch*: (i) I have not looked at that distribution yet. (ii) I have also not examined the phase lag or transformed the observations to the theoretical plane. As you can imagine, there are plenty of things to do with this database!

*Morgan*: Have purely 2nd overtone mode pulsators been detected, and if not, why not?

*Welch*: They *may* be in our database - there are occasional stars with unusual light curve shapes, but I have not yet checked them on a case-by-case basis.

*Kanbur*: (i) To my knowledge, no non-linear models can reproduce double-mode pulsation, so we still have a problem to explain the switch in period ratio from $P_1/P_F$ to $P_2/P_1$ from theory. (ii) Does the steep rise in the $\phi_{21} - P$ plot seen in galactic Cepheids at 10 days occur at the same period for your LMC data?

*Welch*: (i) Not my problem. (ii) Yes, it occurs at very close to 10 days and the difference between the LMC and galactic plots is not obvious.

*Christensen-Dalsgaard*: Have you looked for additional frequencies in the data, beyond the two dominant modes?

*Welch*: Yes, and there are a few that appear to show additional frequencies. The analysis of these stars would benefit from the more complete set of reductions now available (3 years).

*Goldreich*: What fraction of light curves in the Cepheid region of the H-R diagram is not interpretable? For example, could there be non-radial pulsators?

*Welch*: A very small percentage in the Cepheid region. At fainter magnitudes there are more, but I can't give you meaningful percentages from memory.

*Bono*: Did you try to transform the colours of your sample to a temperature scale?

*Welch*: No, but this may be done using published work once the MACHO data are standardised.

*Moskalik*: Calibrating period ratios in terms of metallicity is not so simple. For fixed M and L the period ratios *increase* with decreasing metallicity. But, evolution theory theory tells us that the M-L relation depends on Z. The change of luminosity predicted from evolutionary theory causes the period ratios to *decrease* with decreasing Z. So, the net result is a difference between these competing effects. Thus, calibrating Z from period ratios depends on the evolutionary calculations, not only on the envelope pulsation models.

*Astrophysical Applications of Stellar Pulsation*
*ASP Conference Series, Vol. 83, 1995*
*R. S. Stobie and P. A. Whitelock, eds.*

# HIPPARCOS Data on Pulsating Stars[1]

Catherine Turon
*Observatoire de Paris-Meudon, DASGAL / CNRS URA 335, 92195 Meudon Cedex*

Floor van Leeuwen
*Royal Greenwich Observatory, Cambridge, CB3 0EZ, UK*

**Abstract.** The preliminary astrometric and photometric results obtained from the first 30 months of HIPPARCOS data are presented. Examples of the results obtained for the various categories of pulsating variables included in the HIPPARCOS programme are given. The perspectives opened by GAIA (Global Astrometric Interferometer for Astrophysics), the concept for an astrometric interferometer submitted in September 1994 to the ESA Horizon 2000+ survey committee, are briefly exposed.

## 1. Introduction

The ESA HIPPARCOS satellite was launched by Ariane in August 1989. Despite the failure of its apogee boost motor, scientific data were obtained from November 1989 to March 1993, i.e., during 37 months. In the course of the data analysis, several intermediate catalogues were constructed by each of the two Consortia (NDAC and FAST) involved in the HIPPARCOS data reduction. The preliminary catalogues obtained from the analysis of the first 30 months of data were merged into a unique 'H30' solution, providing mean values of the positions, parallaxes and proper motions for about 107 000 stars, from a total observing programme of 118 000 stars. All statistics on astrometric results given below are based on this preliminary 'H30' solution. The photometric results are from the analysis performed by the NDAC Consortium, as the merging between the two Consortia is not yet available.

Comprehensive descriptions of the HIPPARCOS mission, the astrometric data analysis within NDAC and FAST, the photometric aspects, and the contents of the observing programme are described respectively in Perryman et al. (1992), Lindegren et al. (1992), Kovalevsky et al. (1992), Grenon et al. (1992), Turon et al. (1992). The construction of 'H30' and the properties of this intermediate HIPPARCOS catalogue are described by Kovalevsky et al. (1995)

[1]Based on observations made with the ESA HIPPARCOS satellite, and on work performed within the INCA, NDAC and FAST Consortia

and Lindegren et al. (1995); the HIPPARCOS photometric database and its application to variability in van Leeuwen et al. (1995).

## 2. HIPPARCOS preliminary astrometric results

Although the final HIPPARCOS catalogue will be more accurate and more complete than this provisional version, H30 already gives a fair appraisal of the final mission results. For stars with HIPPARCOS magnitudes in the range 8–10 mag, i.e., for the bulk of the observing programme, the mean values for the formal standard errors of the five astrometric parameters in H30 are as follows: 1.2 mas and 1.0 mas for $\alpha \cos\delta$ and $\delta$ respectively; 1.5 mas for the parallax; 1.7 mas/yr and 1.4 mas/yr for $\mu_\alpha \cos\delta$ and $\mu_\delta$ respectively (Kovalevsky et al. 1995, Lindegren et al. 1995). These errors vary with the ecliptic latitude (by a factor of about 1.5 from the ecliptic plane to high latitudes), and with the apparent magnitude (by a factor of about 3 from the brightest to the faintest stars).

## 3. HIPPARCOS preliminary photometric results

The photometric data obtained with the HIPPARCOS satellite are in many ways different from data obtained on the ground, primarily because the HIPPARCOS satellite was designed to be an astrometric satellite, and the photometric data form a by–product. Nevertheless, the photometric data obtained are of high quality, and much attention has also been paid to achieve the best possible photometric results. This section gives a brief overview of characteristics of the HIPPARCOS photometry. For more detailed information we refer to Van Leeuwen et al. (1995).

### 3.1. Satellite observing mode

The HIPPARCOS payload was based on what could be called a 'double telescope'. Two regions of the sky, separated by 58°, were observed simultaneously and projected on one focal plane, where there was situated a very precise modulating grid of 0.9 by 0.9 degrees on the sky. The satellite operated in scanning mode, such that the two sky regions seen through the telescope described a great circle on the sky. The rotational period of the satellite was $2^h40^m$, causing a star to pass across the grid in 20 seconds. During one rotation of the satellite, the rotation axis of the satellite was moved by approximately 0.4 degrees as a result of its own rotation under an angle of 43° around the direction to the Sun. This precession of the rotation axis completed 6.4 rotations per year, thus providing a complete scan of the sky in at least two different directions every 6 months.

One effect of this scanning law, as it is usually referred to, is that observations on the ecliptic poles are evenly spaced in time, observations at around ecliptic latitudes of ± 47° come often in clumps, that can last up to 2 days, while observations at lower ecliptic latitude are generally scarce but may also contain one or two clumps. The sky coverage is further complicated due to perigee passages, occultations and van Allen belts crossings, during which no observations were possible. On average, observations were obtained for between 60 and 70

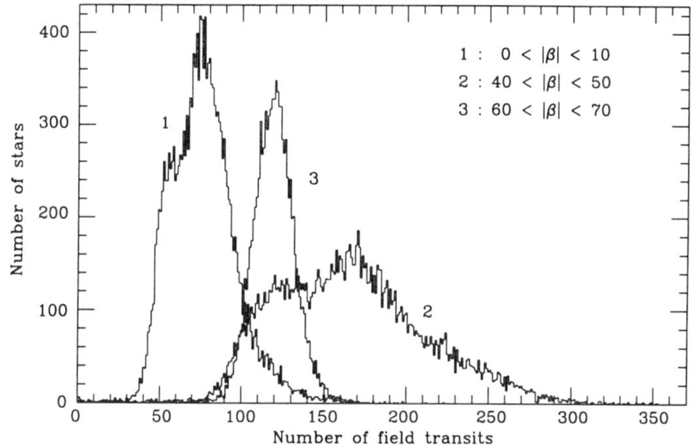


Figure 1. Histograms of the number of observations per star at different ecliptic latitudes

percent of the time. Figure 1 shows histograms of the number of observations per star at different ecliptic latitudes.

A problem unique to the HIPPARCOS photometric data is the superposition of an image from 'the other field of view'. This problem is greatly reduced due to the use of an image dissector tube (IDT from here on). The IDT allows a photomultiplier to observe only a very small region (30″) behind the modulating grid. This still gives the occasional superposition, but these are recognised afterwards and the relevant data is either corrected (for small disturbances) or removed. The IDT also introduced, however, its own complications to the data reductions. The device contains a relatively large number of small lenses allowing it to focus on a small area of the modulating grid. These lenses, when in outer space, get affected by radiation, which changes the transmission as a function of wavelength throughout the mission. This effect could only be calibrated properly when the mission was finished, and the final photometric system has been defined by a passband that represents the situation at the start of 1992. All data were reduced to that passband as far as possible. A full description of these calibrations and the determination of the passband and its evolution is given by Grenon (1995).

### 3.2. Satellite signal

The photometric data obtained with the HIPPARCOS satellite are derived from the modulated signal recorded by the IDT: we used the zero–level and the amplitude of the first harmonic of the modulation. The zero–level, called the DC component, contains the stronger signal, but for fainter stars it is influenced by a background signal that has not been measured independently. The modulation amplitude (AC component) is free of background influences but is much weaker. In the case of close binaries (separations less than 12″) the AC component is also sensitive to the orientation of the binary system with respect to the modu-

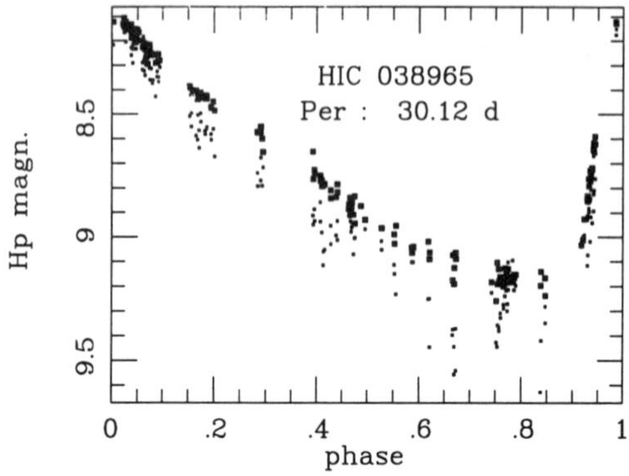


Figure 2. HIPPARCOS observations of the Cepheid AQ Pup; the differences between DC (filled squares) and AC (small dots) photometry indicate the presence of a component

lating grid. Thus, by comparing reduced DC and AC based photometric data, it becomes possible to detect duplicity (for more details see Mignard et al, 1995). As an example, Figure 2 shows the reduced HIPPARCOS observations of the Cepheid AQ Pup in both DC (filled squares) and AC (small dots).

The size of the sensitive area of the IDT also means that double stars with separations between 8″ and 20″ are often rather noisy in their DC component: one of the double star's components would be situated near the edge of the aperture where the sensitivity of the IDT reduces sharply.

The IDT samples the light behind the grid at 1200 Hz. Data are collected over a period of 2.133 seconds (2560 sampling intervals). During that time interval each star travels across 11 % of the grid. Within the time interval, between one and 10, average 4 to 5, stars are observed almost simultaneously. In the photometric reductions these so–called frame transits are used as input data, allowing the calibration of large scale sensitivity corrections across the modulating grid. In the final output, however, the reduced frame transits belonging to one crossing of the grid for a star are combined to what is called a 'Field Transit'. These field transits form the basis of the photometric data base that will become available alongside the HIPPARCOS astrometric catalogue. During the development and implementation of the photometric reduction process, many tests have been carried out to verify that no undue biases and offsets had developed in the final product. This was further enhanced by the independent reductions done by the two data reduction consortia, which allowed extensive comparisons of reduced data.

In the final HIPPARCOS astrometric catalogue, the main photometric data will be based on the DC component, generally referred to as $Hp$. The analysis has involved sophisticated modelling of the background signal. This background signal contains contributions from the diffuse galactic light, the zodiacal light, and the radiation environment (van Allen radiation belts due to the unfortunate elliptical orbit in which the satellite got trapped). Orbits during which

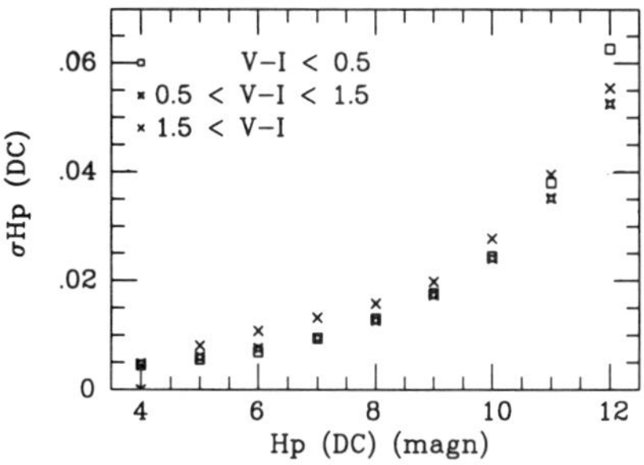


Figure 3. Standard deviations obtained for the individual photometric measurements as a function of magnitude, for three colour ranges

no convergence of the attitude was obtained have been used to calibrate and test the background model. The observed standard deviations obtained for the individual photometric measurements (field transits) are shown in Figure 3 as a function of magnitude for three colour ranges.

Additional photometric information, though of less accuracy and only for stars brighter than 10th magnitude, is provided by the Star Mapper detectors. Here, light is split into a $B_T$ and a $V_T$ channel, resembling the Johnson system (see Grossmann et al., 1995). The information obtained during the mission from the star mapper photometry has been used to improve the knowledge of colour indices, needed in the reductions of both photometric and astrometric data. Figure 4 shows $\delta$ Cephei as observed in the broad-band $Hp$ and $B_T - V_T$.

## 4. HIPPARCOS observing programme

### 4.1. Overall contents

The HIPPARCOS observing programme has been defined, iteratively, on the basis of 210 scientific proposals submitted to the European Space Agency, while making an optimal use of the observing possibilities of the satellite. It was published as the 'HIPPARCOS Input Catalogue' (HIC, Turon et al. 1992b, 1992c and 1994) including the most up–to–date and homogeneous ground-based information related to the 118 000 preselected programme stars. Half of the catalogue consists of a basic list of 'bright' stars, with an apparent limiting magnitude varying with galactic latitude and spectral type. The catalogue is virtually complete for stars brighter than $V = 7.9 + 1.1\ |sin\ b|$ for spectral types earlier than G5 (included), and for stars brighter than $V = 7.3 + 1.1\ |sin\ b|$ for spectral types later than G5. The 60 000 fainter stars have been selected from the proposals on grounds of scientific merit.

### 4.2. Pulsating variable stars

Among the proposals, 32 deal with variable stars, of which 21 more specifically deal with pulsating variable stars. The scientific objectives of these proposals relate to many areas of Stellar Physics: determination of intrinsic parameters

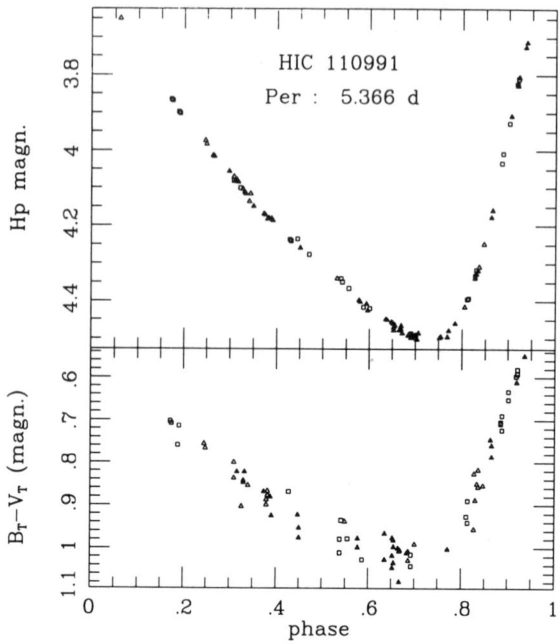


Figure 4. $\delta$ Cephei as observed in the broad-band $Hp$ and $B_T - V_T$

(absolute magnitude, mass, diameter, etc.), internal structure models (understanding of instability strip boundaries), mass transfer, age determination, relations between age, absolute magnitude and metallicity, etc.; Galactic Physics: statistical parallaxes, galactic potential and mass distribution, population and galactic evolution studies (relation of kinematics and metallicity versus age), etc.; Cosmic Distance Scale determination: cluster membership for Cepheids, zeros of period-colour-luminosity relations, kinematics of pulsating variable stars, etc.

5 830 known or suspected variable stars are included in the HIPPARCOS observing programme. Among these are 1 704 pulsating, 504 eclipsing, 223 rotating, 207 eruptive, 16 cataclysmic and 2 X-ray variables. Periods were known for 2 031 stars, and amplitudes for 2 585 stars. For 244 large amplitude stars, satellite observations made use of ephemerides, prepared in advance in order to obtain correct observing time allocation (Mennessier et al. 1992).

## 5. Examples of photometric results on pulsating variable stars

### 5.1. $\delta$ Scuti and Dwarf Cepheids

The $\delta$ Scuti and dwarf Cepheids are relatively common stars, providing a number of examples well within the HIPPARCOS horizon for parallax measurements. The HIPPARCOS distance calibrations will clarify some of the uncertainties in luminosity for stars like AI Vel and SX Phe. HIPPARCOS cannot contribute much to the discussion of the light variations of these multiple mode variables.

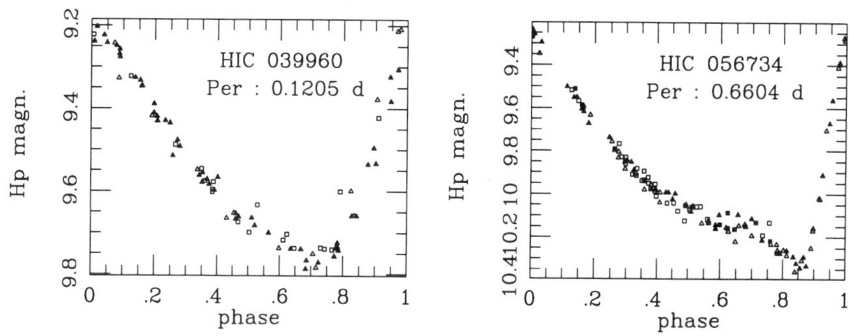


Figure 5. Light curve of the $\delta$ Scuti star SZ Lyn (left), and of the RR Lyrae star SU Dra (right)

For those $\delta$ Scuti stars with a dominant pulsation mode we are, however very well able to establish a pulsation period for the duration of the mission, and compare this with what is available in the literature. Figure 5 (left) shows the data for one of the few single period $\delta$ Scuti stars in our sample: SZ Lyn (HIC 39960).

## 5.2. RR Lyrae stars

There are only a few RR Lyrae stars close enough to obtain a marginal parallax measurement with HIPPARCOS. The HIPPARCOS photometry worked out very well for RR Lyraes. The usually single, short periods received good random coverage and a good chance to obtain accurate periods over the mission time span. In combination with data from the literature this allows a systematic study for period changes. Figure 5 (right) shows an example of HIPPARCOS measurements for the star SU Dra (HIC 56734). Among the original selection of RR Lyrae stars in the HIPPARCOS Input Catalogue there are at least 8 misidentifications (too bad a priori coordinates) and two suspected RR Lyrae stars that turned out to be more likely $\beta$ Lyr or W UMa stars (IV Pav and BX Dra).

## 5.3. $\delta$ Cephei stars

Like the RR Lyrae stars, there are only very few Cepheids with marginal distance determinations in the HIPPARCOS catalogue. These lie generally in the expected range. The photometric coverage of the short period Cepheids is generally good, but it deteriorates towards longer periods, when observations become rather clumped together and coverage is usually poor. For double mode Cepheids like TU Cas, the coverage is too scarce to derive reliable information on its various frequencies. One new Cepheid, HIC 23768, was discovered among the HIPPARCOS stars. Figure 6 (left) shows the accumulated data for this star (spectral type G5), folded with a period of 3.29483 days.

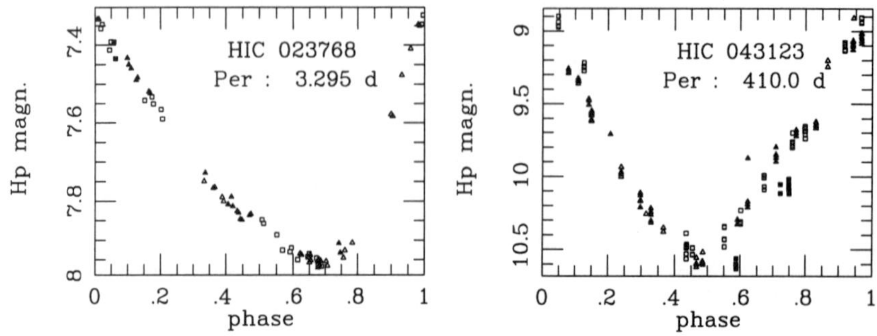


Figure 6. Light curve of a new Cepheid (left), and of a new semi regular- red giant (right)

### 5.4. Irregular K-giants and Miras

The photometric coverage of the long period semi regular and irregular variables is in general poor. In many cases the presentation of the HIPPARCOS photometry will be supported by calibrated data obtained by the AAVSO, to show the HIPPARCOS data points in a wider context. Where the HIPPARCOS data may well be useful is in the study of secondary short term variations, which are occasionally well covered (for up to 2 days) and of an accuracy much better than can be achieved with the AAVSO data. Many new semi regular and irregular K and M giants have already been found in the HIPPARCOS photometric data. Figure 6 (right) shows an example of one such discovery (HIC 43123).

## 6. Astrometric results on pulsating variable stars

The preliminary results obtained in the H30 solution for the trigonometric parallaxes of all types of pulsating variable stars are given in Table 1. The astrometric data for $\delta$ Scuti, $\beta$ Cephei, slow irregular, and semiregular variables will be especially improved. In addition, distance determinations will be obtained for two SX Phœnicis and one ZZ Ceti variables. An HR diagram drawn for $\delta$ Scuti stars using astrometric *and* photometric data from the satellite, immediately shows some misclassifications among these stars. Good accuracies will also be obtained for a few RR Lyræ and Cepheids. The periods of the Cepheids with relative precisions of the H30 parallax better than 30% range from 1.9 to 9.8 days. Final results will slightly improve these statistics.

Besides these direct improvements, indirect important improvements will come from a new calibration of photometric and spectroscopic distances of spectral types F and G, from an excellent direct distance determination of about 20 open clusters of various metallicities, from a much more reliable calibration of the HR diagram main sequence versus metal and helium contents, from a much

Table 1. Pulsating variables in H30. Relative precisions on parallax

| Type | Observed stars | Stars in H30 | Stars in clusters | (sigma $\pi$ / $\pi$) in H30 < 0.05 | 0.05 – 0.10 | 0.10 – 0.20 | 0.20 – 0.30 |
|---|---|---|---|---|---|---|---|
| $\alpha$ Cyg | 21 | 19 | 2 | 0 | 0 | 0 | 0 |
| $\beta$ Cep | 68 | 61 | 10 | 0 | 3 | 16 | 9 |
| Cep | 30 | 25 | 0 | 0 | 1 | 2 | 0 |
| CW | 30 | 25 | 0 | 0 | 0 | 0 | 0 |
| $\delta$ Cep | 250 | 235 | 6 | 0 | 1 | 2 | 2 |
| $\delta$ Sct | 140 | 126 | 14 | 15 | 43 | 30 | 15 |
| L | 290 | 278 | 7 | 2 | 10 | 36 | 54 |
| M | 239 | 178 | 0 | 0 | 0 | 4 | 5 |
| PV Tel | 3 | 3 | 0 | 0 | 0 | 0 | 3 |
| RR | 180 | 151 | 0 | 0 | 0 | 0 | 3 |
| RV | 11 | 10 | 0 | 0 | 0 | 0 | 0 |
| SR | 435 | 399 | 8 | 0 | 7 | 29 | 51 |
| SXP | 6 | 6 | 0 | 0 | 1 | 1 | 0 |
| ZZ | 1 | 1 | 0 | 1 | 0 | 0 | 0 |

more convincing definition of the variable star's cluster membership, and from the discovery of many new periodic variable stars.

## 7. Perspectives opened by GAIA

One of the likely cornerstone missions planned by ESA for the period 2010 to 2020 is a follow up for HIPPARCOS, capable of measuring parallaxes to accuracies of 0.02 mas down to 15th magnitude, and 3 to 5 times more accurate for brighter stars. The mission could do a complete survey of the sky down to that magnitude, and would measure absolute distances to several globular clusters (47 Tuc, $\omega$ Cen, M3, M5, M15, etc.), tens of open clusters and all observable galactic pulsating stars. A possible concept for such a mission, GAIA (Lindegren & Perryman, 1994), includes 6 channel (intermediate bandwidth) photometry (accuracies down to 12th magnitude better than 1 milli-mag, and a few hundredth of a magnitude at 16th mag.) and an additional survey with reduced (1 mas accuracy) positional accuracy for all objects down to 19th or 20th magnitude. GAIA is supported by the extensive experience obtained with the HIPPARCOS mission.

The proposed mission length is 5 years (minimum), which provides an average of 200 to 300 observations per star for each of the 50 million target stars. It will, for example, be able to detect any RR Lyrae star up to a distance of about 7 kpc, (expect a few thousand) and assign a meaningful distance to all; and to provide a good distance determination of the brightest Cepheids in the Magellanic Clouds. Using the multi–colour photometry, reddening corrections can be determined as well as surface gravity values, effective temperatures and metallicities.

This mission would be capable of establishing the influence of metallicity on absolute magnitudes, the relation to age of various forms of pulsation and in general finally provide the solid foundation that is needed for the developments in astronomy and astrophysics that can be expected for the next century.

## 8. Conclusion

The contribution of HIPPARCOS to pulsating variable star studies is multifold: homogeneous photometric observations; the expected discovery of many new periodic variables (already found up to now, one Cepheid, twenty eclipsing binaries, many semi-regular and irregular red giants), and thousands of microvariables; direct high accuracy distance determination for a small number of stars of almost each type of variable; much more reliable calibrations of photometric and spectroscopic distances, of magnitude versus metallicity and helium contents relations can also be expected.

**Acknowledgments.** We are grateful to D. Morin, M.J. Penston, D.W. Evans, F. Mignard and M. Frœschlé for their help during the preparation of this paper.

## References

Grenon, M., Mermilliod, M., & Mermilliod, J.C. 1992, A&A, 258, 88

Grenon, M., et al. 1995, A&A,

Grossmann, et al. 1995, A&A,

Kovalevsky, J., Falin, J.L., Pieplu, J.L., Bernacca, P.L., Donati, F., Frœschlé, M., Galligani, I., Mignard, F., Morando, B., Perryman, M.A.C., Schrijver, H., van Daalen, D.T., van der Marel, H., Villenave, M., Walter, H.G., et al. 1992, A&A, 258, 7

Kovalevsky, J., Lindegren, L., Frœschlé, M., van Leeuwen, F., Perryman, M.A.C., Falin, J.L., Mignard, F., Penston, M.J., Petersen, C., Donati, F., Hering, R., Høg, E., van der Marel, H., Schrijver, H., & Walter, H.G. 1995, A&A,

Lindegren, L., Høg, E., van Leeuwen, F., Murray, C.A., Evans, D.W., Penston, M.J., Perryman, M.A.C., Petersen, C., Ramamani, N., Snijders, M.A.J., Söderhjelm, S., Andreasen, G.K., Cruise, A.M., Elton, N., Lund, N., & Poder, K. 1992, A&A, 258, 18

Lindegren, L. & Perryman, M.A.C., September 1994

Lindegren, L., Röser, S., Schrijver, H., Perryman, M.A.C., van Leeuwen, F., Mignard, F., Falin, J.L., Lenhardt, H., Kovalevsky, J., Frœschlé, M., Bernacca, P.L., & Lattanzi, M. 1995, A&A,

Mennessier, M.O., Barthès, D., Boughaleb, H., Fighueras, F., & Mattei, J.A. 1992, A&A, 258, 99

Mignard, F. et al. 1995, A&A,

Perryman, M.A.C., Høg, E., Kovalevsky, J., Lindegren, L., Turon, C., Bernacca, P.L., Crézé, M., Donati, F., Grenon, M., Grewing, M., van Leeuwen, F., van der Marel, H., Murray, C.A., Le Poole, R., & Schrijver, H. 1992, A&A, 258, 1

Turon, C., Gómez, A., Crifo, F., Crézé, M., Perryman, M.A.C., Morin, D., Arenou, F., Nicolet, B., Chareton, M., & Egret, D. 1992a, A&A, 258, 74

Turon, C., et al. 1992b, ESA SP-1136

Turon, C., et al. 1992c, Bull. Inf. CDS, 41, 9

Turon, C., Morin, D., Arenou, F., Perryman, M.A.C., & Priou, D., 1994, CD-ROM version of the HIPPARCOS Input Catalogue, Turon C. ed.

van Leeuwen, F., et al. 1995, in preparation

## Discussion

*Nemec*: I do not see SX Phe in your $M_V$, B-V diagram at $M_V = +3.04$, (B-V) = 0.24 (Nemec, Nemec & Lutz 1994 AJ 108,222). [Laughter] Is your diagram censored or are our models so much in error?

*Turon*: The diagram is not censored! Answer to your question about your models in 1996!

*Bradley*: You do have a ZZ Ceti star in your list? I assume the star is G226-29, the closest and brightest one. I predict the parallax of G226-29 is 95.5 mas.

*Turon*: Yes, there is a ZZ Ceti star in the HIPPARCOS observing programme. Answer to your prediction in 1996!

*Astrophysical Applications of Stellar Pulsation*
*ASP Conference Series, Vol. 83, 1995*
*R. S. Stobie and P. A. Whitelock, eds.*

# Cepheid Binaries: Frequency and Mass Ratios [1]

Nancy Remage Evans

*Department of Physics and Astronomy, York University, 4700 Keele St., North York, Ontario, M3J 1P3, Canada*

**Abstract.** Ultraviolet spectroscopy with the IUE satellite has been used to survey all Cepheids brighter than $8^{th}$ magnitude to identify blue companions. The derived binary frequency is 21% with blue companions which implies that 34% would have been found from a large radial velocity survey. A second study has used deep IUE exposures of Cepheids showing orbital motion to determine the distribution of mass ratios in Cepheid binaries. Low mass companions are strongly preferred, especially when corrections for incompleteness are included.

## 1. Introduction

Accurate identification of Cepheids with blue companions is difficult from the ground for all but the brightest companions. This has hindered the comparison of the Cepheid binary frequency with that of B stars, and also the investigation of any effects a companion might have on pulsation (amplitude, period variation). It has also proved extremely difficult to extract information about the distribution of mass ratios for binary systems from, for instance, even very large samples of single-lined spectroscopic binaries. Since many cool Cepheids have hot companions, the use of ultraviolet spectroscopy allows us to make a detailed study of the companions. Using this approach, we obtain fresh insight into both these problems.

## 2. Binary frequency

In order to confidently identify blue companions, we have used the International Ultraviolet Explorer (IUE) satellite to observe all Cepheids brighter than 8.0 magnitude from 2000 to 3200 Å (Evans, 1992)

We find that 21% of the 76 Cepheids have blue companions. When stars known to have companions from other techniques such as orbital motion are included, the percentage rises to 29%. Tables have been provided to identify Cepheids with and without blue companions. If we make a statistical correction to derive the percentage of stars which would be found to be binary if extensive radial velocities were available, the percentage becomes 34%. This is actually larger than the fraction of B star binaries (corrected to include only stars with

[1] IUE Guest Observer

orbital periods longer than a year), but we attribute this to more complete detection of Cepheid systems.

## 3. Mass ratios

We have similarly obtained a deep IUE exposure from 1200 to 2000 Å for Cepheids with orbital motion (Evans 1995). A mass can be inferred from the spectral types from these spectra. This mass can be combined with a mass for the Cepheid from a mass-luminosity relation to provide the distribution of mass ratios for 20 binary systems. Low mass companions are strongly preferred. Using the distributions of orbital period, velocity amplitude, and eccentricity, we have derived corrections for incompleteness in the sample. These corrections increase the frequency of low mass companions. We stress that we have been particularly careful that all the stars in the sample clearly have orbital motion, so that the preference for low mass companions cannot be attributed to single stars inadvertently included.

The distribution of mass ratios is important observational information for comparison with star formation calculations. In the case of the Cepheid sample, the results refer to intermediate mass primaries with binary periods longer than a year. This is, of course, a spectroscopic rather than a dynamic determination of mass ratios, however, the technique is powerful and mass ratios $M_2$ / $M_1$ as small as 0.26 have been found.

**Acknowledgments.** Financial support was provided by a Natural Sciences and Engineering Research Council (Canada) grant to NRE. Computing facilities were furnished by Dr. J. J. Caldwell at York University.

## References

Evans, N. R. 1992, ApJ, 384, 220

Evans, N. R. 1995, ApJ, May, in press

*Astrophysical Applications of Stellar Pulsation*
*ASP Conference Series, Vol. 83, 1995*
*R. S. Stobie and P. A. Whitelock, eds.*

# Cepheid Radii and Distances

C. D. Laney and R. S. Stobie

*South African Astronomical Observatory, PO Box 9, Observatory, 7935, Western Cape, South Africa*

Laney & Stobie (1995) (LS95) have determined Baade-Wesselink radii for 49 galactic Cepheids using BVRI and JHK photometry. Four of the possible magnitude-colour combinations were studied in detail – namely, (V,B–V), (V,V–$I_C$), (K,V–K) and (K,J–K). In that paper and in the present note we have used the maximum likelihood method of Balona (1977) as recently modified to allow for the nonlinear effects of the non-negligible radius excursions of Cepheids.

One major advantage of infrared photometry is that the light variation at K is dominated by the change in surface area, unlike the light variation at V which is dominated by temperature changes. Another is that (as shown both empirically and by model atmospheres) one may use a J–K or V–K colour index to calculate the surface brightness at K without having to worry very much about the effects of varying surface gravity or Doppler broadening velocity, while the same is definitely not true when optical colour indices are used to calculate the surface brightness at V.

From radius displacement curves it is clear that the systematic and random errors are least in the case of the infrared solutions, (K,V–K) and (K,J–K). (V,B–V) and (V,V–$I_C$) solutions have large systematic and random errors. Furthermore, "optical" radius determinations appear quite sensitive to whether the phase of the ascending branch is excluded from the solution. "Infrared" solutions show no such effect.

The most accurate period-radius relation was derived by averaging the (K,V–K) and (K,J–K) radius solutions. (V,B–V) and (V,V–I) solutions gave systematically incorrect radius determinations because of the effects of surface gravity and microturbulence variations throughout the pulsation cycle. (V,V–$R_J$) radius solutions (calculated for about half of our sample) showed systematic errors essentially identical to those seen in (V,V–$I_C$) solutions.

Given the availability of radial velocities contemporaneous with our JHK data for LMC and SMC Cepheids, we have calculated "infrared" radii for as many Cepheids as possible, despite the limited quality and limited precision of the data compared to our galactic sample. Contemporaneous radial velocities were available in the literature. Radius solutions were considered successful only if the surface brightness coefficients were within 0.6 of the expected value at a given period. No solutions for Cepheids with periods longer than 100 days were successful. For 5 LMC and 4 SMC Cepheids (HV 12815, 879, 2338, 2827, 5497, 837, 824, 11157, and 829), radii could be determined, and the adopted radii are plotted in Fig. 1, together with our "adopted" galactic Cepheid radii. The LMC and SMC Cepheid radii show no systematic difference from the galactic Cepheid

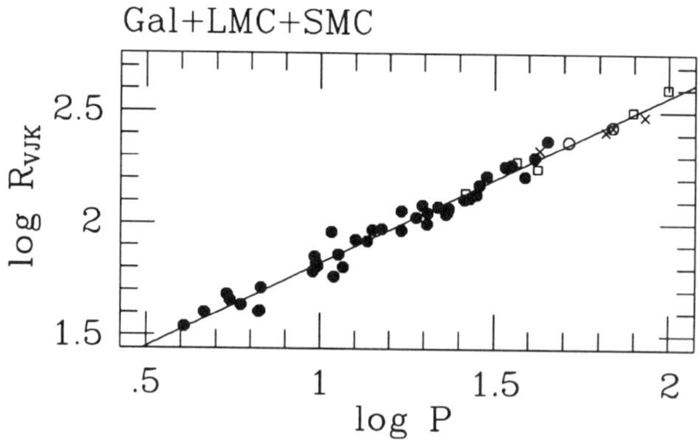


Figure 1. $\log R - \log P$ relation for Cepheids in the Galaxy (circles), LMC (squares) and SMC (crosses).

radii and a combined Galaxy/LMC/SMC period-radius relation

$$\log R = 1.821 + 0.747(\log P - 1) + 0.005(LMC) - 0.009(SMC) \quad \sigma = 0.043$$
$$\pm 0.008 \pm 0.023 \qquad \pm 0.024 \qquad \pm 0.027$$

is the same as within the errors as the (galactic) relation given in LS95, and suggests that (again within the errors) the galactic and MC *bolometric* PL relations are not significantly different.

Absolute magnitudes for 37 Cepheids (V) and 40 Cepheids (JHK) have been calculated using the tables for surface brightness given by Hindsley & Bell (1989, 1990). Combining these with the dereddened mean magnitudes for 45 LMC and 47 SMC Cepheids with VJHK photometry enables period-luminosity relations in each colour to be derived with effective "distance moduli" for the LMC and SMC. Corrected for the effects of mean difference in temperature at a given period and for the differences in blanketing between galactic and Cloud Cepheids (as in Laney & Stobie 1994), we find that the corrected VJHK moduli are in agreement, with means of 18.58±0.04 for the LMC and 19.00±0.04 for the SMC. These mean values are in remarkably good agreement with the moduli of 18.53 and 18.94 we found using the galactic Cepheids in clusters and associations together with the same sample of Cloud Cepheids (Pleiades modulus 5.57).

## References

Balona, L.A. 1977, MNRAS, 178, 231

Hindsley, R.B., & Bell, R.A. 1989, ApJ, 341, 1004

Hindsley, R.B., & Bell, R.A. 1990, ApJ, 348, 673

Laney, C.D., & Stobie, R.S. 1994, MNRAS, 266, 441

Laney, C.D., & Stobie, R.S. 1995, MNRAS, in press

*Astrophysical Applications of Stellar Pulsation*
*ASP Conference Series, Vol. 83, 1995*
*R. S. Stobie and P. A. Whitelock, eds.*

# Common Misconceptions about the Galactic Cluster Calibration of the Cepheid PLC Relation

David G. Turner
*Saint Mary's University, Halifax, Nova Scotia, B3H 3C3, Canada*

Comments in the literature in recent years reveal a variety of misconceptions about the role and value of galactic clusters for calibrating the Cepheid period-luminosity-colour (PLC) relation. This paper attempts to clarify some of these points using personal experiences gained from studies of the sparse galactic clusters and groups in which some classical Cepheids are located.

## 1. Interstellar reddening

Reddening is often blamed for problems originating from other causes. Three features of reddening are sometimes overlooked: (i) no universal reddening law applies to all fields, (ii) mean reddening laws are not applicable to samples of dispersed objects, and (iii) patchy extinction is the norm.

The effect of ignoring such points can be seen in the precision of older estimates of distance modulus for the Cepheid calibrating clusters. For clusters reddened by $E_{B-V} \approx 0.5$, a spread in reddening slopes $E_{U-B}/E_{B-V}$ for cluster B-stars from 0.65 to 0.85 (typical of the galactic plane) results in a dispersion of $\pm 0.19$ in distance modulus if they are dereddened using a common reddening law. Typical uncertainties in zero-age main-sequence (ZAMS) fits under such conditions are probably no better than $\pm 0.15$, while the colour spread of individual Cepheids in the instability strip accounts for an additional scatter of $\pm 0.08$ (or more). The expected scatter in the Cepheid period-luminosity (PL) relation amounts to $\pm 0.25$, typical of what is observed (Gieren & Fouqué 1993). Similar techniques used with Strömgren photometry produce no better results. Adoption of individual reddening corrections for cluster stars and use of reddening laws specific for each field can reduce the dispersion in cluster distance moduli to $\pm 0.05$ or less. With dispersion in the instability strip included, the scatter in the Cepheid PL relation should be no larger than about $\pm 0.09$. Gieren & Fouqué (1993) expressed surprise that the dispersion for cluster calibrators from the author's program should be so small, but it is exactly the result expected from studying clusters in a detailed, homogeneous fashion. Recent results from the program have also led to an estimate of $2.1 \pm 0.2$ for the colour term of the PLC relation from galactic calibrators, consistent with values obtained for Cepheids in Magellanic Cloud clusters. Metallicity spread in the calibrating clusters is apparently not an obvious problem — yet.

Variations in the reddening law also affect the delineation of the Cepheid instability strip. For example, Fernie's (1990a) photometric reddenings for bright Cepheids are a poor match to space reddenings obtained for the same variables; photometric colour excesses for 13 Cepheids with reliable space reddenings differ in the mean (space – photometric) by $\Delta E_{B-V} = -0.01 \pm 0.06$ s.d. The scatter is large, being largest for the longest period Cepheids. The wedge-shaped instability strip delineated by Fernie (1990b) using photometric reddenings contrasts with a tighter strip defined by colour excesses tied to space reddenings and reddening-independent indices (Turner 1995).

Incidentally, luminosities derived for cluster Cepheids do not depend upon $R$, the ratio of total to selective extinction. Cluster distances do, but luminosities for member Cepheids are tied only to luminosities inferred for similarly-reddened cluster ZAMS members, which are independent of $R$. The scatter in ZAMS-fitted absolute magnitudes varies from 5× to 13× the scatter in $(B-V)_o$ for main-sequence B and A stars, so uncertainties of ±0.01 in $(B-V)_o$ for careful cluster studies translate into uncertainties of ±0.05 to ±0.13 in $M_V$ for individual ZAMS stars; samples of 15 or more ZAMS stars reduce the scatter in mean cluster distance moduli to less than ±0.03 (~ 1% in distance). Statements that cluster distance moduli cannot be determined to better than ±0.2 (9% in distance) are therefore not realistic appraisals of the current state of the art.

## 2. Association Cepheids

While a number of long period Cepheids are coincident with OB associations, main-sequence companions of 20-day Cepheids must have spectral types later than B2; such Cepheids cannot be physically related to the youngest association subgroups. Membership in older subgroups of associations is possible, but such subgroups are more spatially dispersed and more challenging to study than open clusters. Yet positive results have come from such studies, including the detection of a previously unrecognised cluster containing the long period Cepheid WZ Sgr. For distances in excess of 2 kpc the dispersion in distance modulus expected from line-of-sight distance spread in associations is less than ±0.06 (Turner 1988), so association Cepheids can be reliable calibrators.

## 3. The ZAMS calibration

The standard ZAMS is tied to the distance of the Hyades cluster. The shape at the faint end is defined by the main sequence for Praesepe, the zero-point is established by the Hyades, and a normalisation to solar metallicity is made using Pleiades stars. Van Leeuwen's (1983) determination of the Pleiades distance using spectroscopic parallaxes is an example of various techniques also used to derive the Hyades distance, but cannot replace the Hyades calibration. Spectroscopic parallaxes use trigonometric parallaxes for nearby stars, and are just as susceptible to systematic error as more direct techniques. The zero-point uncertainty is still a problem to be addressed.

## References

Fernie, J. D. 1990a, ApJS, 70, 153

Fernie, J. D. 1990b, ApJ, 354, 295

Gieren, W. P., & Fouqué, P. 1993, IAU Coll., 139, 72

Turner, D. G. 1988, ASP Conf. Ser., 4, 178

Turner, D. G. 1995, in preparation

van Leeuwen, F. 1983, Ph.D. Thesis, Leiden

*Astrophysical Applications of Stellar Pulsation*
*ASP Conference Series, Vol. 83, 1995*
*R. S. Stobie and P. A. Whitelock, eds.*

# The $H_0$ Key Project: Cepheids in NGC925, M101 and M100

S. M. G. Hughes

*Royal Greenwich Observatory, Madingley Road, Cambridge, UK*

**Abstract.** As part of the Extragalactic Distance Scale Key Project, the *Hubble Space Telescope* has been used to identify Cepheids in M100, M101 and NGC925, and to measure distances derived from the Cepheid PL relation. For M100, the distance of 17.1 ± 1.8 Mpc has been used to infer a preliminary value for $H_0$ of $\sim$ 80 km/s/Mpc, which brings the age of the Universe derived from the standard model of the Big Bang into conflict with the ages of the oldest stars.

## 1. Introduction

To date the quest for $H_0$ has been plagued by systematic errors in at least some of the various methods for measuring distances to galaxies beyond Virgo (and hence to galaxies whose Hubble flow velocities are significantly greater than their peculiar velocities). The aim of the Key Project is to provide a sample of 18 galaxies with well-determined distances derived from Cepheids detected and measured with the *Hubble Space Telescope's* WFPC2, which can then be used to calibrate the secondary distance indicators: the IR Tully-Fisher relation (IRTF), Surface Brightness fluctuations, the luminosity functions of Planetary Nebulae and Globular Clusters, the Expanding Photospheres of Type II SNe (EPM), and Type Ia SNe. The ultimate goal of the Key Project is use these calibrations to measure $H_0$ to an (external) accuracy of 10%.

## 2. Cepheids in NGC 925

NGC925 is a member of the NGC1023 group, which contains galaxies suitable for PNLF, SBF and SNII calibrations. While analysis is still in progress, of 5540 stars, 80 seem to be Cepheids (Silberman et al. 1995).

## 3. Cepheids in M101

M101 was selected as a test for any metallicity effects in the Cepheid PL relation, as M101 has a very steep metallicity gradient. Two fields have been targeted in M101, at radii of 1.7 and 7.9 arcmin, with a metallicity difference of a factor $\sim$ 5 between them. The outer field was observed mainly with WFPC1, and these have now been reduced (Kelson et al. 1995), with 29 Cepheids being found from a total of more than 23,000 stars. Although the calibration of the photometry is still preliminary (based on a transformation from WFPC2 at $-88^o$ to WFPC2 at $-76^o$), the current figures indicate a large extinction of $A_V = 0.45$ mag, and a reddening-free distance of $6.5 \pm 0.6$ Mpc. This compares to ground-based distances of 7.2 Mpc by Cook et al. (1986) derived from 2 Cepheids, and 7.6 Mpc by Cohen (1993) derived from a further 2 Cepheids. SN1970G gives an EPM distance of $7.4^{+1.0}_{-1.5}$ Mpc (Schmidt et al. 1994). The inner field is currently being calibrated (Stetson 1995), and will be very important to see if/how the Cepheid PL relation is affected by metallicity.

## 4. Cepheids in M100

The *HST* results for M100 were reported by Freedman et al. (1994), who detected more than 20 Cepheids out of 40,000 stars, and derived an extinction to M100 of $A_V = 0.15$ mag, and a reddening-free distance of $17.1 \pm 1.8$ Mpc. This compares to IRTF distances of $18.4 \pm 2.2$ Mpc (Pierce & Tully 1988) and $14.5 \pm 2.7$ Mpc (Pierce 1994), and an EPM distance to SN1979C of $15 \pm 4$ Mpc (Schmidt et al. 1994).

An estimate of $H_0$ is derived by assuming the M100 distance is representative of the distance to the Virgo cluster (a large uncertainty of 3.4 Mpc is added to account for the large extent of Virgo). The uncertainty in Virgo's Hubble flow velocity is avoided by using the well-measured distance between the Virgo and Coma clusters of $77.1 \pm 3.1$ Mpc, giving $H_0 = 77 \pm 16$ km/s/Mpc.

$H_0 \sim 80$ km/s/Mpc in the standard (Einstein-de Sitter) model ($\Omega = 1$), gives an age for the Universe of $T_0 = 2/(3H_0) = 8$ Gyr, *inconsistent with oldest star ages of* 14± *2 Gyr.* This conflict with stellar ages can only be resolved by either: (1) allowing a lower density of $0.1 < \Omega < 0.3$, which would then give an age $T_0 = 12$ Gyr ($\Lambda = 0$), consistent (barely) with stellar ages, (2) allowing $\Lambda > 0$, or (3) concluding that theories of stellar and galactic evolution are incorrect.

**Acknowledgments.** These results are due to the hard work of the rest of the $H_0$ Key Project team: Laura Ferrarese, Holland Ford, Wendy Freedman, John Graham, Bob Hill, John Hoessel, John Huchra, Garth Illingworth, Robert Kennicutt, Dan Kelson, Nancy Silberman, Barry Madore, Jeremy Mould, Abhijit Saha, and Peter Stetson.

## References

Cook, K.H., Aaronson, M., & Illingworth, G. 1986, ApJ, 301, L45

Cohen, J.G. 1993, BAAS, 25, 818

Freedman, W.L., et al. 1994, Nature, 371, 757

Kelson, D.D., et al. 1995, ApJS, submitted.

Pierce, M.J., & Tully, R.B. 1988, ApJ, 330, 579

Pierce, M.J. 1994, ApJ, 430, 53

Schmidt, B.P., et al. 1994, ApJ, 432, 42

Silberman, N., et al. 1995, in preparation

Stetson, P.B., et al. 1995, in preparation

*Astrophysical Applications of Stellar Pulsation*
*ASP Conference Series, Vol. 83, 1995*
*R. S. Stobie and P. A. Whitelock, eds.*

# EROS: Short Period Cepheids in the Bar of the LMC

J.P. Beaulieu
*Institut d'Atrophysique de Paris, 98 bis Boulevard Arago, 75014 PARIS, FRANCE*

**Abstract.** We present Fourier analysis at 490 nm for 87 Cepheid light curves in the bar of the Large Magellanic Cloud. The photometry has been obtained through the EROS project whose main purpose is the search for baryonic dark matter in the Galactic halo. The very high quality of the photometry and the good phase coverage enable us to obtain accurate Fourier decomposition of the light curves. As a consequence we confirm that the so-called s-Cepheids are overtone pulsators and represent about 30 % of our sample, whereas the Classical Cepheids are fundamental pulsators. We have identified a number of features in plots of amplitude ratios and phase differences versus period. These features are usually related to resonances between different pulsation modes. If we accept that Fourier decomposition techniques applied to light curves yield positions for the centres of resonances, then it seems that resonances occur in LMC Cepheids at the same period as in Galactic Cepheids for fundamental pulsators, whereas they appear to be at shorter periods for overtone pulsators.

## 1. Introduction

EROS (Experience de Recherche d'Objets Sombres) is a collaboration between French astronomers and particle physicists to search for baryonic dark matter in the galactic halo in the form of compact objects through microlensing effects (Paczyǹski, 1986) on stars in the Large Magellanic Cloud (LMC).

A compact object in the Galactic halo, passing close enough from the line of sight to a background star in the LMC, induces an increase in the apparent brightness of the star. This phenomenon occurs owing to the alignment of the observer, the deflector and the background star. It is achromatic, symmetric in time and transient. Assuming a standard corotating halo, the time scale, $\tau_0$, of an event is given by the relation $\tau_0 = 70\sqrt{\frac{M}{M_\odot}}$ days, where $M$ is the mass of the deflector. Therefore in order to be sensitive to a wide range of mass for compact objects, two complementary approaches have been developed.

The first involves photographic monitoring of 6.4 million stars over a 5 degrees x 5 degrees area using the ESO Schmidt telescope. Plates are taken no more than nightly in two colours, $B_J$ and $R_C$. Isolated stars can be found down to magnitude 20 with a typical photometric uncertainty of 0.3 magnitude. For the period 1990-1994, about 380 plates in blue or red have been taken and the light curves for all the objects have been obtained.

The second approach uses a 0.4 m f/10 reflecting telescope and a mosaic of 16 buttable CCDs covering an area of 1 degree x 0.4 degree centred in the bar of the LMC. Between December 1991 and April 1994 about 15 000 images were taken in two broad bandpass filters $B_E$ and $R_E$ centred respectively on 490 and 670 nm. We have obtained 120 000 light curves, with as many as 48 points in a given night.

General descriptions of the observing program and the CCD camera can be found in Aubourg et al. (1993a, 1993b) and Arnaud et al. (1994a, 1994b).

It is clear that EROS, like MACHO (see Cook in this volume), is specially designed to search for the fingerprint of baryonic dark matter through microlensing effects. Therefore, the study of variable stars in the LMC is not the primary goal, but a by-product. On the other hand, the extremely large amount of data collected is of great interest for systematic studies of variable stars and stellar populations in this nearby galaxy.

## 2. Search for variable stars

### 2.1. Erratic photometric fluctuations

Various software filters have been designed to search for non-periodic fluctuations like microlensing events. They are extremely powerful for the detection of stars with bursts of brightness or irregular variations. To summarise, they search for 'bumps' in a window of given duration in the stars' photometric time sequences.

### 2.2. Periodic variations

A special effort has been made to search for variable stars in a systematic and automatic way. In order to be able to analyse as much information as possible, a new algorithm has been devised which is able to detect periodicities in light curves of arbitrary shape by use of orthogonal combinations of Fourier harmonics. This technique, developed by Grison (1994), is in fact a generalisation of periodogram analysis (Scargle 1982). For a group of test periods we are searching for the best parameters for a periodic-function model $\sum_{i=1}^{M} A_i \cos(i\omega t) + B_i \sin(i\omega t)$, where the coefficients are determined through the mean square method. It allows calculation of a function representative of the measurements' dispersion relative to the model, and therefore yields a confidence level for each trial period $\omega$. For each selected star the following parameters are kept in our data base: a probability of false detection, the value of the most probable period, and the first coefficients of the Fourier series of the best periodic function fitted to the measurements.

The test results for the Schmidt plates and CCDs are as follows.

SCHMIDT PLATES: The variability limit is comparable with the photometric precision, and periodicities between 0.5 days and one year are detectable. About 10 000 variable stars are expected with a confidence level better than 99% that the variations are not due to measurement fluctuations.

CCD DATA: With the 1991-1992 data we are able to detect amplitude variations of 0.3 times the photometric precision with periodicities as short as one hour. About 2000 variable stars are expected with a confidence level better than 99 %. With the 1991-1994 data, we will be able to detect amplitude variations down to 0.1 times the photometric precision.

## 3. Search for Cepheids in the bar of the LMC

### 3.1. Presentation of the 1991-1992 CCD data set

We have obtained ~2000 images spanning ~100 days in 1991-1992. Roughly equal numbers were obtained in $B_E$ and $R_E$. Photometric reductions were performed with a specially-written program of reconstruction and photometry. We built a catalogue comprising 80 000 stars. Then, we searched for periodic photometric fluctuations using the Grison's periodogram method. We detected about 1000 variable stars, of which 79 are eclipsing binaries (Grison et al., 1995), 102 are Cepheids (Beaulieu et al., 1995), about 20 are RR Lyrae stars, and hundreds are red giant variables. In the next sections we will summarise our analysis of the Cepheids we have detected in the bar of the LMC (Beaulieu et al., 1995).

### 3.2. Cepheid selection

We have plotted a colour-magnitude diagram in which we delineated the limits of the Cepheid instability strip via visual inspection of a sample of light curves. We considered only stars with $B_E < 17.6$ in order to reject RR Lyrae stars. Then we performed a visual inspection of all the light curves in order to avoid eclipsing binaries. The resulting 102 stars form our sample of LMC-bar Cepheids. Only 29 of them were previously recorded in the Harvard catalogue by Payne-Gaposchkin (1971).

Some of the objects present clearly anomalous light-curve shapes, or markedly larger scatter than other stars of similar period. In these objects we searched for the most probable period, fit a Fourier model, subtract the Fourier model to the data and search for a second periodicity. Only one star (EROS2040, HV970) showed a second period. The ratio between the two period is $0.710 \pm 0.001$, which leads us to conclude that this star pulsates in both its fundamental and first-overtone mode. This star has been also classified as 'beat Cepheid,' with the same periods, by Welch using data from the MACHO project (Alcock et al., 1995, Welch in this volume).

In the following, we will concentrate on the 'non-anomalous' stars of our sample with periods less than 10 days.

## 4. Fourier analysis of Cepheid light curves.

In the last decade, many authors have shown that coefficients derived from Fourier decomposition of light curves give a quantitative basis on which to compare observations with non-linear hydrodynamic models of variable stars and results from the amplitude-equation formalism. Fourier coefficients are a powerful tool for mode discrimination and to study the presence and effect of resonances in pulsating stars (e.g., Simon & Lee 1981, the review by Simon 1988 and references therein, Poretti 1994 and references therein, Moskalik et al. 1992, Buchler & Moskalik 1994).

Like earlier authors, we adopt a Fourier decomposition of form

$$X_0 + \sum_{i=1}^{M} X_i \cos(i\omega t + \Phi_i),$$

and define the amplitude ratio

$$R_{k1} = \frac{X_k}{X_1} \quad k > 1$$

and the phase difference

$$\Phi_{k1} = \Phi_k - k\,\Phi_1 \quad k > 1.$$

As shown by Stellingwerf & Donohoe (1986), the amplitude ratio reflects the asymmetry of the variation. In other words, the larger the $R_{k1}$, the greater the departure from a sinusoidal form of the light curve. However, as was shown by earlier authors, the $R_{k1}$ can also be strongly affected by the presence of resonances between the pulsational normal modes.

The phase difference $\Phi_{k1}$ has a qualitative meaning in terms of the width of the light curve at half maximum light (Stellingwerf & Donohoe 1987). A smaller value of $\Phi_{k1}$ implies a more acute light curve. Like $R_{k1}$, $\Phi_{k1}$ can also be strongly affected by the presence of periods resonances among the normal modes of stars.

A type of low amplitude, almost sinusoidal Cepheids are classified in the General Catalogue of Variable Stars (Kholopov 1985) as s-Cepheids. It is suggested that they are fundamental mode pulsators during the first crossing of the instability strip, or first overtone pulsators.

While various terminologies have been employed in the past, we shall adopt this nomenclature. Simon & Lee (1981) were the first to make a mapping of the Hertzsprung progression of the changing form of classical Cepheids light curves through the use of Fourier decomposition. Then, Antonello & Poretti (1986) showed that the Galactic s-Cepheids follow different sequences in the $R_{21} - P$ and $\Phi_{21} - P$ plane. They propose a separation between classical and s-Cepheids based on the $\Phi_{21} - P$ plane, and interpreted the differences as a consequence of different pulsation modes : fundamental radial mode for classical Cepheids, and first overtone radial mode for s-Cepheids. On the other hand Gieren et al. (1990) propose that at least some of the s-Cepheids are fundamental mode pulsators.

In Figure 1, the distinction between the skew, higher amplitude and possibly bumpy light curves of classical Cepheids and the more sinusoidal, lower amplitude ones of s-Cepheids can be seen.

### 4.1. Classification

Mindful of earlier works, we used the $R_{21} - P$ plane to discriminate between classical and s-Cepheids (Figure 2). Let us recall that a low value of $R_{21}$ indicates a more-sinusoidal light curve. Therefore, we defined as s-Cepheids the 30 stars that lie under $R_{21} < 0.3$ for $P < 3$ d and $R_{21} < 0.2$ for $3 < P < 5.5$ d.

A group a six stars with $R_{21} \sim 0.25$ and $4 < P < 6.5$.d. stand apart as of uncertain classification in $R_{21} - P$ plane. We defined them as 'intermediate stars'.

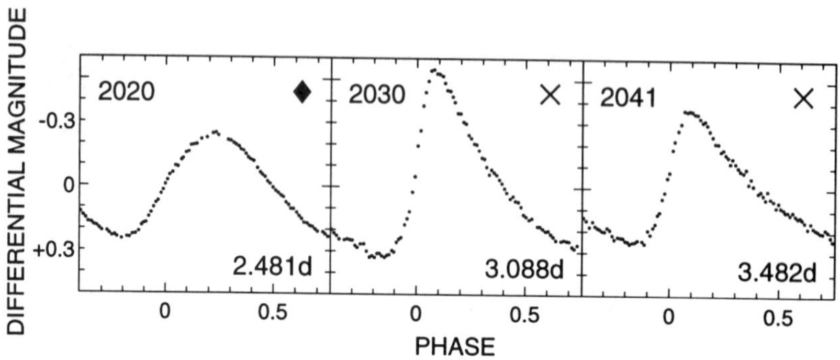


Figure 1. $B_E$ mean light curves for three Cepheids of our sample. The following symbols are used: Classical light curves (diagonal crosses), s-Cepheids (filled diamonds).

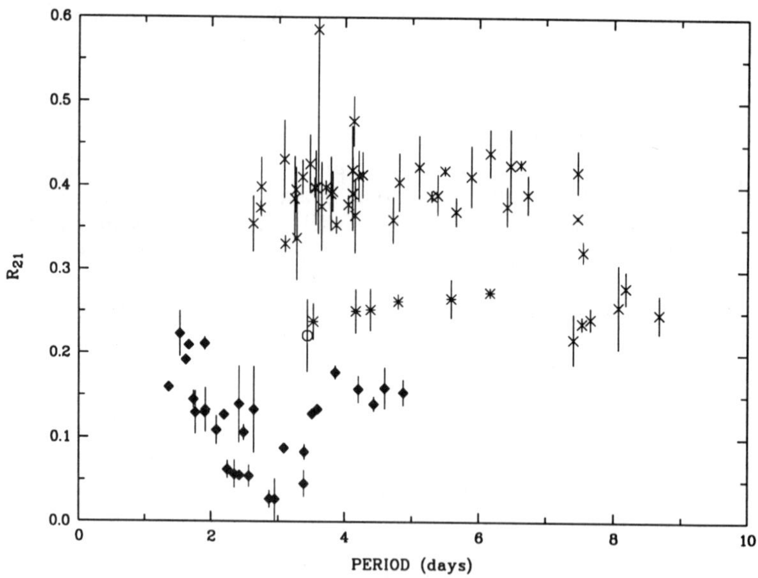


Figure 2. The $R_{21} - P$ diagram for $B_E$ light curves of the 87 objects with $P < 10$ d. The following symbols are used: Classical light curves (diagonal crosses), s-Cepheids (filled diamonds), double-mode Cepheid (open circle) and intermediate objects in the $R_{21} - P$ plane (asterisks).

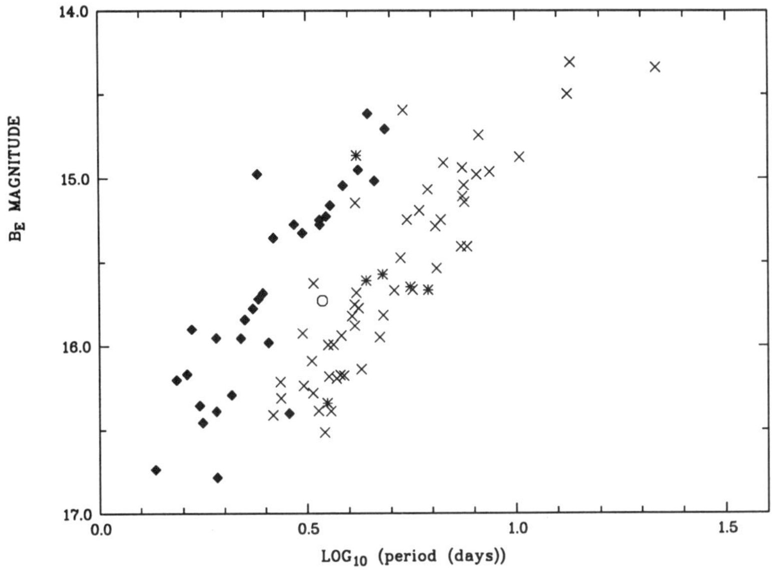


Figure 3. Period-luminosity relation for Cepheids in the $B_E$ filter. The same symbols are used as in Figure 2.

The remaining 51 stars are classified as classical Cepheids.

We find that our classification based on morphological criteria (Fourier component values) is mirrored by a clear separation in two sequences when we plot the period-luminosity relation (Figure 3). We note that at a given period the s-Cepheids are $\sim 1$ mag. brighter. An obvious interpretation is that the upper sequence correspond to first-overtone pulsators, and most of our s-Cepheids pulsate in this mode, while the lower sequence corresponds to fundamental-mode pulsators. Most of our classical Cepheids are pulsating in this mode.

A few stars deviate from the sequence to which they may be supposed to belong according to the classification based on Figure 2. For most of them, the images of the stars are oval, a signature of optical binarity. For some overtone pulsators, another alternative is that some of them are pulsating in the second overtone. However, it is not possible to give a conclusive answer from the available data.

The stars we defined as intermediate stars are a mixed bag. Five of them are fundamental-mode pulsators, one is an overtone pulsator, and one is a beat Cepheid.

## 4.2. Period-luminosity relation

We plotted a period-luminosity diagram where the periods of the overtone pulsators are converted to expected fundamental-mode values through the classical relation $\frac{P(1H)}{P(1F)} \sim 0.71$. The two sequences are now closer, but not coincident. We applied a classical colour correction as defined by Feast (1984, and in this volume) to all the stars. This 'modified-period'-luminosity-colour relation is plotted

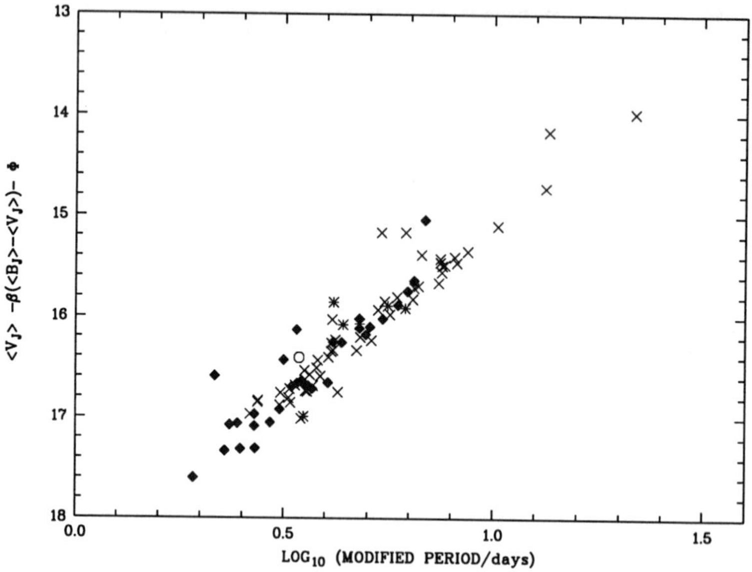


Figure 4. Period-luminosity-colour relation. The periods were modified as described in the text. The ordinate is the numerical value $< B_E > -2.95(< B_E > - < R_E >) + 2.39$. The same symbols are used as in Figures 2 and 3.

in Figure 4. All the stars obey the same period-luminosity-colour relation. This confirms that **the s-Cepheids are overtone pulsators** and that **the same colour correction applies to overtone pulsators and fundamental-mode pulsators**.

### 4.3. $\Phi_{21} - P$ plane for classical Cepheids

The classification defined in $R_{21} - P$, and mirrored in period-luminosity relation, is also observed in the $\Phi_{21} - P$ plane plotted in Figure 5. The classical Cepheids lie along a sequence corresponding to the so-called Hertzsprung progression.

The extremely regular progressions in $R_{21} - P$, $\Phi_{21} - P$ and other Fourier components diagrams for classical Cepheids is now well understood. The break around 10 days, observed for Galactic (Simon & Lee 1981) and Magellanic Cepheids (Andreasen & Petersen 1987) is due to the chance 2:1 period resonance between the fundamental and the second overtone in the normal-mode period spectrum of these stars. It has been understood through the formalism of amplitude equations (Buchler & Goupil, 1984, Klap et al. 1985) and reproduced by the non-linear calculations of Moskalik et al. (1992) with the new OPAL opacities.

With the present data we can only infer that the resonance lies between 8.7 and 10.2 days, as pointed out by earlier authors. A better answer will be obtained via analysis of a bigger sample.

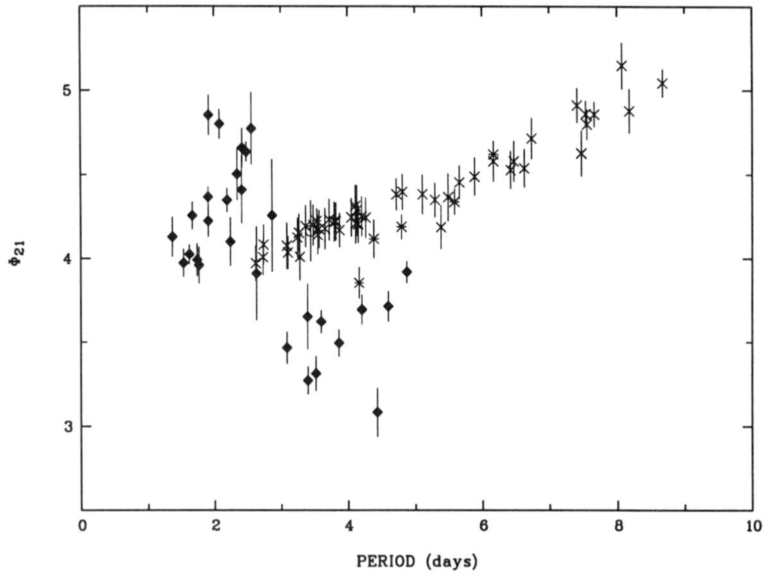


Figure 5. The $\Phi_{21} - P$ diagram for $B_E$ light curves for the 87 objects with $P < 10$ d. The same symbols are used as in Figures 2-4.

## 4.4. $\Phi_{21} - P$ plane for overtone Cepheids

The overtone pulsators fall in two well defined sequences, as was noted by Antonello et al. (1990) for Galactic Cepheids. The upper sequence is characterised by $P < 3.2$ d and $\Phi_{21} > 4.2$ rad, while for the lower sequence $P > 3.2$ d and $\Phi_{21} < 4.0$ rad. With a little calligraphic licence we called these two sequences and their possible link, the 'Z-shape'. We confirmed that all the stars that belong to the Z are overtone pulsators. Therefore, the alternative hypothesis of Gieren et al. (1990) that the stars of the lower part of the Z are fundamental pulsators is refuted.

Mindful of the effect that the 2:1 resonance between the fundamental mode and the second overtone for classical Cepheids around 10 days has on Fourier components, the Z-shape has been interpreted by earlier authors (e.g., Poretti 1994 and references therein) as the signature of a resonance between the first and the fourth overtone.

From our sample we estimate that the Z-shape is centred around $2.7 \pm 0.2$ d, whereas it is observed around 3.2 d in the Galaxy by Poretti (1994). Buchler & Moskalik (1994) observed that in the SMC it should occur at shorter periods than in the Galaxy, but data quality could not precisely yield the resonance centre period. A deeper comparison between the Z-shape obtained for these 3 galaxies would be of great interest for observational study of metallicity effects on this probable resonance, followed by a modelling attempt using non-linear calculations.

### 4.5. Signature of resonances for classical Cepheids around 7.5 days

For the classical Cepheids, a sharp dip feature is observed in the $R_{21} - P$ plane at $\sim 7.5 \pm 0.2$ d. It is weakly visible in $\Phi_{21} - P$, and more and more visible when going to higher-harmonic Fourier components. We think that it may be related to the 3:1 resonance between the fundamental and fourth overtone, like the feature found by Moskalik et al. (1992) in their non-linear calculations around 7.7 d and related to a dip in $\Phi_{41} - P$ found by Simon (1985), and also discussed by Antonello (1994).

## 5. Conclusion

We have Fourier analysed Cepheid light curves in the bar of the LMC using the high quality CCD photometry obtained by the EROS collaboration. We selected 87 Cepheid variables with period less than 10 days, of which 26 were recognised in the Harvard catalogue. We adopted a classification based on morphological criteria using Fourier components and defined as s-Cepheids the stars with low $R_{21}$ value and period less than 5.5 days. One star (EROS 2040, HV 970) is a double mode Cepheid beating in the fundamental and first overtone modes. 6 stars with $R_{21} \sim 0.26$ are classified as "intermediate Cepheids". The rest are classified as classical Cepheids.

The s-Cepheids and the classical Cepheids follow different loci while plotting the period luminosity relation, and that leads us to conclude that the s-Cepheids are overtone pulsators (about 30 % of the sample ) whereas the classical Cepheids are fundamental pulsators.

The features exhibited in the $R_{21} - P$ and $\Phi_{21} - P$ plane are similar to those of galactic Cepheids. The 2:1 resonance between the fundamental mode and the second overtone for classical Cepheids occurs between 8.7 and 10 days as already pointed out by earlier authors. A sharp dip feature is observed for classical Cepheids in the $R_{21} - P$ plane around 7.5 days and we think it can be related to the 3:1 resonance between the fundamental and the fourth overtone for classical Cepheids. (This is the first detection of this resonance in the LMC, and it is found at nearly the same period as in the Galaxy.)

The Z-shape for overtone pulsators near 3 days, interpreted as the signature of the 2:1 resonance between the fourth and the first overtone is clearly seen, but is observed at a lower period ($2.7 \pm 0.2d$) than for the galaxy (3.2 d). This may be due to the lower metallicity of the LMC.

We thus find that the position of the resonance in overtone pulsators is shifted in the LMC, whereas for fundamental pulsators the values are similar to galactic ones.

**Acknowledgments.** This work has been done in collaboration with P. Grison, J. Pritchard and W. Tobin. I thank J. Percy for inviting me to IAU 155 Colloquium. I would like to thank M. Feast, R. Ferlet, M.J. Goupil, D. Sasselov and A. Vidal-Madjar for useful discussions.

## References

Alcock, C., et al. 1995, AJ, in press

Andreasen, G.K., & Petersen, J.O. 1987, A&A, 180, 127

Antonello, E., & Poretti, E. 1986, A&A, 169, 149

Antonello, E., Poretti, E., & Reduzzi, L. 1990, A&A, 236, 138

Antonello, E. 1994, A&A, 282, 835

Arnaud, M. et al. 1994a, Experimental Astron., 4, 265

Arnaud, M. et al. 1994b, Experimental Astron., 4, 279

Aubourg, E., et al. 1993a, The Messenger, 72, 20

Aubourg, E., et al. 1993b, Nature, 365, 623

Beaulieu, J.P., et al. 1995, A&A, in press

Buchler, J.R., & Goupil, M.J. 1984, ApJ, 279, 394

Buchler, J.R., & Moskalik, P. 1994, A&A, 292, 450

Feast, M.W. 1984, Magellanic Cloud Cepheids. in: van den Bergh, S., de Boer, K.S. (eds) Proc. IAU Symp. 108, Structure and evolution of the Magellanic Clouds. Reidel, Dordrecht p.157

Gieren, W.P., et al. 1990, AJ, 99, 1196

Grison, P. 1994, A&A, 289, 404

Grison, P., et al. 1995, A&AS, 109, 1

Kholopov, P.N., et al. 1985, General Catalogue of Variable Stars, 4th edition, Vol. I. Nauka, Moscow

Klapp, J., Goupil, M.J., & Buchler, J.R. 1985, ApJ, 296, 514

Paczyǹski, B., 1986, ApJ, 304, 1

Moskalik, P., Buchler, J.R., & Marom, M. 1992, ApJ, 385, 685

Payne-Gaposchkin, C.H. 1971, Smithson. Contr. Astrophys. No. 13

Poretti, E. 1994, A&A, 285, 524

Scargle, J.D. 1982, ApJ, 263, 835

Simon, N.R., 1988, The Fourier decomposition technique: Interpreting the variations of pulsating stars. In: Stalio, R., Willson, L.A. (eds) Pulsation and Mass Loss in Stars. Kluwer, Dordrecht p.27

Simon, N.R., & Lee A. 1981, ApJ, 248, 291

Simon, N.R. 1985, ApJ, 299, 723

Stellingwerf, R.F., & Donohoe M. 1986, ApJ, 306, 183

Stellingwerf, R.F., & Donohoe M. 1987, ApJ, 314, 252

*Astrophysical Applications of Stellar Pulsation*
*ASP Conference Series, Vol. 83, 1995*
*R. S. Stobie and P. A. Whitelock, eds.*

# Interferometry of Cool Pulsating Stars

Chris Haniff

*Mullard Radio Astronomy Observatory, University of Cambridge, Cavendish Laboratory, Madingley Road, Cambridge, CB3 OHE, England*

**Abstract.** The drive for high-angular resolution from the ground has led to the development of optical and near infrared interferometric telescopes analogous to radio synthesis arrays. These now provide a diffraction-limited imaging capability well suited to studies of nearby cool giants and supergiants. In this review I summarise the current status of these instruments, their scientific potential for studies of cool pulsating variables (LPVs), and present some recent results that highlight the promise that such methods hold.

## 1. Introduction

The application of interferometric techniques to astronomical imaging has had a long and successful history, and is exemplified most visibly nowadays in large radio synthesis telescopes such as the VLA & AT. Recently such methods have been transferred to optical and near-infrared (NIR) wavelengths and are reaching fruition in a new generation of separated-element interferometric telescopes. Because of their similarity in implementation, throughout this review any mention of optical interferometry should in fact be taken to refer to both optical and NIR systems. The purpose of this review is in essence threefold. First, to explain why and how interferometric methods have only now expanded into the optical and NIR wavebands. Second, to outline the specific astrophysical questions related to LPVs that can be addressed by such instruments. And finally, to demonstrate, by way of examples, the contribution that imaging interferometry is already making to our understanding of cool pulsating variables today.

## 2. Michelson interferometry in the 1990s

Before discussing the scope of interferometry it is worthwhile stepping back and reviewing exactly what we mean when we talk of an "interferometer". My own feeling is that most optical astronomers, have a reasonably well defined, though outdated, picture of an unwieldy instrument attached to the front of a large telescope, producing an image crossed by fringes, i.e. they envision a classical Michelson stellar interferometer. Only infrequently do they recognise optical interferometers as precise analogues of radio synthesis arrays, where the primary observable of the instrument is not an image at all, but a measurement of the spatial coherence function of the radiation from the source. Hereafter, I shall use the terms spatial coherence function and visibility function interchangeably.

For the sake of completeness, I remind the reader that the visibility function and source brightness distribution are related to each other, in most cases of interest, via a Fourier transform.

In fact it is instructive to use a typical radio synthesis telescope to characterise the features we ought to expect in its optical counterpart:

- Small unit telescopes: these should be smaller than the characteristic scale size of the atmospheric wavefront perturbations, $r_o$. At cm radio wavelengths this is typically many hundreds of metres, and so is often not an important design constraint. However, for optical and NIR observations this restriction implies that $D < 0.1\text{m} - 1.0\text{m}$.

- Long baselines: $B \gg D$. Usually this applies to all the interferometer baselines, with even the shortest spacings being many times the unit telescope size. This implies a lack of sensitivity to large-scale structures on the sky.

- Short exposure times: $T_{exp} <$ the timescale associated with the atmospheric wavefront perturbations, $\tau_o$. This scale is much longer at radio wavelengths than in the optical/NIR regime, where typical exposure times lie in the range 0.01s – 0.1s. In general $T_{exp} \ll$ the total time to observe the source.

- The primary observable is the spatial coherence function or complex visibility function, not an image as is the case for a conventional telescope. In this sense the two processes of data acquisition and image recovery can be treated as quite separate steps, with the production of an image being one of many possible procedures for visualising the interferometer's output.

Given the above, we should expect interferometers to have good spatial resolution, but very poor sensitivity, at least when compared to filled-aperture telescopes. At optical and NIR wavelengths the small values of $r_o$ and $\tau_o$ limit the scope of interferometry to bright astrophysical targets, but the ability to overcome the seeing limit means that interferometer arrays are ideally suited for studying luminous evolved giants and supergiants whose apparent angular sizes are well matched to the resolutions attainable with 10 – 100m baselines.

## 2.1. Developments in optical interferometry

Perhaps the most important recent advances in interferometry have been the development of methods to recover the phase of the visibility function from atmospherically perturbed measurements. In the absence of phase information, unambiguous image reconstruction is usually not possible, and so for many years interferometric methods played only a minor role in optical and NIR astronomy.

Current techniques take advantage of the fact that, while individual measurements of the visibility phase are corrupted by atmospheric turbulence, certain combinations of them are immune to this type of disturbance. More specifically, sums of Fourier phases measured round closed loops of baselines, or equivalently the arguments of the products of the complex visibilities on these baselines, remain good observables. These so-called closure, or bispectral, phases provide access to the true source visibility phases, albeit in the form of linear combinations, and thereby lead to the following important repercussions:

- Imaging becomes feasible: although for a 3-element interferometer (the minimum sized array that permits the use of closure techniques) the fraction of visibility phase information recoverable is relatively small, for arrays with larger numbers of telescopes high-fidelity imaging of complex sources is possible. The use of these closure methods has been the key to the success of radio VLBI (see, e.g., Pearson & Readhead 1984).
- Information on scales smaller than the Rayleigh limit ($\theta_{min} \sim \lambda/B_{max}$) becomes accessible. This does not contradict the usual criterion for resolving power, but reflects the fact that closure phases are sensitive diagnostics for certain source structures. In particular, symmetric components in the source produce no closure phase signal, whereas asymmetric structures can be reliably detected even when present on scales smaller than $\theta_{min}$.

Verification of these two new possibilities for optical interferometry had to wait until the late 1980s, and so the status of optical interferometry is comparable to that of radio synthesis in its early days. Nevertheless, the demonstrable success of interferometric imaging experiments and the, perhaps unexpected, robustness of the closure phase to calibration errors, has meant that spatial interferometry is a far more powerful tool nowadays than it has ever been.

### 2.2. Current interferometer projects

The scientific potential of interferometric imaging can best be gauged by the large investment of both effort and money in the field today. Table 1 provides a brief summary of optical interferometers currently operational, or in the process of construction or planning. Only systems designed for closure-phase imaging are listed, and so a number of scientifically important 2-element instruments, including GI2T and SUSI, are missing. Of those listed, COAST, ISI and IOTA are operational, though the last two of these are still awaiting their third unit telescopes. Further details of all of these interferometers can be found in Breckinridge (1994).

| Name | Location | Type | Operating Bandpass |
|---|---|---|---|
| NPOI | Anderson Mesa, AZ | Imaging/Astrometry | Optical |
| COAST | Cambridge, UK | Imaging | Optical/Near IR |
| CHARA | t.b.d. | Imaging | Optical/Near IR |
| ISI | Mt. Wilson, CA | Imaging | Mid IR |
| IOTA | Mt. Hopkins, AZ | Imaging | Near IR |
| VLTI | Paranal, Chile | Imaging | Optical/Near IR |
| KECK | Mauna Kea, Hawaii | Imaging/Astrometry | Near/Mid IR |

Although these interferometers differ greatly in scope and scale, not surprisingly they share numerous common features:

- Small telescopes in the range 10cm to 2m. Apart from the KECK and VLTI arrays, which will utilise the very large unit telescopes in combination with arrays of smaller outrigger telescopes, most of these interferometers have been designed to operate with aperture sizes of $\sim r_o$.
- Baselines in the range tens to hundreds of metres. This implies resolving powers of between one and four milliarcseconds, sufficient to permit imaging of LPVs out to $\sim$ 1kpc.
- Broad band imaging capabilities, with spectral resolutions from 10 to 1000. Most of these interferometers have use medium and narrow-band filters to define the bandpasses observed. However, given the diagnostic importance of spectral information, beam combination architectures incorporating dispersion are envisioned in all second-generation detector subsystems.

Apart from separated-element arrays, individual monolithic telescopes can also be operated in an interferometric mode. This is accomplished by apodizing the telescope pupil with a mask perforated with a number of appropriately arranged holes, thereby emulating a Fizeau interferometer (see, e.g., Haniff et al. 1987). When implemented on 10m-class telescopes this type of approach can be very attractive for studies of nearby LPVs for which resolutions of the order of ten milliarcseconds are still scientifically valuable.

## 3. Science goals for cool pulsating stars

Despite much observational work, numerous aspects of the theory and physics of Mira variables still remain in dispute. To some extent this is a consequence of their intrinsic complexity, but the paucity of precise observational diagnostics has been a significant impediment to progress. Interferometry offers the potential for addressing a number of pressing issues, some of which are outlined below:

### 3.1. Photospheric angular diameters

Ostensibly one of the most straightforward applications of interferometry is to establish reliable angular sizes for M-type variables and hence determine the effective temperature scale for these stars. This has been hampered in the past by the wealth of atomic and molecular features present in the optical spectra of Miras. These make it difficult to isolate bandpasses that are sensitive only to photospheric emission, and are responsible for a strong wavelength dependence of apparent angular size.

One of the most suitable regions of the spectrum for this type of work lies between 1 and 2$\mu$m (Bessell et al. 1989). However, until recently, most studies have had to rely upon occultation techniques to provide the necessary resolution to achieve accurate results at these longer wavelengths (see Richichi 1994 for a review of these methods). Interferometry now seems well placed to become the method of choice for stellar diameter measurement, offering full sky coverage, high resolution, and of course freedom from the temporal constraints set by the lunar motion. Furthermore, the combination of adaptive optics and interferometry on large monolithic telescopes offers the prospect of measurements shortwards of $0.4\mu$m where alternative bandpasses exist that also probe the deeper continuum layers of the atmosphere (Bessell et al. 1989).

### 3.2. Pulsation modes of Mira variables

The mode of pulsation of Miras has remained controversial for many years, with photometric studies suggesting overtone pulsation, but shock models implicating fundamental mode pulsation in order to reproduce the velocities seen in high-resolution NIR spectra (see Wood, these proceedings, for a discussion of these issues). In principle, selection between these two options ought to be simple, requiring estimates only of the period, $P$, mass, $M$, and radius, $R$, of the star. The combination of these three quantities in the pulsation constant, $Q = P(M/M_{\odot})^{1/2}(R/R_{\odot})^{-3/2}$, would then provide the necessary discriminant. On the assumption of a consensus as to the values of $Q$ expected for fundamental and first overtone pulsation, it seems likely that the availability of a small number of precise interferometric diameter estimates at selected photospheric bandpasses, together with accurate parallax distances, either from HIPPARCOS or modern ground-based measurements (Gatewood 1992), will soon put an end to this impasse.

### 3.3. Testing models of Mira atmospheres

Over the past decade, theoretical modelling of LPV atmospheres has matured greatly. Not content with static plane-parallel models, workers have introduced increasing levels of sophistication in order to better accommodate, for example, the sources of opacity in the cool outer layers of Miras as well as the stratification expected in a periodically perturbed atmosphere (Scholz & Takeda 1987; Bessell et al. 1989; Bessell et al. 1995). With interferometry it has now become possible to check these models directly.

Apart from verifying the predicted variations of apparent diameter with $\lambda$, and locating the regions responsible for the emission lines seen at certain phases of the variability cycle, one of the most interesting possibilities will be to monitor the temporal behaviour of the atmospheric structure. This will require only moderate resolution but high sensitivity and fidelity, particularly if subtle variations are to be detected. Typical of these are the relatively small modulations predicted for the photospheric diameter, which have been estimated to lie in the range 5 – 10% on the basis of bolometric photometry (Tuthill et al. 1994). Dynamical models of Miras (see Wood, these proceedings) often suggest larger amplitudes (30–50% for fundamental models) and so it ought to be straightforward to confirm these predictions with interferometric monitoring.

### 3.4. Surface imaging of Mira variables

The most exciting and challenging role for interferometry will be to reveal the surface morphologies of LPVs directly. Indirect evidence for deviations from circular symmetry and an otherwise uniform brightness distribution has often been cited on the basis of polarimetric studies (Shawl 1974; Boyle et al. 1986) and observations of non-uniform distributions of SiO masers (e.g., Diamond et al. 1994; Greenhill et al. 1995), which trace material close to the stellar surface. Asymmetries on larger scales have also been detected using molecular species

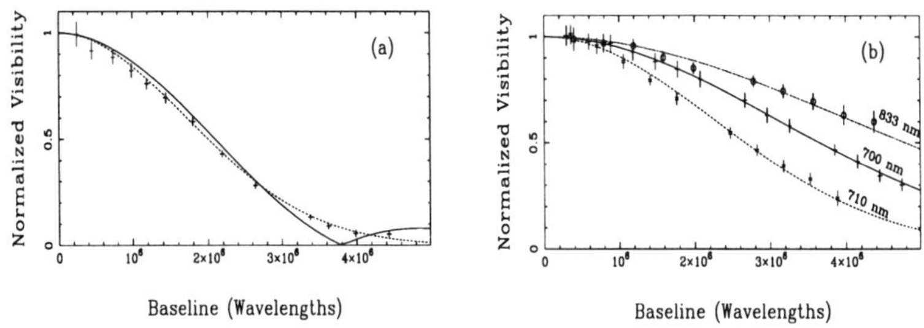


Figure 1. Measured visibility data for W Hya (a) and R Cas (b). In the first panel the data are plotted together with two models: the filled line is a theoretical prediction, while the dotted line is a Gaussian fit. The data for R Cas are typical for Miras, which show large changes in apparent diameter as a function of wavelength.

such as CO (Planesas et al. 1990), and it remains to be seen to what extent these are governed by corresponding asymmetries at the stellar surface.

High quality imaging will demand good Fourier plane coverage and both accuracy and precision in the visibility measurements. Observations within the optical bandpass will be best suited to identify and define the number, location and timescales of any surface phenomena, while those at longer wavelengths will begin to probe hot dust close to the photosphere. Sequences of images should uncover, not only the location of the hottest dust, but also identify when in the pulsation cycle it forms and hence whether or not mass loss in Miras is sporadic or more closely linked to the regular pulsation cycle.

## 4. Results from ground-based programmes

Though I have talked thus far about the future role of interferometry, its notable successes to date deserve some mention. Rather than providing an exhaustive review of all previous studies, for reasons of expediency I have concentrated on an admittedly biased, but hopefully representative, selection of results from our group in Cambridge. These were obtained by a number of my colleagues during interferometric observing campaigns that utilised the 4.2 WHT telescope in La Palma between 1991 and 1994 : I am very grateful to them for allowing me to present their results here.

### 4.1. Atmospheric structure

Measurements of visibility amplitude as a function of baseline have now become almost routine (see, e.g., Benson et al. 1991; Quirrenbach et al. 1993), although in most instances the range of baselines used is small, which can limit the scope of any interpretation. Recently, workers have begun to use more extensive datasets to search for deviations from uniform limb-darkening and circular symmetry. Figure 1a shows data for W Hya obtained at 700nm in July 1992. The measurements appear to fit a smooth Gaussian profile in preference to a

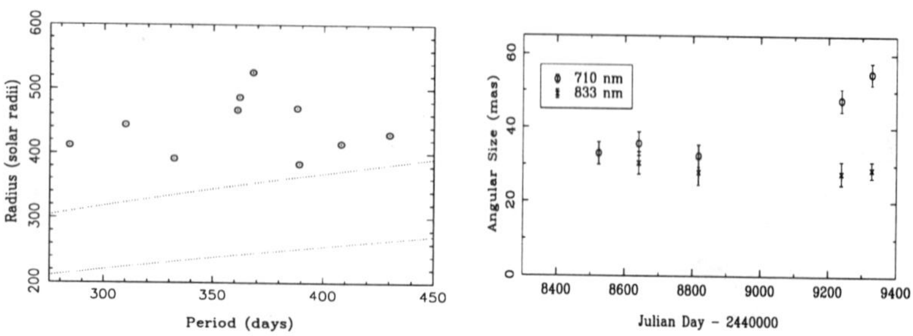


Figure 2. The left hand panel shows photospheric radii for a sample of nearby Miras from Haniff et al. (1995). The dotted lines are predictions for $1M_\odot$ (lower) and $1.5M_\odot$ (higher) fundamental mode pulsators. The other panel shows the measured size variations of *o* Ceti between 1991 and 1993. Note the rapid increase in size towards the end of this period, and its prominence in the 710nm (TiO) bandpass.

theoretical limb-darkened model. This appears to be a general trend for the sample of LPVs we have observed, with the visibility amplitudes at the longer baselines not showing any pronounced nulls, indicative of very smooth brightness profiles.

The $\lambda$-dependence of the apparent diameters of LPVs was first noticed by Labeyrie and his colleagues in the late 1970s (Labeyrie et al. 1977; Bonneau et al. 1982). Nowadays observations are targeted towards selected narrow bandpasses over as wide a range of $\lambda$ as possible so as to permit useful comparisons with models. Figure 1b shows data for the source R Cas: the sequence of wavelengths 833 $\rightarrow$ 700 $\rightarrow$ 710nm corresponds to layers increasingly distant from the centre of the star, and highlights the strength of the wavelength dependence.

As mentioned by Wood (these proceedings) model atmospheres can be used to derive correction factors relating these quasi-monochromatic measurements to the true photospheric size. Tuthill and colleagues (Tuthill et al. 1994; Haniff et al. 1995) have used such models, together with interferometric data and distances based on the (K, log P) relation, to estimate the linear radii for a sample of 10 nearby Miras. Their results are summarised in Figure 2a. Subject to the usual caveats regarding the correction factors used, their data appear to rule out fundamental mode pulsation for this sample of short period ($P \leq 450$days) LPVs. Hopefully, interferometric measurements in less confused regions of the spectrum will soon place this result on a sounder footing.

## 4.2. Variability effects

There has been relatively little interferometric monitoring of the spatial structure of Mira variables. Variations in the shape of Mira itself were first noticed by Karovska et al. (1991) using speckle interferometry, and further studied by Quirrenbach et al. (1992) with the MkIII optical interferometer. Their data showed variability on a timescale of less than a month, but its interpretation was made difficulty by the broad (22nm) bandpass of their instrument. Some of our own results for Mira are shown in Figure 2b. Measurements in pseudo-

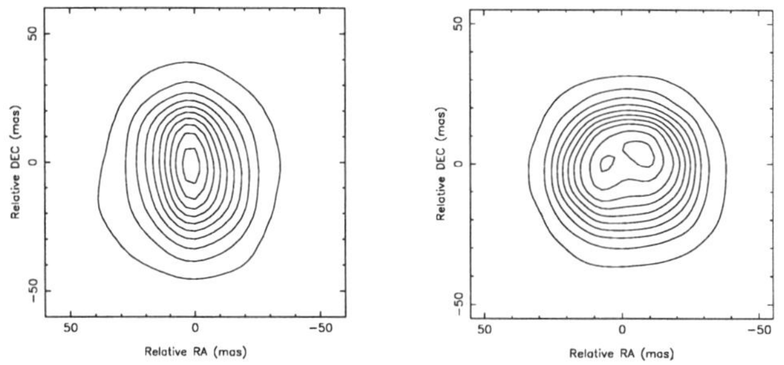


Figure 3. Aperture synthesis images of R Cas (left) and $\alpha$ Ori (right). The data were secured at 710nm and 700nm respectively. North is up and East to the left. Contour levels are from 5 to 95% of the peak flux, in increments of 10%.

continuum bandpasses exhibit relatively small fluctuations, but at wavelengths affected by TiO absorption the apparent size of the star can vary considerably over quite short timescales. More work certainly remains to be done, both in terms of monitoring the smaller modulation of the photospheric radius, as well as identifying any links these larger fluctuations may have with mass-loss and dust formation.

## 4.3. Stellar surface imaging

Probably the most unusual results from spatial interferometry have arisen from imaging experiments (see, e.g., Buscher et al. 1990; Wilson et al. 1992). These have uncovered several especially intriguing phenomena:

- Gross deviations from circular symmetry: these are clearly demonstrated in Figure 3a which shows the nearby LPV R Cas. The highly distorted atmosphere showed little change between July 1992 and September 1993, and is the most extreme example of its type we have observed. The connections, if any, of this puzzling morphology with the pulsation cycle, with deeper surface asymmetries (see below) or with possible stellar rotation all remain to be determined.

- Isolated unresolved hotspots of emission on the stellar surface: although not obvious from the presentation of Figure 3a, for a sample of 10 nearby Miras including R Cas, we have found no cases where a star has exhibited a distortion from circular symmetry that is symmetric. The key to this discovery has been the availability of closure phase information as a diagnostic for asymmetric structure. A clearer demonstration of this type of surface phenomenon is presented in Figure 3b, which shows a map of the M supergiant $\alpha$ Ori. The two hotspots contain 12% and 6% of the total flux respectively, and appear to show little variation in brightness with $\lambda$. For the LPV sample, the hotspots detected have similar fluxes, while for other supergiants we have studied they can sometimes be a factor of two brighter.

The ubiquity of these types of surface and atmospheric disturbances is quite remarkable: observations at different epochs and wavelengths have consistently revealed them for almost every M giant and supergiant we have studied with sufficient resolution. One possibility is that we are seeing the effects of large scale convective zones on the stellar surface (Schwarzschild 1975). The number, size, and luminosity of the hotspots are certainly consistent with such an hypothesis, and could go some way towards explaining the low-amplitude chaotic fluctuations seen in the light curves of many of these sources.

## 5. Conclusions

Interferometric imaging is already providing a unique tool for cool variable star research today, and the advent of long-baseline interferometers looks set to reveal a range of unsuspected phenomena. Two areas appear particularly fruitful. Observations in the infrared, targeted towards the interface between the stellar surface and circumstellar environment, will help to understand the geometry and dynamics of mass loss, as well as the link between mass loss and pulsation, and will call for long-baseline instruments. At the other extreme, the combination of adaptive optics and interferometry on 10m-class telescopes offers the possibility of very narrow-band ($R \sim 1000$) diffraction-limited imaging of nearby LPVs, isolating individual atmospheric layers and monitoring their evolution in time.

**Acknowledgments.** It is a pleasure to thank the members of our group in Cambridge involved in our aperture synthesis programme, in particular John Baldwin & Peter Tuthill, and also Michael Feast & Michael Scholz for their useful input. I am very grateful to the Royal Society for their financial support.

## References

Benson, J.A., Dyck, H.M., Ridgway, S.T., Dixon, D.J., Mason, W.L., & Howell, R.R. 1991, AJ, 102, 2091

Bessell, M.S., Brett, J.M., Scholz, M., & Wood P.R. 1989, A&A, 213, 209

Bessell, M.S., Scholz M., & Wood P.R. 1995, A&A, Submitted

Bonneau, D., Foy, R., Blazit, A., Labeyrie, A. 1982, A&A, 106, 235

Boyle, R.P., Aspin, C., Coyne, G.V., & McLean, I.S. 1986, A&A, 164, 310

Breckinridge, J.B. 1994, Proc SPIE, 2200

Buscher, D.F., Haniff, C.A ., Baldwin, J.E., & Warner, P.J. 1990, MNRAS, 245, 7p

Diamond, P.J., Kemball, A.J., Junor, W., Zensus, A., Benson, J., & Dhawan, V. 1994, ApJ, 430, L61

Gatewood, G. 1992, PASP, 104, 23

Greenhill, L.J., Colomer, F., Moran, J.M., Backer, D., Danchi, W.C., & Bester, M. 1995, ApJ, submitted

Haniff, C.A., Mackay, C.D., Titterington, D.J., Sivia, D., Baldwin, J.E., & Warner, P.J. 1987, Nature, 328, 694

Haniff, C.A., Scholz, M., & Tuthill, P.G. 1995, MNRAS, In press

Karovska, M., Nisenson, P., & Papaliolios, C. 1991, ApJ, 374, L51
Labeyrie, A., Koechlin, L., Bonneau, D., Blazit, A., & Foy, R. 1977, ApJ, 218, L75
Pearson, T.J. & Readhead, A.C.S. 1984, ARA&A, 22, 97
Planesas, P., Kenney, J.P.D., & Bachiller, R. 1990, ApJ, 364, L9
Quirrenbach, A., Mozurkewich, D., Armstrong, J.T., Johnston, K.J., Colavita, M.M., & Shao, M. 1992, A&A, 259, L19
Quirrenbach, A., Mozurkewich, D., Armstrong, J.T., Buscher, D.F., & Hummel, C.A. 1993, ApJ, 406, 215
Richichi, A. 1994, in Very High Angular Resolution Imaging, J.G.Robertson & W.J.Tango, Kluwer: Dordrecht, 71
Scholz, M., & Takeda, Y. 1987, A&A, 186, 200 (erratum: A&A, 196, 342)
Schwarzschild, M. 1975, ApJ, 195, 137
Shawl, S.J. 1974, PASP, 86, 843
Tuthill, P.G., Haniff, C.A., Baldwin, J.E., & Feast, M.W. 1994, MNRAS, 266, 745
Wilson, R.W., Baldwin, J.E, Buscher, D.F., & Warner, P.J. 1992, MNRAS, 257, 369

## Discussion

*Chapman*: Are you confident that your visibilities at the highest spatial frequencies are not affected by instrumental effects, i.e. could instrumental effects explain the offsets between the data and limb-darkened models?

*Haniff*: Yes we are. Our calibration procedures are quite extensive, and at the longest baselines the visibilities stay so high that it's difficult to see how large enough errors can have arisen to erase the nulls in the visibility function.

*Kawaler*: Have you looked at any unresolvable stars intentionally and, if so, what do these sources look like?

*Haniff*: Yes, a standard part of our procedure is to observe stars known to have small apparent diameters (e.g., early type stars) so as to calibrate our optical system. For these we see reductions in visibility consistent with atmospheric decorrelation, and closure phases close to zero, as expected for unresolved targets.

*Hughes*: Are the positions and changes in positions of the hotspots consistent with planetary orbits?

*Haniff*: Thus far we have been unable to monitor the change in position of any individual feature: what we see are either different features at different epochs, or the vanishing of a single feature between observations several months apart. Given these limitations, there does not appear to be any evidence yet of a systematic variation in the hotspot parameters that one could associate with planetary motion.

*Smith*: Your resolved features in $\alpha$ Ori may be related to the "long periods" discovered in this and other M supergiants and giants on the basis of their lightcurves. Over the last decade we (Smith et al. these proceedings) have found that both the H-$\alpha$ and CaII fluxes are correlated with these long periods. Our belief is that these periods could arise from an analogue to supergranulation in this and related stars.

*Haniff*: That's an interesting point. With our current poor temporal sampling we have not yet been able to determine any long term trends in the features detected on the stellar disk. However it would certainly be interesting to see if we could correlate our images with your data at the relevant epochs.

*Astrophysical Applications of Stellar Pulsation*
*ASP Conference Series, Vol. 83, 1995*
*R. S. Stobie and P. A. Whitelock, eds.*

# SUSI's Potential for Cepheid Studies

A.J. Booth, J. Davis, and R.R. Shobbrook

*Chatterton Astronomy Department, School of Physics, University of Sydney, NSW 2006, Australia*

**Abstract.** The Sydney University Stellar Interferometer (SUSI) is the first instrument with sufficient resolving power to measure the angular sizes of a large number of stars over the full range of spectral classes. It can also resolve the orbits of spectroscopic binaries, allowing masses and distances to be determined directly. One of the major programs envisaged for the instrument is a study of pulsating stars, in particular Cepheids.

## 1. Introduction

SUSI is a very long baseline optical amplitude interferometer with a north-south baseline array of eleven 14cm apertures, which can be combined in pairs to give baselines between 5m and 640m. The longest baseline results in a limiting resolution of ~50 micro arc sec at $\lambda$=442nm. Currently, baselines up to 160m are equipped, giving a resolution of 300 micro arc sec. 14cm apertures sample single $r_0$ (Fried, 1965) sections of wavefront, which are actively corrected for tip-tilt error and optical path length variations. Aperture size and the tip-tilt servo place an ultimate limit of B~8 on the system. At present, observations are limited to very bright stars by the path length tracking servo, but a new detector should allow observations to V~4, with further improvements later. Currently, spectral bandwidths of a few nanometers in the range $\lambda\lambda$400–500 are used. For a full description of the instrument see Davis et al. 1995.

## 2. Distance determinations for Cepheids using SUSI

SUSI will measure the angular size change of Cepheids due to their radial pulsation. Combining this with a linear size change obtained by integrating the velocity curve will yield a distance to the star, which is purely geometric and independent of other calibrators. This impacts on the zero point of the period luminosity relation, as other methods are all indirect or model dependent.

It may also be possible to improve the *accuracy* of the zero point of the period luminosity relation. Accuracies < 2% can be achieved in angular size measurements, giving a typical accuracy for the change in angular size of a Cepheid of ~10%. If this is the dominant uncertainty then the distance can be obtained to the same accuracy (distance modulus $\simeq \pm$0.2mag). Some 20 southern Cepheids of V $\leq$ 6.5 will be accessible to SUSI, thus a statistical improvement in accuracy to ~3% ($\pm$0.07mag) should be possible.

In practice, several factors can bias the result of the distance determination. The p-factors to convert the observed Doppler shifts into pulsation velocities must be carefully computed. For this level of accuracy it appears necessary to do fully dynamical calculations for the pulsating atmosphere (Sasselov & Karovska, 1994). A further complication is line asymmetries. It is well known that Cepheid metal lines display moderate asymmetry that changes through the pulsation cycle (Butler, 1993), and this must reflect velocity changes with atmospheric depth (Albrow & Cottrell, 1995). Thus, not only is it difficult to integrate the velocity curve unambiguously, but there will be some uncertainty in relating the velocity at the depth of line formation to that at the depth of continuum formation, which is essentially what is measured by SUSI. Such uncertainties amount to velocities of order $1\,\mathrm{km\,s^{-1}}$, which corresponds to a distance error of about 10%. In the most extreme cases line splitting can occur (Sasselov & Lester, 1990), making velocity assignment very difficult.

Limb darkening affects both the p-factors (Sasselov & Karovska, 1994) and angular diameters. SUSI observations yield a diameter of an equivalent uniform disc, not a true stellar diameter. Fortunately the resulting uncertainties are small; in any case, SUSI observations can probe limb darkening by measurement of angular size at different continuum wavelengths.

## 3. Other fundamental data obtainable from SUSI observations

In addition to distances, SUSI can be used to accurately determine other fundamental properties that will enhance our understanding of these stars. Angular diameters combined with total fluxes give effective temperatures (accurate to a few percent), and temperature changes during the cycle. From distances and angular sizes come linear sizes with similar accuracies. In addition to limb darkening investigations, observations inside broad metal and hydrogen lines can be combined with continuum observations to probe atmospheric structure and refine models. SUSI can also be used to determine masses and distances of stars in spectroscopic binaries (including some Cepheids) by resolving the orbit and thereby determining its angular size and inclination.

**Acknowledgments.** SUSI is supported by the Australian Research Council, the University of Sydney, the Pollock Memorial Fund, and the Science Foundation for Physics within the University of Sydney.

## References

Albrow, M.D., & Cottrell, P.L. 1995, MNRAS, submitted

Butler, R.P. 1993, ApJ, 415, 323

Davis, J., Tango, W.J, Booth, A.J., Minard, R., ten Brummelaar, T., Owens, S.M., & Shobbrook, R.R. 1995, MNRAS, in preparation

Fried, D.L. 1965, J. Opt. Soc. Am., 55, 1427

Sasselov, D.D., & Karovska, M. 1994, ApJ, 432, 367

Sasselov, D.D., & Lester, J.B. 1990, ApJ, 362, 333

*Astrophysical Applications of Stellar Pulsation*
*ASP Conference Series, Vol. 83, 1995*
*R. S. Stobie and P. A. Whitelock, eds.*

# Detection of Weak Periodic Signals from Irregularly Spaced Observations

Jörg Pfleiderer and Martin Mössner[1,2]

*Inst. of Astronomy, Leopold-Franzens-University, Innsbruck, Austria*

**Abstract.** We perform time series analysis on irregularly spaced data with large gaps. Our algorithm is based on multidimensional nonlinear optimisation. Particular emphasis is given to the selection and deletion of frequencies. The final result is checked by various statistical tests.

**1 Method.** The algorithm performs in an infinite loop, consisting of the steps: Selection, adjustment, and deletion of frequencies. The algorithm stops when the standard deviation of the given data is reached. The computed results are tested by applying various statistical tests.

A set of $M$ data $\mathbf{m} = (m_j)_j$, such as time-dependent brightness data, is to be approximated by $\mathbf{s} : s_j = A_0 + \sum_{i=1}^{N} A_i \cos(2\pi f_i t_j - \varphi_i)$. The amplitudes $\mathbf{A}$, frequencies $\mathbf{f}$, phases $\varphi$, as well as the number N of frequencies considered are unknown. The data and, perhaps, the times contain observational errors. The problem is how to choose $N$ and $\mathbf{f}$ and then to compute $\mathbf{A}$ and $\varphi$.

**1.1 Selection of frequencies.** Our main criterion for selecting new frequencies is the power function $p(f)$ of the data minus the signal expected from the last iteration. Since $p(f)$ is rapidly oscillating, high resolution is necessary for a clear image. Optimal resolution is given by $\Delta f = 0.1 \frac{1}{T}$ with $T = t_M - t_1$.

In order to overcome local oscillations we smooth $p(f)$ by computing the maximum within small windows. From this we choose the largest contributions and add them to the frequency vector of the last iteration. The new frequency vector will be used as initial guess in the adjustment step. If the number of frequencies gets too large only the strongest contributions will be accepted.

**1.2 Adjustment of frequencies.** Given a start frequency vector $\mathbf{f}$, we use a damped Newton solver (implementation of Deuflhard 1974), to minimise $\| \mathbf{g}(\mathbf{f}) \|_2$

with $$g_j(\mathbf{f}) = a_0(\mathbf{f}) + \sum_{i=1}^{N} a_i(\mathbf{f}) \cos(2\pi f_i t_j) + \sum_{i=1}^{N} b_i(\mathbf{f}) \sin(2\pi f_i t_j) - \mathbf{m}_j.$$

For this we need the amplitudes $a_0$, $a_i$, $b_i$ as function of $\mathbf{f}$. The derivatives with respect to $\mathbf{f}$ are computed by numerical differentiation.

The amplitudes for a particular $\mathbf{f}$ can be computed as the solution to the overdetermined linear system $\mathbf{A} \cdot \mathbf{a} \approx \mathbf{m}$ with $\mathbf{A} = (1, \cos(2\pi f_i t_j), \sin(2\pi f_i t_j))_{j,i}$ and $\mathbf{a} = (a_0(\mathbf{f}), a_i(\mathbf{f}), b_i(\mathbf{f}))_i$. In the full rank case, an unique $\mathbf{a}$, the so-called least

[1]email: joerg.pfleiderer@uibk.ac.at and martin.moessner@uibk.ac.at

[2]This work was supported by the Austrian Research Foundation under grant P8568-PHY and by the Austrian Academy of Sciences (Space Research / National Programs).

squares solution (LS), exists, which minimises $\|\mathbf{Aa}-\mathbf{m}\|_2$. For solving the linear system, we use either the complete orthogonal factorisation (COF, e.g., Golub & van Loan 1991) or the total least squares method (TLS, van Huffel & Vandewalle 1991). We do not use normal equations since this leads to bad conditioning. TLS has the advantage that it considers errors both in **A** and **b**.

**1.3 Deletion of frequencies.** In practical examples, it happens that the strongest power-amplitude is caused by aliasing effects. Nevertheless, if the correct frequencies are chosen, amplitudes for aliased frequencies get small and therefore can be deleted. In such cases we insert some test values and perform one or two Newton iterations. Frequencies which are *very* closely spaced are substituted by their mean value.

**1.4 Grid search.** The main break criterion is the norm of the residual vector. Since it shows strong oscillations in each component of the frequency vector, most optimisation techniques tend to catch one of the local minima instead of the absolute minimum. The typical frequency for these oscillations is well known ($\Delta f = \frac{1}{2T}$). Therefore, we do, after each major iteration, some grid search in each component of **f**.

**1.5 Statistical package** (see Dougherty 1990). We begin by computing standard statistics, such as mean, s.d., etc. To test the randomness of the residual vector, we apply the signed-rank or runs test and the Anderson-Darling test (Mössner et al. 1995). For estimating the accuracy of the chosen frequencies we compute linear confidence intervals. The reliability of weak signals can be tested by Fisher's test.

**2 Results.** The method was assessed by test-data and real astrophysical data: Delta Scuti star $\theta^2$ Tauri. The results of an early version of our method are given in Mössner & Pfleiderer (1993). The results of $\theta^2$ Tauri compare well to the results of Breger (1989) but contain additional periods.

## References

Breger, M., et al. 1989, "Multiple close frequencies of the Delta Scuti star $\theta^2$ Tauri. II. The second multi-site campaign", A&A, 214, 209–219

Deuflhard, P. 1974, "A Modified Newton Method for the Solution of Ill-Conditioned Systems of Nonlinear Equations with Applications to Multiple Shooting", Numerische Mathematik, 22, 289–315

Dougherty, E.R. 1990, "Probability and statistics for the engineering, computing and physical sciences", Prentice Hall International, London

Golub, G.H., & van Loan, Ch.F. 1991, "Matrix Computations", The Johns Hopkins University Press, Baltimore

Mössner, M., Netzer, N., & Pfleiderer, J. 1995, "Efficient computing of the Anderson-Darling Test Statistic", submitted

Mössner, M. & Pfleiderer, J. 1993, "Automatic Analysis for Time-Series with Large Gaps", Proc. of the 5th ESO/ST-ECF Data Analysis Workshop (P.J. Grosbøl and R.C.E. de Ruijsscher, eds.), ESO, 197–202

van Huffel, S. & Vandewalle, J. 1991, "The Total Least Squares Problem. Computational Aspects and Analysis", SIAM, Philadelphia

*Astrophysical Applications of Stellar Pulsation*
*ASP Conference Series, Vol. 83, 1995*
*R. S. Stobie and P. A. Whitelock, eds.*

# A New Asymptotic Treatment of $g$-modes of a Star

P. Smeyers, T. Van Hoolst, I. De Boeck[1] and L. Decock

*Instituut voor Sterrenkunde, Katholieke Universiteit Leuven, Celestijnenlaan 200B, 3001 Heverlee, Belgium*

## 1. Introduction and basic equations

An asymptotic representation of low-frequency, linear, isentropic $g$-modes of a star is developed without the usual neglect of the Eulerian perturbation of the gravitational potential. Our asymptotic representation is based on the use of asymptotic expansions adequate for solutions of singular perturbation problems (see, e.g., Kevorkian & Cole 1981).

Linear, isentropic oscillation modes with frequency different from zero are governed by a fourth-order system of linear, homogeneous differential equations in the radial parts of the radial displacement $\xi(r)$ and the divergence $\alpha(r)$. These equations take the form

$$\frac{d^2\alpha}{dr^2} + K_2(r)\,\frac{d\alpha}{dr} + \left[\frac{\sigma^2}{c^2} + K_3(r) + \frac{K_1(r)}{\sigma^2}\right]\alpha = -K_4(r)\,\frac{d\xi}{dr}, \tag{1}$$

$$\frac{d^2\xi}{dr^2} + \frac{4}{r}\frac{d\xi}{dr} - \frac{\ell(\ell+1)-2}{r^2}\,\xi = \frac{d\alpha}{dr} - \left[\frac{c^2}{g}\,\frac{K_1(r)}{\sigma^2} - \frac{2}{r}\right]\alpha. \tag{2}$$

The symbols have their usual meaning. $N^2$ is the square of the frequency of Brunt-Väisälä. The functions $K_1(r)$, $K_2(r)$, $K_3(r)$, $K_4(r)$ depend on the equilibrium model, e.g.,

$$K_1(r) = \ell(\ell+1)\,\frac{N^2}{r^2}, \tag{3}$$

We introduce the small expansion parameter

$$\epsilon = |\sigma|.$$

and assume, for the sake of simplification, $N^2$ to be positive everywhere in the star so that the star is everywhere convectively stable.

## 2. Method

Because of the singularities appearing in the differential equations at the centre and the surface of the star, it is necessary to make a distinction between the region sufficiently far from both singular points, and the two regions near the singular points.

[1]Research Assistant of the Belgian National Fund for Scientific Research

At distances sufficiently large from the star's centre and surface, we consider Eq. (1) as a differential equation governing a linear oscillator with a high frequency and a small damping. By using a two-variable expansion procedure, we seek asymptotic solutions in terms of a fast and a slow variable in the radial direction. Next, we seek solutions for the other second-order differential equation in terms of the same variables.

We treat the regions near the centre and the surface of the star as boundary layers.

By matching the appropriate asymptotic expansions and imposing the continuity of the gravitational potential and its gradient, we derive the equation determining the eigenfrequencies, and the various constants involved in the solutions.

## 3. Conclusions

- The Cowling approximation ($\Phi' = 0$) is valid in the lowest-order asymptotic approximation. The same eigenfunctions and eigenfrequencies are found as previously derived in the Cowling approximation (see, e.g., Smeyers & Tassoul 1987).
- The subseismic approximation ($P' = 0$) is valid from the centre to a distance sufficiently far from the surface in the lowest-order asymptotic approximation.
- The relative order of the lowest-order approximations of the eigenfunctions of $g$-modes in the boundary layer near $r = 0$ (central), the outer region, and the boundary layer near $r = R$ (surface) are given in the table below. When the lowest-order approximation is equal to zero, the lowest, possibly non-zero order is indicated between braces.

| | central | outer | surface |
|---|---|---|---|
| $\alpha/\xi$ | $\varepsilon$ | $\varepsilon^0$ | $\varepsilon^{-2}$ |
| $\eta/\xi$ | $\varepsilon$ | $\varepsilon^{-1}$ | $\varepsilon^{-2}$ |
| $\rho'/(\rho\xi)$ | $\varepsilon$ | $\varepsilon^0$ | $\varepsilon^{-2}$ |
| $P'/(\rho\xi)$ | $(\varepsilon^3)$ | $(\varepsilon)$ | $\varepsilon^0$ |
| $\Phi'/\xi$ | $(\varepsilon^3)$ | $(\varepsilon)$ | $(\varepsilon^2)$ |

Full details will be published in Astronomy and Astrophysics.

## References

Kevorkian, J., & Cole, J.D. 1981, Perturbation Methods in Applied Mathematics, Springer, New-York

Smeyers, P., & Tassoul, M. 1987, ApJS, 65, 429

*Astrophysical Applications of Stellar Pulsation*
*ASP Conference Series, Vol. 83, 1995*
*R. S. Stobie and P. A. Whitelock, eds.*

# Quadratic and Cubic Couplings of Oscillation Modes of Stars

T. Van Hoolst

*Instituut voor Sterrenkunde, Katholieke Universiteit Leuven, Celestijnenlaan 200B, 3001 Heverlee, Belgium*

## 1. Introduction

The strength of nonlinear interactions of oscillation modes of stars is determined by the amplitudes as well as by the eigenfunctions of the oscillation modes. The intrinsic couplings of modes through their eigenfunctions can be described by coupling coefficients. Here, we concentrate on quadratic and cubic coupling coefficients that describe the nonlinear coupling of modes with itself and are called self-coupling coefficients.

We considered radial and nonradial oscillation modes of polytropic models with degrees of central condensation that correspond to central condensations of main sequence stars to highly condensed evolved stars. We study the influence of the radial order and the degree of the oscillation mode on the self- coupling coefficients.

We adopt the isentropic approximation.

## 2. Coupling coefficients

When the nonlinear interactions of the stellar oscillation modes are taken into account up to the third order in the amplitudes, the time-behaviour of the amplitudes $a_k(t)$ of the linear, isentropic eigenfunction $\vec{\varphi}_k(\vec{r})$ of mode $k$ is governed by the coupled-mode equation:

$$\frac{d^2 a_k}{dt^2} + \omega_k^2 a_k = Q_k^{nm} a_n a_m + C_k^{nmr} a_n a_m a_r$$

(see Van Hoolst & Smeyers 1993, Van Hoolst 1994), where $Q_k^{nm}$ and $C_k^{nmr}$ are quadratic and cubic coupling coefficients and are measures of the strengths of the nonlinear interactions of the modes. Coupling coefficients are defined as the ratios of integrals involving the eigenfunctions $\vec{\varphi}_n(\vec{r})$ of the modes $n$ considered and the oscillatory moment of inertia of the mode $k$. The integrals of the quadratic and the cubic self-coupling coefficients contain the linear, isentropic eigenfunction $\vec{\varphi}_k(\vec{r})$ of mode $k$ three and four times, respectively.

In the oscillation calculations, the generalised isentropic coefficient $\Gamma_1 = (\partial \ln \rho / \partial \ln P)_S = 5/3$. The linear, isentropic eigenfunctions are normalised by setting the radial part of the relative radial displacement equal to one.

We determined dimensionless self-coupling coefficients $Q_k^{kk}/\omega_k^2$ and $C_k^{kkk}/\omega_k^2$ for modes $k$ as well as their isentropic frequencies $\omega_k$ and oscillatory moments of inertia.

## 3. Discussion

### 3.1. Radial order

With increasing radial order $n$, the frequencies and the absolute values of the dimensionless self-coupling coefficients increase whereas the moments of inertia decrease. Specifically, the quadratic self-coupling coefficient is found to increase almost linearly with radial order. An increase of coupling coefficients with increasing radial orders has also been noted by Van Hoolst (1994), and by, e.g., Takeuti et al. (1992) for quadratic coupling coefficients.

### 3.2. Spherical degree

The quadratic and cubic self-coupling coefficients $Q_k^{kk}$ and $C_k^{kkk}$ are proportional to the angular integral $\int |Y_k|^2 Y_k d\Omega$ and $\int |Y_k|^2 (Y_k)^2 d\Omega$, respectively. Here, $Y_k$ is the spherical harmonic associated with the mode $k$, and $d\Omega = \sin\theta d\theta d\phi$ is the infinitesimal solid angle in angular coordinates $\theta$ and $\phi$. The integrals over three spherical harmonics decrease and the integrals over four spherical harmonics increase with increasing degree.

The integral over the radial parts of the eigenfunction and the oscillatory moment of inertia also depend on the degree. We find that the dimensionless quadratic self-coupling coefficient decreases slightly with increasing degree $l$, and that the dimensionless cubic self-coupling coefficient decreases almost linearly with the degree.

### 3.3. Polytropic index

With increasing polytropic index $n_e$, or central mass-concentration, the frequencies increase and the moments of inertia decrease. The absolute values of the dimensionless self-coupling coefficients increase with increasing polytropic index for the lowest-order modes and decrease for the higher-order modes. Around polytropic index 4, this behaviour is sometimes reversed for low-order modes. Generally, the influence of $n_e$ is not very large, so that it can be expected, at least as concerns the density profile, that coupling coefficients for various classes of pulsating stars are very similar to each other and do not differ much from the coupling coefficients for polytropic models.

Full details will be published elsewhere.

## References

Van Hoolst, T. 1994, A&A, 286, 879

Van Hoolst, T., & Smeyers, P. 1994, A&A, 279, 417

Takeuti, M., Yamakawa, F., & Ishida, T. 1992, PASJ, 44, 101

*Astrophysical Applications of Stellar Pulsation*
*ASP Conference Series, Vol. 83, 1995*
*R. S. Stobie and P. A. Whitelock, eds.*

# Asteroseismology of the $\beta$ Cephei Variable 16 (EN) Lacertae

W.A. Dziembowski

*Copernicus Astronomical Centre, Polish Academy of Sciences*
*ul. Bartycka 18, 00–716 Warsaw, Poland*

M. Jerzykiewicz

*Wrocław University Observatory*
*ul. Kopernika 11, 51–622 Wrocław, Poland*

**Abstract.** Various observational data and model calculations are used to identify four pulsation modes detected in this star and to infer its properties. It is concluded that the star's interior must rotate faster than the outer layers.

This object is a single–lined spectroscopic binary and an eclipsing variable. It consists of the well–known $\beta$ Cephei star and an invisible secondary. Three circumstances make the $\beta$ Cephei primary exceptionally attractive as a target for asteroseismology: (1) frequencies of as many as four of its pulsation modes are precisely known from time–series observations, (2) multicolour photometry and radial velocity data show that one of the four modes is radial and, of course, which of them it is, and (3) the orbital and eclipse solutions yield a reliable value of the star's mean density.

Frequencies of the four pulsation modes observed in 16 Lac are $f_1 = 5.9112$, $f_2 = 5.8551$, $f_3 = 5.5033$, and $f_4 = 7.194$ c/d (Jerzykiewicz 1993). The colour–to–light and velocity–to–light amplitude ratios show that the $f_1$ mode is radial (Jerzykiewicz 1993, Cugier et al. 1994). Mean density of the star amounts to $\langle\rho\rangle = 0.0391 \pm 0.0053\ \langle\rho\rangle_\odot$. This value allows us to identity the $f_1$ mode as the fundamental, that is, $p_1$.

The measured value of the fundamental mode frequency determines a one–parameter family of stellar models. We assumed standard Population I composition ($X = 0.7$, $Z = 0.02$) and standard treatment of stellar evolution and pulsations (see Dziembowski & Pamyatnykh 1993 for details). The range of $\log T_{\rm eff}$ consistent with various determinations is 4.357 to 4.333 and corresponds to stellar mass range 9 to 10 $M_\odot$. The range of the star age corresponding to this mass range is 20 to 16 million years and that of the H abundance in the core, $X_c$, is 0.196 to 0.231.

Considering possible identifications of the $f_2$, $f_3$ and $f_4$ modes we limited ourselves to modes of spherical harmonic degree $l \leq 2$. We took into account uncertainty of the rotation rate and the azimuthal order of the modes, $m$. Initially we assumed a uniform rotation and we used for model frequencies, $\nu$, the well–known formula $\nu_m = \nu_0 + m f_{\rm rot}(1 - C)$, where $f_{\rm rot} = V_{\rm rot}/2\pi R$. The centroid frequencies, $\nu_0$, and the Ledoux constants, $C$, were calculated for the models.

For the surface rotational velocity, $V_{\rm rot}$, we adopted 40 km/s. This number is equal to a mean of two published $V_{\rm rot} \sin i$ values divided by $\langle \sin i \rangle$.

The frequency $f_2$ may be identified with the $m = 0$ modes whose frequencies are independent of the rotation as long as the effects of centrifugal force are ignored. There are two such possibilities. One is $l = 2$, $p_0$. This implies the following parameters: $\log T_{\rm eff} = 4.358$, $M = 9.9\ M_\odot$. The other is $l = 1$, $g_1$, formally implying a mass somewhat above the allowed upper limit of 10 $M_\odot$. Nevertheless, in view of uncertainties, we regard this identification as tenable. Discrimination between these two $m = 0$ modes should be possible with precise measurements of the light and colour phases for the $f_2$ term. Closer to the lower mass limit, the $l = 1$, $m = -1$, $g_1$ is also a possibility. More accurate values of $T_{\rm eff}$ would help discriminate between this and the two previous options.

Neither $f_3$ nor $f_4$ frequencies may be identified with any of the centroid frequencies in our models. A conservative identification of $f_4$ is $l = 2$, $m = -2$, $p_1$. This identification requires that the mean value of $f_{\rm rot}$ is higher than that assumed in the calculated rotational splitting. The required increase ranges from 21% for the 9 $M_\odot$ model to 24% for 10 $M_\odot$ model. The discrepancy may be due to an underestimate of $V_{\rm rot}$ or to an inward increase of $f_{\rm rot}$.

The latter interpretation is supported by two possible identifications of $f_3$. The more plausible one seems to be $l = 2$, $m = -2$, $p_0$. It requires a still larger value of the mean $f_{\rm rot}$. The increase above the adopted value ranges from about 61% at 10 $M_\odot$ to 73% at 9 $M_\odot$. This mode, being in fact a mixed $\rm p_0/g_1$ mode, samples much more of the star's deep interior than does $\rm p_1$. Thus, higher mean $f_{\rm rot}$ value required implies a significant inward increase of the rotation rate.

The other identification for $f_3$ is $l = 2$, $m = 2$, $g_2$. This mode samples preferentially the innermost part of the radiative interior where the hydrogen concentration gradient occurs. At $M = 9\ M_\odot$ this identification requires a 63% increase of $f_{\rm rot}$ over the adopted value. The required increase growth rapidly with mass and, for instance, at $M = 9.5\ M_\odot$ it is 224%.

Let us point out that we could discriminate between the two identifications of $f_3$ if we knew the sign of $m$, but this requires a difficult analysis of the line–profile observations. However, regardless of which identification is correct, the conclusion that the interior of EN Lac rotates significantly faster than the outer layers is unavoidable within the adopted assumptions.

**Acknowledgments.** This research was supported by KBN grants Nr 2 P304 013 07 and Nr 2 P304 001 04. MJ gratefully acknowledges a travel grant from the Government of the Republic of South Africa, WAD that from the IAU.

## References

Cugier, H., Dziembowski, W. A., & Pamyatnykh, A. A. 1994, A&A, 291, 143

Dziembowski, W. A., & Pamyatnykh, A. A. 1993, MNRAS, 262, 204

Jerzykiewicz, M. 1993, Acta astr., 43, 13

*Astrophysical Applications of Stellar Pulsation*
*ASP Conference Series, Vol. 83, 1995*
*R. S. Stobie and P. A. Whitelock, eds.*

# Uncertainties in the Position of the $\beta$ Cephei Instability Strip in the HR Diagram

A.A. Pamyatnykh, W.A. Dziembowski, P. Moskalik

*Copernicus Astronomical Center, Polish Academy of Sciences*
*ul. Bartycka 18, 00–716 Warsaw, Poland*

**Abstract.** We discuss the sensitivity of the theoretical B star instability domains to the heavy element abundance $Z$, the adopted metal mixture, the assumed overshooting from stellar convective cores and the choice of the opacity data.

A very good agreement between results of stability surveys (Dziembowski & Pamyatnykh, 1993; Gautschy & Saio, 1993; Dziembowski, Moskalik, & Pamyatnykh, 1993) conducted for models employing the OPAL opacities (Iglesias, Rogers, & Wilson, 1992) with observational data on B type pulsators leaves no doubt that the long searched driving mechanism has been correctly identified. This is the usual $\kappa$ mechanism acting in the layer at $T \approx 2 \times 10^5$ K where a pronounced bump in the metal opacities occurs. The good agreement concerns not only $\beta$ Cep stars but also SPB (53 Per) stars which are B3-B9 type objects exhibiting light variations with periods of 1 to 4 d. The whole domain in the H-R diagram where the instability driven by the metal opacity bump occurs we call *the $\beta$ Cephei instability strip.* This is in an analogy with *the $\delta$ Cep instability strip*, where stellar pulsations are driven by the opacity mechanism acting in the HeII ionization zone.

The purpose of our research was to examine prospects for using occurrence of B type pulsators in a stellar system as a diagnostic of the system properties as well as a test of the stellar evolution theory. The main advantage over the similar efforts in the case of $\delta$ Cep instability strip is that the convection plays no role in B star pulsations. We conduct a similar stability survey to those described in our earlier papers. However, in addition to original OPAL opacities, we use their updated version as well as OP (Seaton et al. 1993) data, which allowed us to assess uncertainties introduced by various approximations adopted in the opacity calculations. Moreover, we considered cases of lower metal abundance and models with convective overshooting.

The position of the Blue Edge of the $\beta$ Cephei instability domain is determined by the metallicity, primarily by the abundance of the iron-group elements (Fe, but also Cr, Mn, Ni). An observational determination of the Blue Edge in a stellar system, therefore, yields a probe of the system's metallicity. With the OP tables $Z = 0.015$ is sufficient to explain the excitation of almost all cluster $\beta$ Cephei variables. With the original OPAL data the corresponding $Z$ values was above 0.02. Most recent data, based on calculations with additional 7 elements included, lower that number but still there is some 0.003 difference in the inferred $Z$-value depending on the choice of the opacity tables (OP *vs.* OPAL).

The instability domain extends to red well beyond the Main Sequence band. However, since the Post-Main Sequence evolution is very fast we identify the red edge of the instability strip with that of the MS band (TAMS). Position of TAMS in the HR diagram depends on the overall metallicity $Z$ and on the amount of the convective overshooting assumed in the models. Both, increase in $Z$ and the overshooting distance moves TAMS toward red. The latter effect is more significant. The observed position of the red edge constrains the overshooting length. Standard models without convective overshooting explain very well the position of the $\beta$ Cephei stars in the HR diagram. On the contrary, allowing even small overshooting results in a sizable extension of the instability domain toward lower $T_{\rm eff}$ and longer periods, where no $\beta$ Cephei stars are found.

With decreasing $Z$ the $\beta$ Cephei instability domain shrinks down and finally vanishes. In the case of OP opacities this happens at $Z \approx 0.012$.

The observed range of $\beta$ Cep star luminosities and corresponding masses is significantly narrower than that predicted by the theory. In our survey we found that the short-period mode instability begins at $M \approx 7.5M_{\odot}$, while the minimum mass for the observed $\beta$ Cep stars is about $9M_{\odot}$. We do not know whether this discrepancy reflects some inadequacy in the stellar opacity calculation or a difference in the nonlinear development of the instability.

There is no upper luminosity limit of the instability domain. Kiriakidis, Fricke & Glatzel (1993) showed that the blue edge, which initially moves away from ZAMS with mass increase, at $M \approx 20M_{\odot}$ turns back and reaches it at about $70M_{\odot}$. We have confirmed this result and found that the instability strip is the narrowest in terms of $T_{\rm eff}$ at $M \approx 18M_{\odot}$. In Sterken & Jerzykiewicz (1993) list of $\beta$ Cep stars there is only one (V986 Oph) occuring in the upper part of the instability domain. The apparent paucity of hot and luminous $\beta$ Cep calls for explanation.

The high order g-mode instability occurs in a wide range of temperatures and it is quite robust. It persists at lower metal abundance ($Z \approx 0.006$ for the OP opacities) than that of the p modes excited in $\beta$ Cep stars. The use of the OP instead of OPAL opacities promotes the instability, extending it to somewhat higher degrees (longer periods).

**Acknowledgments.** This research was supported by KBN grant Nr 2 P304 013 07. WAD and PM gratefully acknowledge travel grants from the IAU.

## References

Dziembowski, W. A., Moskalik, P., & Pamyatnykh, A. A. 1993, MNRAS, 265, 588

Dziembowski, W. A., & Pamyatnykh, A. A. 1993, MNRAS, 262, 204

Gautschy, A., & Saio, H. 1993, MNRAS, 262, 213

Iglesias, C. A., Rogers, F. J., & Wilson, B. G. 1992, ApJ, 397, 71

Kiriakidis, M., Fricke, K. J., & Glatzel, W. 1993, MNRAS, 264, 50

Seaton, M. J., Yu Yan, Mihalas, D., & Pradhan, A. K. 1994, MNRAS, 266, 805

Sterken, C,. & Jerzykiewicz, M. 1993, Space Sci. Rev., 62, 95

*Astrophysical Applications of Stellar Pulsation*
*ASP Conference Series, Vol. 83, 1995*
*R. S. Stobie and P. A. Whitelock, eds.*

# A Search for Variable B Stars in the Northern Open Clusters

Andrzej Pigulski, Mikołaj Jerzykiewicz and Grzegorz Kopacki

*Wrocław University Observatory, ul. Kopernika 11, 51–622 Wrocław, Poland, Electronic mail: pigulski@ii.uni.wroc.pl*

**Abstract.** In April 1994 we started a program of searching for variable B-type stars in young open clusters. The equipment consisted of a 60-cm reflecting telescope and a CCD camera with a Johnson's B filter. Up to date we obtained nearly 1500 CCD frames of the central regions of four clusters: IC 4996, NGC 7235, NGC 7510, and NGC 654. In the range from B 0 to B 5, the photometric accuracy we achieve allows detection of short-period variability with amplitudes of about 5 mmag or greater. In this note we present the short-period variable stars we found in NGC 7235 and NGC 7510.

## 1. Introduction

With the discovery of the opacity-driven pulsational instability in models of B-type stars (see, e.g., Moskalik, these proceedings), the importance of finding $\beta$ Cephei variables and SPB stars in open clusters has shifted from mapping the instability domain in the HR diagram to (1) providing information about the clusters themselves and (2) identifying objects for detailed asteroseismological studies. The first possibility may lead to constraining the age and metal content of the clusters, the second will ultimately yield data for comparison with the theory of the interiors of hot stars.

## 2. Equipment and reductions

All observations were obtained in the Białków station of the Wrocław University Observatory. We used the Photometrics Star I camera with the Thomson TH 7883 384 × 576 CCD chip, mounted in the Cassegrain focus of a 60-cm reflecting telescope. The field was $4' \times 6'$. Most exposures were taken through a Johnson's $B$ filter. The exposure times ranged from 40 to 90 seconds.

The initial processing of the frames included corrections for bias, dark- and flat-fields. Stellar magnitudes were then derived by means of Stetson's (1987) Daophot II profile-fitting software. In the next step, differential magnitudes were calculated between pairs of stars. Variable stars were picked out at this stage. In the last step, we derived differential magnitudes once more, now using three nearby comparison stars for each variable we found.

## 3. NGC 7235

There are 376 frames, obtained on 13 nights in the interval from JD 2449516 to JD 2449545. Out of the 80 brightest stars investigated so far, two show periodic variations. The numbers we assigned to them are 198 and 167.

#198. This is #8 in the photoelectric sequence of Hoag et al. (1961). Our B observations can be represented as a sum of two sinusoids with periods equal to $P_1 = 0\overset{d}{.}20305$ and $P_2 = 0\overset{d}{.}17800$. The amplitudes (half-ranges) amount to $12.6 \pm 0.6$ mmag and $10.1 \pm 0.6$ mmag, respectively. We conclude that the star is a $\beta$ Cephei variable.

#167 (47″ north and 18″ west of the second star in the photographic sequence of Hoag et al. 1961) exhibits a variation typical of the W Ursae Majoris eclipsing systems. The period is $0\overset{d}{.}2800$, implying component masses close to one solar mass. On the other hand, the star's $B$ magnitude is equal to 16.2. If the star were a member of NGC 7235, this would translate into $M_V = 0\overset{m}{.}2$, corresponding to spectral type of B 9. We conclude that #167 is a foreground object.

## 4. NGC 7510

This cluster has been the subject of a recent CCD study by Sagar & Griffiths (1991, henceforth SG). Our observations consist of 284 frames, taken on 11 nights in the interval from JD 2449547 to JD 2449569. About 100 brightest stars were checked for variability. Two turned out to be short-period variables.

SG 516 has a sinusoidal light-curve with a period of $0\overset{d}{.}1022$ and an amplitude equal to $11.2 \pm 1.2$ mmag.

According to Hoag et al. (1961), the photographic $U - B$ and $B - V$ of the star are equal to 0.58 and 0.81, respectively. Using these data and the cluster's mean $E_{U-B}$ and $E_{B-V}$, we get $(U - B)_0 = -0.19$ and $(B - V)_0 = -0.25$. These numbers are clearly inconsistent. Note, however, that the ratio of the observed colour indices is very nearly equal to the slope of the reddening line in the two-colour diagram. Thus, a consistent pair of de-reddened colour indices would be obtained if SG 516 were assumed to be an early A-type star. We conclude that SG 516 is a foreground $\delta$ Scuti variable.

SG 101. Our observations can be represented by a period of $0\overset{d}{.}659$, or its 1 c/d alias, equal to $0\overset{d}{.}3976$. The data are insufficient to exclude either value.

**Acknowledgments.** This research was supported by KBN grant Nr 2 P304 001 04. M.J. gratefully acknowledges a travel grant from the Government of the Republic of South Africa.

## References

Hoag, A. A., Johnson, H. L., Iriarte, B., Mitchell, R. I., Hallam, K. L., & Sharpless, S. 1961, Publ. U.S. nav. Obs., 2nd ser., Vol. 17, part 7

Sagar, R., & Griffiths, W. K. 1991, MNRAS, 250, 683

Stetson, P. B. 1987, PASP, 99, 191

*Astrophysical Applications of Stellar Pulsation*
*ASP Conference Series, Vol. 83, 1995*
*R. S. Stobie and P. A. Whitelock, eds.*

# On the H$\alpha$ Emission Phase of $\beta$ Cep

L. Kaper

*ESO, Karl-Schwarzschild-Str. 2, Garching bei München, Germany*

P. Mathias

*Observatoire de la Cote d'Azur, BP 229, F-06304 Nice Cedex 4, France*

**Abstract.** During the past five years, we monitored the H$\alpha$ line of $\beta$ Cep. The strength of the H$\alpha$ emission, first discovered in 1990, is linearly declining with time. Several options are explored, but a satisfactory explanation for the recurring phases of H$\alpha$ emission is not yet found.

## 1. Introduction

We report on the H$\alpha$ emission phase that is presently observed for the pulsating B2 III star $\beta$ Cep ($P_{\rm pul} = 4^h34^m$). The H$\alpha$ emission, which was first observed (Mathias et al. 1990) in July 1990 (and absent in 1985), is superposed on the Stark-broadened photospheric absorption profile and has a Gaussian shape. The H$\alpha$ emission reached maximum strength at the end of 1990 and is presently decreasing in strength, but still well pronounced.

A similar H$\alpha$ emission phase was observed in 1956 (Wilson & Seddon 1956) and apart from that, several cases of incipient H$\alpha$ emission have been reported during this century. The characteristic "$\beta$ Cep" pulsation seems not to be affected by the presence of the H$\alpha$ emission (see also Telting et al., this issue).

## 2. Discussion

The origin of the observed H$\alpha$ emission is unclear. As a first option, the H$\alpha$ emission could be caused by a Be-star-type equatorial disk. Be stars and $\beta$ Cep stars share the same region in the HRD and the H$\alpha$ emission officially classifies $\beta$ Cep as a Be star. It has been suggested (e.g., Balona 1990) that the two phenomena are mutually exclusive. However, transitions from Be to $\beta$ Cep stars (and vice versa) or Be stars with $\beta$ Cep characteristics have been observed. Although $\beta$ Cep might be a pole-on Be star, the H$\alpha$ emission is much weaker than observed for some "classical" Be stars. Secondly, Pigulski & Boratyn (1992) showed that $\beta$ Cep has a close companion in an eccentric orbit of 92 years. Bedding (1993) reported a similar discovery for the B star $\delta$ Sco, for which H$\alpha$ emission has been detected (Coté & Van Kerkwijk 1993). Both $\beta$ Cep and $\delta$ Sco are candidates for having a circumstellar cloud (Gurzadyan & Rustambekova 1987), which may perhaps be formed during periastron passage. The orbital period, on the other hand, does not correspond to the recurrence time-scale of

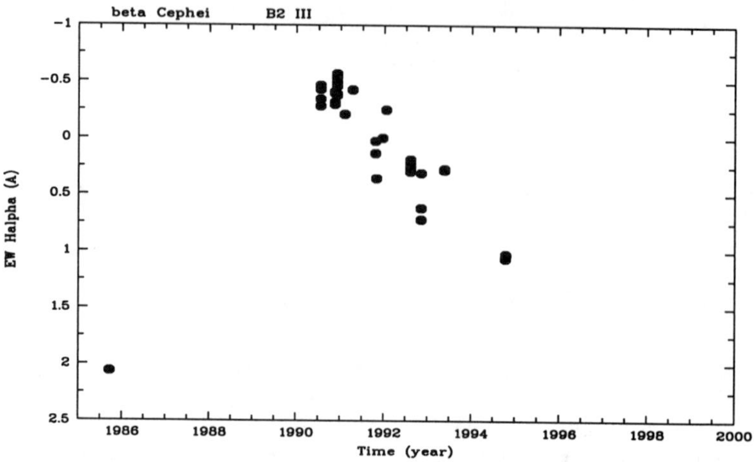


Figure 1. The H$\alpha$ emission of $\beta$ Cep is gradually declining in strength. The EWs have been corrected for the presence of telluric lines and the C II lines around 6580 Å. We expect that the H$\alpha$ emission phase will end around 1998.

H$\alpha$ emission. Thirdly, Chapellier (1986) suggested that a 30 years cycle might be present in the radial-velocity curve's amplitude, with minima around 1920 and 1950. The apparent change in radial velocity amplitude could be due, however, to the multi-periodicity in $\beta$ Cep's pulsational behaviour (Aerts et al. 1994). Lastly, Henrichs et al. (1993) proposed that the periodic change in EW found in the UV resonance lines of $\beta$ Cep is caused by modulation of the stellar wind by an oblique dipolar magnetic field at the stellar surface. The period of 12 days would then correspond to the stellar rotation period. One might speculate that a changing magnetic field configuration has consequences for the distribution of wind material close to the star, and thus the amount of H$\alpha$ emission.

## References

Aerts, C., et al. 1994, A&A, 286, 109

Balona, L.A. 1990, MNRAS, 245, 92

Bedding, T.R. 1993, AJ, 106, 768

Chapellier, E. 1986, A&AS, 64, 275

Coté, J., & Van Kerkwijk, M.H. 1993, A&A, 274, 870

Gurzadyan, G.A., & Rustambekova, S.S. 1987, Ap&SS, 134, 1

Henrichs, H.F., et al. 1994, in: IAU Coll. 139, Eds. Nemec & Matthews, p. 186

Mathias, P., Gillet, D., & Kaper, L. 1990, in ESO Workshop No. 36, Ed. D. Baade

Pigulski, A., & Boratyn, D.A. 1992, A&A, 253, 178

Wilson, R., & Seddon, H. 1956, The Observatory 76, 145

*Astrophysical Applications of Stellar Pulsation*
*ASP Conference Series, Vol. 83, 1995*
*R. S. Stobie and P. A. Whitelock, eds.*

# A Period Analysis of the Optical Line Variability of $\beta$ Cephei [1]

J. Telting

*Astronomical Institute 'Anton Pannekoek', University of Amsterdam, Kruislaan 403, NL-1098 SJ Amsterdam, Netherlands*

C. Aerts[2], and P. Mathias[3]

*Instituut voor Sterrenkunde, Katholieke Universiteit Leuven, Celestijnenlaan 200 B, B-3001 Heverlee, Belgium*

## 1. Introduction

Recently Aerts et al. (1994, Paper I) discovered multi-periodicity in the variations of optical lines of $\beta$ Cephei. They identified the frequency of the radial pulsation ($f_1$ = 5.25 cycles/day), and additionally found 2 frequencies ($f_2$ = 5.38 c/d, $f_3$ = 4.92 c/d) close to the main frequency. The amplitude of these variations was found to be much smaller than that of the radial pulsation. Aerts et al. (1994) attributed these small variations at the newly found frequencies to non-radial pulsation modes of the star, but labelled the mode-identifications as uncertain.

As a follow up on this work, we investigated the whole frequency range that can be analysed with the data set described in Paper I. For four absorption lines (Si III $\lambda$ 4552, 4567, 4574 and O II $\lambda$ 4591) we examined variations in EW and the velocity moments of the absorption lines (yielding the velocity shift, line width and skewness). We also investigated the variations of the normalised intensity for each individual wavelength bin of the spectra.

## 2. Results

We find, apart from the already known frequencies, two other frequencies ($f_4$ = 5.08 c/d, $f_5$ = 5.42 c/d) around the main frequency, which have variations exceeding the noise level. These frequencies clearly show up in the CLEANed periodograms of the first velocity moment and of the intensity variations across the lines. Furthermore, we find a peak in the periodograms of the moments at 0.17 c/d, corresponding to a period of 6 days.

Henrichs et al. (1993) and Veen et al. (1995) identify the frequency of variations detected in UV wind lines with the rotation period of a magnetic dipole,

[1] Based on observations obtained at the Observatoire de Haute-Provence, France

[2] Senior Research Assistant, Belgian National Fund for Scientific Research

[3] Present Address: Observatoire de la Côte d'Azur, Département Fresnel, BP 229, F-06304 Nice Cedex 4, France

which implies a rotation period of the star of 12 days. We find that the spacings between the main frequency $f_1$ and the frequencies $f_3$, $f_4$, and $f_5$ are multiples of the frequency detected in the UV wind lines. This means that we can identify the spacing between the frequencies (i.e., $f_1 - f_3 \approx 0.33$ c/d, $f_1 - f_4 \approx 0.17$ c/d, $f_5 - f_1 \approx 0.17$ c/d), as derived from the optical Si III and O II lines, as multiples of the rotation period.

We find that for the intensity variations across the lines, the pattern of frequencies ($f_2$, $f_3$, $f_4$ and $f_5$) around the frequency $f_1$ of the main (radial) pulsation, repeats itself around every detected harmonic of the main frequency, with the same frequency spacing as around the main frequency.

## 3. Discussion

The pattern of frequency combinations of $f_1$ and $f_2$ (i.e., $f_1$, $f_2$, $f_1+f_1$, $f_1+f_2$, $2f_1+f_1$, $2f_1+f_2$, etc.), as detected in the intensity variations across the lines, can be modelled as a multi-periodic star with l=m=0 and l=m=2 spheroidal modes (Paper I). This model also gives a good description of the observed power and phase distribution across the absorption lines. At present we are investigating if the power and phase distribution of the frequencies $f_3$, $f_4$ and $f_5$ can be modelled as the result of (non-)radial pulsations.

The fact that the spacings of the detected frequencies are multiples of the rotation frequency suggests that the observed variations with frequencies $f_3$, $f_4$, and $f_5$ might be due to modulation effects. The similarities between the variations of UV wind lines and optical lines suggest that either a common phenomenon in wind and photosphere (e.g., a magnetic field) causes these similarities or that one type of variation (wind variability) may be caused by the other (pulsations).

Further investigations will hopefully show whether the variations with frequencies $f_3$, $f_4$, and $f_5$ in the spectra of $\beta$ Cephei are best described as (non)-radial pulsations, or as the result of modulation caused by the combination of rotation, pulsations and the presence of a magnetic field.

## References


Aerts, C., Mathias, P., Gillet, D., & Waelkens, C., 1994, A&A 286, 109

Henrichs, H.F., Bauer, F., Hill, G.M., Kaper, L., Nichols Bohlin, J.S., & Veen P.M., 1993, Proceedings IAU Coll. 139, p. 186, Cambridge Univ. Press

Veen, P.M., Henrichs, H.F., et al. 1995, in preparation

*Astrophysical Applications of Stellar Pulsation*
*ASP Conference Series, Vol. 83, 1995*
*R. S. Stobie and P. A. Whitelock, eds.*

# Search for Pulsational Instability among h Persei Stars

J. Krzesinski
*Mt. Suhora Observatory, Cracow Pedagogical University, ul. Podchorążych 2, 30-084 Kraków, Poland*

**Abstract.** The northern-hemisphere double cluster h and $\chi$ Persei is one of the richest young open clusters. At least half of the brightest stars in the cluster appear to be variables and most of them are Be stars. Till now there was no evidence for the presence of $\beta$ Cephei stars in the cluster. It is remarkable that the similar age southern cluster NGC 3293 and other clusters belonging to the same age group are documented as clusters in which $\beta$ Cephei stars have been observed. In this work we report finding one suspected $\beta$ Cephei-like object, basing on our own observations.

## 1. Introduction.

A new thorough search for pulsational instability among h Persei stars has been undertaken during last months of 1994. The search was focused on three types of variable stars: $\beta$ Cephei, Supergiant and Slowly Pulsating B stars, which should pulsate due to the same excitation mechanism, i.e., $\kappa$-effect acting in the metal opacity bump (Dziembowski et al. 1993). The CCD observations of the central region of the h Persei cluster were carried out with 60-cm telescope (equipped with an autoguider) at Mt. Suhora Observatory. The field of view of a CCD camera, equal to 4x6 arc minutes, allowed us to take images of nearly a hundred stars in the same frame, but only about 20 of them were stars, which fall into instability strip predicted by the theory. Exposures were taken through the V filter close to Johnson's V filter and with integration times from 50 up to 150 seconds. To derive instrumental stellar magnitudes for the brightest separated (or semi-separated) stars visible in frames, an aperture method was used. The accuracy of a single measurement was better than 0.02 magnitude for 11-th magnitude stars.

## 2. Results.

Among 40 stars taken into analysis we have found one suspect $\beta$ Cephei-like object with P=0.172 days, which was identified to be the Oo 692 h Persei, B0 type (9.45 mag. in V) and luminosity class V star. This type corresponds to the $\log(T_{\rm eff})$ value equal to 4.48 and $\log(L/L_{\odot}) = 4.72$. It means that the Oo 692 star has mass $\sim$17 $M_{\odot}$ and can be placed at the edge of an instability strip, determined for the metal abundance of Z=0.02.

There is no doubt that this star is a fast rotator (v*sin(i)=135 km/sec, Waelkens et al. in preparation), thus its behaviour (if pulsations are real) would be very important in connection with the problem of damping of pulsations in fast rotators.

The Oo 963, B2 IV (m=10.7 mag. in V) star has been also taken into analysis as a $\beta$ Cephei suspect from a list of $\beta$ Cephei suspects published by C. Sterken and M. Jerzykiewicz (1992), but we cannot confirm this type of variability. If this star is a variable, its amplitude of variations should be less than 0.01 magnitude.

## References

Dziembowski, W.A., Moskalik, P., & Pamyatnykh, A.A. 1993, MNRAS, 265, 588.

Waelkens, C., Aerts, C., Simeons, K., De Mey, K., & Denoyelle, J. 1995, in preparation.

Sterken, C., & Jerzykiewicz, M. 1993, Space Sci. Rev., 62, 95.

*Astrophysical Applications of Stellar Pulsation*
*ASP Conference Series, Vol. 83, 1995*
*R. S. Stobie and P. A. Whitelock, eds.*

# NRP Modes in a Be Star Zeta Oph

E. Kambe

*Department of Geoscience, National Defense Academy, Yokosuka, Kanagawa 239, Japan*

R. Hirata and M. Kato

*Dept of Astronomy, Kyoto University, Kyoto 606-01, Japan*

H. Ando

*National Astronomical Observatory, Mitaka 181, Japan*

E. J. Kennelly and G. A. H. Walker

*Department of Geophysics and Astronomy, University of British Columbia, Vancouver, BC, V6T1Z4, Canada*

S. Štefl

*Astronomical Institute, Academy of Sciences of the Czech Republic, CZ-251 65 Ondrejov, Czech Republic*

A. E. Tarasov

*Crimean Astrophysical Observatory, Nauchnyj, Crimea, 334413, Ukraine*

**Abstract.** Two large sinusoidal variations with periods of 3.337 hrs and 2.018 hrs and other smaller variations have been detected from the period analysis of HeI $\lambda$ 6678 spectra obtained by our multi-site campaign for $\zeta$ Oph in May, 1993. The resultant periodgram is considerably improved (aliasing free) from those in previous publications. The periodicity may be consistent with previous studies except ambiguities of aliasing. It is surprising that almost all detected periods have a common superperiodicity of about 10.05 hrs. We emphasise the importance of multi-site campaign for the study of line-profile variations (lpv) in early-type stars.

The precise determination of the periodicity of modes of lpv in Be and related stars is very important to infer the modal nature of the oscillations and their possible physical link to Be phenomenon. Unfortunately, it is not easy to obtain good temporal coverage data from an observing site alone. So, we have carried out a couple of multi-site campaigns of Be stars since 1993 (Hirata 1994). Here we present a preliminary result of an analysis of spectroscopic observation which is a part of a multi-site campaign of $\zeta Oph$ in May, 1993. The detailed result including photometric observations and others will be published elsewhere soon.

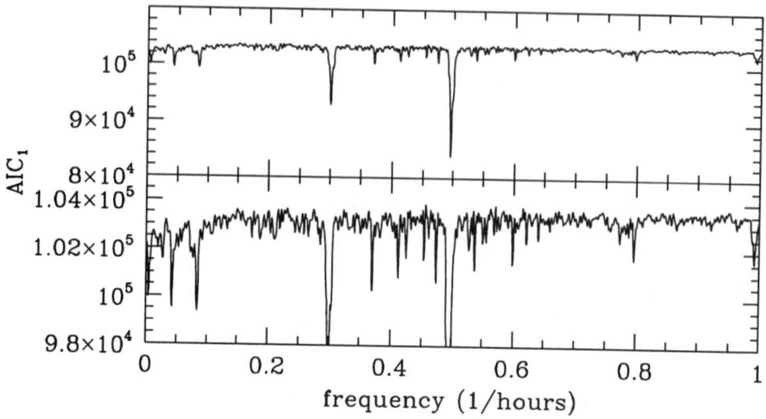


Figure 1. The periodgram of one sinusoid fitting. See text for detail.

The spectroscopic data were monitored at four sites (Okayama, Crimea, ESO/CAT, and Dominion) and 411 HeI $\lambda$ 6678 spectra were obtained in 15 nights during the campaign. The resolution of the spectra (slit width) is 0.1A-0.33A and the typical signal to noise ratio of the spectra is 300-500.

The result of the period analysis is shown in Fig. 1. The AIC$_1$ corresponds to the residual of the least-square fitting of one sinusoid to the data in temporal domain (see Kambe et al. 1993 for the period analysis in detail). The periodgram is much more aliasing free than in previous studies. The lower panel of the figure is the magnified diagram of AIC$_1$, where less significant peaks are seen with aliasings of the two main periods. The longer periods are likely to be due to the systematic errors of the reduction. In summary, we could detect at least two strong periodicities in the data, i.e., 2.018hr and 3.337hr, with other possible periods of 1.008 hr, 1.256 hr, and 1.668 hr. All these periods have a common superperiod of about 10.05 hrs.

The result may consistent with previous studies by Kambe et al. 1993 and Reid et al. 1993 except for an ambiguity of aliasing, in which periodicity of 2.43 hrs and 3.33 hrs are detected as lpv with two largest amplitudes. If the true periodicity is 2.018 hrs instead of 2.43 hrs, it is clear that multi-site campaigns are very important to establish the true periods of the lpv.

The extensive discussion of the new periodicity will be published soon. Although we still have not made any satisfactory understanding of the lpv in Be stars, multi-site campaigns will bring us some significant progress in understanding the feature and cause of lpv in Be stars and other related stars.

## References

Hirata, R. 1994, Be Newsletter, No. 28, p.11

Kambe, E., Ando, H., & Hirata R. 1993, A&A, 273, 435

Reid, A.H.N., et al. 1993, ApJ, 417, 338

*Astrophysical Applications of Stellar Pulsation*
*ASP Conference Series, Vol. 83, 1995*
*R. S. Stobie and P. A. Whitelock, eds.*

# A First Search for Rapid Variations of Stellar and Circumstellar Balmer Discontinuity in Be Stars

S. Štefl

*Astronomical Institute, Academy of Sciences, Ondřejov, Czech Republic*

D. Baade

*European Southern Observatory, Garching bei München, Germany*

J. Cuypers

*Belgian Royal Observatory, Brussel, Belgium*

**Abstract.** Spectrophotometric observations of the Be star $\eta$ Cen obtained in May, 1993 show smooth variations of both the stellar and circumstellar Balmer discontinuity (BD) on a time scale of hours. Simultaneous photometry and high-resolution spectroscopy suggest possible correlations with the large-amplitude brightness and line-profile variations.

## 1. Introduction

Simultaneous monitoring of the southern Be star $\eta$ Cen by means of flux calibrated spectra ($\lambda\lambda$ 340–510 nm, resolving power 5,000, temporal sampling 3–5 minutes), high-resolution line profiles (Si III $\lambda$455.3 and 456.2 nm, resolving power 65,000, temporal sampling 15–20 minutes) and *uvby* magnitudes was carried out with three telescopes of the La Silla Observatory. The aim was to search for the rapid BD variations predicted by Štefl et al. (Štefl, S., Baade D., Harmanec, P., Balona L.A. 1995, A&A, 294, 135), and to investigate their correlations, if any, with the known rapid light and line-profile variations. Due to exceptionally bad weather, good quality data were obtained during one night only (May 10/11, 1993). The new results reported here are therefore preliminary.

## 2. Results

The BCD parameters $D$ and $d$ (height of the photospheric and circumstellar BD, respectively, measured at the common wavelength $\lambda_0$ = 370nm; for the definition see, e.g., Divan L.: 1979, IAU Coll. 47, p. 247 and references therein) characterizing the Balmer discontinuity were measured. The main results are:

- Variations of the photospheric BD were detected at high confidence level (peak-to-peak amplitude 4% of the ambient continuum flux).
- Small variations (1% amplitude) of the *circumstellar* BD (in emission) were also found.

- Scattered observations from two more nights show that the variations of the BD parameters probably follow the 0.6424-day photometric period.

The range in $u-b$ amounts to $0\overset{m}{.}02$. However, unlike 1992 when the peak-to-peak amplitude was $0\overset{m}{.}03$ (Štefl et al. 1995), the 1993 observations do not cover the phases of light and $u-b$ colour maximum. Therefore, the true amplitudes of the BD variations may be larger than observed so far.

Ultra-rapid light variations observed after phase $0\overset{P}{.}5$ may represent an intrinsic flickering. No such flickering was observed in 1992 (Štefl et al. 1995).

Almost no Si III line-profile variations were detected in the first half of the night. Profiles with large asymmetric wings extending to more than –1000 km/s appear in the phase interval $0\overset{P}{.}5$–$0\overset{P}{.}6$ and were followed by strong absorption bumps arising in the far violet wings and moving to the line centre.

Two components can be distinguished in the variability of the $1^{st}$ moment of the Si III 455.3 line: i) the smooth (probably phase dependent) variations described by Štefl et al. (1995), ii) an abrupt decrease during the phase interval $0\overset{P}{.}45$–$0\overset{P}{.}60$ when the extended violet wings were present.

If fixed phase relations between the rapid variations of the 3 quantities studied (Balmer discontinuity, line profiles, and light and colour) are assumed, our observations suggest the following correlations:

- The photospheric BD height may correlate with brightness and low-order line-profile variations which maintain a stable period of 0.6424 days.
- The variability of the circumstellar BD coincides with the occurrence of rapid photometric flickering and of strong narrow features moving across the Si III line profiles. The two high-frequency phenomena appear when the circumstellar Balmer height $d$ is low.

## 3. Discussion

BD variations are detectable from ground-based data. Our observations of $\eta$ Cen represent the first direct spectroscopic evidence of rapid BD variations in early-type stars. Possibly, not only the stellar Balmer discontinuity but the circumstellar one as well varies with phase of the general photospheric variability.

The sensitivity of height and position of the BD to temperature and density, respectively, and possible phase relations between variations of BD, line profiles, and light and colour may permit, for the first time, an analysis of rapid variations to be performed separately for photosphere and circumstellar envelope.

The narrow moving features in the Si III lines, photometric flickering, and the low state of the circumstellar Balmer jump may all be connected with processes in the circumstellar gas. We note that these phenomena were observed shortly after the development of the highly asymmetric Si III profiles which persisted for less than 1 hour. If the latter are due to mass outflow and we may assume an expansion velocity of 1000 km/s and $R_\star = 4.0\ R_\odot$ (Štefl et al. 1995), the processes giving rise to these rapid variations would take place within about $1\ R_\star$ of the photosphere.

**Acknowledgments.** The authors thank Dr. J. Zorec for constructive comments to the manuscript.

*Astrophysical Applications of Stellar Pulsation*
*ASP Conference Series, Vol. 83, 1995*
*R. S. Stobie and P. A. Whitelock, eds.*

# Rapid Photometric and Spectroscopic Variability of the Be Star DX Eri

S. Štefl

*Astronomical Institute, Academy of Sciences, Ondřejov, Czech Republic*

L. A. Balona

*South African Astronomical Observatory, P.O. Box 9, Observatory 7935, Cape, South Africa*

**Abstract.** Short-period variations of high-resolution He I 667.8 nm line profiles obtained on 5 nights are analysed and compared to extensive *uvby* photometry obtained during 1986 – 1993. The separation of short-period and medium-term light variations was done for 4 seasons. All prewhitened magnitudes can be phased with a single period of 1.2564 d. The radial velocity and the first moment of the He I 667.8 line follows the photometric period with a semi-amplitude of 42 km $s^{-1}$. The large amplitude is very difficult to understand in terms of current models of non-radial pulsation or star spots.

## 1. Introduction

We analysed 280 new *uvby* observations of the Be star DX Eri (56 Eri=HR 1508) obtained at ESO, La Silla and SAAO during 1991 – 1993 together with those published by Balona et al. (1992). In addition, 28 high-resolution He I 667.8 nm profiles were obtained with the ESO 1.4-m telescope during 5 nights in November 1991. The resolving power is 70 000, temporal sampling 60 – 80 minutes and typical S/N ≈ 200.

The purpose is to derive the period of the rapid variations, to determine the colour variation and to model the relationship between the photometric and spectroscopic parameters by means of non-radial pulsations and star spots. In this way we hope to discriminate between the two models.

## 2. Results

Random medium-term (with a characteristic time scale of about 10 days) and short-period light variations combine in a complicated pattern. In order to separate the strictly periodic, variation characteristic of many Be stars (the $\lambda$ Eri phenomenon), it was necessary to employ a smoothing procedure.

---

[1]Based on data collected at the ESO and SAAO.

Time series analysis of smoothed $b$-magnitudes in individual seasons gives significant periods in the interval 1.241 – 1.369 d. We cannot determine whether the differences reflect intrinsic variations in the period or whether they are caused by imperfect prewhitening. Nevertheless, all smoothed $b$-magnitudes in the interval 1986 – 1993 are well phased with one single period of 1.256 d.

The colour-index variations corresponding to the periodic component were derived from the prewhitened *uvby* magnitudes obtained in 1991 and 1993. The following conclusions can be drawn regarding this component:

- All significant variations are in phase. The semi-amplitude of the brightness variation ( $0\overset{m}{.}08$ – $0\overset{m}{.}10$ ) is comparable in all passbands and seasons.
- The light curves are not symmetrical: light maxima occur near phase $0\overset{P}{.}60$.
- The $m_1$ index is constant within the errors, but the amplitude of $b-y$, $u-b$ and $c_1$ varies from season to season.
- The star is bluest when brightest.

The line-profiles show low-degree line-asymmetry variations. The radial velocity (RV) measured in the line wings follow the photometric 1.256 d period with a semi-amplitude of 41 km s$^{-1}$. The RV minimum is shifted by $0\overset{P}{.}25$-$0\overset{P}{.}30$ with respect to light minimum. The first line moment follows the same curve with a lower amplitude and larger scatter.

The large radial velocity amplitude cannot be understood as a result of the pulsational displacement alone. Statistical results obtained by comparing the photometric periods and projected rotational velocities show that the period in the rotating frame is at least 4 times larger than the observed period (a three standard deviation result; Balona 1990). This, coupled with the large RV amplitude leads to an unrealistically high fractional radius variation. If pulsation is involved, its main effect must be a change in temperature due to compression and decompression.

The net result is that the NRP model is observationally similar to a large star spot which drifts slowly across the photosphere. However, a star spot model cannot account for the large amplitude either, since the expected semi-amplitude is only about 20 km s$^{-1}$.

This study is a second attempt, after that of $\eta$ Cen (Štefl et al. 1995), to derive relations between the rapid photometric and spectroscopic variations. In $\eta$ Cen the medium-term variations occur on a considerably longer time-scale. An interesting difference between the two stars is that the minima of light and RV coincide for $\eta$ Cen, but are shifted by a quarter of the period in DX Eri.

## References

Balona, L.A. 1990, MNRAS, 245, 92

Balona, L.A., Cuypers, J., & Marang, F. 1992, A&AS, 92, 533

Štefl, S., Baade, D., Harmanec, P., & Balona, L.A. 1995, A&A, in press

*Astrophysical Applications of Stellar Pulsation*
*ASP Conference Series, Vol. 83, 1995*
*R. S. Stobie and P. A. Whitelock, eds.*

# The Line Profile Variations of 60 Cyg (HD 200310)*

P. Koubský[1], D. Ballereau[2], J. Chauville[2], and J. Horn[1†]

[1]*Astronomical Institute, Academy of Sciences of Czech Republic, CZ-251 65 Ondřejov, Czech Republic*
[2]*Observatoire de Paris, Section d'Astrophysique de Meudon, F-92195 Meudon Principal Cedex, France*

**Abstract.** Results of the preliminary investigation of the monitoring of line-profile variations of the Be star 60 Cyg are reported. At least three apparent absorption or emission bumps moving from blue to red were detected.

## 1. Introduction

60 Cygni (HD200310, V1931 Cyg) is a variable Be star (sp. type B1 Ve, v sin i = 300 km/s). Variable radial velocity was reported by various authors. Harmanec et al. (1986) from the coincidence of photometric and spectroscopic periods concluded that 60 Cyg might be a SB1 with a period 2.48257 days.

Helium profiles on spectra taken at Ondřejov and OHP have shown variable asymmetries reminiscent of "moving bumps" observed in OB stars . In this paper we present the first detection of moving subfeatures in the HeI 447.1 nm line of 60 Cyg.

## 2. Variability of the HeI 447.1 line profile

Spectra were obtained in August 1994 with the Aurélie/T152 instrumentation at OHP by D.B. Two helium lines were observed - 447.1 and 667.8 nm. Mean profiles for the particular night were then subtracted from the individual spectra. Results of this subtraction during the night (JD 49570.5) are shown in Fig. 1. Features moving from blue to red through the HeI 447.1 line are identified by numbers: 1, 2, 3 and 4. It was not possible to construct similar figure for series covering the HeI 667.8 line. The reason might be instrumental (lower S/N ratio) and/or it might reflect the influence of the Be star envelope. The "moving bumps" were measured for radial velocity (displacement from the centre of HeI 447.1 line). The bump crossing rate is about 80 km/s/hr (in both nights where 447.1 line was observed. This is in good agreement with other stars (c.f., e.g., $\theta$ CrB - Hubert et al. 1990 or $\eta$ Cen - Leister et al. 1994).

[1†]Jiři Horn died December 13, 1994

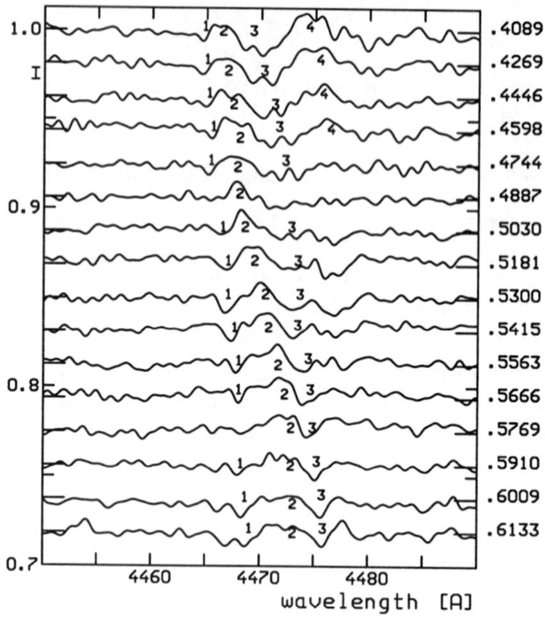


Figure 1. Residuals formed by subtracting the mean night (JD 49570) profile from the individual profiles. Prominent features are identified by numbers. Profiles are labelled by fraction of Julian date.

## 3. Conclusions

High resolution, high S/N spectra of the Be star 60 Cyg revealed the presence of features moving from blue to red across the HeI 447.1 profile. Due to the limitation of only two short data strings we have not made any attempt to explain these variations.

It should be stressed that 60 Cyg is a very interesting object worth of systematic study. The open questions are:

- character of light variations

- correlation between the short term variation of lines like HeI 447.1 and emission components to H$\alpha$ or HeI 667.8 line.

* Based on observations collected at the Haute Provence Observatory

## References

Harmanec, P., et al. 1986, IBVS, No 2912

Hubert, H., et al. 1990, A&A, 230, 131

Leister, N.V., et al. 1994, A&A, 287, 789

Slettebak, A. 1982, ApJS, 50, 55

*Astrophysical Applications of Stellar Pulsation*
*ASP Conference Series, Vol. 83, 1995*
*R. S. Stobie and P. A. Whitelock, eds.*

# Two Modes of Pulsation in the Rapidly Rotating O9.5 Dwarf, HD 93521

A.H.N. Reid[1], D. Bohlender[2], C.T. Bolton[3], A.W. Fullerton[4], I.D. Howarth[1], C. Manz[3], K.C. Smith[1], and R.H.D. Townsend[1].

**Abstract.** We present preliminary results of a time-series analysis on the He II $\lambda$4541Å, Si III $\lambda\lambda$4552, 4567, 4575Å and H$\beta$ $\lambda$4861Å lines of the late O-type dwarf, HD 93521. The spectra show line-profile variations (lpv) – 'bumps' and 'dips' – of amplitude $\sim$ 1% of the continuum, moving systematically redwards through the photospheric absorption lines. The analysis implies non-radial pulsation (NRP) as the origin for the lpv.

## 1. Introduction

HD 93521 is a rapidly rotating ($v_e \sin i$ = 400 km s$^{-1}$, Lennon et al. 1991) O9.5 main-sequence star. Analysis of the UV P-Cyg profiles suggest that the stellar wind of HD 93521 forms a wind-compressed disc (Howarth & Reid 1993; Bjorkman et al. 1994). Fullerton (1990) identified lpv at $\sim$ 1% of the continuum. Subsequent time-series analyses (Fullerton, Gies & Bolton 1991, 1994; Howarth & Reid 1993) suggested that the photospheric velocity field of the star is dominated by a single prograde sectorial NRP ($\ell = -m = 9 \pm 1$).

## 2. Time-series analysis

New data were obtained in March 1994 with the WHT. They were analysed by using the Fourier technique of Baade (1988) and Gies & Kullavanijaya (1988). This technique incorporates the CLEAN algorithm (Roberts et al. 1987) which iteratively removes the aliasing associated with the sampling window. Two periods, $P_1$ = 1.76-hours and $P_2$ = 2.90-hours, were detected in the Si III triplet and in H$\beta$. An equatorial origin for the modulation is indicated by the absence of a detectable period in He II (confined to the polar regions because of rotationally-induced gravity darkening). Coupled with the cylindrical symmetry, implied by the sinusoidal phase across the line profile, an interpretation in terms of prograde

[1]Dept. Physics and Astronomy, University College London, Gower Street, London WC1E 6BT, United Kingdom.

[2]Canada-France-Hawaii Telescope Corporation, P.O. Box 1597, Kamuela, Hawaii, USA 96743.

[3]David Dunlap Observatory, P.O. Box 360, Richmond Hill, Ontario, Canada L4C 4Y6.

[4]Max-Planck-Institut für Astrophysik, Karl-Schwarzschild-Str. 1, D-85740 Garching bei München, Germany

sectorial NRP is suggested, with mode identifications of $P_1$: $\ell = -m = 9 \pm 1$; and $P_2$: $\ell = -m = 5 \pm 1$.

## 3. Discussion

The amplitude of variation across the line profile indicates the ratio of the horizontal-to-vertical pulsation velocity, $k$. For the 2.90-hour period in Si III, $k \sim 0.0$-$0.2$ (by comparison with NRP models, Reid 1994). However, in H$\beta$, $k \sim 1$-$2$.

Lee & Saio (1990) have shown that NRP-generated temperature variations, when coupled to rapid rotation, may significantly affect the appearance of lpv. For a line which is temperature-sensitive, NRP with large $k$ will produce lpv indicative of small $k$. For lines which are weakly affected by temperature variations only NRP velocity variations are observed (Reid 1994), thus preserving $k$.

The equivalent width of Si III $\lambda$4552Å is very sensitive to temperature variations of a few kK in HD 93521. However, the strength of H$\beta$ remains relatively constant. Thus, H$\beta$ should be more sensitive to velocity variations and therefore, $k$. This may explain the discrepancy in the sinusoidal semi-amplitudes seen between the Si III line and H$\beta$ for the 2.90-hour period.

**Acknowledgments.** We thank Miss Despina Savva for her assistance with part of the data reduction. AHNR, KCS and RHDT acknowledge the financial assistance of UK's PPARC.

## References

Baade, D. 1988, in IAU Symposium 132: The Impact of Very High S/N Spectroscopy on Stellar Physics, G. Cayrel de Strobel & M. Spite, Dordrecht: Kluwer, 193

Becker, S.R., & Butler, K. 1990, A&AS, 84, 95

Bjorkman, J.E., Ignace, R., Tripp, T.M., & Cassinelli, J.P. 1994, ApJ, 435, 416

Fullerton, A.W. 1990, Ph.D. Thesis, University of Toronto

Fullerton, A.W., Gies, D.R., & Bolton, C.T. 1991, BAAS, 23, 1379

Fullerton, A.W., Gies, D.R., & Bolton, C.T. 1994, in IAU Symposium 162: Pulsation, Rotation & Mass Loss in Early-Type Stars, L. Balona, H. Henrichs, & J.M. Le Contel, Dordrecht: Kluwer, 114

Gies, D.R., & Kullavanijaya, A. 1988, ApJ, 326, 813

Howarth, I.D., & Reid, A.H.N. 1993, A&A, 279, 148

Lee, U., & Saio, H. 1990, ApJ, 349, 570

Lennon, D.J., Dufton, P.L., Keenan, F.P., & Holmgren, D.E. 1991, A&A, 246, 175

Reid, A.H.N. 1994, Ph.D. Thesis, University of London

Roberts, D.H., Lehàr, J., & Dreher, J.W. 1987, AJ, 93, 698

*Astrophysical Applications of Stellar Pulsation*
*ASP Conference Series, Vol. 83, 1995*
*R. S. Stobie and P. A. Whitelock, eds.*

# Eta Orionis

K. De Mey, C. Aerts, H. Van Winckel, C. Waelkens

*Instituut voor Sterrenkunde, Katholieke Universiteit Leuven, Celestijnenlaan 200B, 3001 Leuven, Belgium*

## 1. Introduction

Eta Orionis is an interesting multiple system of early type stars. It consists of a visual triple system (A, B, C), a spectroscopic triple system (Aabc) and an eclipsing binary Aab. A is a triple system consisting of a 7.989 day eclipsing binary Aab and a gravitationally bound third component Ac with orbital period of about 9.5 years. All components of the pair AB have spectral types in the range O9-B3 (C. Waelkens & P. Lampens 1988). The radial velocity curve is used to constrain the orbital elements and properties of the binary Aab.

The second component Ab is a B-type star that displays rapid absorption line profile variability. Regular patterns are easily visible in the data.

## 2. Observations

All high resolution spectra were obtained with ESO's CES Spectrograph linked to the 1.4m CAT telescope in December 1993. Mean exposure times of 2 minutes yielded signal-to-noise ratios of about 400. We obtained 36 and 40 spectra of Si III-4552 on 2 successive nights HJD 2449338.5 and HJD 2449339.5. The spectra were reduced using the MIDAS software package.

The Si III-profiles of Ab vary strongly over our observing run. The spectra are corrected for the orbital motion.

## 3. Line profile variations (LPV)

To demonstrate the reality of the LPV, the longest nightly sequences of observations are used. We subtracted the nightly mean profile from the observed ones. The residuals have been combined into two-dimensional gray-scale pictures. This diagram reveals a blue-to-red movement of spectral sub-features (bumps at about $\pm$ 1% of the continuum) across the rotationally broadened line profiles. The time scale for a sub-feature to traverse the width of the line profile is of the order of hours. The bumps extend over a large fraction of the profile. At least 2 bumps are visible at any time. Small but significant equivalent width variations are also observed.

A detailed frequency analysis of the equivalent width of the secondary component has been carried out for the full data set using the phase dispersion minimisation method. The analysis reveals that the most significant frequency is 7.54 cycles/day.

Mode-identifications were obtained by means of the moment method (Aerts et al. 1992). *If* the line profile variations are due to stellar pulsation, they cannot be due to an axisymmetric nor to a radial mode as derived from the moment variations. Assuming pulsation, the best result is l=4, m=−3.

## References

Aerts, C., De Pauw, M., & Waelkens, C. 1992, A&A, 266, 294

Waelkens, C., & Lampens, P. 1988, A&A, 194, 143

*Astrophysical Applications of Stellar Pulsation*
*ASP Conference Series, Vol. 83, 1995*
*R. S. Stobie and P. A. Whitelock, eds.*

# Second-Order Rotational Effects on Observations of $\beta$ Cephei Stars

C. A. Engelbrecht (jr)

*Department of Physics, University of South Africa, PO Box 392, Pretoria 0001, South Africa*

Practically all published models of $\beta$ Cephei ($\beta$ Cep) stars assume a spherical profile, despite the fact that many actual $\beta$ Cep stars display high rotation velocities. As a result of the centrifugally altered profile of the surface of a rapidly rotating star, the surface area as well as the surface normal of an arbitrary element on the stellar surface will be modified. Numerical integration over the observed surface elements produces the observed luminosity and radial velocity of such a star. The effect of including the centrifugal distortion of a rotating model $\beta$ Cep star on its observed characteristics was studied for inclination angles ($i$) of 0°, 22°.5, 45°, 67°.5 and 90° and equatorial rotation speeds of 0, 80, 160, 240, 320 and 400 km/s, for pulsation modes of order $l = 0$ to 4.

Previous studies of the ratios between *radial velocity variations* and *magnitude variations* by Balona & Stobie (1979) indicated that these ratios are independent of $i$. Results of the present study for the radial mode and the sectoral quadrupole mode are shown below. In each case, the bottom panel shows the results obtained assuming no centrifugal distortion, while the top panel shows what is obtained when centrifugal distortion is included. The increasing dependence of these amplitude ratios on $i$ with increasing rotation speed is clear. Both graphs used an adiabatically-derived value of −17.5 for the ratio of proportional flux variation to proportional radius variation (commonly called the $f$-ratio), and the spread of the graphs was found to vary with $f$, although the trend remained similar.

The overall results obtained may be summarised as follows: Provided that accurate values of $f$ are used, the method of calculating amplitude ratios of light and velocity variations may be used to determine either equatorial rotation speed or angle of inclination of a star, provided that the pulsation mode of the star has been identified and the complementary parameter ($i$ and rotation speed respectively) are known. This application becomes increasingly useful as the rotation speed increases, provided that proper account (second-order in $\Omega$) is taken of centrifugal distortion of stellar profiles.

Spectral line profiles were also modelled for the above mentioned range of parameters. The main conclusion was that rotation speeds determined from rotational broadening are significantly underestimated (by up to 25%) if centrifugal distortion of the stellar profile is disregarded, at least for rotation speeds higher than 250 km/s. This agrees with a finding of Hardorp & Strittmatter (1968).

E. C. Reynhardt is thanked for travel assistance.

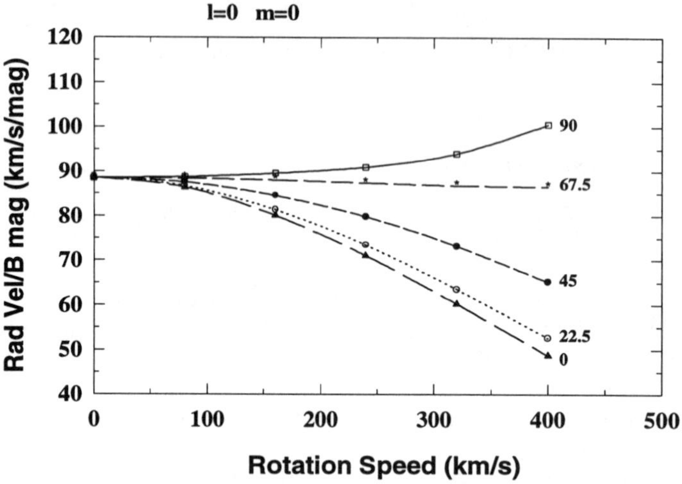


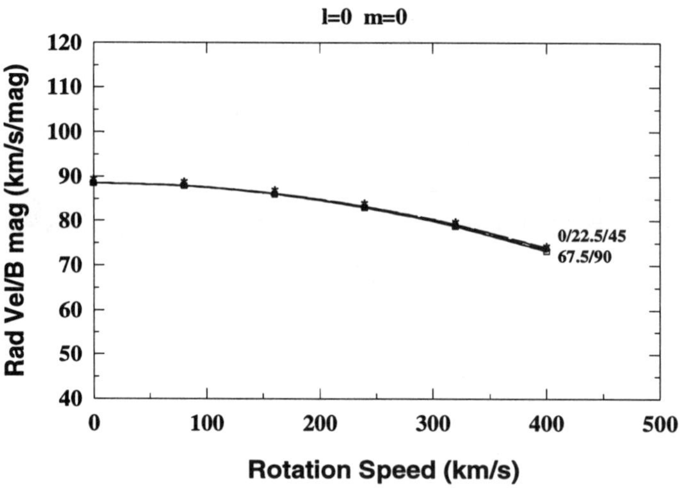


## References

Balona, L.A., & Stobie, R.S. 1979a, MNRAS, 187, 217
Balona, L.A., & Stobie, R.S. 1979b, MNRAS, 189, 649
Hardorp, J., & Strittmatter, P.A. 1968, ApJ, 153, 465

*Astrophysical Applications of Stellar Pulsation*
*ASP Conference Series, Vol. 83, 1995*
*R. S. Stobie and P. A. Whitelock, eds.*

# The Evolutionary Status of $\lambda$ Boo Stars

E. Paunzen, M. Gelbmann, U. Heiter, F. Kupka, R. Kuschnig, W. W. Weiss

*Inst. f. Astronomy, Türkenschanzstr. 17, A-1180 Vienna, Austria*

**Abstract.** We summarise the history of the $\lambda$ Boo criterion and our strategy to determine the evolutionary status of this group of stars.

## 1. Introduction

$\lambda$ Boo stars present a considerable challenge to our understanding of stellar atmospheres, structure and evolution. They are rather fast rotators ($v \sin i >$ 50 km s$^{-1}$), have no measurable magnetic field, 7000 K $< T_{\rm eff} <$ 9000 K, and $3.5 < log\, \rm g < 4.2$. CNO slightly is over-abundant, but most of the heavy elements are under-abundant up to 1/100 relative to the Sun. From dynamic arguments these metal poor stars belong to Pop. II.

Two mechanisms are discussed for generating the $\lambda$ Boo phenomenon:

• *Accretion* of metal depleted gas from a remnant protostellar cloud which would generate *unevolved* $\lambda$ Boo stars at the ZAMS (Charbonneau 1991). A flux-excess (marginally detected by IRAS) corroborate this model.

• *Diffusion and mass loss* proposed by Michaud & Charland (1986) produce $\lambda$ Boo stars only at the end of the main-sequence lifetime. A mass-loss rate of about $10^{-13} M_{\odot} y^{-1}$ can lead to the observed under-abundances after $10^9$ years, thus implying *evolved* $\lambda$ Boo stars.

## 2. Classification criteria

Unfortunately, the group of $\lambda$ Boo stars, which was introduced by Morgan et al. (1943), still is not well defined in contrast to, e.g., the Am or CP2 stars. Morgan described the spectrum of the prototype, $\lambda$ Boo, as weak in Ca II K lines, hydrogen lines being similar to A0 V, and an otherwise featureless classification spectrum. Later research helped to improve the classification criteria by including, among others, spectral features in the UV (Baschek, Gray, Faraggiana, Hauck, Holweger, Slettebak).

Morgan et al. (1943) refers to only one $\lambda$ Boo star, Faraggiana et al. (1990) list 20 members and Renson et al. (1990) already 101 candidates. A critical assessment of these lists, however, resulted in only 40 probably true $\lambda$ Boo stars (Paunzen et al. 1995). Evidently, a larger sample of well defined $\lambda$ Boo stars is needed for a statistically sound treatment.

## 3. Survey for $\lambda$ Boo stars

Based on $uvby\beta$ and Geneva photometry we selected stars in the galactic field and in various galactic clusters of different age which populate the parameter space typical for $\lambda$ Boo stars (Weiss et al. 1994). This procedure resulted in about 1800 candidates which by and large will be checked with classification spectroscopy. Presently, 120 candidates have been observed at OHP, UTSO (Chile) and Monterey (California) and 2 new $\lambda$ Boo stars have been found so far.

The problem of border-line cases for membership is frequently encountered and only high resolution spectroscopy can decide about inclusion to the group of $\lambda$ Boo stars. In addition, similar investigations are needed for undisputed $\lambda$ Boo stars in order to improve the significance of the abundance pattern assumed to be typical for these stars. Presently, we are studying 5 $\lambda$ Boo stars (for 29 Cyg see paper on *Fine analysis of pulsating CP stars* in this volume).

All together, a catalogue of stars which belong with high probability to the group of $\lambda$ Boo stars has been compiled (Paunzen et al. 1995) and presently contains 40 entries, thus doubling the catalogue of Gray (1988).

## 4. Asteroseismology of $\lambda$ Boo stars

A very powerful tool to determine the evolutionary status of $\lambda$ Boo stars and hence to decide between the two competing theories can be provided by asteroseismology (Weiss et al. 1994). We initiated therefore a photometric survey for pulsating $\lambda$ Boo stars at ESO, CTIO, SAAO and Lowell observatory which up to now resulted in the discovery of 10 new pulsating stars.

For an asteroseismic inversion it is necessary to determine a set of at least 4 frequencies (Breger, this volume). Such a frequency spectrum can only be obtained from the ground with a multi-site campaign. The first two campaigns are presently organised for April to June '95 at SAAO, New Zealand and CTIO.

**Acknowledgments.** Financial support for this international collaboration was provided by the FwF (*P 8776-PHY*) and Stadt Wien ($\lambda$ Boo Sterne). We acknowledge contributions to our research by R. Gray (Boone), P. North and B. Hauck (Lausanne), and A. Torres-Dodgen (Monterey).

## References

Charbonneau, P. 1991, ApJ, 372, L33

Farragiana, R., Gerbaldi, M., & Böhm, C. 1990, A&A, 235, 311

Michaud, G., & Charland, Y. 1986, ApJ, 311, 326

Morgan, W. W., Keenan, P. C., & Kellman, E. 1943, Atlas of Stellar Spectra, Univ. Chicago

Renson, P., Faraggiana, R., & Böhm, C. 1990, Bull. Inform. CDS 38, 137

Gray, R. O. 1988, AJ, 95, 220

Paunzen, E., Weiss, W. W., Heiter, U., & North, P. 1995, preprint

Weiss, W. W., Paunzen, E., Kuschnig, R., & Schneider, H. 1994, A&A, 281, 797

*Astrophysical Applications of Stellar Pulsation*
*ASP Conference Series, Vol. 83, 1995*
*R. S. Stobie and P. A. Whitelock, eds.*

# Fine Analysis of Pulsating CP Stars

F. Kupka, M. Gelbmann, U. Heiter, R. Kuschnig, W. W. Weiss

*Institute for Astronomy, University of Vienna, A-1180 Vienna, Austria*

T. A. Ryabchikova

*Institute for Astronomy, Russian Academy of Sciences, Moscow, Russia*

**Abstract.** We present fine analysis results for roAp and $\lambda$ Boo stars.

## 1. Introduction: goals and methods

Temperature and pressure structure of cool pulsating CP stars have an important role as boundary conditions for pulsation models. We perform a fine analysis of pulsating CP stars in order to determine the atmospheric structure with special respect to the differences between pulsating and *non* pulsating stars. In addition, possibly existing characteristic abundance patterns, which could be used as a discrimination between pulsating and *non* pulsating stars, are studied.

To achieve these goals we first perform a fine analysis for various cool CP2 stars based on line profile fitting of synthetic spectra with the codes SYNTH and ROTATE and on equivalent width calculations using WIDTH9 (Kurucz 1993, modified by V. Tsymbal). The results obtained are basic stellar parameters, abundance patterns, and estimations of the magnetic surface fields. These results will be compared with 'similar' stars which do *not pulsate* according to state-of-the-art observing techniques. As a first step towards self consistency, the initial fine analysis, which is based on model atmospheres from ATLAS9 (Kurucz 1993) with scaled solar abundances, will be repeated based on specific opacity distribution functions and model atmospheres for these stars.

## 2. Observations and results

We use high resolution spectra for wavelength regions in the visual and near infrared. Details will be given in publications related to the particular objects.

Up to now we have been working on the fine analysis of four roAp stars based on solar scaled model atmospheres. The analysis of $\alpha$ Cir and BI Mic has been finished. For $\gamma$ Equ and the line variable DO Eri we have preliminary results. Table 1 gives the relative abundance for the investigated elements. Note the characteristic under-abundance for CNO, the over-abundance of rare earth elements (REE, La to Lu), and the complex pattern for iron peak elements. The most interesting results are the indication from surface gravity that BI Mic is still on the ZAMS and the pronounced over-abundance of Co for all roAp stars. For DO Eri, Co joins the REE in their variation in phase with the magnetic field, whereas Mg, Fe, and Ni show an anti-phase variation. The strikingly different abundance pattern of the pulsating $\lambda$ Boo star 29 Cyg is not very surprising. For comparison the solar abundances according to Anders & Grevesse (1989) are also given in the table.

Table 1. Element abundances $\log(N_x/N_{tot})$ and basic parameters

| | roAp BI Mic HD 203932 | roAp $\alpha$ Cir HD 128898 | roAp $\gamma$ Equ HD 201601 | roAp DO Eri HD 24712 | roAp DO Eri HD 24712 | $\lambda$ Boo 29 Cyg HD 192640 | Sun - |
|---|---|---|---|---|---|---|---|
| C | -4.09 | -4.00 | -4.40 | - | - | -3.48 | -3.48 |
| N | -4.60 | -4.40 | -4.50 | - | - | - | -3.99 |
| O | -3.81 | -3.80 | -3.80 | - | - | - | -3.11 |
| Na | -5.76 | -5.90 | -5.90 | - | - | - | -5.71 |
| Mg | -4.18 | -4.46 | - | -4.50 | -5.60 | -6.26 | -4.46 |
| Al | -5.28 | - | - | - | - | - | -5.57 |
| Si | -4.37 | -4.20 | -4.05 | -4.70 | -4.70 | - | -4.49 |
| Ca | -5.09 | -5.15 | -5.40 | -5.39 | -5.13 | -6.98 | -5.68 |
| Sc | -9.40 | -9.80 | -9.80 | - | - | -10.74 | -8.94 |
| Ti | -6.71 | -6.87 | -6.80 | -7.06 | -7.29 | -8.60 | -7.05 |
| V | -7.24 | -7.65 | -7.65 | - | - | - | -8.04 |
| Cr | -5.45 | -5.55 | -5.75 | -5.85 | -5.70 | -8.27 | -6.37 |
| Mn | -6.18 | -6.00 | -6.25 | -6.69 | -7.11 | -6.65 | -6.65 |
| Fe | -4.23 | -4.50 | -4.50 | -4.86 | -5.07 | -6.37 | -4.37 |
| Co | -5.83 | -5.50 | -6.00 | -5.94 | -5.58 | - | -7.12 |
| Ni | -5.73 | -6.15 | -6.15 | -6.24 | -6.57 | - | -5.79 |
| Sr | -7.47 | -7.25 | -7.25 | - | - | -10.84 | -9.14 |
| Y | -8.04 | -8.50 | -8.33 | -8.23 | -8.06 | - | -9.80 |
| Zr | -8.90 | -9.00 | -9.10 | - | - | - | -9.44 |
| Ba | -9.20 | -10.30 | -9.10 | -9.13 | -9.27 | -11.41 | -9.91 |
| La | -9.72 | -10.32 | - | -10.14 | -9.55 | - | -10.82 |
| Ce | -9.13 | -9.40 | -9.40 | -9.15 | -8.82 | - | -10.49 |
| Pr | -11.20 | -10.40 | -10.50 | -10.04 | -9.51 | - | -11.33 |
| Nd | -9.24 | -9.30 | -9.22 | -9.21 | -8.59 | - | -10.54 |
| Sm | -9.52 | -9.50 | -9.70 | -9.65 | -8.98 | - | -11.04 |
| Eu | -8.92 | -9.40 | -9.20 | -9.80 | -9.20 | - | -11.53 |
| Gd | -9.28 | -9.45 | -9.25 | -9.26 | -8.80 | - | -10.92 |
| Dy | - | -10.00 | -9.65 | -9.59 | -9.13 | - | -10.94 |
| Er | -9.70 | - | - | -9.84 | -9.71 | - | -11.11 |
| $T_{\rm eff}$ [K] | 7750 | 7900 | 7750 | 7250 | 7250 | 7800 | 5777 |
| log(g) | 4.6 | 4.2 | $< 4.5$ | 4.3 | 4.3 | 4.0 | 4.44 |
| $B$ [kG] | - | $\sim 1.5$ | - | $+0.4min$ | $+1.5max$ | - | - |
| $v \sin(i)$ | 12.5 | 12.5 | $< 4.5$ | $< 5$ | $< 5$ | 80 | - |
| $v_{micro}$ | $< 0.5$ | 1.5 | $\sim 1$ | 1.0 | 1.0 | 3.0 | 1.5 |

The typical intrinsic errors of the abundances range from 0.1 to 0.3, in some cases to 0.5 (depending on the number of lines). Errors for $T_{\rm eff}$ are less than 200 K, for $\log(g)$ less than 0.15. Velocities are given in [km s$^{-1}$] with relative errors of about 10% .

**Acknowledgments.** We acknowledge N. Piskunov for providing software to us and G. Mathys for observations. The project is supported by FWF (*P 8776-PHY*), BM f. Wiss. Forsch. and the Österr. Akad. Wiss. (*Asteroseismologie*).

## References

Anders, E., & Grevesse, N. 1989, Geochimica et Cosmochimica Acta, 53, 197

Kurucz, R. L. 1993, Smithsonian Astrophys. Observatory, CDROM 13

*Astrophysical Applications of Stellar Pulsation*
*ASP Conference Series, Vol. 83, 1995*
*R. S. Stobie and P. A. Whitelock, eds.*

# Spatial Integration Grids in Line Profile Modelling of Nonradially Pulsating Magnetic Stars

R.M. Fensl, G. Könighofer, M.J. Stift

*Institut für Astronomie, Türkenschanzstr. 17, Wien, Austria*

## 1. The philosophy of the resolution function

For an optimum grid we postulate a fractional change of the emergent Stokes parameters between two adjacent surface points not to exceed some fixed value (say 20 %) at any frequency point. The contribution from a local Stokes $IQUV$ profile to the observed integrated profile is a complicated non-linear function of rotational and pulsational Doppler shifts respectively, of Zeeman splitting, magnetic field direction relative to the line-of-sight, of the magnetic field azimuth in the observer's frame and of limb darkening. The run of $IQUV$ with visible disk coordinates $(x, y)$ can be represented by curved surfaces $f(x, y)$; these are different for each Stokes parameter and frequency point. The optimum integration grid then has to ensure that the maximum change $\Delta f(x, y)$ over all frequencies and Stokes parameters does not exceed the above-mentioned limit.

Expressing the average fractional change in the value of the Voigt function by an average change in normalised (to the Doppler width of the line) wavelength difference, we quadratically sum up the individual contributions to the total difference in adjacent $IQUV$ intensities as enumerated above to obtain a piecewise resolution function. Inverse equidistant interpolation in the *cumulative* resolution function yields the desired optimum grid (for details see Fensl 1995).

## 2. Application to the Oblique Pulsator model and conclusions

The Oblique Pulsator (Kurtz 1990) explains the observed short-term variations (4-15 min) of roAp stars by invoking non-radial pulsations of low degree $l$ and of order $m = 0$; the pulsation axis is aligned with the magnetic axis which is itself inclined at a non-zero angle to the rotation axis. We have made calculations for this simple geometry with a code based on Stift's (1985) Stokes code but including linear polarisation and magneto-optical effects. The code is written in Ada83 making the manipulation of Stokes, magnetic and coordinate vectors, and of transformation matrices particularly easy and transparent (operator overloading, reusable packages).

Fig. 1 (top row) shows Stokes $IQUV$ profile for three different models at three different rotational phases (low degree zonal modes $l \leq 5, m = 0$ only). Note the wealth of non-spurious structure in $QUV$, perfectly resolved with 260 grid points! On account of the simple geometry of the problem the improvement over fixed grids is not overly dramatic but we find that the excellent accuracy obtained with our grid – measured by $\Sigma$ the square root of the cumulated variances of the deviations in $IQUV$ from the reference profile – is *always maintained*

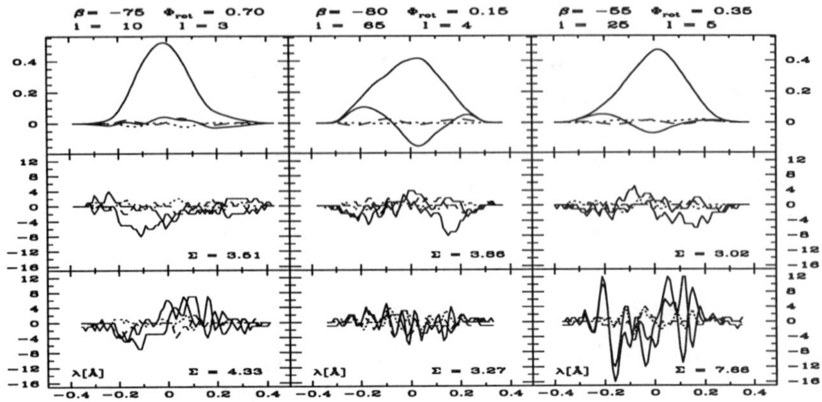


Figure 1. **Top:** *IQUV* profiles (full, dashed, dotted, full) for a Fe II $\lambda\,6150$ line, assuming a Zeeman triplet, $v \sin i = 7\,\mathrm{km/s}$, $v_{\mathrm{puls}} = 10\,\mathrm{km/s}$, $\Phi_{\mathrm{puls}} = 0.0$, $H_{\mathrm{pole}} = 0.4\,\mathrm{T}$ and using a (9000, 4.0, 0.0) Kurucz atmosphere. Inclination $i$, obliquity $\beta$, pulsation degree $l$ (with $m = 0$) and rotational phase $\Phi_{\mathrm{rot}}$ are given above. **Middle:** Differences between profiles calculated with 260 adaptive points and reference profiles (7000 points); 100 times exaggerated scale. *IQ* (full, dotted) above, *UV* (dotted, full) below. **Bottom:** Same for fixed grid of 204 equally sized visible elements. For $\Sigma$ the numerical accuracy in $10^{-4}$ see text.

(middle rows). Fixed grids can sometimes yield comparable results but one can never rely on that (bottom rows).

The main advantage of adaptive grids lies in the guaranteed resolution without the need to provide for huge safety margins in the number of grid points. Additionally, in the large majority of cases – particularly those with more complex magnetic geometries – we obtain considerable gains in accuracy at constant grid size and CPU time. This becomes possible only because the grid adapts to the velocity and magnetic field distributions particular to the magnetic geometry and pulsational pattern at a given rotational and pulsational phase. Our extension of the method of Stift (1985) proves to deal very effectively and accurately with linear polarisation and with non-radial pulsation. We would contend that at present adaptive grids as outlined above constitute the best way to guarantee reliable and accurate Stokes profile modelling in the presence of macroscopic velocity and of inhomogeneous magnetic fields, using a minimum of computational resources.

An Ada83 package for establishing an adaptive grid is available upon request from stift@astro.ast.univie.ac.at.

## References

Fensl, R.M. 1995, A&A, in press

Kurtz, D.W. 1990, ARA&A, 28, 607

Stift, M.J. 1985, MNRAS, 217, 55

*Astrophysical Applications of Stellar Pulsation*
*ASP Conference Series, Vol. 83, 1995*
*R. S. Stobie and P. A. Whitelock, eds.*

# Consequences of Rapid Rotation on Mode Identification

F. Soufi[1], M. J. Goupil[1], W.A. Dziembowski[2], R. Sienkiewicz[2].

[1] *Observatoire de Paris, URA 335, 92190 Meudon, France*

[2] *Copernicus Astronomical Center, 00-716 Warsaw, Poland*

The present work deals with the effect of rotation on the adiabatic oscillation frequencies of stellar models. Here, rotation, as described in a perturbation theory, is said to be rapid in the sense that second order perturbation effects have to be taken into account. Rotation has two second order effects on the oscillation frequencies: departure from equidistance within a given multiplet split by rotation and an overall shift of the multiplet itself. The first effect has been investigated for realistic models of $\delta$ Scuti stars in uniform rotation (Dziembowski & Goode, 1992), whereas H. Saio's study includes both effects for a polytrope of index 3 in uniform rotation (Saio 1981). Both effects are investigated for realistic models of $\delta$ Scuti stars. Results for the overall shift for a 2 $M_{\odot}$ model in rapid uniform rotation ($\mathbf{v}$ up to $\sim$ 200 km/s) are outlined below. The rotation rate corresponding to this surface velocity is $\Omega = 1.44\ 10^{-4}$ rad/s.

The relevant eigensystem of linear adiabatic equations is expanded with respect to the parameter $\Omega/\omega_0$ with $\omega$ the eigenfrequency. In the notation used by H. Saio, the second order correction $\delta\omega_{n,l,m}$ to the eigenfrequeny $\omega_{n,l,m}$ can be written as:

$$\delta\omega_{n,l,m} = \underbrace{\omega_{0n,l}\left(\frac{\Omega}{\omega_{0n,l}}\right)^2 (X_1 + X_2 + Z)}_{\text{overall shift}} + \omega_{0n,l}\left(\frac{\Omega}{\omega_{0n,l}}\right)^2 m^2 (Y_1 + Y_2)$$

where $n$,$l$,$m$ are the radial order, the degree, and the azimuthal order of the eigenmode respectively. All perturbative frequencies depend on the eigenmode $(n, l)$ and on the associated eigenfrequency $\omega_{0n,l}$ of the non rotating star. The second order correction which is induced by the centrifugal force is :

$$\delta\omega_{centrifug,n,l,m} = \omega_{0n,l}\left(\frac{\Omega}{\omega_{0n,l}}\right)^2 \left[(X_2 + Z) + m^2\ Y_2\right]$$

The $Z$ contribution comes from structure perturbations induced by the spherical part of the centrifugal force while $X_2$ and $Y_2$ are due to those induced by the non spherical part of the same force. $Z$ and $X_2$ appear in the correction which is independent of $m$ and therefore leads to an overall shift of each given $n, l$ multiplet split by rotation(first order effect).

A Chandrasekhar-Milne expansion (Tassoul, 1978) can be used to obtain an analytical expression for the $Z$ contribution in terms of the structural quantities of the non rotating star. The results thereby obtained for a polytrope of index 3 are in good agreement with those obtained by H. Saio (1981). However this method applied to realistic models of stars does not take into account the effect of rotation on the evolution of the star, and leads to values of $Z$ which significantly differ from *numerical* $Z$ values directly computed from a 'pseudo' rotating model, i.e., a model which is evolved with the spherically symmetric part of the centrifugal force included from the start. Therefore, in what follows the overall shift of the oscillation frequencies is computed with numerical values of $Z$. As an illustration, resulting $m = 0$ eigenfrequencies in a given frequency range is displayed in Figure 1 for a 2 $M_\odot$ model having burned 32% of its central hydrogen.

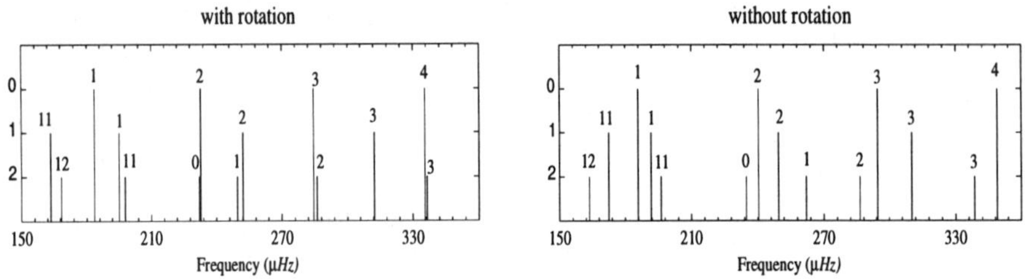


Figure 1. Effect of rotation on the oscillation frequencies. Numbers refer to radial order n, different arbitrary amplitudes are assigned to modes of different degree, which is specified in ordinate.

The overall shift clearly leads to a frequency spectrum for a rotating star which greatly differs from the one obtained when rotation is disregarded: for instance, the order of appearance of the frequencies $(n, 2)$ and $(n, 0)$ is reversed. Also pairs of very close frequencies appear much more often for the rotating model. Quantitatively here, the smallest separation between two frequencies is 0.6 $\mu Hz$, most of the time it is larger than 1 $\mu Hz$. A three weeks campaign of observation gives a resolution in frequencies which is about 0.55 $\mu Hz$ so that these patterns are resolved. Close pairs are indeed often seen in observed spectra (E. Michel, private communication). The present investigation leads therefore to a new understanding of observational frequency spectra as it provides a possible explanation for the closest observed pairs otherwise difficult to explain with nonrotating models.

As a conclusion, not taking into account rotation effects for most rapid $\delta$ Scuti stars could lead to a misleading mode identification. Detailed discussion and results will be found in a paper that is to be submitted. Note further that the above frequency range includes not only eigenfrequencies of pure p-modes modes but also those of mixed modes, i.e., modes which have amplitude both near the convective core and at the surface of the star. As the shift should be different for these modes if the core is rotating faster than the surface, our next step will be to investigate depth dependent rotation effect.

## References

Dziembowski, W., & Goode, P. 1992, ApJ, 394, 670
Saio, H. 1981, ApJ, 244, 299
Tassoul, J.-L. 1978, Theory of rotating stars (Princeton Univ. Press, Princeton)

*Astrophysical Applications of Stellar Pulsation*
*ASP Conference Series, Vol. 83, 1995*
*R. S. Stobie and P. A. Whitelock, eds.*

# Light Variability of 28 Aql

William Wehlau and Amelia Wehlau
*Astronomy Department, University of Western Ontario, London, Ontario N6A 3K7, Canada*

## 1. Introduction

28 Aquilae (HR 7331, V1208 Aql) was first announced as a $\delta$ Scuti type variable by Breger in 1969 with a period of about 0.15 day. It was clear that the light variability could not be adequately represented as a sinusoidal variation with just one frequency. Since then several observers have provided additional light curves and new analyses of the light variation. In this paper we discuss observations which we have made and an analysis of the data available since 1966.

Our objective was to find the frequencies of components which would represent the light variation of 28 Aql over the 28 years for which the star has been observed.

## 2. Observations

Using the 30 cm reflector of the Hume Cronyn Memorial Observatory we observed 28 Aql in 1979, 1980, 1981, and 1993 in blue light. The light curves for JD 2444140 and 2449190 are shown below. Comparison stars were HD 180868 and HD 181383. Magnitude differences are referred to HD 181383. Nights on which the scatter in the observations, due to poor transparency, appeared to be excessive were discarded. Observations on 21 nights were retained for analysis.

Observations have been obtained in a number of years by Breger (1966), Breger, Hutchins, & Kuhi (1967), Moreno (1967), Pena & Warman (1978), and by Reed & Welch (1983).

## 3. Analysis

The periodogram programs described by Matthews and Wehlau, and by Wehlau and Leung were used to identify the frequencies present in the data. Spectra and window functions were produced for individual years, as well as for the entire data set. In searching for the frequencies present, differences between blue and yellow light curves were frequently ignored. Of course, the amplitudes and phases will depend on the colour, so that prewhitening should be done with values determined from one colour only.

To identify the frequencies present we calculated the periodogram of the data for each year. As an example, the periodogram and window function for 1978 (Pena & Warman's data) is shown in Figure 1. The periodograms for the different years were combined to form an average periodogram. The largest amplitude was obtained for a frequency of about 6.68 c/d, so a periodogram

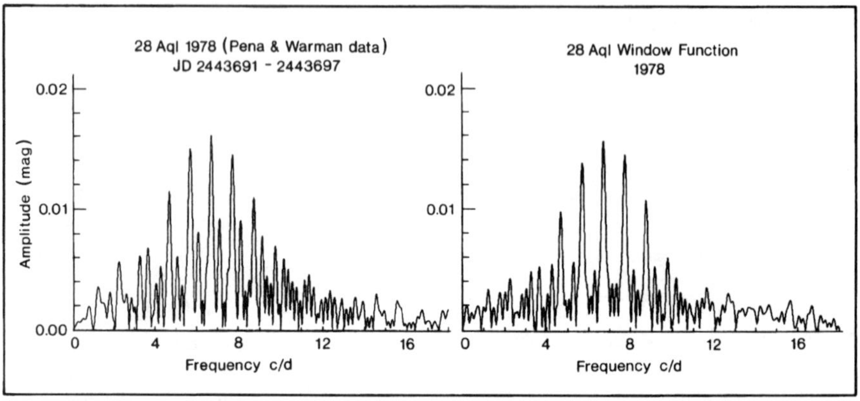


Figure 1. Periodogram and window function for 1978 data.

of the entire data set was calculated near this frequency. After prewhitening for this frequency, another average periodogram was calculated which indicated the presence of another component with a period near 7.1 c/d. The search for a third frequency did not produce any peak in the periodogram significantly above the noise.

Synthetic light curves were computed for each light curve using the two components found.

## 4. Results

The frequency with the largest amplitude was present in every year for which observations were available. The frequency, approximate amplitude, and phase (for $JD_0$ = 2444075.0) were

| Frequency | Standard Dev. | Amplitude | Phase |
|---|---|---|---|
| 6.68072 | ± 0.00006 | 0.0181 | 6.227 rad. |
| 7.12466 | | 0.0063 | 5.906 |

Comparison of the observations with the synthetic light curves computed with these elements show deviations which are greater than can attributed to observational error.

## References

Breger, M. 1969, AJ, 74, 166

Breger, M., Hutchins, J.H., & Kuhi, L.V. 1976, ApJ, 210,163

Moreno, M.A. 1980, Rev. Mexicana Astron. Astrof., 5, 19

Pena, J.H., & Warman, J. 1979, AJ, 84, 1046

Reed, L.G., & Welch, G.A. 1988, AJ, 95, 1510

*Astrophysical Applications of Stellar Pulsation*
*ASP Conference Series, Vol. 83, 1995*
*R. S. Stobie and P. A. Whitelock, eds.*

# Analysis of the Asteroseismological Data of Six $\delta$ Scuti Stars in the Open Cluster NGC 6134

Nathalie Audard, Hans Kjeldsen, Søren Frandsen

*Institut for Fysik og Astronomi, Aarhus Universitet, DK 8000 Aarhus C, Denmark*

**Abstract.** Six $\delta$ Scuti stars have been detected in the open cluster NGC 6134 (Frandsen et al. 1995). A first mode identification attempt shows that several radial modes of radial order $n = 0$, 1, 2 or 3 might have been observed. We may also be dealing with $g$-modes and/or modes split by rotation.

## 1. Theoretical isochrones

We use the photometric data obtained by Frandsen et al. (1995).

Models are computed with the CESAM code (Morel, 1993). It turns out that isochrones constructed with convective-core overshooting over 0.2 pressure scale height provide a better fit than those derived from standard models.

## 2. Asteroseismology of stars in the open cluster NGC 6134

Table 1 gives the ID number of the variables, their observed frequencies and periods $P$, and the corresponding pulsation constants $Q = P\ \sqrt{M/R^3}$ (where $M$ is the stellar mass and $R$ the radius), as well as the amplitude of the modes (columns 1 to 5 respectively).

We compare these $Q$ values to those provided by Stellingwerf (1979): Stellingwerf's model number 2.1 for the star 9, number 4.2 for the stars 87, 159 and 397, number 1.1 for the star 161, and number 2.2 for the star 348. Table 1 shows the resulting possible identification (last column).

The first four frequencies of star 9 are so close that they could be modes split by rotation. But since the $Q$ values for these modes are larger than those of Stellingwerf's fundamental mode, they may also be $g$-modes. For the same reason, the lower frequencies of the stars 159 and 348 may also be $g$-modes.

## 3. Conclusion

Our main results are the following:

• Convective-core overshooting is needed in order to make isochrones which fit the observations.

• A mode identification attempt shows that several of the observed oscillations may be fundamental, first, second and third radial modes.

• Some oscillations having very close frequencies could be $g$-modes and/or modes split by rotation. The observation of $g$-modes would be of great interest for studying the deep stellar interior and the importance of convective-core overshooting.

However, we must keep in mind that the $Q$ values derived from the observations are subject to uncertainties in the calculation of the effective temperature, mass and radius. Moreover, rotation is not taken into account in our computations, and may lead to a different identification of the modes.

Table 1. Asteroseismological data of the six $\delta$ Scuti stars in NGC 6134: ID numbers, frequencies, periods, pulsation constants $Q$, amplitudes and possible mode identification.

| ID | Frequency ($\mu$Hz) | Period (days) | $Q$ (days) | Amplitude (mmag) | Possible mode identification $(n, \ell)$ |
|---|---|---|---|---|---|
| 9 | 59.3 | 0.1958 | 0.0395 | 6.4 | $g$-mode |
| | 61.2 | 0.1891 | 0.0383 | 6.7 | $g$-mode |
| | 63.5 | 0.1823 | 0.0369 | 9.7 | $g$-mode |
| | 65.6 | 0.1764 | 0.0357 | 8.1 | $g$-mode |
| | 89.8 | 0.1289 | 0.0261 | 4.3 | (2,0) |
| 87 | 162.4 | 0.0713 | 0.0319 | 0.8 | |
| | 200.5 | 0.0577 | 0.0258 | 2.0 | (2,0) |
| | 220.5 | 0.0525 | 0.0235 | 0.8 | |
| 159 | 116.1 | 0.0997 | 0.0384 | 4.5 | $g$-mode |
| | 139.3 | 0.0839 | 0.0320 | 5.9 | |
| | 244.8 | 0.0478 | 0.0182 | 4.1 | (4,0) |
| 161 | 87.3 | 0.1326 | 0.0224 | 6.2 | |
| 348 | 76.4 | 0.1515 | 0.0363 | 4.2 | $g$-mode |
| | 133.5 | 0.0867 | 0.0208 | 6.2 | (3,0) |
| 397 | 168.1 | 0.0689 | 0.0317 | 1.7 | |
| | 324.3 | 0.0357 | 0.0164 | 1.6 | |

## References

Frandsen, S., Balona, L.A., Viskum, M., Koen, C., & Kjeldsen, H. 1995, *"Multisite CCD Photometry of δ Scuti stars in the open cluster NGC 6134 (I STACC Campaign)"*, to be submitted to A&A

Morel, P. 1993, in *"Inside the Stars"*, Weiss W.W., Baglin A. (eds), ASP Conf. Ser., vol. 40, p. 445

Stellingwerf, R.F. 1979, ApJ, 227, 935

*Astrophysical Applications of Stellar Pulsation*
*ASP Conference Series, Vol. 83, 1995*
*R. S. Stobie and P. A. Whitelock, eds.*

# Asteroseismology in Open Clusters

S. Frandsen and M. Viskum

*Institute of Physics and Astronomy, Aarhus University, Denmark*

M.M. Hernandéz and J.A. Belmonte

*Instituto de Astrofisica de Canarias, Tenerife, Spain*

## 1. A stellar evolution test

Within two groups or networks we are trying to build the observational fundamentals for better tests of stellar evolution. The first group is **STACC** (**S**mall **T**elescope **A**rray with **C**CD **C**ameras) and the second the EU HCM program **ANTENA** (involving institutes in Spain, France and Denmark).

The first information needed is photometric data. Colour-Luminosity diagrams and time series photometry must be obtained. The time series require extensive campaigns in order to give the high quality data needed for asteroseismology. Ideally one should have time series of several quantities (see Balona & Stobie 1979, Mantegazza et al. 1994).

The second type of observations is of spectroscopic nature. It is needed to ascertain membership of the stars in a given cluster and to check binarity. Also rotational velocities must be observed to help the attempt to identify the modes in $\delta$ Scuti stars. Finally some information about metallicity and peculiarity is required.

With a complete set of data for a cluster, many normal degrees of freedom disappear and a very stringent test of several aspects of stellar models becomes feasible.

In addition some specific questions about $\delta$ Scuti stars can be addressed like the identification of modes, the excitation and the period and amplitude changes reported (Breger 1990).

## 2. A definition of a good open cluster

Distance

- $d \geq 1$Kpc: The angular size of the cluster on the sky must fit into the field of view of the telescope/CCD system typically available.
- $d \leq 2$Kpc: The precision of the time series studies needed prevents small telescopes from being used for distant clusters. Crowding becomes worse and influences the measurements even if large telescopes are used. Spectroscopic information becomes very expensive in terms of telescope allocation: Long exposures on large telescopes.

Age

- $8.5 < \log t < 9.0$ Younger clusters only have main sequence $\delta$ Scuti stars with very low amplitudes and therefore hard to locate and hard to obtain good precision for. Older clusters simply have turnoff below the instability region. The best clusters have a population of giants, which is useful, when the age and metallicity of the cluster have to be determined.

Metal abundance

- $0.3 \geq [Fe/H] \geq -0.3$ Although it probably does not remove any cluster from the list of targets a nearly solar abundance is preferable. But, open clusters normally satisfy this demand without problem.

## 2.1. Open clusters under study

Table 1 presents a range of clusters with measurements of very different extent from just a few frames up to thousands in the case of NGC 6134. The lack of variables in some of the clusters does not necessarily reflect a true deficit of variables. It is as much a function of the amount of data available for the clusters.

Table 1. Open clusters with CCD measurements.

| Cluster | Decl. | Dist.(pc) | log(Age) | Comments[a] |
|---|---|---|---|---|
| NGC 2099 | +32° | 1350 | 8.48 | |
| NGC 2194 | +13° | 2700 | 8.57 | |
| NGC 2251 | +8° | 1550 | 8.48 | |
| NGC 2355 | +13° | 1600 | 8.99 | |
| NGC 2660 | −47° | 2680 | 9.22 | 4 $\delta$'s, 1 EB |
| NGC 6134 | −49° | 900 | 8.90 | 7 $\delta$'s, 1 BS |
| NGC 7062 | +46° | 1700 | 8.69 | 3-5 $\delta$'s |
| NGC 7209 | +46° | 900 | 8.50 | |
| NGC 7226 | +55° | 2100 | 8.34 | 2 $\delta$'s, 1 EB |
| NGC 7245 | +54° | 1850 | 8.19 | 2 $\delta$'s, 1 EB, 1 W UMA |
| NGC 7654 | +61° | 1600 | 8.23 | 1 $\delta$, 1 EB, 1? |

[a]The explanation of signatures is EB: Eclipsing Binary, BS: Blue Straggler, $\delta$: $\delta$ Scuti star, W UMA: Contact binary

In order to succeed in the search the procedures described by Kjeldsen & Frandsen (1992), Gilliland & Brown (1994) and Frandsen et al. (1995) should be followed in detail.

## References


Balona, L. A., & Stobie, R. S. 1979, MNRAS, 189, 649

Breger, M. 1990, in "Confrontation between stellar pulsation and evolution", (eds.) C. Cacciari & G. Clementini, ASP Conf. Ser. 11, 263

Gilliland, R. L., & Brown, T.M. 1994, ARA&A, 32, 37

Frandsen, S., Jones, A., Kjeldsen, H., et al. 1995, A&A, in press

Kjeldsen, H., & Frandsen, S. 1992, PASP, 104, 413

Mantegazza, L., Poretti, E., & Bossi, M. 1994, A&A, 287, 95

*Astrophysical Applications of Stellar Pulsation*
*ASP Conference Series, Vol. 83, 1995*
*R. S. Stobie and P. A. Whitelock, eds.*

# Radial pulsation among $\delta$ Sct stars

E. Rodríguez, A. Rolland, P. López de Coca, S. Martín
*Instituto de Astrofísica de Andalucía, CSIC, Apdo. 3004, 18080 Granada, Spain*

**Abstract.** It has been pointed out by earlier authors that radial pulsation seems to be the only one type present in high amplitude $\delta$ Sct stars, at least for monoperiodic stars. In order to confirm or not this point, we have collected all the multicolour data available for these stars in the Strömgren and Johnson photometric systems. Then, the type of pulsation has been analysed on the basis of the phase shifts and amplitude ratios between observed light and colour variations. The results indicate that all the stars analysed are radial pulsators.

It has been pointed by earlier authors that two clearly different groups of pulsators seems to coexist among the $\delta$ Sct stars. First, variables with low amplitudes of pulsation (visual amplitudes smaller than roughly $0.^m1$). The majority of these variables show complex pulsation frequency spectra where a number of simultaneously excited nonradial (and radial) modes are present. Second, variables with high amplitudes of pulsation (visual amplitudes greater than roughly $0.^m3$). In most of them only one or two frequencies appear excited, and both single and double mode stars seem to show radial pulsation.

In all these double mode high amplitude stars, radial pulsation has been found from comparison between observed and theoretical period ratios. Moreover, the method based on "amplitude ratios versus phase shifts" between different filters, in *uvby* photometry, was applied by earlier authors to a large sample of monoperiodic high amplitude $\delta$ Sct and SX Phe stars leading to radial pulsation for all the variables analysed. A similar method has also been shown in UBV photometry. Thus, radial pulsation seems to be the only one type present in high amplitude $\delta$ Sct stars, at least for mono-periodic and double mode pulsators.

In order to confirm or not this point for mono-periodic stars, we have collected all the simultaneous or nearly simultaneous multicolour data available for these stars in the Strömgren and Johnson photometric systems. The data have been considered for Population I and also Population II (SX Phe type) stars. Then, the type of pulsation has been analysed on the basis of these phase shifts and amplitude ratios between observed light and colour variations. Table 1 lists the stars analysed with their periods and full visual amplitudes. In all cases, the results lead to radial pulsation according to the corresponding diagrams of "amplitude ratios versus phase shifts". In addition to mono-periodic variables, some double mode stars (AE UMa, RV Ari, BP Peg, V1719 Cyg and SX Phe) were considered and radial pulsation was also found in good agreement with their period ratios. It is also true for the two first modes of AI Vel. Radial pulsation was also found for the three mono-periodic stars of medium amplitude ($0.^m1 \leq \Delta V < 0.^m3$) considered: V1162 Ori, $\rho$ Pup and V393 Car.

Table 1. Stars analysed.

| Star | Period (days) | ΔV (mag) | source | Star | Period (days) | ΔV (mag) | source |
|---|---|---|---|---|---|---|---|
| Pop. I | | | | Pop. I | | | |
| ZZ Mic | 0.0672 | 0.35 | 23,24,25 | V567 Oph | 0.1495 | 0.33 | 6,7 |
| GP And | 0.0787 | 0.52 | 1 | CW Ser | 0.1892 | 0.43 | 15,16 |
| V1162 Ori | 0.0787 | 0.20 | 6 | V798 Cyg | 0.1948 | 0.46 | 4 |
| KU Cen | 0.0800 | 0.47 | 6 | BS Aqr | 0.1978 | 0.44 | 10,11 |
| AE UMa(F) | 0.0860 | 0.44 | 1 | RY Lep | 0.2254 | 0.35 | 1 |
| EH Lib | 0.0884 | 0.50 | 17,18,19,20,21,22 | TV Lyn | 0.2407 | 0.44 | 1 |
| RV Ari(F) | 0.0931 | 0.43 | 2 | DE Lac | 0.2537 | 0.32 | 12,13 |
| BE Lyn | 0.0959 | 0.39 | 1 | DH Peg | 0.2555 | 0.50 | 14 |
| YZ Boo | 0.1041 | 0.42 | 26,27,28,29 | UY Cam | 0.2670 | 0.34 | 1 |
| BP Peg(F) | 0.1095 | 0.41 | 1 | V1719 Cyg(1) | 0.2673 | 0.31 | 4 |
| AI Vel(F) | 0.1116 | 0.67 | 3 | V1719 Cyg(2) | 0.2138 | | 4 |
| AI Vel(1) | 0.0862 | | 3 | Pop. II | | | |
| SZ Lyn | 0.1205 | 0.51 | 29,34,35,36,37,38 | SX Phe(F) | 0.0550 | 0.41 | 1 |
| AD CMi | 0.1230 | 0.30 | 1 | SX Phe(1) | 0.0428 | | 1 |
| ρ Pup | 0.1409 | 0.09 | 8,9 | KZ Hya | 0.0595 | 0.80 | 1 |
| V393 Car | 0.1413 | 0.19 | 5 | CY Aqr | 0.0610 | 0.71 | 1 |
| RS Gru | 0.1470 | 0.56 | 1 | DY Peg | 0.0729 | 0.54 | 1 |
| DY Her | 0.1486 | 0.51 | 30,31,32,33 | XX Cyg | 0.1349 | 0.80 | 1 |

1 Garrido et al. 1990, A&A 234, 262
2 Rodríguez et al. 1992, A&AS 93, 189
3 Watson R.D. 1988, Ap&SS 140, 255
4 Poretti et al. 1988, A&A 199, 191
5 Helt B.E. 1984, A&AS 56, 457
6 Poretti et al. 1990, A&A 228, 350
7 Powell et al. 1990, PASP 102, 1131
8 Balona et al. 1983, South African Astr. Obs. Circ. 7, 19
9 Dravins et al. 1977, A&A 54, 381
10 Elst E.W. 1976, A&AS 23, 419
11 Meylan et al. 1986, A&AS 64, 25
12 Fitch et al. 1966, Comm. Lunar Planet. Observ. 71, 3
13 Paczynski B. 1965, Acta Astron. 15, 115
14 Tifft W.G. 1964, ApJ 139, 451
15 Gieren et al. 1975, Inf. Bull. Var. Stars No. 1055
16 Bookmeyer et al. 1977, Rev. Mex. Astron. Astrof. 2, 235
17 Joner M.D. 1986, PASP 98, 651
18 Broglia et al. 1977, A&AS 29, 321
19 Terzan et al. 1974, A&AS 16, 155
20 Garrido et al. 1979, A&AS 36, 51
21 Mahdy et al. 1980, Comm. Konkoly Obs. No. 74
22 Karetnikov et al. 1979, Perem. Zvezd. 21, 199
23 Churms et al. 1961, Observatory 920, 25
24 Balona et al. 1978, MNRAS 184, 11
25 Chambliss C.R. 1971, ApJ 165, 365
26 Szeidl et al. 1981, Comm. Konkoly Obs. No. 75
27 Broglia et al. 1957, Cont. Ob. Ast. Mil.-Mer. No. 102
28 Joner et al. 1983, PASP 95, 433
29 Langford W.R. 1976, Ph. D. Thesis Brigham Univ.
30 Breger et al. 1978, PASP 90, 754
31 Hardie et al. 1961, ApJ 133, 71
32 Spinrad H. 1959, ApJ 130, 539
33 Broglia et al. 1955, Cont. Ob. Ast. Mil.-Mer. No. 66
34 Soliman et al. 1986, Comm. Konkoly Obs. No. 88
35 Broglia P. 1963, Cont. Ob. Ast. Mil.-Mer. No. 222
36 Barnes et al. 1975, AJ 80, 48
37 Duerbeck H.W. 1976, Inf. Bull. Var. Stars No. 1971
38 Szeidl B. 1983, Comm. Konkoly Obs. No. 84

*Astrophysical Applications of Stellar Pulsation*
*ASP Conference Series, Vol. 83, 1995*
*R. S. Stobie and P. A. Whitelock, eds.*

# New WET Observations of the $\delta$ Scuti Star CD−24 7599: amplitude variability and discovery of 13 pulsation modes

G. Handler[1], D. O'Donoghue[2], D. A. H. Buckley[3], O. Giovannini[4], S. O. Kepler[4], J. L. Provencal[5], M. A. Wood[6], J. C. Clemens[7], M. S. O'Brien[7], S. J. Kleinman[8], T. K. Watson[8], R. E. Nather[8], D. E. Winget[8], A. Kanaan[8], A. Nitta[8], M. H. Montgomery[8], E. Klumpe[8], P. A. Bradley[9], D. J. Sullivan[10], K. Wu[11], E. Leibowitz[12]

## 1. Introduction

The variability of CD−24 7599 ($V$=11.48 mag) was discovered by JCC during observing run XCOV7 of the Whole Earth Telescope (WET, Nather et al. 1990) network in February, 1992. The star was observed as an additional target and 117 hours of high-quality temporal spectroscopic observations were obtained.

Our analysis of these data revealed the presence of 7 independent pulsation modes between 27.0 and 38.1 cycles per day (313 – 441 $\mu$Hz) with semi-amplitudes of 2.1 – 10.2 milli-modulation amplitudes (mma). We showed that peaks at linear combination frequencies detected in the power spectra were *not* due to eigenmodes excited to visible amplitude by resonant mode coupling.

The examination of subsequent uvby$\beta$ photometry showed the star to be located on the lower main sequence, near the observed hot border of the classical instability strip. The complete analysis was published by Handler et al. (1995), and the reader is referred to this paper for more information. Due to the suspected presence of several further pulsation modes with photometric amplitudes around 1 mma, CD−24 7599 was re-observed with WET in May, 1994.

---

[1]Institut für Astronomie, Türkenschanzstraße 17, A-1180 Wien, Austria

[2]Department of Astronomy, University of Cape Town, Rondebosch 7700, South Africa

[3]South African Astronomical Observatory, P. O. Box 9, Observatory 7935, South Africa

[4]Instituto de Fisica, Universidade Federal do Rio Grande do Sul, 90049 Porto Alegre-RS, Brazil

[5]Department of Physics and Astronomy, University of Delaware, Newark, DE 19716, USA

[6]Department of Physics and Space Sciences, Florida Institute of Technology, 150 W. University Boulevard, Melbourne, FL 32901, USA

[7]Department of Physics and Astronomy, Iowa State University, Ames, IA 50211, USA

[8]Department of Astronomy and McDonald Obs., University of Texas, Austin, TX 78712, USA

[9]X-2 MS B-220, Los Alamos National Laboratory, Los Alamos, NM 87545, USA

[10]Department of Physics, Victoria University of Wellington, New Zealand

[11]RCfTA, School of Physics, University of Sydney, NSW 2006, Australia

[12]Department of Physics and Astronomy, University Tel Aviv, Tel Aviv 69978, Israel

## 2. New WET observations and their analysis

CD−24 7599 was included as 3rd priority target in WET run XCOV 10. Due to good weather, we acquired 86 hours of new measurements during a 14-day time span. Frequency analysis of the new data set confirmed the presence of the previously detected pulsation modes. However, conspicuous amplitude variations occurred within the two years. Moreover, when analysing both data sets together, we were able to establish the presence of 6 further pulsation modes (with frequencies between 315 and 390 $\mu$Hz and amplitudes between 0.9 and 1.3 mma), bringing the total number to 13.

## 3. The importance of CD−24 7599 for $\delta$ Scuti star asteroseismology

CD−24 7599 is a good (maybe the best) candidate for the application of asteroseismology to $\delta$ Scuti stars because:

- Of course, due to its large number of detected pulsation frequencies. It is the $\delta$ Scuti star with the most observed pulsation modes at this time.
- Its unevolved state excludes the presence of a large number of mixed modes.
- Only 2 or 3 radial overtones appear to be excited. Therefore, we expect to observe many almost completely rotationally split multiplets.
- Its relatively large photometric amplitudes allow us to determine accurate phase shifts to constrain the $\ell$ value of many modes by using time-series multi-colour photometry.
- Its amplitude variations may help us to discover even more pulsation frequencies in the future.

What can be said at the present stage?

- We cannot find closely spaced triplets, quintuplets, or septuplets in the amplitude spectrum, suggesting the star is not a slow rotator.
- A regular frequency spacing (about 26 $\mu$Hz) of the 13 modes may already be detected. This could be due to rotational splitting or to the near degenerate frequencies of modes with the same $k$, but different $\ell$.
- The knowledge of the star's projected rotational velocity is of great importance. A measurement will be obtained soon. Collaboration of the Vienna group with W. Dziembowski and colleagues will enable us to calculate specific pulsational models for CD−24 7599 and other $\delta$ Scuti stars.

## References

Handler, G., Breger, M., Sullivan, D. J., et al. 1995, A&A, submitted

Nather, R. E., Winget, D. E., Clemens, J. C., Hansen, C. J., & Hine, B. P. 1990, ApJ, 361, 309

*Astrophysical Applications of Stellar Pulsation*
*ASP Conference Series, Vol. 83, 1995*
*R. S. Stobie and P. A. Whitelock, eds.*

# Asteroseismology in Praesepe: the case of EP Cnc

J.A. Belmonte, A. Claret and F. Pérez Hernández

*Instituto de Astrofísica de Canarias, 38200 La Laguna, Tenerife, Spain*

**Abstract.** On the 1992 STEPHI IV campaign, the multi-periodic $\delta$ Scuti stars BN Cnc and BU Cnc of the Praesepe cluster were monitored during a three-week, three-continental run (Belmonte et al. 1994). Five and six frequency peaks, respectively, were recorded as a result of the campaign. In a subsequent paper (Pérez Hernández et al. 1995, hereafter Paper I), an attempt to identify these frequency peaks as pulsation modes was done. An important point is that both stars belong to the same cluster and, hence, more severe constraints than for isolated stars can be imposed. Recently, Breger et al. (1994) have reported results on EP Cnc, another $\delta$ Scuti in Praesepe. This star is also multi-periodic, showing three frequency peaks above the noise level. In the present work we extend our analysis to this star.

## 1. Discussion

As a matter of fact, EP Cnc is more luminous than BN and BU Cnc and presumably all the hydrogen in its convective core has been depleted, as opposite to BN Cnc and BU Cnc. In Paper I, the model parameters were chosen in order to fit the observed frequencies of BN Cnc and BU Cnc. The range of ages considered were also obtained from the theoretical analysis of the observed frequencies. Although for EP Cnc (as well as for BN Cnc and BU Cnc), other models can also fit the observed frequencies, to be consistent, we shall assume the same parameters for the three stars. The models corresponding to each star are constrained by the observed magnitudes and effective temperatures. In this sense, a mass between $2.3M_\odot$ and $2.4M_\odot$ can be estimated for EP Cnc.

Frequencies are computed in the linear and adiabatic approximation. Rotation effects are considered up to second order, namely

$$\nu_{nlm} - \nu_{nl}^{(0)} = -m(1 - C_{nl})\nu_{rot} + (X_{nl} + m^2 Y_{nl})\frac{\nu_{rot}^2}{\nu_{nl}^{(0)}} \quad , \tag{1}$$

where $\nu_{nl}^{(0)}$ is the unperturbed frequency and $\nu_{rot}$ the rotation rate of the star.

As for BN and BU Cnc, we start the identification of the frequency peaks by looking for possible radial modes. This can be done by comparing the observed period ratios with the theoretical ones. For EP Cnc, evolutionary stages before and after the red hook have been considered. In contrast to BN and BU Cnc, for which two radial modes ($n = 5$ and $n = 6$) can be present, for EP Cnc only one of the frequency peaks can be a radial model, either $n = 2$ or $n = 3$. The fundamental mode $n = 1$ is unlikely. It is interesting to note that this latter fact is due to the inclusion of the frequency shift caused by rotation and not to the restrictive set of stellar models considered here.

There are at least two non-radial modes. When considering all the theoretical modes with $l = 1$ and $l = 2$, it is found that tens of them are present in the observed frequency range. However, not all of them are of the same nature. The $l = 2$ modes can be regarded as $g$-modes, while for the $l = 1$ modes there are both $g$-modes and modes which are trapped predominantly in the envelope. Envelope-trapped modes have more relative energy in the upper layers than $g$-modes and rotational splitting parameters closer to those of $p$-modes. We note that this situation is typical for these kind of stars (see e.g. Dziembowski & Goode 1992 for the case of $\theta^2$ Tau).

In the observed frequency range, and for our set of models, there is just one $(n, l = 1)$ envelope-trapped mode. In principle it can be thought that this $p$-like mode is the more likely to be detected since, for a given energy, it has the largest surface amplitude. However the (unknown) excitation mechanism could give a substantially larger energy to some of the $g$-modes.

On the other hand, the rotational frequency splittings of the g and envelope-trapped modes are rather different. If there were no $g$-modes, then we found that $\nu_2$= 67.0 $\mu$Hz is a radial mode with $n = 2$ and $\nu_1$=63.3 $\mu$Hz and $\nu_3$=75.4 $\mu$Hz correspond to an $l = 1$ envelope-trapped mode split by rotation. The difference between $\nu_3$ and $\nu_1$ is 12.1$\mu$Hz, while from the observed value of $V \sin i$ and our models, it follows that $\nu_{rot} > 7.2\mu$Hz. If they correspond to $m = \pm 1$, then $C_{nl} > 0.16$. This is compatible with an envelope-trapped mode.

It is interesting to consider a frequency peak at 72.9 $\mu$Hz, although, following Breger et al. (1994), it is below the noise level. If $g$-modes are not present in the spectrum then it must correspond to the $m = 0$. From eq. (1) the value of $Y_{nl}$ can be estimated, yielding $Y \sim -4$, which is in the range of values expected for envelope-trapped modes. Hence, even in this case, the presence of $g$-modes is not necessary.

## References

Belmonte, J.A., Michel, E., Alvarez, M., Jiang, S.Y., Chevreton, M., Auvergne, M., Liu, Y.Y., Goupil, M.J., Baglin, A., Roca Cortés, T., Mangeney, A., Dolez, N., Valtier, J.C., Massacrier, G., Sareyan, J.P., Schmider, F.X., & Vidal I. 1994, A&A, 283, 121

Breger, M., Martin, B., Garrido, R., Jiang, S.Y., Li, Z.P., Hube, D.P., Stich, J., Ostermannn, W., & Paparo M. 1994, A&A, 281, 90

Pérez Hernández, F., Claret, A., & Belmonte, J.A. 1995, A&A, in press

Dziembowski, W.A., & Goode, P.R. 1992, ApJ, 394, 670

*Astrophysical Applications of Stellar Pulsation*
*ASP Conference Series, Vol. 83, 1995*
*R. S. Stobie and P. A. Whitelock, eds.*

# The Variable Star Population in the Center of 47 Tuc as defined by *HST* Observations[1]

Ronald L. Gilliland, Peter D. Edmonds, Larry Petro, Abhijit Saha, and Michael Shara

*Space Telescope Science Institute, 3700 San Martin Drive, Baltimore, MD 21218 USA*

**Abstract.** The high stellar densities at the centre of some globular clusters provide prolific breeding grounds for exotic stars, e.g., the 11 msec pulsars in 47 Tuc and numerous blue stragglers. The same high density of stars as viewed on the sky makes ground-based photometry problematic for variable star detection. Even a very few tight binaries can be of fundamental importance for the dynamic evolution of the cluster core. We present results of a continuous 40-hour sequence of U-band CCD exposures acquired with WF/PC on HST. The high spatial resolution coupled with a stable PSF allows nearly Poisson limited time series on some 20,000 stars. The 1000 second sampling and long coverage provide excellent sensitivity to W UMa and $\delta$ Scuti variations. Light curves are shown for selected variables and the relative frequencies of binary and variable stars are discussed.

## 1. Introduction and observations

A matter of considerable importance to the dynamical evolution of globular clusters is whether or not the centrally condensed (and hence presumably more massive than most stars) blue stragglers are primarily single stars, or binaries. Variability monitoring for $\delta$ Scuti (usually called SX Phe for Pop. II stars) or W UMa modulation provides important clues to the status (single or binary) of the blue stragglers. We find numerous examples of $\delta$ Scuti, W UMa and eclipsing binary variables in 47 Tuc. Variability in the blue stragglers is clearly over-represented relative to the general population.

Our observations were the first to exploit the continuous viewing zone (CVZ) of *HST* for a substantial science program. Over 5-7 September 1993 we obtained a nearly continuous sequence of 99, 1000s exposures with the Planetary Camera (0.043 arcsec pixels) of WF/PC using the F336W (U) filter.

We have discussed the observations and data analysis details in Gilliland et al. (1995). Details concerning variable star detection, and completeness statistics for binary systems will appear in Edmonds et al. (1995).

---

[1]Based on observations with the NASA/ESA *Hubble Space Telescope* obtained at the Space Telescope Science Institute, which is operated by Association of Universities for Research in Astronomy, Incorporated, under NASA contract NAS5-26555.

## 2. Binary star detections

Our time series with 99 points over 38.5 hours have single measure precisions of about 3-4 mmag for the brightest stars near $m_U$ ~15 to 0.2 mag near $m_U$ ~20. The precisions are well modelled by Poisson statistics and readout noise (plus a 'sky' term from the overlapping spherical aberration halos).

The W UMa systems are of particular interest, since our observations are useful for providing a nearly complete census within the 66×66 arcsec region observed. Above $m_U = 19$ our detection sensitivity (Edmonds et al. 1995) averages 77% for a statistically representative distribution of W UMa light curves (Rucinski 1994). We detect 3 W UMa systems among 7700 stars brighter than $m_U = 19$, and thus predict the existence of 4 W UMas in this sample. We have also detected at least 4 other binary systems through eclipses.

## 3. δ Scuti (SX Phe) detections

We have not yet finalised our U, B, V photometry solutions, and thus will not present colours for the δ Scuti variables. Our search of 7700 stars for δ Scuti-like oscillations turned up 4 stars all of which are contained within the blue straggler domain of $m_V \leq 16.5$, $0.0 \leq B - V \leq 0.6$ (there are 1325 stars in total with $m_V \leq 16.5$, 50 within this preliminary blue straggler domain). Variability at the level of stars A, B & C would have been detected in any of the several thousand stars. The lower amplitude of variable D would only be detectable in favourable cases and our search for variables has not yet been completed at these lower signal levels (nor is its statistical significance established).

Table of δ Scuti Properties

| Star | V | Amp. (mag) | Frequency (mHz) |
|---|---|---|---|
| A | 15.5 | 0.075 | 0.1830 |
| B | 15.1 | 0.06,0.05 | 0.1133,0.1471 |
| C | 15.8 | 0.038 | 0.2078 |
| D | 14.4 | 0.003 | 0.210 |

Object B clearly displays a double mode; with a period ratio of 0.770 ± 0.002 this falls within the narrow range (0.768 to 0.778) expected for the fundamental and first overtone period ratio for Pop II stars (Peterson & Jørgensen 1972). Consistency checks on membership, cluster distance, and parameters of the variable stars can be obtained by comparison with calibrated period-luminosity relations (Nemec et al. 1994); to be discussed in a journal paper.

## References

Edmonds, P.D., et al. 1995, in preparation

Gilliland, R.L., et al. 1995, ApJ, in press (July issue)

Nemec, J.M., Linnell Nemec, A.F., & Lutz, T.E. 1994, AJ, 108, 222

Peterson, J.D., & Jørgensen, H.E. 1972, A&A, 17, 367

Rucinski, S.M, 1994, PASP, 106, 462

*Astrophysical Applications of Stellar Pulsation*
*ASP Conference Series, Vol. 83, 1995*
*R. S. Stobie and P. A. Whitelock, eds.*

# Insights into the Pulsational Behaviour of $\delta$ Sct Stars

L. Mantegazza

*Università di Pavia, Dipartimento di Fisica Nucleare e Teorica, Via Bassi 6, 27100 Pavia, Italy*

E. Poretti

*Osservatorio di Brera, Via E. Bianchi 46, 22055 Merate, Italy*

## 1. Introduction

Extensive photometric monitoring of bright $\delta$ Scuti stars, made in the last years, allowed us to get reliable frequency determinations also in very complicated cases. Table 1 gives a full description of our photometric targets. Recently, we realized that combining photometry with simultaneous high resolution spectroscopy, in order to study line profile variations, offers the possibility to perform a much more reliable mode identification. Hence, the photometric runs of FG Vir, X Cae, HD 2724 were paired with spectroscopic ones obtained at the Coudé Auxiliary Telescope (ESO, La Silla, Chile). The stellar spectra are centred at 4508 Å and they cover a range of 37.6 Å; the resolving power is about 50000-60000 and the linear dispersion is 2.4 Å/mm.

## 2. New results

*X Caeli* – Already observed in the $V$ colour in 1989, X Caeli was reobserved both phometrically and spectroscopically in 1992. The frequency analysis of the new photometric data confirmed the presence of a dominant mode (7.39 c/d, amplitude 37 mmag), two modes with intermediate amplitudes (6.04 c/d, amplitude 7 mmag, and 7.46 c/d, amplitude 5 mmag), many other modes with smaller amplitude, the first Fourier harmonic of the three strongest modes and their nonlinear coupling terms. The amplitude ratios and phase shifts between $V$ and $B-V$ curves suggest that the three strongest modes are nonradial with $\ell$=1 or $\ell$=2.

The availability of light, colour and radial velocity curves allowed us to perform a preliminary application of the Baade-Wesselink method to the dominant mode. There is an acceptable solution for $\ell$=2 only, from which we also obtain a mean stellar radius of $R = 5.0 \pm 0.4\ R_{\odot}$, $f = 18.9 \pm 1.5$ (ratio of flux to radius variation), $\psi = 100.5^o \pm 3.5^o$ (phase difference between the radius and the flux variations), $\epsilon = 0.0052 \pm 0.0004$ (the projected radial semi-amplitude), $A = 2.48 \pm 0.26$ (the gradient of the surface brightness–colour relationship).

In the stellar spectra obtained in 1992, a narrow absorption core is visible at the bottom of the spectral lines. This feature is stable with respect to the stellar reference frame; 9 additional spectra taken in November 1994 confirm its stability over an interval of years. Therefore, we could be in presence of a circumstellar shell.

*HD 16439* – This star was observed at Merate Observatory in 1988-89. Two frequencies at 17.1266 c/d and 10.1273 c/d were clearly identified; no additional peak was visible. HD 16439 was re-observed at Merate Observatory in December 1994 and January 1995. The presence of the main frequency at 17.13 c/d was confirmed, while the difference in height between the 10.13 c/d peak and its alias at 11.13 c/d is reduced. Moreover, a new peak at 10.48 c/d, not observed in the 1988-89 dataset, is clearly identified in the frequency analysis. Since the two datasets were obtained in the same conditions (site, length of the observations, number of data, ...), the evidence for a change in the frequency content between 1989 and 1994 seems well established. Moreover, the similar height of the peaks at 10.13 and 11.13 c/d makes one suspect that other undetectable terms are present in this spectral region.

Table 1. Summary of the photometric observations carried out by the Merate group

| Star | Site | Year | N | Survey [hours] | Reference |
|---|---|---|---|---|---|
| HD 37819 | Merate | 1986 | 462 | 32 | A&A 1987, **181**, 273 |
| HR 1225 | ESO | 1987 | 705 | 38 | A&A 1989, **220**, 144 |
| $o^1$ Eri | ESO | 1987 | 710 | 38 | A&A 1989, **220**, 144 |
| HR 547 | ESO | 1987 | 462 | 22 | A&A 1989, **220**, 144 |
| HD 16439 | Merate | 1988-89 | 1020 | 54 | A&A 1990, **230**, 91 |
| HD 16439 | Merate | 1994-95 | | | work in progress |
| HD 101158 | ESO | 1989 | 1234 | 62 | A&A 1991, **245**, 136 |
| X Cae | ESO | 1989 | 1013 | 54 | A&A 1992, **255**, 153 |
| X Cae | ESO | 1992 | 1634 | 100 | work in progress |
| 44 Tau | Merate | 1989-90 | 2434 | 117 | A&A 1992, **256**, 113 |
| HD 66853 | ESO | 1991 | 1390 | 100 | A&A 1994, **281**, 66 |
| HD 66853 | Merate | 1991 | 642 | 43 | A&A 1994, **281**, 66 |
| HD 224639 | ESO | 1991 | 2567 | 120 | A&A 1995, in press |
| HD 224639 | ESO | 1994 | | | work in progress |
| HD 18878 | Merate | 1991-92 | 2915 | 150 | A&A 1993, **274**, 811 |
| HD 19279 | Merate | 1991-92 | 2679 | 150 | A&A 1993, **274**, 811 |
| $\beta$ Cas | Merate | 1986 and 1992 | 925 | 30 | A&AS 1994, **108**, 55 |
| FG Vir | ESO | 1992 | 792 | 53 | A&A 1994, **287**, 95 |
| HD 2724 | ESO | 1993 | | | work in progress |
| AZ CMi | Merate | 1994 | | | work in progress |

*Astrophysical Applications of Stellar Pulsation*
*ASP Conference Series, Vol. 83, 1995*
*R. S. Stobie and P. A. Whitelock, eds.*

# Simultaneous Photometric and Spectroscopic Observations of the unusual Variable F-type stars HD 224638 and HD 224945.

L. Mantegazza[1], E. Poretti[2], M. Bossi[2] and F.M. Zerbi[1]

[1] *Dipartimento di Fisica Nucleare e Teorica, Università di Pavia, Italy*

[2] *Osservatorio Astronomico di Brera, Italy*

## 1. Introduction

The stars HD 224638 and HD 224945 belong to the new class of F0 V spectral type stars which show an unusual kind of variability for that region of the HR diagram (Mantegazza et al. 1993).

The variability of these two stars has been discovered and studied by Mantegazza, Poretti & Zerbi (1994). They show small amplitude light variations with characteristic time scales of the order of one day. The light curves are not periodic and it is difficult to satisfactorily fit them even with several periodic terms.

In order to explain our 1991 *B* colour observations we tentatively suggested that in both stars two close periodic terms were present with a characteristic frequency of about 0.8 c/d, and that these terms had a double wave shape for HD 224638 and a triple wave one for HD 224945.

Three high resolution spectrograms of HD 224638 also showed that probably there were line profile variations.

At the moment two competitive models are the most plausible in order to explain these variations: non-radial *g*-mode pulsations or spots carried across the visible stellar disk by rotation.

## 2. New observations

In order to better understand the phenomenology of these stars we performed simultaneous *B*,*V* (13 nights, Oct. 23 – Nov. 6, 1994) and high resolution spectroscopic (5 nights, Oct. 24 – 29, 1994) observations with the 50 cm and CAT telescopes respectively at La Silla (ESO). The spectrograms cover the region between 4490 and 4525 Å with a resolution of 60000 and have a mean S/N ratio at a continuum level of about 250.

A preliminary data analysis allows us to establish the following facts:

- The light curves have a frequency content very similar to that of previous observations in 1991. For example for HD 224638 we find 5 terms with frequencies of 1.66, 1.37, 1.56, 0.96 and 1.69 c/d in the 1994 data and of 1.62, 1.37, 1.70, 0.89 and 0.81 c/d in the 1991 data.
- The frequency analysis of colour index data indicates that the colour index varies approximately in phase with the light.

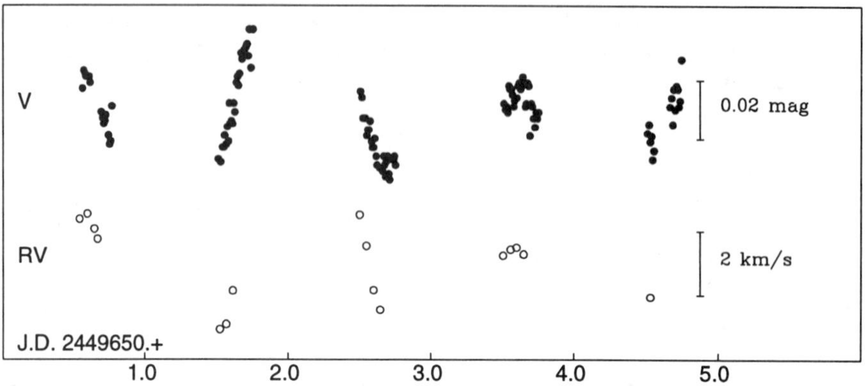


Figure 1. V and radial velocity curves of HD 224638

- There are large variations in the shapes of line profiles, which are also detectable in the course of the same observing night.
- The frequency analysis of the time-series of the first 5 moments of the line profiles (Balona 1986), indicates that they have approximately the same frequency content as does the photometric data.
- The full amplitudes of the radial velocity variations ($m1$ moment) are of about 3.5 km/s for both stars. Fig. 1 shows that for HD 224638 these variations are in phase with the light curves. This behaviour is the opposite to that observed in radial and non-radial $p$-mode pulsators in the instability strip.

## 3. Conclusion

The observed light, colour and line profile variations could in principle be explained both by the pulsation and the spot model. However the spot model can hardly explain both the observed amplitudes in radial velocities and the presence of several periodicities stable on a baseline of a few years. In order to definitely confirm this last point it would be of paramount importance to perform a multi-site campaign on these stars, due to the difficulty of deconvolving unambiguously from the spectral window time-series with periods of the order of 1 day.

## References

Mantegazza, L., Poretti, E., Antonello, E., & Zerbi, F.M. 1993, in: Inside the Stars, W.W. Weiss & A. Baglin, ASP Conf. Ser. 40, 733

Mantegazza, L., Poretti, E., & Zerbi, F.M. 1994, MNRAS, 270, 439

Balona, L. 1986, MNRAS, 219, 111

*Astrophysical Applications of Stellar Pulsation*
*ASP Conference Series, Vol. 83, 1995*
*R. S. Stobie and P. A. Whitelock, eds.*

# Slowly Variable Early F-type Stars

L.A. Balona

*South African Astronomical Observatory, P.O. Box 9, Observatory 7935, Cape, South Africa*

K. Krisciunas

*Joint Astronomy Centre 660 N. A'ohōkū Place, University Park, Hilo, Hawaii 96720, USA*

**Abstract.** A third periodicity is found in the early F-type star $\gamma$ Dor which probably rules out an interpretation in terms of differential rotation of a spotted star. One of the two periods in 9 Aur is not visible in the radial velocity. This is not explained by the star spot model, but may indicate that the two periodicities are nonradial pulsations of different spherical harmonic degrees.

## 1. Introduction

In recent years there has been growing evidence for the existence of a new class of variable among main-sequence early F-type stars. These stars show periodic light variations of many hours – too long to be attributed to $p$-mode pulsations characteristic of $\delta$ Scuti stars. They are all early-F dwarfs on, or beyond, the red edge of the $\delta$ Scuti instability strip. Few of these stars have been well studied, but whenever sufficient observations are available, they are found to be multiperiodic.

## 2. Observations

By far the best studied member of this class is $\gamma$ Dor, an F1V star in which two sinusoidal variations with periods $P_1 = 0.7570$ and $P_2 = 0.7334$ d are present (Balona, Krisciunas & Cousins 1994). For two weeks in 1994, November, we obtained photometry of $\gamma$ Dor from SAAO (LAB) and CTIO (KK). Observations from two sites are required to ensure that $P_1$ and $P_2$ in $\gamma$ Dor are separate periodicities. The latest observations eliminate the possibility that only one period is present. Balona et al., (1994) found marginal evidence for a third periodicity at $P_3 = 0.678$ d, not too different from the well known periods at $P_1$ and $P_2$ d. Our most recent data again show the presence of $P_3$ and we conclude that is most probably real. This component was not present before 1993.

Krisciunas et al. (1995) provide significant evidence that the photometric variations of 9 Aur are due to nonradial $g$-mode pulsations of low spherical harmonic order. Their radial velocity data, obtained by R. Griffin with CORAVEL at Haute Provence, exhibit variations entirely consistent with only the longer of the two periods of 9 Aur. The preliminary conclusion is that $P_2$ comes from an $\ell = 1$ or 2 spherical harmonic, while $P_1$ comes from a higher order harmonic.

A multi-longitude campaign on 9 Aur was carried out in 1994 December and 1995 January. These observations show 9 Aur exhibiting the largest photometric amplitudes seen to date. It ranged in brightness by 0.090 in the $V$-band, and by 0.124 in the $B$-band. As with $\gamma$ Dor, the $B$–$V$ colours of 9 Aur phase up with the $V$-band light curve in the sense that the bluest colour (i.e., hottest temperature) occurs when the star is brightest.

## 3. Star spots or nonradial pulsations?

There are only two viable hypotheses to explain the light variations in these stars: star spots or nonradial pulsations. Since they are multiperiodic, the star spot model presumes differential rotation. In order to understand the rather large frequency differences, Mantegazza, Poretti & Zerbi (1994) assume that the true periods are actually harmonics and/or aliases of the observed periods. Since there is little or no observed power at the fundamental frequencies, the diametrically opposed spots must be on the same latitude and must have equal sizes.

In the case of $\gamma$ Dor the situation is simple, since the periods are close enough to be understood in terms of differential rotation. However, the detection of a third period implies an implausibly large differential rotation. In the case of 9 Aur, it is difficult to understand why one of the periods is not visible in the radial velocity observations.

In a study of $\gamma$ Dor, Balona et al. (1994) found that spots would have to be larger than $20^o$ in radius and more or less on the same latitude. The spots then interact at every beat period (about 23.5 d), in contradiction to the fact that the two principal periods have been extremely stable over the last 12 years with little or no change in amplitude.

In our opinion, there seems little doubt that there is a special region of pulsational instability for early F dwarfs which is not at present understood. Current models do not predict pulsational instability of low-degree $g$-modes in these stars. It is therefore unlikely that the $\kappa$ mechanism which is responsible for the $p$-mode instability in $\delta$ Scuti stars will also apply in these stars.

## References

Balona, L.A., Krisciunas, K., & Cousins, A.W.J. 1994, MNRAS, 270, 905

Krisciunas, K., Griffin, R.F., Guinan, E.F., Luedeke, K.D., & McCook, G.P. 1995, MNRAS, in press

Mantegazza, L., Poretti, E., & Zerbi, F.M. 1994, MNRAS, 270, 439

*Astrophysical Applications of Stellar Pulsation*
*ASP Conference Series, Vol. 83, 1995*
*R. S. Stobie and P. A. Whitelock, eds.*

# On the Mechanism of Mode Selection in Rapidly-Oscillating Magnetic Ap-Stars

I. W. Roxburgh and S. V. Vorontsov[1]

*Astronomy Unit, Queen Mary and Westfield College, Mile End Road, London E1 4NS, UK*

**Abstract.** We address the problem of the preferential mode excitation in magnetic Ap-stars by estimating the pulsation energy losses in the atmosphere through the generation of the Alfvenic waves.

Most oscillations of magnetic Ap stars are identified as dipole ($\ell = 1$) modes, with the symmetry axis aligned with magnetic axis. The preferential excitation of this mode might be provided by some selective properties of the excitation mechanism itself (Dolez & Gough 1982, Shibahashi 1983, Dziembowski 1984, Dziembowski & Goode 1985, Dolez et al. 1988). An alternative explanation related with the magnetically-induced lateral inhomogeneity of the acoustic cut-off frequency, was considered by Shibahashi (1991).

It was first demonstrated by Roberts & Soward (1983) and Campbell & Papaloizou (1986), that wave transformation in the magnetic atmosphere can provide a significant dissipation of the acoustic mode energy. For acoustic waves which propagate almost vertically near the surface, the generation of Alfvenic waves requires the magnetic field to have both a horizontal and vertical component. We thus expect the acoustic energy losses to be localised in two belts which are symmetric around the magnetic equatorial plane. These simple qualitative arguments suggest that $\ell = 1$, $m = 0$ mode with maximum amplitudes in polar regions might suffer smaller energy losses than $\ell = 1$, $m = 1$ mode. Our quantitative estimates are based on the perturbation analysis developed by Roberts & Soward (1983). Their result for the relative loss of the acoustic energy flux in the polytropic atmosphere of index $n$ can be written as

$$\tilde{R}^2 \propto B_x^2 \left(B_z^2\right)^{\frac{1}{n+2}}, \qquad (1)$$

where $B_x$ and $B_z$ denote the horizontal and the vertical component of the magnetic field. For a centred dipole field,

$$\tilde{R}^2 \propto \sin^2\theta \left(\cos^2\theta\right)^{\frac{1}{n+2}}. \qquad (2)$$

Note that the perturbation analysis of Roberts & Soward (1983) breaks down when $B_z \ll B_x$, i.e., near the magnetic equator. The correct analysis remains to be developed; it is reasonable to expect, however, that a more accurate description of $\tilde{R}(\theta)$ in the equator region will not change the qualitative conclusions.

---

[1]Institute of Physics of the Earth, Moscow 123810, Russia (permanent address)

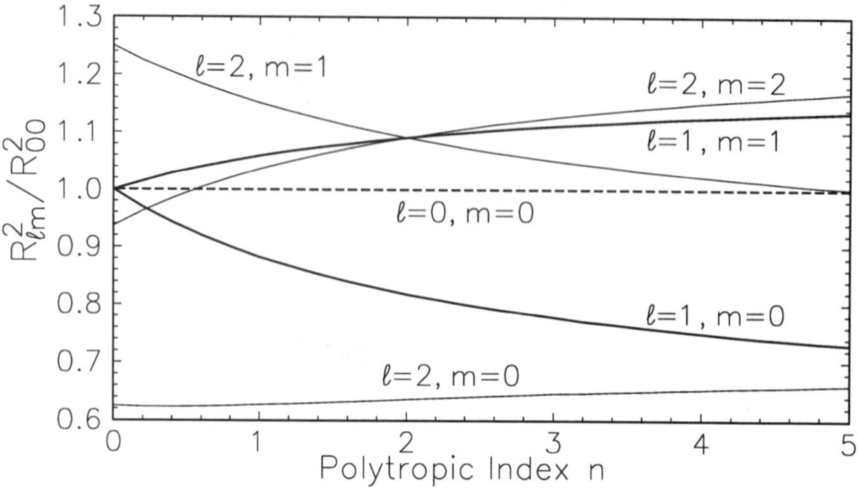


Figure 1. Energy dissipation compared with that of $\ell = 0$ mode.

The integral energy losses are estimated by the integration of $\tilde{R}^2$ with pulsation amplitudes. The energy losses of $\ell = 1$, $m = 0$ mode appear to be smaller than those of $m = 1$ mode for all $n > 0$ (Fig. 1). The energy losses are also smaller than those of a radial mode, which is in agreement with the observational evidence that dipole, but not radial modes are predominantly excited. It is interesting that for $\ell = 2, m = 0$ mode the energy losses appear to be even smaller, so that it is even more competitive for being excited. The last conclusion, however, is more sensitive to the correct treatment of the magnetic effects near the equator.

## References

Campbell, C. G., & Papaloizou, J. C. B. 1986, MNRAS, 220, 577

Dolez, N., & Gough, D. O. 1982, in *Pulsations in Classical and Cataclismic Variables*, eds J. P. Cox and C. J. Hansen, Boulder, JILA, p.248

Dolez, N., Gough, D. O., & Vauclair, S. 1988, in *Advances in Helio- and Asteroseismology*, (eds.) J. Christensen-Dalsgaard & S. Frandsen, Reidel, Dordrecht, p.291

Dziembowski, W. 1984, in *Theoretical Problems of Stellar Stability and Oscillations*, (eds.) A. Noels & M. Gabriel, Univ. de Liege, Liege, p.346

Dziembowski, W., & Goode, P. R. 1985, ApJ, 296, L27.

Roberts, P. H., & Soward, A. M. 1983, MNRAS, 205, 1171

Shibahashi, H. 1983, ApJ, 275, L5.

Shibahashi, H. 1991, in *Challenges to Theories of the Structure of Moderate-Mass Stars*, (eds.) D. Gough & J. Toomre, Springer-Verlag, Berlin, p.393

*Astrophysical Applications of Stellar Pulsation*
*ASP Conference Series, Vol. 83, 1995*
*R. S. Stobie and P. A. Whitelock, eds.*

# The Cape Rapidly Oscillating Ap Star Survey

Peter Martinez

*Department of Astronomy, University of Cape Town, Rondebosch 7700, South Africa, peter@physci.uct.ac.za*

**Abstract.** The Cape rapidly oscillating Ap star survey is a systematic search for rapidly oscillating Ap stars in the southern hemisphere. To date, 12 new roAp stars have been discovered. This paper describes the current state of the survey.

## 1. The Cape Survey

magnetic chemically peculiar A-type stars which pulsate non-radially with periods in the range 6–15 min and peak-to-peak photometric Johnson $B$ amplitudes $\leq 16$ mmag. The observed properties of these stars are reviewed in detail by Martinez & Kurtz elsewhere in these proceedings. Since May of 1990 Don Kurtz and I have searched for roAp stars systematically using the telescopes of the South African Astronomical Observatory at Sutherland. Candidates were drawn mostly from the Michigan Spectral Catalogue, which covers the declination range $-12^\circ \leq \delta \leq -90^\circ$, with the balance being drawn from other sources in an attempt to extend the survey to the equator. To date, $uvby\beta$ observations of 563 southern Ap stars have been obtained. The goal of these observations is to allow us to establish the photometric limits of the roAp phenomenon and to produce a catalogue of Strömgren photometry for all the cool southern Ap stars. New roAp stars are detected and studied using non-differential high-speed photometry. The technique and its potential pitfalls have been discussed in detail by Martinez (1993a). As of this writing 147 candidates have been observed in this mode at least once. Each star is observed for 1–2 hr in Johnson $B$ light and the resulting light curve is Fourier analysed to search for coherent oscillation frequencies. The observations are repeated on several different nights to sample different (unknown) rotation phases and thereby to increase the likelihood of obtaining observations during a phase when one of the pulsation poles dominates the visible hemisphere. Most of our candidates do not yet have satisfactory phase coverage. For a discussion of the null results of the survey, refer to the paper by Martinez & Kurtz (1994b).

been highly successful. Prior to the start of this survey, only 14 roAp stars, discovered over a period of 12 years, were known. During its first four years, the Cape Survey yielded 12 confirmed new roAp stars (Martinez et al. 1991, 1994a; Kurtz & Martinez 1994). Figure 1 shows a preliminary $[c_1]$ versus $(b-y)$ diagram for the Cape Survey Ap SrCrEu stars. The dots represent the stars for which Cape Survey $uvby\beta$ photometry exists. The filled circles represent the roAp stars. The open circles are stars with $uvby\beta$ photometry for which

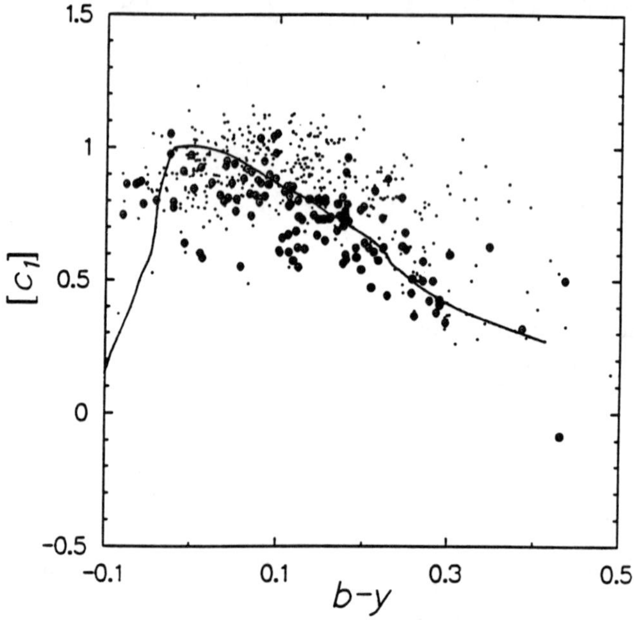


Figure 1. The $[c_1]$ versus $(b-y)$ diagram for the Cape Survey stars.

we obtained null results during searches for rapid oscillations. The solid line is the empirical main sequence relation for the chemically normal B-, A- and F-type stars. The distribution of open circles in Figure 1 shows that the Cape Survey has accomplished its objective of searching for rapid oscillations in the Ap SrCrEu stars over a wide range of temperature and luminosity. We also see that the roAp stars are confined to the coolest Ap SrCrEu stars and, for a given $(b-y)$, they are confined to the smaller values of $[c_1]$. Martinez (1993b) examines and discusses various two-colour diagrams for the roAp stars in more detail and deduces the following photometric limits for the roAp phenomenon: $2.69 \leq \beta \leq 2.88$, $0.08 \leq (b-y) \leq 0.31$, $0.19 \leq m_1 \leq 0.33$, $-0.12 \leq \delta m_1 \leq 0.02$, $0.46 \leq c_1 \leq 0.88$ and $-0.31 \leq \delta c_1 \leq 0.04$.

## References

Kurtz, D.W. & Martinez, P. 1994, Inf. Bull. Variable Stars No. 4013

Martinez, P., Kurtz, D.W. & Kauffmann, G. 1991, MNRAS, 250, 666

Martinez, P. 1993a, in: D. Kilkenny, E. Lastovica, E. & J.W. Menzies (eds.), *Precision Photometry*, South African Astronomical Observatory, Cape Town, p. 134

Martinez, P. 1993b, Ph.D. Thesis, University of Cape Town

Martinez, P. & Kurtz, D.W. 1994a, MNRAS, 271, 118

Martinez, P. & Kurtz, D.W. 1994b, MNRAS, 271, 129

*Astrophysical Applications of Stellar Pulsation*
*ASP Conference Series, Vol. 83, 1995*
*R. S. Stobie and P. A. Whitelock, eds.*

# Velocity Variations of the roAp Star $\gamma$ Equulei: Probing a puzzling pulsator

Jaymie M. Matthews and Sandra Scott

*Department of Geophysics & Astronomy*
*University of British Columbia, Vancouver, V6T 1Z4, Canada*

**Abstract.** We present a pulsational RV curve for $\gamma$ Equ whose amplitude is about 200 m/s and period is roughly 11 minutes. These data, combined with rapid photometry and line-profile variations, are being used to identify the nonradial mode(s) and study mode growth/decay in this unusual star.

## 1. The long and short of $\gamma$ Equ

The peculiar magnetic CP2 star HD 201601 ($\gamma$ Equulei) represents two extremes of stellar variability. On the one hand, based on its very slow magnetic variations (between about +500 and −1000 G), it is believed to have a rotation period *longer than 75 years!* At the same time, this object is also a rapidly oscillating Ap (roAp) star, with high-overtone nonradial pulsations of periods near 12 minutes.

The rapid radial velocity (RV) variations associated with roAp pulsations are challenging to measure – with amplitudes of less than a few × 100 m/s. Among the more than two dozen roAp stars known, there have been only two firm detections of RV oscillations: HR 1217 (Matthews et al. 1988) and $\gamma$ Equ (Libbrecht 1988). The latter was based only on low-level peaks in a Fourier periodogram of an RV time series.

Because RV-to-light amplitude ratios and phase lags are a potential mode diagnostic, we have undertaken a major spectroscopic analysis of $\gamma$ Equ.

## 2. Observations and analysis

In collaboration with W.H. Wehlau (UWO), G.A.H. Walker (UBC) and S. Yang (UVic), we obtained over *1620 individual high-resolution spectra* ($R \simeq 60,000$) of $\gamma$ Equ during 19 – 23 August 1988, using the CFHT coudé spectrograph and 1872 Reticon detector. Each exposure was about 30 sec long and synchronised with the best estimate of the star's photometric oscillation period, sampling 20 phases per cycle. These exposures were binned in pulsational phase to improve S/N and the binned spectra cross-correlated to measure relative velocity shifts. Superposed on each spectrum was a mercury emission line from an arc source at the spectrograph pupil. The apparent shifts in this line were used to gauge instrumental stability and remove drifts. The resulting RV curve for the night of 23 Aug 1988 UT (which shows the largest amplitude) is presented in Fig. 1.

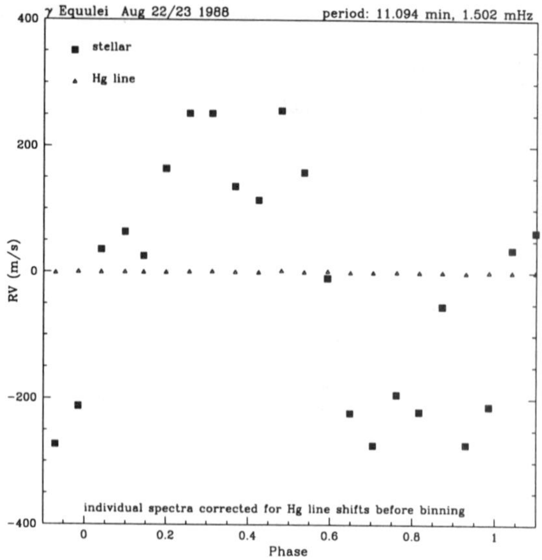


Figure 1. The phased RV curve. Solid symbols are the stellar data; open triangles, the mercury line fiducial to check instrumental stability.

## 3. Continuing surprises from $\gamma$ Equ

The stellar RV curve in Fig. 1 is surprising in two ways: (1) its period, which differs significantly from the dominant photometric period observed on the same night, and (2) its relatively "large" amplitude of about 250 m/s. The latter is a particular puzzle, given an upper limit of about 0.2 mmag on any light oscillations at the same frequency. Based on RV-to-light amplitude ratios predicted for nonradial pulsators, we would naively expect a photometric amplitude of 2 – 4 mmag. The mystery is compounded by the fact that Libbrecht's (1988) Fourier analysis of his spectral time series yielded an RV amplitude of only 20 m/s. Weiss & Schneider (1984) set an upper limit around 50 m/s for $\gamma$ Equ.

The data from our entire four-night run confirm the findings of previous observers that the oscillations of $\gamma$ Equ undergo amplitude modulation (or mode evolution) on timescales of a day or less. This is unlikely to be rotational modulation as seen in other roAp stars, if the rotation period of $\gamma$ Equ is indeed as long as decades. The star also shows evidence of profile variability in some lines – the first time this has been seen in an roAp star. We are currently interpreting these data.

## References

Libbrecht, K.A. 1988, ApJ, 330, L51

Matthews, J.M., et al. 1988, ApJ, 324, 1099

Weiss, W.W. & Schneider, H. 1984, A&A, 135, 141

*Astrophysical Applications of Stellar Pulsation*
*ASP Conference Series, Vol. 83, 1995*
*R. S. Stobie and P. A. Whitelock, eds.*

# The Cepheid Data Base

L. N. Berdnikov

*Sternberg Astronomical Institute, 13 Universitetskij prosp., Moscow 119899, Russia*

**Abstract.** Since 1981, we have been regularly carrying out photoelectric observations of Cepheids. By now nearly 20450 observations in UBVRI have been acquired. These observations are the basis of our Cepheid data base; presently we have collected on magnetic media near 88000 photoelectric as well as near 18000 radial velocity measurements. We shall continue to append to this Cepheid data base our new observations as well as published data.

## 1. Introduction

It is very important today to have the photometric and radial velocity measurements of Cepheid variables in electronic form. These data can be used for study of the Cepheids themselves as well as for study of the Galaxy structure.

We began the work of creation of Cepheid data base in 1980.

## 2. Structure of Cepheid data base

The collection of the observational data is managed in two ways. Firstly, there are photoelectric data on the Cepheids observed by us, and secondly, published photoelectric observations and radial velocity measurements are collected. At the present moment nearly 20450 observations in UBVRI system have been obtained, and the information from 442 published papers has been gathered.

Both the photoelectric observations and radial velocities catalogues have been converted to machine readable form. The first catalogue contains nearly 88000 photoelectric observations obtained for the both classical and type W Vir Cepheids and several related stars in various photometric systems, while the second one includes approximately 18000 radial velocity measurements.

Also, there is a machine readable version of the reference – bibliographic catalogue, which consists of two files. The one file contains bibliographic information on publications accessible to us which are placed in alphabetical order of the authors. There is following information for each paper: the first line is the key words one beginning with symbols "******"; the second line contains the paper's number, authors and year of the publication; the third line is the paper's title; the fourth line contains the reference. If there are rows of continuing characters they are shifted on two positions to right. For example:

```
****** CEP    PHE    ABUND
   73. Diethelm R., 1986.
       Physical  parameters  of  pulsating  variables with periods
         between   one   and   three   days.  I.  Photometry  and
         metallicities.
       Astron. and Astrophys. Suppl., 1986, v.64, p.261-274.
```

The other file contains the cross-referencing in the papers Cepheid variable names, placed after paper's number. This catalogue allows one to find the necessary literature by key words and/or by Cepheid's name.

Moreover, a series of FORTRAN programs has been written, which are used for modernisation and maintenance of the catalogues data.

## 3. Discussion

The Cepheid data bank is used in a series of the Cepheid studies. Some results had already been published, for example: "The catalogue of light curves parameters, distances and space coordinates of classical Cepheids" (Berdnikov, 1987a), "The distribution of Cepheids in the Galaxy" (Berdnikov, 1987b), "The groupings of the Cepheids in the Galaxy" (Berdnikov, Efremov, 1989), "Light curves and stability of the periods of the components of CE Cassiopeiae" (Berdnikov, 1990), "Light curves of double-mode Cepheids" (Berdnikov, 1992), "Isolines of surface density and z-coordinates of Cepheids" (Berdnikov, Efremov, 1993), "A study of pulsation stability of seven northern Cepheids with the longest periods" (Berdnikov,1994), "A study of period changes of small amplitude Cepheids in Aql, CMa, Car and Cas" (Berdnikov L.N., Pastukhova E.N., 1994a), "A study of period changes of small amplitude Cepheids in Cen, Cru, Cyg, Gem, Lac, Lyr, Mon, Nor and Oph" (Berdnikov L.N., Pastukhova E.N., 1994b), etc.

At present, a new version of the atlas of light and colour curves of the all photoelectrically observed in UBVRI system Cepheids has been created.

In the near future it is planned to create the Cepheid light curves atlas for all filter-measurements in all photometric systems.

We shall continue to append our Cepheid data base with our new observations as well as with published data.

## References

Berdnikov, L.N. 1987a, Peremennye Zviozdy, 22, 505

Berdnikov, L.N. 1987b, Pis'ma Astron. Zh., 13, 110

Berdnikov, L.N. 1990, AZh, 67, 798

Berdnikov, L.N. 1992, Pis'ma Astron. Zh., 18, 654

Berdnikov, L.N. 1994, Pis'ma Astron. Zh., 20, 285

Berdnikov, L.N., & Efremov, Yu.N. 1989, AZh, 66, 537

Berdnikov, L.N., & Efremov, Yu.N. 1993, Pis'ma Astron. Zh., 19, 957

Berdnikov, L.N., & Pastukhova, E.N. 1994a, Pis'ma Astron. Zh., 20, 567

Berdnikov, L.N., & Pastukhova, E.N. 1994b, Pis'ma Astron. Zh., 20, 829

*Astrophysical Applications of Stellar Pulsation*
*ASP Conference Series, Vol. 83, 1995*
*R. S. Stobie and P. A. Whitelock, eds.*

# Estimating Cepheid Metallicities: a new approach

E. Zsoldos

*Konkoly Observatory, Budapest XII, P.O. Box 67, 1525 Hungary*

**Abstract.** A relation between the Fourier decomposition parameters of Cepheid light curves and metallicities in the form [Fe/H]=$f(\{R_{i1}, \varphi_{i1}\})$ is presented. It gives [Fe/H] values with prediction errors 0.16–0.22 dex depending on the number of Fourier parameters used in the relation.

## 1. Introduction

A new method has been introduced to estimate RR Lyrae metallicities from the light curves by Kovács & Zsoldos (1995). It is based on the assumption that the light curve shape depends only on a finite (and small) number of parameters, e.g., mass, luminosity, chemical compositions, etc. Here I apply this method to Cepheids, another application is given by Jurcsik & Kovács (1995). A full description of the method is given by Kovács & Zsoldos (1995).

## 2. The data

I used 58 stars with well covered *UBV* light curves and with photometric metallicities given by Harris (1981). The light curves were fitted with an 8th- or higher order Fourier-sum.

The overall accuracy of the fits is about $0\overset{m}{.}02$–$0\overset{m}{.}04$, the relative error of the Fourier components is low ($< 10\%$), reaching 50% only in some cases for $A_7 - A_8$.

## 3. The relation

Four cases were studied, quadratic fits with 2–5 parameters. The most important parameters (i.e., appearing in all four cases) are $\varphi_{31}$ (a metallicity indicator in RR Lyrae stars, too!) and $R_{51}$. The prediction error decreased from 0.22 to 0.16 as the number of parameters increased from 2 to 5. The equation for the five parameter quadratic case is:

$$\begin{aligned}
[\mathrm{Fe/H}] &= 0.334 + 0.109p_4 + 3.360p_5 + 34.211p_1^2 - \\
&- 7.386p_1p_2 - 111.617p_1p_3 + 2.146p_1p_4 + \\
&- 0.675p_2^2 - 0.535p_2p_4 + 15.875p_2p_5 + \\
&+ 1.717p_3p_4 + 208.527p_3p_5 - 8.961p_4p_5 - \\
&- 255.853p_5^2
\end{aligned}$$

where $p_1 = A_1 - 0.371$, $p_2 = \varphi_{21} - 2.887$, $p_3 = R_{31} - 0.151$, $p_4 = \varphi_{31} + 1.054$ and $p_5 = R_{51} - 0.070$ (the phases refer to a sine decomposition).

Some coefficients were omitted from the equation because their relative errors were larger than 60%. Their omission did not change the prediction accuracy, it still remains 0.16.

## 4. Conclusions

Cepheid light curves show a similar type of correlation with metallicity as the RR Lyrae stars do. Though three of the five parameters are identical, it is interesting to note the lack of period dependence in the case of Cepheids.

The main uncertainty of the result lies in the uncertainties of the metallicities. There are few spectroscopic [Fe/H] determinations and they usually do not pay attention to the phase. The correlation between the existing spectroscopic data and Harris' metallicities is not as good as one would wish.

Figure 1 compares the three- and five parameter fits of Cepheids and the five parameter fit of RR Lyrae stars.

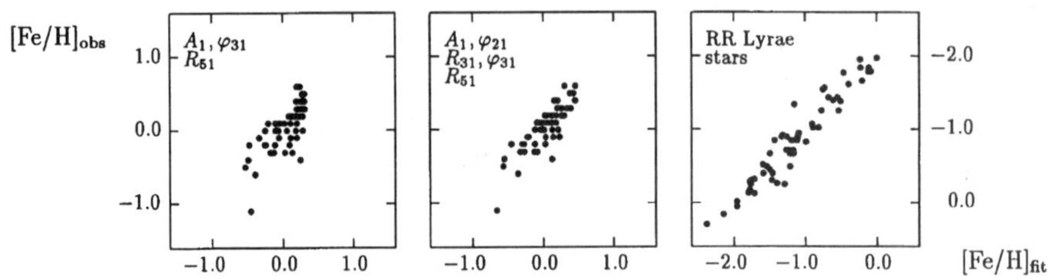


Figure 1. Three- and five parameter fits of Cepheids and the five parameter fit of RR Lyrae stars respectively.

**Acknowledgments.** It is a pleasure to thank Géza Kovács, Béla Szeidl and Johanna Jurcsik for their useful comments. This work has been supported by OTKA grant T–014183.

## References

Harris, H.C. 1981, AJ 86, 707

Jurcsik, J., & Kovács, G. 1995, these proceedings

Kovács, G., & Zsoldos, E. 1995, A&A 293, L57

*Astrophysical Applications of Stellar Pulsation*
*ASP Conference Series, Vol. 83, 1995*
*R. S. Stobie and P. A. Whitelock, eds.*

# The Chemical Composition of S-Cepheids and Double-Mode Cepheids

I.A. Usenko[1], V.V. Kovtjukh[1], S.M. Andrievsky[1], V.G. Klochkova[2], and G.A. Galazutdinov[2]

[1] *Department of Astronomy, Odessa State University, Shevchenko Park, Odessa, 270014, Ukraine*

[2] *Special Astrophysical Observatory, Russian Academy of Sciences, Nizhny Arkhyz, Stavropol' Territory, 357147, Russia*

The main aim of this work is to consider the evolutionary status of s-Cepheids (SC), - small-amplitude Cepheids with smooth sinusoidal light, colour and radial velocity curves. The unusual character of these objects could be explained in two ways: (1) SCs are crossing the Cepheid instability strip for the first time and they are first (or high) overtone pulsators (Efremov 1970); (2) SCs have close early-type companions, which affect gravitationally the Cepheid He II zone. Such an effect can decrease the Cepheid pulsational activity (Lloyd Evans, 1968).

According to theoretical calculations, low helium abundance in Cepheid atmospheres, which are crossing the instability strip for the first time, could produce small-amplitude oscillations. On the other hand, double-mode Cepheids (DMC) show oscillations with two (or three) periods: fundamental with large-amplitude asymmetrical light, colour and radial velocity curves, and first overtone with smooth sinusoidal small-amplitude ones. Nevertheless, detailed spectral and photometric investigations permit the detection of companions for the brightest SCs, and support the second hypothesis.

In order to solve these problems we obtained during 1990-1995 high-resolution photographic and CCD spectra of some SCs and DMCs for further abundance analysis of these stars. Our observational programme included 14 SCs (V473 Lyr, SU Cas, EU Tau, DT Cyg, V526 Mon, SZ Tau, V1334 Cyg, $\alpha$ UMi, FF Aql, V1162 Aql, V924 Cyg, V440 Per, V636 Cas, Y Oph) and 4 DMCs (TU Cas, EW Sct, BQ Ser, VX Pup).

Spectral observations of SCs and DMCs were made with: (1) The Main Stellar Spectrograph fed by the 6-m telescope SAO RAS at the reciprocal dispersion 9 Å/mm and 14 Å/mm. (2) The CCD-echelle spectrograph LYNX fed by the 6-m telescope SAO RAS (Panchuk et al. 1993) with resolution power about 25000. (3) The CCD-spectrograph AURELLIE fed by the the 1.52-m telescope Haute Provence Observatory with resolution power about 11000.

All spectra were reduced on the automatic computing system of the Crimean Astrophysical Observatory. For reducing CCD-echelle LYNX and AURELLE spectra we used DECH 20 code (Galazutdinov, 1992). We have carried out the LTE analysis and the atmosphere models for the programme stars were interpolated from Kurucz (1979) (for SC) and Kurucz (1992) (for DMC) grids.

The comparative results of spectroscopic investigations have shown: (1) Carbon is under-abundant for most SCs. This can be explained by the reduction of carbon's abundance after dredge-up (Luck & Lambert, 1985). It is accompanied by nitrogen overabundance with normal oxygen content. This agrees with Luck & Lambert (1985) about unaltered oxygen content in the supergiants after the first dredge-up. (2) The majority of SCs shows abundances for $\alpha$- and Fe- group elements, close to solar ones (with small deficiency for EU Tau, V526 Mon and V924 Cyg). DMCs indicate a greater deficiency for these elements. (3) The majority of s-process elements show some overabundance or close to solar for SCs, except EU Tau, V526 Mon and V924 Cyg, mentioned above. It is noticeable especially for SU Cas, DT Cyg and FF Aql. Probably, all these stars have passed through red-giant stage and they could not be crossing the instability strip for the first time. (4) Iron abundance data for DMC strongly support the existence of $P_1/P_0$ - [Fe/H] relation, which was discussed in Andrievsky et al. 1993). From theoretical models of DMC pulsation (e.g., Moskalik et al. 1992; Christensen-Dalsgaard 1993) the period ratio $P_1/P_0$ can depend both on the detailed chemical composition and on the pulsation period $P_0$. Theoretical analyses based on the 'old' Los Alamos opacities focussed on the dependence on period, whereas models using the 'new' Livermore opacities show a small systematic change of $P_1/P_0$ ratio with period in the period interval relevant for the DMC. Since the revised models show a strong dependence of $P_1/P_0$ ratio on the composition parameters (mainly through the opacity) it is necessary to know the detailed chemical composition in order to perform precise comparisons of observed and theoretical period ratios. (5) Carbon deficiency, nitrogen overabundance and some overabundance of s-process elements in SC atmospheres indicate that these objects are not crossing the Cepheid instability strip for the first time, and therefore they are not overtone pulsators. (6) Two SC objects, V1162 Aql and V636 Cas, showed the agreeable surprise: their CNO abundances are solar-like. Consequently, we have detected Cepheids, crossing the Cepheid instability strip for the first time. These stars have the lower $T_{eff}$ value (5500 K) and later spectral type (G0-G5) compared with other SC. It should be noted that, according to Berdnikov (1994), V1162 Aql has a very asymmetrical light curve and, probably it is a normal classical Cepheid. (7) The peculiarity of s-Cepheid light, colour and radial velocity curves could be explained by the presence of the companion.

## References

Andrievsky, S.M., et al. 1993, MNRAS, 265, 257

Christensen-Dalsgaard, J. 1993, Inside The Stars, ASP Conf. Ser. 40, 483

Efremov, Yu.N. 1970, Pulsating Stars (in Russian), Moskow, Nauka, 64

Galazutdinov, G.A. 1992, Preprint SAO RAS, 92

Kurucz, R.L. 1979, ApJS, 40, 1

Kurucz, R.L. 1992, In: The Stellar Populations of Galaxies, IAU Symp. 149, 225

Lloyd Evans, T. 1968, MNRAS, 141, 109

Luck, R.E., & Lambert, D.L. 1985, ApJ, 298, 782

Moskalik, P., Buchler, J.R., & Marom, A. 1992, ApJ, 385, 685

Panchuk, V.E., et al. 1993, Soviet Ast. Lett., 19, 1069

*Astrophysical Applications of Stellar Pulsation*
*ASP Conference Series, Vol. 83, 1995*
*R. S. Stobie and P. A. Whitelock, eds.*

# Models of the Double-mode Cepheids in the Large Magellanic Cloud

J. O. Petersen[1] and J. Christensen-Dalsgaard[2]

[1] *Theoretical Astrophysics Center, and*
*Niels Bohr Institute for Astronomy, Physics and Geophysics,*
*Astronomical Observatory, Øster Voldgade 3, DK 1350 Copenhagen K,*
*Denmark, (E-mail: oz@astro.ku.dk)*
[2] *Theoretical Astrophysics Center, and*
*Institute of Physics and Astronomy, Aarhus University,*
*DK 8000 Aarhus C, Denmark, (E-mail: jcd@obs.aau.dk)*

**Abstract.** Very recently the MACHO Collaboration published detailed data for 45 double-mode Cepheids in the Large Magellanic Cloud. Here we compare these data with new stellar pulsation models. We conclude that the new period ratios are in good agreement with a metal content $Z = 0.005 - 0.01$ in the Large Magellanic Cloud.

## 1. Introduction

The double-mode Cepheids pulsating in two radial modes have been crucial for testing of Cepheid models (e.g., Andreasen & Petersen 1988, Christensen-Dalsgaard 1992). These studies have been based on observations of 15 double-mode Cepheids known in the Galaxy. Very recently the MACHO Collaboration (Alcock et al. 1995) published detailed data for 45 double-mode Cepheids in the Large Magellanic Cloud, 30 oscillating in the fundamental mode and the first overtone and 15 oscillating in the first two overtones.

## 2. Model series and comparisons with observations

We have calculated precise periods of linear, adiabatic oscillations for the fundamental mode and the first few overtones of series of Cepheid models generally situated in the middle of the Cepheid instability strip using OPAL opacities. We use two mass-luminosity relations: ML1 is the BIT mass-luminosity relation (Becker et al. 1977), while ML2 approximately fits calculations including a moderate amount of overshooting from convective cores. Comparing model series with somewhat changed parameters (e.g., increased effective temperature at constant luminosity or a change in the mixing-length parameter, $\alpha$, from our standard value of 2 to 1), we find that effects from these changes are minor compared to the effects of changes in the metal content, $Z$, and the mass-luminosity relation. We therefore concentrate on comparing the observed period ratios with results for models with varying $Z$, for the mass-luminosity relations ML1 and ML2.

For $\Pi_1/\Pi_0$ we find that the computed period ratios show a substantial dependence on both $Z$ and the $M-L$ relation. The 14 DMCs in the Galaxy correspond to models with $Z$ in the interval 0.01–0.02, and we cannot distinguish between ML1 and ML2 from the available period ratios. It is interesting that the average $Z$ for the relatively young DMCs in the Milky Way appears to be somewhat smaller than the the solar value of $Z = 0.017 - 0.020$.

The LMC stars are nicely described by models with $Z$ in the range 0.005–0.010, except for one star apparently indicating $Z \simeq 0.015$. For periods larger than about 3 d (18 stars) we find that $\Pi_1/\Pi_0$ is mainly determined by $Z$. But for periods smaller than about 3 d (12 stars) we now find that the change from ML1 to ML2 can have effects on the period ratio that are larger than those of a change in $Z$ by a factor of 2. Given our lack of knowledge about the $Z$-value, we cannot choose between these possibilities.

LMC contains 15 Cepheids oscillating in the first and second overtones, providing a new test of the theoretical models. For periods smaller than about 1 d (where five stars are observed), we find a separation after $Z$ in the computed $\Pi_2/\Pi_1$, similar to the separation in $\Pi_1/\Pi_0$ for $\Pi_0 \geq 3$ d. In the period interval $\Pi_1$ = 1.0–1.5 d, where the bulk of the 1OT/2OT variables are found, the theoretical $\Pi_2/\Pi_1$ depends sensitively both on $Z$ and the assumed $M-L$ relation. For $\Pi_1 \geq 1.5$ d we now find that $\Pi_2/\Pi_1$ is mainly determined by the $M-L$ relation. Thus the accurately known $\Pi_2/\Pi_1$ is determined by different effects in different period intervals.

## 3. Conclusion

Our results confirm that period ratios for double-mode Cepheid models computed with the OPAL opacities, and for masses corresponding to evolution calculations, are broadly consistent with the observed values.

Comparison of double-mode Cepheids in the Galaxy and the Large Magellanic Cloud shows a difference in the average $\Pi_1/\Pi_0$ corresponding to a lower heavy-element abundance $Z$, by roughly a factor of 2, in the LMC, and this result is also in agreement with the observed values of $\Pi_2/\Pi_1$.

**Acknowledgments.** We wish to thank Doug Welch for sending us the data for the LMC double-mode Cepheids by e-mail. Support from the Danish National Research Foundation via TAC is gratefully acknowledged.

## References

Alcock, C., Allsman, R.A., Axelrod, T.S., Bennett, D.P., Cook, K.H., Freeman, K.C., Griest, K., Marshall, S.L., Peterson, B.A., Pratt, M.R., Quinn, P.J., Reimann, J., Rodgers, A.W., Stubbs, C.W., Sutherland, W., & Welch, D.L. 1995, preprint (astro-ph/9411061)

Andreasen, G.K., & Petersen, J.O. 1988, A&A, 192, L4

Becker, S.A., Iben, I., & Tuggle, R.S. 1977, ApJ, 218, 633

Christensen-Dalsgaard, J. 1992, in: "Inside the Stars", IAU Coll. 137, eds. W.W. Weiss, A. Baglin, ASP Conf. Ser., 40, 483

*Astrophysical Applications of Stellar Pulsation*
*ASP Conference Series, Vol. 83, 1995*
*R. S. Stobie and P. A. Whitelock, eds.*

# Cepheids in Binary and Multiple Systems

László Szabados

*Konkoly Observatory, P.O. Box 67, H-1525 Budapest XII, Hungary*

**Abstract.** Compilation of the list of Cepheids which are members in binary and multiple systems is in progress. During this project SB-nature of a number of Cepheids was revealed. The frequency of binaries among Cepheids exceeds 60 per cent.

Pulsating stars belonging to binary (multiple) systems offer special opportunity for determining the physical parameters of the components (e.g., Evans, 1994). Cepheids are particularly important because of their rôle in calibrating the cosmic distance scale. In view of their importance, the list of Cepheids with companion(s) is being compiled. The list contains some basic parameters for the Cepheids and their companion(s) as well as key references. The table supplemented with additional remarks will be published in the near future.

The use of novel and more accurate observational techniques (e.g., IUE-study, extensive radial velocity measurements) has resulted in discovering companions to an unexpectedly large percentage of Cepheids. However, the earlier observational material is not obsolete. It is worth mentioning that Joy's (1937) classical radial velocity study is still invaluable because the comparison of the modern data with Joy's early (therefore obviously less accurate) measurements often helps revealing variability in the $\gamma$-velocity. Such recent discoveries include SB-nature of Cepheids TZ Mon, CR Ori and YZ Sgr.

Another fact to be emphasised is that Joy's data were collected during a decade-long-interval (or even longer), and the effect of the orbital motion appears as an extra scatter in the radial velocity curves. RU Sct may be a member in a spectroscopic binary system based on this criterion.

The orbital effect has been discovered in other cases, as well. VZ CMa, previously known as having a blue companion (Stobie & Balona, 1979; Szabados, 1993), has a variable $\gamma$-velocity. Moreover, the comparison of two recent observational series on SV Vul (Gorynya et al., 1992; Bersier et al., 1994) shows that this long-period Cepheid also belongs to an SB-system. The detailed discussion on these (and some more) recently discovered SB-Cepheids is left for a future paper (Szabados, 1995).

Another useful criterion for pointing out companion(s) to Cepheids is the relatively large radial velocity amplitude with respect to the amplitude of the light variation (the AR-method – Szabados, 1993). Based on this parameter, UZ Cas and AK Cep are not solitary Cepheids, either.

Although there are well over a hundred items in the list of Cepheids belonging to binary and multiple systems, a big number of companions have not been discovered yet, as is seen in Figure 1. This histogram shows the number of classical Cepheids as the function of the mean V-magnitude (divided into

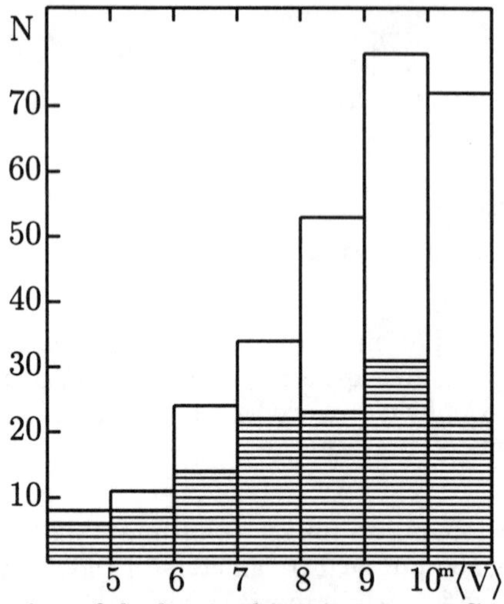


Figure 1. Fraction of the known binaries among Cepheids. The number is plotted against the mean V brightness for one magnitude wide intervals. Hatched area: known binaries; white rectangles: rest of the sample. Note the high occurrence of binaries among the brightest Cepheids.

one-magnitude wide intervals). The hatched area represents Cepheids known to have companions in each bin. It is clearly visible that the occurrence of binaries exceeds 60 % among the brighter Cepheids. The decreasing frequency of occurrence towards the fainter stars is the result of an obvious selection effect. The observed frequency of binaries among Cepheids of 10th visual magnitude is still about 40 %, a larger value than used to be assumed. Nevertheless, dozens of new binaries remain to be discovered among Cepheids brighter than 11th magnitude. For modern instrumentation these stars are not faint any more.

**Acknowledgments.** Financial assistance from the OTKA T7438 and the Hungarian Academy of Sciences is gratefully acknowledged.

## References

Bersier, D., Burki, G., Mayor, M., & Duquennoy, A. 1994, A&AS, 108, 25

Evans, N. R. 1994, ApJ, 436, 273

Gorynya, N. A., Irsmambetova, T. R., Rastorgouev, A. S., & Samus, N. N. 1992, Pis'ma v Astr. Zhu., 18, No. 9, 777

Joy, A. H. 1937, ApJ, 86, 363

Stobie, R. S., & Balona, L. A. 1979, MNRAS, 189, 641.

Szabados, L. 1993, in "New Perspectives on Stellar Pulsation and Pulsating Variable Stars", Proc. IAU Coll. No. 139; eds.: J. M. Nemec & J. M. Matthews, Cambridge Univ. Press, p. 406

Szabados, L. 1995, to be published

*Astrophysical Applications of Stellar Pulsation*
*ASP Conference Series, Vol. 83, 1995*
*R. S. Stobie and P. A. Whitelock, eds.*

# The Use of Maximum Light in Cepheid Distance Estimators

S.M. Kanbur

*Department of Physics and Astronomy, University of Glasgow*

M.A. Hendry

*Astronomy Centre, University of Sussex*

**Abstract.** Cepheid PLC relations are usually constructed using mean light. We show that the use of maximum and mean light leads to a distance estimator for the LMC with 30% less scatter than existing relations.

## 1. Introduction

Simon, Kanbur & Mihalas (1993) showed that the range of Galactic Cepheid temperatures at maximum light is about half that at mean light. In order to use this in constructing Cepheid PLC relations, Kanbur & Hendry (1995, hereafter KH) derived the following relation between mean and maximum light for a Cepheid,

$$\log P - \alpha M_{\rm v} + \beta M_{\rm vmax} + \gamma (B-V)_{\rm max} = \text{constant}, \tag{1}$$

where $\alpha$, $\beta$, $\gamma$ are constants and the other symbols have their usual meanings. Using the well known relation between apparent and absolute magnitude, we obtain an estimate of distance as

$$\widehat{\log D} = 0.2(m_{\rm vmax} - \hat{M}_{\rm vmax} - 25), \tag{2}$$

which gives the distance in Mpc. Our model for $\hat{M}_{\rm vmax}$ is given by a modified form of equation (1),

$$\hat{M}_{\rm vmax} = a + b \log P + c(B-V)_{\rm max} + d(M_{\rm vmean} - < M_{\rm vmean} >), \tag{3}$$

where all terms on the right hand side are directly observable and in particular the final term is just numerically equal to $m_{\rm vmean} - < m_{\rm vmean} >$, the mean apparent magnitude minus the sample average of the mean magnitudes of a group of equidistant Cepheids. We can also formulate a PLC(max) estimator by using

$$\hat{M}_{\rm vmax} = a + b \log P + c(B-V)_{\rm max}, \tag{4}$$

in equation 2, the justification again coming from equation (1).

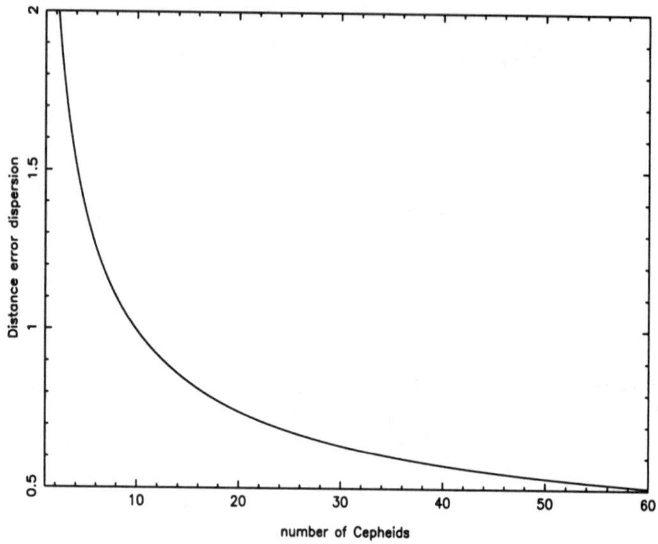


Figure 1. Rms percentage distance error derived from equation 3 divided by the percentage error of that derived from equation 4 as a function of $n$, the number of distant Cepheids.

## 2. Results and discussions

We have tested the distance estimators given by equations 2 and 4 using data for the LMC obtained by Martin, Warren & Feast (1979). The statistical techniques used do take account of correlations between the variables and are described in KH and Hendry & Simmons (1994). We found that the PLC(max) estimator given by equation 4 has about 10 percent less scatter than a PLC estimator using mean light. In Figure 1, we plot the rms percentage error derived from equation 3 divided by the percentage distance error of the PLC relation at maximum light - equation 4. It can be seen that when the number of Cepheids in the distant sample exceeds about 30, we obtain a significant reduction in scatter when using equation 3. These results may have interesting consequences for the cosmic distance scale.

## References

Hendry, M.A., & Simmons, J.F.L. 1994, ApJ, 435, 515

Kanbur, S.M., & Hendry, M.A. 1995, A&A, submitted

Martin, W.L., Warren, P.R., & Feast, M.W. 1979, MNRAS, 188, 139

Simon, N.R., Kanbur, S.M., & Mihalas, D. 1993, ApJ, 414, 310

*Astrophysical Applications of Stellar Pulsation*
*ASP Conference Series, Vol. 83, 1995*
*R. S. Stobie and P. A. Whitelock, eds.*

# Independent Distance Determinations to Milky Way Cepheids in Open Clusters: A progress report on four stars

Wolfgang P. Gieren[1], Jaymie M. Matthews[2]
Jean-Claude Mermilliod[3] & Douglas Welch[4]

[1] *Pontificia Universidad Católica de Chile, Santiago, Chile*
[2] *University of British Columbia, Vancouver, Canada*
[3] *Université de Lausanne, Switzerland*
[4] *McMaster University, Hamilton, Canada*

**Abstract.** We have undertaken a programme to calibrate the Cepheid PL relation zero-point by obtaining distances of Cepheids in open clusters and associations via the visual surface brightness technique. Results are now available for four stars (SZ Tau, CF Cas, CV Mon and DL Cas) and others are currently under analysis. Preliminary results suggest the 'ZAMS-fitting' distances to the host clusters are systematically smaller than those we derive from Cepheid surface brightnesses.

## 1. Introduction

One of the most nagging contributors to uncertainty in the intergalactic distance scale remains the calibration of Galactic Cepheids. We are currently deriving distances to Cepheids in open clusters and associations independent of ZAMS-fitting and/or H$\beta$ calibrations.

Our approach is to apply the "visual surface brightness" method (Barnes & Evans 1976) – a variant of the classical Baade-Wesselink technique – using new high-quality radial velocity measurements supplemented by velocity and photometric data from the literature. This allows us to compare our results with the period-luminosity (PL) calibration of Gieren, Barnes & Moffett (1993), who obtained distances to 100 Galactic Cepheids by the same method. Our ultimate goal is to calibrate the zero point of the PL relation to within a few $\times$ 0.01 mag, compared to current accuracies of $\pm 0.1$ (cf. Gieren & Fouqué 1993).

The present uncertainty is not surprising, given the many problems associated with ZAMS fitting to derive distances of Galactic clusters containing Cepheid variables. Strong visual extinction (varying across the cluster field), main-sequence bands which are often poorly defined in observed colour-magnitude diagrams (CMDs), contamination by field stars, and the difficulty in finding a ZAMS fiducial appropriate for a cluster are among the most serious sources of error.

## 2. DL Cas in NGC 129; CF Cas in NGC 7790

As an eight-day Cepheid which is both a component of a spectroscopic binary and a member of the open cluster NGC 129, DL Cas should be a very accurate calibrator of the PL relation and Cepheid mass. From 160 high-precision ($\sigma < 1$ km/s) radial velocity observations made with the CORAVEL and DAO spectrometers - including 67 new unpublished data - we have obtained both the orbital and pulsational velocity curves of this binary Cepheid (Gieren et al. 1994). This body of RV data makes DL Cas one of the best observed Cepheids in our Galaxy. The radius of 66.0 ± 3.5 $R_\odot$ we derive indicates that DL Cas is a fundamental-mode pulsator, removing any ambiguity in mode identification.

Containing three Cepheids and an eclipsing binary, the open cluster NGC 7790 has the potential to be a Rosetta Stone for refining the zero-point of the extragalactic distance scale. Unfortunately, ZAMS fitting of this relatively sparse, heavily reddened cluster is fraught with difficulty, and even modern determinations of its distance modulus based on comparable data and techniques differ by over 0.3 mag. To provide an independent calibration of the distance to NGC 7790, we have performed a surface brightness analysis of one of its Cepheids, CF Cas ($P \simeq 4.88$ days). Our new measurements almost double the published RV data for this star (Matthews et al. 1995, submitted).

## 3. Checking the ZAMS calibration

With our results for four stars, we have taken the first step towards checking for any systematic differences between the ZAMS-fitting and surface brightness calibrations. Our initial findings are summarised below. We have another four stars in our main programme and several other possible candidates, which should allow us to make a realistic comparison of the two zero-point systems, tied in to the extensive study of field Cepheids by Gieren, Barnes & Moffett (1993).

| Cepheid | log P (d) | Distance (parsecs) | $(m-M)_{\rm ZAMS}$ | $(m-M)_{\rm SB}$ |
|---|---|---|---|---|
| SZ Tau | 0.4982 | 580 ± 040 | 8.68 | 8.82 |
| CF Cas | 0.6875 | 3230 ± 160 | 12.34 – 12.65 | 12.55 |
| CV Mon | 0.7307 | 2160 ± 150 | 11.22 | 11.67 |
| DL Cas | 0.9031 | 2034 ± 110 | 11.11 | 11.54 |

## References

Barnes, T.G., & Evans, D.S. 1976, MNRAS, 174, 489

Gieren, W.P., Barnes, T.G., & Moffett, T.J. 1993, ApJ, 418, 135

Gieren, W.P., & Fouqué, P. 1993, AJ, 106, 734

Gieren, W.P., et al. 1994, AJ, 107, 2093

Matthews, J.M., et al. 1995, AJ, submitted

*Astrophysical Applications of Stellar Pulsation*
*ASP Conference Series, Vol. 83, 1995*
*R. S. Stobie and P. A. Whitelock, eds.*

# Uncertainties in the Galactic Cepheid Distance Scale

Thomas G. Barnes III
*McDonald Observatory, University of Texas at Austin, Austin, TX 78712*

Thomas J. Moffett
*Physics Department, Purdue University, W. Lafayette, IN 47907-1396*

**Abstract.** We have examined the uncertainties involved in using the visual surface brightness technique on Galactic Cepheid variables. The random error in a single Cepheid distance measurement is well determined to be ±8%. An upper limit to the systematic uncertainty is shown to be ±6% in distance. These combine for a single Cepheid to a typical uncertainty of ±10% and, for samples larger than ten Cepheids, to a typical uncertainty of less than ±6% in distance.

## 1. Introduction

Using the visual surface brightness technique on 100 Cepheids, Gieren, Barnes, & Moffett (1993) determined the Galactic Cepheid PL relation. This relation is an independent check on the PL relation determined from the cluster-fitting technique. Thus, it is important to understand the uncertainties in the visual surface brightness method. There are two questions to address: (1) what is the individual uncertainty in a Cepheid distance determined by the surface brightness technique (the random error), and (2) what is the uncertainty of the method itself (the systematic error)?

## 2. Random uncertainties

The random uncertainties are rather straightforward to understand. The surface brightness analysis involves fitting photometrically determined angular diameters to spectroscopically determined linear diameters, as discussed by Gieren et al. The fitting process calculates an uncertainty for each Cepheid's distance from the quality-of-fit, which is an internal random error. From our analyses of 100 Cepheids, the median uncertainty in distance for an individual Cepheid is found to be ±7.8%, *i.e.*, ±8%.

## 3. Systematic errors

Estimating the systematic uncertainty in distance measurements is rather more difficult. Hindsley & Bell (1989) carefully investigated the systematic errors in the surface brightness method. The ones appropriate to the Gieren et al. analysis are (1) the choice of $E(B-V)$ (0.7–2.2%), (2) the choice of $A_V/E(B-V)$ (0.8–1.1%), (3) the choice of $E(B-V)/E(V-R)$ (3.1–3.6%), (4) phasing problems between the radial velocities and the photometry (4.0–4.7%), and (5) the choice of correction factor from radial velocity to pulsational velocity (2.0%). These combine to a typical value of ±6.1% uncertainty in distance.

The best way to estimate systematic uncertainty is to inter-compare fully independent methods of determining the same quantity. The Galactic Cepheid PL relation may be determined by means of surface brightness analyses and by means of the *completely independent* method of cluster fitting. The surface brightness analysis gives the Galactic Cepheid PL relation

$$\begin{array}{rll} M_V = & -1.371 & -2.986 \log P. \\ & \pm.095 & \pm.094 \end{array}$$

The cluster fitting method gives a relation (Gieren & Fouqué 1993)

$$\begin{array}{rll} M_V = & -1.329 & -2.911 \log P. \\ & \pm.134 & \pm.122 \end{array}$$

These independent methods of determining Cepheid distances differ by less than 35% of the combined sigma in slope and by less than 20% of the combined sigma in zero point. A straightforward interpretation is that there is no difference between the results and thus there can be *no significant systematic error in either of them.* In the worst case, the systematic difference between the relations is 0.12 mag, as measured at the middle of the period distribution. If this difference were wholly due to systematic error in the surface brightness technique, its systematic error would be 5.7% in distance. This is close to the systematic error estimated by Hindsley & Bell, ±6.1%. We adopt the conservative stance that ±6% is a reasonable upper limit for the systematic uncertainty in visual surface brightness distances.

## 4. Conclusions

We have shown that the random error in a single Cepheid distance measurement by the surface brightness method is ±8%. We have also shown that the systematic uncertainty is less than ±6% in distance. These combine for a single Cepheid to a typical uncertainty of ±10% in distance and for samples larger than ten Cepheids, appropriate to studies of other galaxies, to a typical uncertainty of less than ±6% in the distance.

## References

Gieren, Barnes, & Moffett 1993, ApJ, 418, 135

Gieren, & Fouqué 1993, AJ, 106, 734

Hindsley, & Bell 1989, ApJ, 341, 1004

*Astrophysical Applications of Stellar Pulsation*
*ASP Conference Series, Vol. 83, 1995*
*R. S. Stobie and P. A. Whitelock, eds.*

# Comparison of the Galactic Cepheid Open Cluster and Surface Brightness Distance Scales

Wolfgang P. Gieren

*Pontifica Universidad Católica de Chile, Casilla 104, Santiago 22, Chile*

Pascal Fouqué

*DESPA Observatoire de Paris-Meudon, F-92195 Meudon Cedex France and European Southern Observatory, Casilla 19001, Santiago 19, Chile*

**Abstract.** We compare the best determined Baade-Wesselink (BW) period-luminosity (PL) relation for 100 galactic Cepheids to the PL relation derived for 32 open cluster and association Cepheids by the ZAMS-fitting method. Eighteen stars in common lead to the conclusion that BW and ZAMS-fitting distances can agree to better than 0.1 mag, after proper allowance of systematic effects in both methods.

Gieren & Fouqué (1993; = GF) have calibrated the ZAMS-fitting distance scale on 32 galactic Cepheids which are presumably all members in open clusters and associations. They used as a reference cluster the Pleiades assuming $\mu_0 =$ 5.57 (van Leeuwen 1983) in spite of recent indications (Gatewood et al. 1990, Feast 1991) that the Pleiades distance is more likely to be ~5.70. For details on how the true cluster distance moduli were derived, see GF.

Among the 32 Cepheids with ZAMS-fitting distances, 18 have now BW distance determinations from the surface brightness method (see Gieren et al. 1993). For four of these stars (CF Cas, CV Mon, SZ Tau and UY Per) we have derived new improved distances; the underlying new data will be published elsewhere. These distances and the corresponding absolute magnitudes are given in Table 1.

Comparing these data with the ZAMS-fitting moduli, it is found that the BW moduli are, on average, 0.23 ± 0.10 mag larger. There are five Cepheids for which the discrepancy between the determinations from both methods is larger than 0.6 mag; omitting these stars as yet unreliable for a comparison of both methods of distance determination, the mean difference between BW and ZAMS-fitting moduli reduces to 0.15 ± 0.06.

We conclude that the agreement between both methods is encouraging, if we regard possible systematic errors in each method: ZAMS-fitting is affected by absorption uncertainties, metallicity differences among clusters, Pleiades zero-point and location of the Cepheid into the sparse associations. The BW method is affected by metallicity, surface gravity, and microturbulence differences among Cepheids; it also exhibits systematic differences according to which colour is adopted to predict surface brightness. Adopting infrared colours which are less affected by secondary effects, and which reduce the BW radii (Laney & Stobie 1995), and hence the distances by several percent, we may safely conclude that ZAMS-fitting and BW distance moduli agree within 0.1 mag, for the galactic sample. Applying the calibration to extragalactic samples is another matter as, for example, the limited range of metallicity observed in our Galaxy needs extrapolation if one goes to the Magellanic Clouds.

Table 1. Metallicity-corrected surface brightness distances and absolute magnitudes of 18 open cluster and association Cepheids

| Cepheid | log P (days) | distance (pc) | $\sigma$ (pc) | $M_V^{SB}$(mag) | $\mu_0^{SB}$ (mag) | $\mu_0^{ZAMS}$† (mag) | $\mu_0^{SB} - \mu_0^{ZAMS}$ (mag) |
|---|---|---|---|---|---|---|---|
| EV Sct | 0.4901 | 1370 | 110 | -2.66 | 10.69 | 10.88 | -0.19 |
| SZ Tau | 0.4982 | 580 | 40 | -3.22 | 8.82 | 8.68 | +0.14 |
| QZ Nor | 0.5782 | 1740 | 190 | -3.23 | 11.20 | 11.13 | +0.07 |
| CF Cas | 0.6880 | 3130 | 160 | -3.15 | 12.48 | 12.68 | -0.20 |
| UY Per | 0.7296 | 1560 | 110 | -2.57 | 10.97 | 11.70 | -0.73 |
| CV Mon | 0.7307 | 2160 | 150 | -3.67 | 11.67 | 11.22 | +0.45 |
| V Cen | 0.7399 | 920 | 50 | -3.92 | 9.82 | 9.11 | +0.71 |
| BB Sgr | 0.8220 | 770 | 90 | -3.41 | 9.42 | 9.05 | +0.37 |
| U Sgr | 0.8290 | 640 | 40 | -3.66 | 9.04 | 8.95 | +0.09 |
| DL Cas | 0.9031 | 2030 | 110 | -4.20 | 11.54 | 11.12 | +0.42 |
| ζ Gem | 1.0065 | 410 | 70 | -4.18 | 8.05 | 7.75 | +0.30 |
| VY Car | 1.2767 | 2600 | 150 | -5.38 | 12.08 | 11.42 | +0.66 |
| RZ Vel | 1.3096 | 1870 | 110 | -5.32 | 11.36 | 11.24 | +0.12 |
| WZ Sgr | 1.3394 | 2610 | 210 | -5.59 | 12.08 | 11.27 | +0.81 |
| SW Vel | 1.3700 | 2750 | 140 | -5.18 | 12.20 | 12.00 | +0.20 |
| T Mon | 1.4317 | 1770 | 100 | -5.77 | 11.24 | 11.02 | +0.22 |
| RS Pup | 1.6172 | 2580 | 300 | -6.51 | 12.06 | 11.28 | +0.78 |
| SV Vul | 1.6532 | 2330 | 160 | -6.48 | 11.83 | 11.82 | +0.01 |

† Values from Gieren & Fouqué, AJ, 106, 734, 1993.

**Acknowledgments.** This research was supported by FONDECYT grant No. 1940710 which is gratefully acknowledged.

## References

Feast, M. W. 1991, in Observational Tests of Cosmological Inflation, eds. Shanks et al., (Kluwer, Dordrecht), p. 147

Gatewood, G., Castelaz, M., Han, I., Persinger, T., Stein, J., Stephenson, B., & Tangren, W. 1990, ApJ, 364, 114

Gieren, W. P., & Fouqué, P. 1993, AJ, 106, 734

Gieren, W. P., Barnes, T. B., & Moffett, T. J. 1993, ApJ, 418, 135

Laney, C. D., & Stobie, R. S. 1995, MNRAS, in press

van Leeuwen, F. 1983, PhD Thesis, Leiden University, Netherlands

*Astrophysical Applications of Stellar Pulsation*
*ASP Conference Series, Vol. 83, 1995*
*R. S. Stobie and P. A. Whitelock, eds.*

# Type II? Cepheid Radii and TX Del

C. D. Laney

*South African Astronomical Observatory, PO Box 9, Observatory 7935, Cape, South Africa*

There is often some dispute in the literature as to whether a particular Cepheid (S Vul, for example) belongs to Type I or Type II. $JHK$ photometry has been used to calculate Baade-Wesselink radii for a number of stars always, sometimes or occasionally stated to be Type II Cepheids. Radial velocities have been taken from the literature.

## 1. Type II Cepheids

The $(K, J-K)$ radius of **SW Tau** is $8.30 \pm 0.34$. This is about half of the expected radius for a Type I Cepheid at this period and normal for Type II.

For **V553 Cen**, the relative phasing of radial velocities and infrared photometry has to be done by choosing a relative phasing where the phase shift between the radius displacement curve integrated from the radial velocities and the photometric radius displacements is minimized, and where the surface brightness coefficient $A$ (Balona 1977) is normal for a star with the colour indices of V553 Cen. Uncertainty in the reddening has a negligible effect on the radius solution obtained in this way. I have also considered the degree to which the $K, J-K$ and $K, V-K$ solutions give similar answers, as the relative phasing of the optical photometry and the radial velocities was more straightforward. The $(K, J-K)$ radius is $11.2 \pm 0.4$, normal for a Type II Cepheid.

For **AU Peg**, the data have been phase shifted as described for V553 Cen, but the required value is very small. Only the last 18 radial velocities in Harris, Olszewski, & Wallerstein (1984) have been used, as the phase jitter gives excessive scatter in $RV_{puls}$ if all the available RV data are included. The $(K, J-K)$ radius obtained is about $11.8 \pm 1.0$, indicating that this is a BL Her star. This is as expected given the short orbital period, high galactic latitude, and IR colour excess.

For **Kappa Pav**, relative phasing is not a problem, and the $K, J-K$ and $K, V-K$ radii are very similar. The $(K, J-K)$ radius is $29.06 \pm 0.85$, again about half the expected radius of a Type I Cepheid of this period.

The available data strongly suggest that **AL Vir** is a binary W Vir star with a period somewhat less than 1900 days. This considerably complicates the solution, but the $(K, J-K)$ radius of $34 \pm 2$ allows the exclusion of any possibility that this halo object belongs to Type I.

## 2. Type I Cepheids

For **T Ant**, phasing can be established from the sharp photometric temperature maximum in both JHK and optical photometry. The $(K, J-K)$ radius of $46.64 \pm 1.31$ is normal for a Type I Cepheid with a period of 5.9 days. Its Type II status assigned by Harris (1985) seems erroneous.

The existence of contemporaneous radial velocities and JHK photometry for **CT Car** means that the $(K, J-K)$ radius of $94.4 \pm 2.8$ unambiguously excludes any possibility that this star is anything but a classical Cepheid.

Phase coverage with good radial velocities is rather incomplete for **GY Sge**, and the phasing is more ambiguous than would be desirable. Nevertheless, the $(K, J-K)$ solution for the GY Sge phase interval for which RV data are available gives a radius of about $233 \pm 12$, a normal Type I value.

For **S Vul**, older data (heavily smoothed) have been used to patch the gap in the modern data, and phasing is unambiguous. The $(K, J-K)$ and $(K, V-K)$ radii are 271 and 272, respectively, with an uncertainty of about 14. S Vul is about the same radius as Type I Cepheids of similar period in the Magellanic Clouds.

## 3. Type X?

**TX Del** is a well known binary Cepheid with an orbit (Harris & Welch 1989) whose size seems to compel Type II status. I have used their pulsational radial velocity curve, phase shifted so that the surface brightness is normal for a Cepheid with the observed colour indices. The $(K, J-K)$ radius obtained is $47.6 \pm 5.4$, and the $(K, V-K)$ radius is considerably larger regardless of the phasing. Agreement between the $(K, J-K)$ and $(K, V-K)$ radii can only be obtained if the phasing is set so that the surface brightness coefficients are highly abnormal, and in that case the radius becomes about 75. This might be explained as the effect of a bright red companion, but the type of companion required would produce a J-K much redder for the observed B-V than is the case. TX Del is *normal* in the ratio of these two colours. Alternatively one might restrict the radius solution to the descending branch. This gives a $(K, V-K)$ radius of 45.6, about the same as for $(K, J-K)$ (which is not much affected by the restriction). TX Del appears to be far too large (and bright) to be a normal Type II Cepheid. A second difference between Type I and Type II Cepheids also highlights the oddity of TX Del. The amplitude in $J-K$ is always more than half the $K$ amplitude for Type I, typically much less for Type II. TX Del is the only star yet tested which has a $J-K$ amplitude *larger* than its $K$ amplitude.

TX Del almost seems to need a type of its own.

## References

Balona L. A. 1977, MNRAS, 178, 231

Harris H. C. 1985, AJ, 90, 756

Harris H. C., Olszewski E. W., & Wallerstein G. 1984, AJ, 89, 119

Harris H. C., & Welch D. L. 1989, AJ, 98, 981

*Astrophysical Applications of Stellar Pulsation*
*ASP Conference Series, Vol. 83, 1995*
*R. S. Stobie and P. A. Whitelock, eds.*

# Baade-Wesselink Analysis of Type II Cepheids

J. Vinkó and Z. Balog

*Department of Optics, JATE University, Szeged, H-6720 Hungary, e-mail: vinko@physx.u-szeged.hu*

**Abstract.** New physical parameters of selected Cepheid variables classified as CW or CEP (Type II) in GCVS are presented. The analysis is based on published light and radial velocity curves. Two of the stars turned out to be classical Cepheids.

The short period metal-rich variables form an interesting group among Type II Cepheids (also called metal-rich BL Her-type stars). Their kinematic properties indicate that they differ from the normal disk population, because of e.g., their high galactic latitude. On the other hand, their chemical abundances are much closer to Pop. I composition. Moreover, from theoretical point of view, it is not clear yet how these stars can evolve into the instability strip and become pulsating stars.

It is known that Type I and II Cepheids are difficult to distinguish (Harris, 1985), because absolute physical parameters are not easy to obtain due to mainly the faintness of Type II Cepheids. Recently some new radial velocity curves have become available (e.g., from Gorynya et al., 1992) which enabled us to compute Baade-Wesselink radii for five selected Type II Cepheids which are relatively well observed and have nearly solar metallicity, $[Fe/H]$ close to 0. The resultant radius values were used to calculate other physical parameters via theoretical period-mass-radius and mass-luminosity relations for both populations.

All of these results are collected in Table 1, while Fig. 1 shows the position of the variables in the $HR$-diagram (filled symbols). Fig. 1 also contains reference data for both Pop. I and Pop. II objects: some known Cepheids in galactic open clusters and several short period BL Her-type stars in globular clusters (open squares and circles, respectively).

The conclusions are the following. TX Del and V351 Cep are probably misclassified as CW (Type II Cepheid), because they are too large and hence too bright to be Type II Cepheids. These results strongly suggest their Pop. I nature, because they are in the region of Classical Cepheids. Moreover, V351 Cep is probably an s-Cepheid which supports its Pop. I classification. TX Del is actually a binary Cepheid with a low mass function (Harris & Welch, 1989), which gives $M_2(\text{min}) = 1.1\ M_\odot$ for the companion. The Roche-lobe filling factor in this case becomes f = 0.44. The Pop.I nature of TX Del is also supported by its infrared light curves (Laney, 1995). Note that the galactic z-distance of TX Del is much larger than most of the classical Cepheids.

IX Cas is also close to the regime of classical Cepheids; however, its physical parameters indicate later evolutionary state (e.g., its mass is too low). BL Her, the prototype of this class of variables, is clearly among Pop. II objects. AU

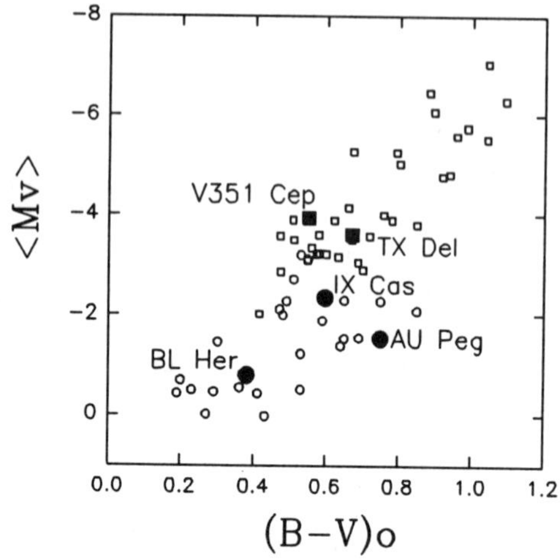


Figure 1. The position of the variable in the HR diagram. Open symbols represent reference Cepheids of both types (see text).

Peg, another binary Cepheid, is peculiar in many respects (e.g., Vinkó, Szabados & Szatmáry, 1993). Referring to Table 1, it is more massive than most of the Pop. II Cepheids (which usually have $M < 1M_{\odot}$). On the other hand, AU Peg does not resemble any classical Cepheids, therefore its classification is uncertain.

Table 1. Inferred physical parameters of the program stars.

| Star | $R/R_{\odot}$ | $M/M_{\odot}$ | $M_v$ | $d(kpc)$ | $z(kpc)$ |
|---|---|---|---|---|---|
| BL Her | 9.4 | 0.8 | -0.78 | 1.43 | 0.47 |
| AU Peg | 20.0 | 2.3 | -1.60 | 1.28 | 0.48 |
| IX Cas | 24.0 | 0.5 | -2.29 | 4.81 | 0.99 |
| TX Del | 48.0 | 5.9 | -3.54 | 2.98 | 1.22 |
| V351 Cep | 49.6 | 6.3 | -3.89 | 2.58 | 0.03 |

This work has been supported by the Hungarian OTKA Grant #F7318.

## References


Gorynya, N.A., Irsmambetova, T.R., Rastorgouev, A.S., & Samus, N.N. 1992, Pisma A.Zh., 18 No.9. 777

Harris, H.C. 1985, AJ, 90, 756

Harris, H.C., Welch, D.L. 1989, AJ, 98, 981

Laney, C.D. 1995, in this proceedings

Vinkó, J., Szabados, L., & Szatmáry, K. 1993, A&A, 279, 410

*Astrophysical Applications of Stellar Pulsation*
*ASP Conference Series, Vol. 83, 1995*
*R. S. Stobie and P. A. Whitelock, eds.*

# Turbulence Variations in Cepheids

David Bersier and Gilbert Burki

*Geneva Observatory, 1290 Sauverny, Switzerland*

## 1. Introduction

From radial velocity measurements obtained with a cross-correlation technique, the variation of turbulence during the pulsation cycle is studied for a sample of 40 Cepheids. We will propose a new way to separate classical and s-Cepheids. More complete results will appear in a forthcoming paper (Bersier & Burki 1995)

## 2. The residual width

The radial velocities have been measured with the spectrometer CORAVEL (Baranne et al. 1979), whose cross-correlation function (CCF) is fitted with a Gaussian, giving the radial velocity $V_r$, the width $\sigma_{obs}$, the depth $H$ and the continuum, normalised to 1. The pulsation broadens the lines and thus also the CCF. In the Gaussian approximation one can write

$$\sigma_{obs}^2 = \sigma_{inst}^2 + \sigma_{puls}^2 + \sigma_{res}^2 \tag{1}$$

where $\sigma_{obs}$ is the observed width, $\sigma_{inst}$ is the instrumental width, $\sigma_{puls}$ is the additional width caused by the pulsational velocity field and $\sigma_{res}$ contains all the other effects (turbulence, rotation, magnetic field, etc.). To be less affected by the noise in the data, a Fourier series has been fitted to each curve of $\sigma_{obs}$. With numerical simulations, one is able to synthesise the additional Doppler width due to pulsation, with a high accuracy. The instrumental width being well known for CORAVEL, the computation of $\sigma_{res}$ is then straightforward. One then has a curve in phase for $\sigma_{res}$. From this curve, we determined the maximum residual broadening $\sigma_{max}$ (observed at or very close to minimum radius), and the width $\sigma_0$ that the star would have if it did not pulsate. As shown by Bersier & Burki (1995), $\sigma_0$ is slightly higher than the mean value of $\sigma_{res}$.

## 3. The link between $\sigma_{res}$ and turbulence

Luck & Bond (1989) gave atmospheric parameters for $\sim$ 160 supergiants and Cepheids, in particular the microturbulence. We searched in this sample all the stars that have also been measured with CORAVEL and we calculated $\sigma_{res}$ according to eq. (1) with $\sigma_{puls} = 0$ for supergiants. We then quadratically subtracted the instrumental width to obtain $\sigma_{res}$. Figure 1a) shows the relation between the microturbulence $\xi_t$ and $\sigma_{res}$. The equation of the straight line is $\xi_t = 0.286\,\sigma_{res} + 1.49$.

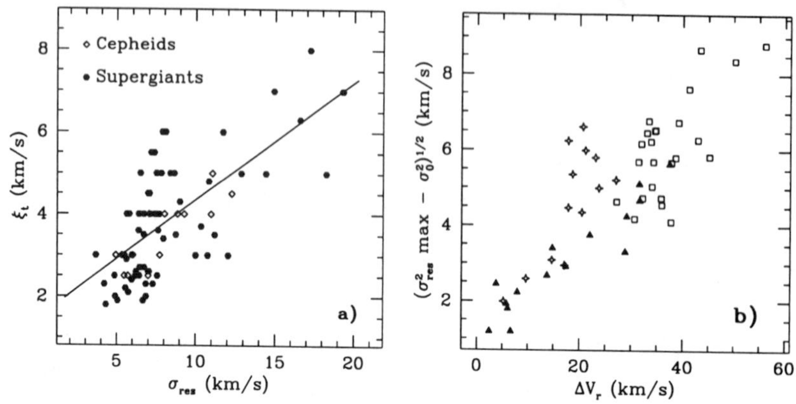


Figure 1. **a)** The relation between the microturbulence $\xi_t$ and the residual width. The solid line is the regression for supergiants and Cepheids. **b)** The correlation between $\sigma^2_{max} - \sigma^2_0$ and the amplitude $\Delta V_r$. Squares are for classical Cepheids, stars for s-Cepheids and triangles for V473 Lyr

## 4. Separation between classical and s-Cepheids

The quantity $(\sigma^2_{max} - \sigma^2_0)^{1/2}$ is viewed as the maximum excess turbulence that the star has compared to what it would have without pulsation. There is a clear separation in amplitude between classical and s-Cepheids. In Fig. 1b) are also plotted the points computed for the exceptional Cepheid V473 Lyr, whose amplitude varies by a factor 15 on about 1200 days (Burki 1994). The s-Cepheids are at relatively low amplitude whereas classical Cepheids are more or less along a straight line. The cause of this separation is not clear. One can imagine that for the s-Cepheids, as the amplitude increases, more and more energy is transferred in small-scale motions (turbulence) than in large-scale motion (re-expansion of the atmosphere), whereas for classical Cepheids, $\sigma^2_{max} - \sigma^2_0$ remains proportional to the amplitude. This could be related to the fact that most s-Cepheids are probably overtone pulsators. This way to separate classical and s-Cepheids gives new informations on the difference between these two classes.

**Acknowledgments.** This work has been partly supported by the Swiss National Science Foundation.

## References

Bersier, D., & Burki, G. 1995, A&A, in preparation

Bersier, D., Burki G., Mayor M., & Duquennoy A. 1994, A&AS, 108, 25

Burki G. 1994, in: C. Sterken, & M. de Groot (eds), The Impact of Long Term Monitoring on Variable Star Research, Kluwer Academic Press, Dordrecht, The Netherlands, 247

*Astrophysical Applications of Stellar Pulsation*
*ASP Conference Series, Vol. 83, 1995*
*R. S. Stobie and P. A. Whitelock, eds.*

# Formation Depths of Spectral Lines in Cepheids

Michael D. Albrow[1] and P. L. Cottrell
*Mount John University Observatory, Department of Physics and Astronomy, University of Canterbury, Private Bag 4800, Christchurch 8001, New Zealand*

## 1. Introduction

There has been a number of observational programmes that have endeavoured to investigate the atmospheric velocity fields in Cepheids (e.g., Sanford 1956, Wallerstein et al. 1992, Butler 1993). These studies measured the radial velocities of lines of different strength, excitation and ionisation potential as these provide an indication of line formation at different levels in the atmosphere. From these measurements, the presence of velocity gradients can be inferred, but determination of the magnitude of such gradients requires knowledge of the spectral line depth of formation. Through dynamical modelling we are endeavouring to ascertain what is actually being measured in the above observational programmes.

## 2. Contribution function

The function giving the relative contribution of different atmospheric layers to a spectral line flux depression,

$$C_U(\log_{10}\tau_0) = \frac{(ln10)\tau_0}{\kappa_0}\int_0^1 \kappa_l(I_c - S_l)e^{-\tau/\mu}d\mu, \tag{1}$$

has been derived by Albrow & Cottrell (1995). This function has optical depth as its independent variable but also changes with wavelength. The most meaningful way to display the flux contribution across a spectral line is to plot a series of contours in what is known as the 'flux contribution contour' diagram (Albrow 1994, Albrow & Cottrell 1995).

Using 'snapshot' models of Cepheid atmospheres we have investigated the effects of velocity fields on the depths of formation of spectral lines and how radial velocities measured from these lines relate to the velocities in the model. In most cases the correlation between the atmospheric velocity and the centre of absorption is good, with the contours tending to follow the velocity line. The relationship tends to worsen as line strength is increased.

An interesting phenomenon that occurs with a changing velocity gradient is the presence of spatially separated absorption centres. These are more prevalent in stronger lines but are still present at certain phases in weak lines. In Fig. 1, for

[1]New Zealand Science and Technology Postdoctoral Fellow; Visiting Fellow, SAAO

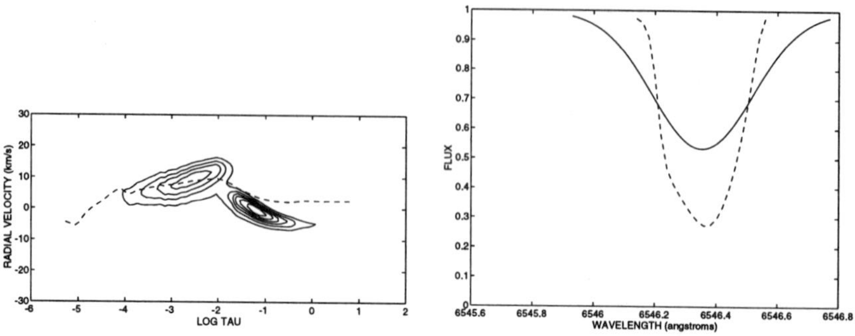


Figure 1. (a) Flux contribution contour diagram for an Fe I line with $\log gf = -2$ and $\chi = 2.76$eV. The atmosphere is taken from the bump phase of a model of an 11 day Cepheid. The velocity (divided by a projection factor) is shown as a dashed line. (b) Flux profiles for the spectral line from (a). Dashed line, raw (unbroadened) flux; solid line, with radial-tangential macroturbulence $\zeta = 10$km s$^{-1}$.

example, the velocity near $\log \tau \sim -2.5$ (the formation depth of the core of an unshifted line) is sufficient to move the absorption at this depth by $\sim 10$ km s$^{-1}$ in wavelength space. The projection effect means that this shift is not the same amount across the stellar disc (more near the centre of the disc, less towards the limb) and the peak magnitude of the contribution function in the shifted 'core' is much less than if the atmosphere was static. Interior to this, near $\log \tau \sim -1$ (the formation depth of the wing of the spectral line in a static atmosphere), the projected velocity is close to zero ($\sim 3$ km s$^{-1}$). The blue (negative radial velocity) wing of the spectral line will be formed as in a static atmosphere. There is little differential Doppler shift across the stellar disc to smear out the contribution function in wavelength space and the flux contribution is actually greater here than at the 'core' depth. The red (positive radial velocity) wing of the spectral line makes no contribution at this depth because it is masked by the shallower Doppler-shifted 'core' absorption. The observed (real) core of the spectral line, which appears with a radial velocity of $\sim 5$ km s$^{-1}$, contains contributions from the blue wing of the unshifted line and from the redshifted 'core'. The structure inherent in the line profile is unobservable because of the effects of turbulent broadening.

## References

Albrow, M. D. 1994, PhD thesis, University of Canterbury

Albrow, M. D., & Cottrell, P. L. 1995, MNRAS, submitted

Butler, R. P. 1993, ApJ, 415, 323

Sanford, R. F. 1956, ApJ, 123, 201

Wallerstein, G., Jacobsen, T., Cottrell, P. L., Clark, M., & Albrow, M. 1992, MNRAS, 259, 474

*Astrophysical Applications of Stellar Pulsation*
*ASP Conference Series, Vol. 83, 1995*
*R. S. Stobie and P. A. Whitelock, eds.*

# The Centre-of-Mass Velocity of a Radially Pulsating Star: Insights from NLTE Models

Dimitar D. Sasselov

*Harvard-Smithsonian Center for Astrophysics, Cambridge, MA 02138*

**Abstract.** The problem of deriving the centre-of-mass velocity of a radially pulsating star is reexamined. New observations of line asymmetry and Non-LTE radiation hydrodynamics point to a systematic effect of about 1 km s$^{-1}$ in Cepheids.

## 1. Introduction

Accurate measurements (better than 1 km s$^{-1}$) of the centre-of-mass velocity of a pulsating star, commonly referred to as $\gamma$-velocity, are needed both for kinematic studies (e.g., Pont et al. 1994) and for determining radii and distances using the Baade-Wesselink (BW) method (Gautschy 1987).

The standard method to measure the $\gamma$-velocity from an observed radial velocity curve requires the integral

$$\int (v_{rad} - \gamma)\, d\phi \;=\; 0,$$

when extended over a whole cycle. Thus path conservation is a basic assumption in the BW method (Gautschy 1987). From this also follows the constancy with phase of the projection factor, $p \equiv V_{\rm pulsation}/V_{\rm radial}$. The $p$-factor is used in a BW solution to convert from observed radial velocities to pulsation velocities.

Karp (1975) studied the systematics in Cepheid $\gamma$-velocities from hydrodynamic models with *gray* atmospheres. He found no significant systematic effect above 0.4 km s$^{-1}$. We reexamine the problem with more realistic models and in view of new observational evidence.

## 2. The new observational evidence and non-LTE models

Discrepant asymmetries of the photospheric spectral lines in Cepheids and RR Lyraes indicate that path conservation *may not* be valid. Naively one expects the asymmetry of spectral lines to be a simple function of radial velocity over the cycle. What is observed in Cepheids is very different – large asymmetry during contraction, much reduced asymmetry during expansion (Butler 1993; Albrow & Cottrell 1994; Sabbey et al. 1995).

The discrepant line asymmetries are reproduced by the radiation hydrodynamics NLTE models of Cepheids (Sasselov & Raga 1992), as caused by the varying depth of line formation and coupled opacity changes (see Sabbey et al.

1995 for details). It is crucial that the atmosphere is *not gray* and that the NLTE radiative transfer is coupled to the hydrodynamics. Otherwise none of these spectral line effects are treated properly, as noted by Karp (1975) himself and shown further by Albrow (1994) with line-blanketed LTE synthetic spectra on a separately computed first-order hydrodynamic model after Wood (1974).

## 3. Practical considerations

While the systematic effects discussed here have been suspected to exist for more than three decades now, there has been no direct empirical evidence to help determine their magnitude. Therefore current use of radial velocity curves involves an implicit assumption of zero average line asymmetry over a cycle. Sabbey et al. (1995) showed that this assumption is not good at the 5−10% level of accuracy.

As a solution, we suggest that the line asymmetry curves be used *in conjunction with* the radial velocity curves of Cepheids. First, the radial velocity curve should be measured from Doppler shifts of selected photospheric lines (see Butler 1993 for lists) with a knowledge of the spectral line profiles. For example, some methods measure Doppler shifts from the minimum of an absorption line profile, others take a weighted mean, etc. Second, the asymmetry curve should be obtained for the same list of photospheric lines. If the asymmetry curve is not symmetric, i.e., the average is not zero, corrections to both the $\gamma$-velocity and $p$-factor must be made. In the case of deriving a phase-dependent $p$-factor, this can be accomplished directly from the observations. One needs to quantify the asymmetry curve in terms of Doppler shift (in wavelength or velocity), depending on which method was used above to measure the radial velocities, e.g., see Fig. 3 in Sabbey et al. (1995). In the case of the $\gamma$-velocity, the deviation of the asymmetry curve from a zero-average (in units of Doppler shift) would provide most of the needed correction. Part of this correction depends on the details of line formation and is thus model dependent. Its model dependence can be reduced substantially by observing the asymmetry curves of a large number of photospheric lines with different excitation potentials.

## References

Albrow, M. D. 1994, Ph.D. Thesis, University of Canterbury

Albrow, M. D., & Cottrell, P. L. 1994, MNRAS, 267, 548

Butler, R. P. 1993, ApJ, 415, 323

Gautschy, A., 1987, Vistas in Astronomy, 30, 197

Karp, A. 1975, ApJ, 201, 641

Pont, F., Mayor, M., & Burki, G. 1994, A&A, 285, 415

Sasselov, D., & Raga, A. 1992, in: Cool Stars, Stellar Systems, and the Sun, (eds.) Giampapa & Bookbinder, ASP Conf. Ser. 26, p.549

Sabbey, C. N., Sasselov, D. D., Fieldus, M. S., Lester, J. B., Venn, K. A., & Butler, P. 1995, ApJ, June 10

Wood, P. R. 1974, ApJ, 190, 609

*Astrophysical Applications of Stellar Pulsation*
*ASP Conference Series, Vol. 83, 1995*
*R. S. Stobie and P. A. Whitelock, eds.*

# Accurate CCD Light Curves of Cepheids in the Large Magellanic Cloud

E. Poretti[1], L. Mantegazza[2], E. Antonello[1]

[1] *Osservatorio di Brera, Via E. Bianchi 46, 22055 Merate, Italy*

[2] *Università di Pavia, Dipartimento di Fisica Nucleare e Teorica, Via Bassi 6, 27100 Pavia, Italy*

## 1. Introduction

In the past years, in a series of papers (Antonello & Poretti 1986, Antonello et al. 1990, Mantegazza & Poretti 1992, Poretti 1994) the Fourier decomposition was successfully applied to the light curves of galactic Cepheids with P< 8 d. The separation of these Cepheids into two groups was evident. The first, more numerous (195 stars after the latest inclusion) group is constituted by the classical Cepheids, i.e., the Cepheids which follow the Hertzsprung progression: they occupy a narrow region in the $\phi_{21} - P$ plane and are also characterised by a $R_{21}$ ratio larger than 0.20. The second group, containing 30 objects, is constituted by Cepheids which deviate from the Hertzsprung progression, describing a "Z" crossing the classical sequence in the $\phi_{21} - P$ plane; their $R_{21}$ values are smaller than 0.20. The relation between stars of the upper and lower sequence of the "Z" is confirmed by the $\phi_{31} - P$ plane, where these stars describe an unique sequence. As first suggested by Antonello & Poretti (1986), the splitting of Cepheids into two groups can be explained by a different pulsation mode: while the classical Cepheids are pulsating in the fundamental radial mode, the other Cepheids are pulsating in the first overtone radial mode.

## 2. Observations of Cepheids in the LMC

As a natural extension of this work, the same method should be applied to extragalactic Cepheids. For the Cepheids belonging to the same galaxy, we directly observe a separation between fundamental and first overtone objects since for a given period the former are fainter than the latter. The Magellanic Clouds objects were considered as the first target; photographic light curves were available, but their precision was not adequate for our purposes. Hence, we planned to obtain CCD measurements of the Cepheids in the Large Magellanic Cloud. 21 Cepheids with periods between 2.4 and 4.0 d were selected and about 20 CCD measurements for each star were obtained at the Dutch 90–cm telescope at La Silla (Chile) in December 1993. The reduction of the 450 science images has been completed and it shows that the requested precision has been achieved: the standard deviations of the fits are usually <0.010 mag and also the measurements of the constant stars as bright as the Cepheids show a standard deviation better than 0.010 mag. Figure 1 shows the measurements of four LMC Cepheids with the fitted light curve.

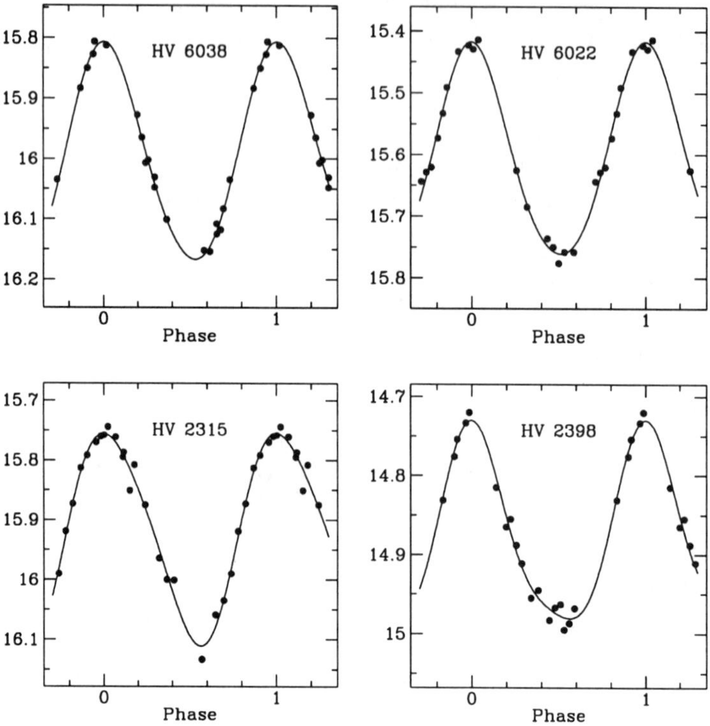


Figure 1. LMC Cepheids light curves: HV 6038 ($P$=2.796 d), HV 6022 ($P$=3.651 d), HV 2315 ($P$=2.412 d), HV 2398($P$=3.276 d).

About 30–40 stars were surveyed in each stellar field and six new Cepheids were discovered. The collected data are not sufficient to uniformly cover the light curves of the whole sample and, in some cases, no reliable Fourier decomposition can be performed. For this reason, another 600 images of the same stellar fields were obtained in November 1994. When the data reduction of these images is also completed, we are confident we will obtain very precise values of the Fourier coefficients for the whole sample. The preliminary values confirm the subdivision of the Cepheids into two well–separated subgroups, in agreement with the characteristic observed in the Galaxy.

## References

Antonello, E., & Poretti, E. 1986, A&A, 169, 149

Antonello, E., Poretti, E., & Reduzzi, L., 1990, A&A, 236, 138

Mantegazza, L., & Poretti, E. 1992, A&A, 261, 137

Poretti, E. 1994, A&A, 285, 524

*Astrophysical Applications of Stellar Pulsation*
*ASP Conference Series, Vol. 83, 1995*
*R. S. Stobie and P. A. Whitelock, eds.*

# Baade-Wesselink Analysis of a Sample of Equal Period SMC Cepheids. A progress report.

Jesper Storm[1], Bruce W. Carney[2], Wendy Freedman[3], and Barry F. Madore[4]

[1]*European Southern Observatory,* [2]*Univ. of North Carolina,* [3]*The Observatories,* [4]*IPAC*

**Abstract.** We present a progress report on a project to investigate the differences in absolute physical parameters for a sample of five equal period SMC Cepheids from a Baade-Wesselink analysis of the stars.

## 1. Introduction

With the recent observations of Cepheids in Virgo galaxies (Pierce et al. 1994 and Freedman et al. 1994) the Period-Luminosity (PL) relation is finally being applied to objects at cosmologically interesting distances. As the PL relation is a statistical relation for an ensemble of stars and not an exact physical relation, it is important to understand how this relation can be affected by differences in the sample. This is particularly important when applying the relation to distant galaxies where the properties of the stellar populations are not well known but might well be significantly different from the local and well studied ones. As demonstrated by Mateo (1992) the ensembles obtained from one young cluster to the next in the LMC can lead to significantly different slopes of the PL relation. The reason for this discrepancy is not entirely clear, but abundance differences might provide part of the explanation. In an attempt to understand better the physical properties of Cepheids, in particular with respect to differences in abundances, we will apply the Baade-Wesselink (BW) method to a sample of five SMC Cepheids of similar period. The BW analysis gives the absolute physical properties for the individual stars and we will investigate how the physical parameters for the stars, at a fixed period, change as a function of color and luminosity. Comparing these results with similar results for galactic stars will enable us to assess the effect of abundance variations on luminosity which could be important for the application of the PL relation to stars in other galaxies.

## 2. The sample

To obtain a good baseline in metallicity with respect to the galactic Cepheids it is preferable to observe SMC Cepheids ([Fe/H] $\approx -0.65$) over LMC Cepheids. Also, the SMC has a lower dust content than the LMC, and thus the reddening corrections are in general smaller and more uniform.

Three of the stars (HV1328, HV1333, and HV1335) are in the south-eastern part of the SMC, whereas HV822 and HV1345 are located in the southern end of the bar of the SMC. Although the SMC possess a significant depth the BW method will provide distances that we hope will be able to disentangle such problems.

All the stars have periods close to 15 days, short enough to limit the problems with the model atmospheres for the very extended envelopes, and at the same time long enough to reduce the contamination of the sample by overtone pulsators.

| ID | HV822 | HV1328 | HV1333 | HV1335 | HV1345 | |
|---|---|---|---|---|---|---|
| Period | 16.742 | 15.8396 | 16.289 | 14.96 | 13.4767 | (days) |

## 3. The data

We have collected about 2000 BVRI CCD frames at Las Campanas and Cerro Tololo for these stars and additional stars in the LMC. The magnitudes have been derived using DoPHOT1.0 (Schecter, Mateo & Saha, 1993) modified to run on a SUN workstation. A database package has been developed within IRAF to keep track of the magnitudes of all the objects in the frames. This package enables us to achieve the optimal relative calibration of the frames and to find and study the other variable stars in the field as well.

The radial velocities have been derived from high resolution echelle, low signal to noise, spectra obtained at Las Campanas supplemented with data from ESO. These data have been reduced using IRAF and the package `xcor`.

## 4. The Baade-Wesselink analysis

The temperature is the most critical parameter for the BW analysis. Fry & Carney are currently determining excitation temperatures for galactic Cepheids in open clusters, and they find good agreement between their temperatures and temperatures from the (B–V) colour index combined with the Buser & Kurucz (1992) model atmospheres. Thus we can use the model atmosphere data for the temperature determination and follow the recipe used for the RR Lyrae stars described in Storm et al. (1993).

## References

Buser R., & Kurucz R.L.; 1992, A&A, 264, 557

Freedman, W., et al.; 1994, Nature, 371, 757

Mateo, M., 1992, PASP, 104, 824

Pierce, M.,J., et al.; 1994, Nature, 371, 385

Schecter, P.L., Mateo, M., and Saha, A.; 1993, PASP, 105, 1342

Storm, J., Carney, B.W., & Latham, D.W.; 1994, A&A, 290, 443

*Astrophysical Applications of Stellar Pulsation*
*ASP Conference Series, Vol. 83, 1995*
*R. S. Stobie and P. A. Whitelock, eds.*

# Cepheid Distances to Nearby Galaxies

Wolfgang P. Gieren, Pablo Candia

*Pontifica Universidad Católica de Chile, Casilla 104, Santiago 22, Chile*

Pascal Fouqué

*DESPA Observatoire de Paris-Meudon, F-92195 Meudon Cedex France and European Southern Observatory, Casilla 19001, Santiago 19, Chile*

Gisela Hertling

*Pontifica Universidad Católica de Chile, Casilla 104, Santiago 22, Chile*

Eduardo Brieva

*Observatorio Astronómico Nacional, Apartado Aéreo 2584, Bogotá, Colombia*

**Abstract.** We have determined new Cepheid distances on a homogeneous system to nine nearby galaxies with Cepheid observations in the V bandpass, using the extensive calibration of the galactic Cepheid PL(V) relation of Gieren, Barnes & Moffett (1993). Absorption corrections to the observed distance moduli were taken from the RC3 catalog (de Vaucouleurs et al. 1991). For most of the galaxies, the galactic PL(V) relation of GBM gives a very good fit to the Cepheid data.

## 1. Introduction

Recently Gieren, Barnes & Moffett (1993; hereafter GBM) have published a new galactic calibration of the Cepheid period-luminosity (PL) relation in the V bandpass. This relation, based upon the distances of 100 Cepheids derived via the surface brightness method (Barnes & Evans 1976), is the most extensive calibration of the PL relation currently available. Using this calibration which is assumed to be universal, and applying it to Cepheids in other galaxies for which V light curves have been observed, a homogeneous set of internally consistent distances for these galaxies can be derived.

## 2. Data and distance determination

We surveyed the literature for galaxies for which V light curves for a substantial number of Cepheid variables are available. This defined a sample of nine galaxies including M81, IC 4182 and NGC 5253 for which Hubble Space Telescope data have recently become available.

For each of these galaxies we plotted the mean V magnitudes of the Cepheids, as given in the original references, versus the logarithm of the pulsation period. We then fitted the GBM PL relation

$$\begin{aligned} M_{\langle V \rangle} &= -1.371 - 2.986 logP \\ & \quad \pm 0.095 \pm 0.094 \end{aligned} \tag{1}$$

to the data. The GBM calibration gives a very good representation of the $\langle V \rangle$ versus $logP$ diagrams for most, albeit not all, of the studied galaxies. On the basis of these plots, we decided the location of the detection limits for each galaxy and omitted those short-period (faint) Cepheids from the samples which appeared to be affected by the Malmquist bias. For each of the remaining Cepheids in the different galaxy samples, a distance modulus $(V - M_V)$ was calculated where $M_V$ came from equation (1). Finally, a mean distance modulus was calculated for each galaxy averaging the individual Cepheid $\mu$ values. In this process, weights of 0, 0.5, and 1 were assigned to the individual Cepheids according to the quality and amplitudes of their light curves. The apparent distance moduli were corrected for interstellar absorption using the $A_B$ values given in the RC3 catalog (de Vaucouleurs et al. 1991) which were transformed to absorption in the V band using $A_V/A_B = 0.78$.

On average the distance moduli we obtain are about 0.3 mag larger than the mean of previous determinations (see de Vaucouleurs 1993). This is partly due to the adopted galactic calibration of the PL(V) relation of GBM, which makes Cepheids ~0.2 mag brighter than the Feast & Walker (1987) calibration from ZAMS-fitting of open clusters, as well as to our adopted absorption corrections. Numerical values will be published elsewhere.

**Acknowledgments.** This work was supported by FONDECYT grant No. 1940710 given to W. Gieren, and by a Colciencias grant given to the Universidad Nacional de Columbia. Both grants are gratefully acknowledged.

## References

Barnes, T. G. & Evans, D. S. 1976, MNRAS, 174, 498

De Vaucouleurs, G. 1993, ApJ, 415 10

De Vaucouleurs, G., de Vaucouleurs, A., Corwin, H. G., Buta, R. J., Paturel, G., & Fouqué, P. 1991, Third Reference Catalogue of Bright Galaxies, Vol. II, Springer-Verlag

Feast, M. W. & Walker, A. R. 1987, ARA&A, 25, 345

Gieren, W. P., Barnes, T. G. & Moffett, T. J. 1993, ApJ, 418, 135

*Astrophysical Applications of Stellar Pulsation*
*ASP Conference Series, Vol. 83, 1995*
*R. S. Stobie and P. A. Whitelock, eds.*

# Cepheids and LPVs in and around Magellanic Cloud Clusters

K.M. Sebo and P.R. Wood

*Mount Stromlo & Siding Spring Observatories, Private Bag, Weston Creek PO, Canberra ACT 2611, Australia*

**Abstract.** V and I band CCD photometry of the young clusters NGC 330 (SMC), NGC 1850, NGC 2058 and NGC 2065 (LMC) has been used to find variable stars in and around the clusters. The variables detected are mostly Cepheids and long-period variables (LPVs). Here we outline some results of these studies.

## 1. Introduction

This study was designed to address a number of problems: firstly, the disagreement between pulsation mass and evolution mass (Cox 1980), which may be resolved by allowing the convective cores of intermediate mass stars to overshoot beyond the classical Schwarzschild limit (Chiosi et al. 1992; Moskalik, Buchler & Marom 1992); secondly, the mass-period relation for LPVs, which can be studied by finding LPVs in Magellanic Cloud clusters; and, thirdly, the Mira pulsation mode, which can be determined by comparing large and small amplitude LPVs in either the clusters or the field.

## 2. Results

Pulsation masses for Cepheids were derived from $<V>$ and $<V> - <I>$ and the $V, V-I, M, P$ relation in Chiosi, Wood & Capitanio (1993). Evolution masses were derived by fitting the Cepheids to the core helium burning loops of the isochrones of Bertelli et al. (1994). These tracks included a moderate amount of main-sequence convective core overshoot ($\lambda$=0.5). The pulsation masses are compared to evolution masses in Figure 1.

For the Cepheids around NGC 1850 (Sebo & Wood 1995), pulsation masses are $\approx$20% smaller than evolution masses for stars with M $\gtrsim$ 5 $M_\odot$. It thus appears that for stars in this mass range, a large amount of main-sequence convective core overshoot ($\lambda \sim 1$) may be needed to bring pulsation and evolution masses into agreement. One of the Cepheids around NGC 1850 was a bump Cepheid and detailed fitting of the light curve with nonlinear pulsation models gave excellent agreement with the pulsation mass. Most of the Cepheids in the field around NGC 330 are less massive (M $\sim$ 3 $M_\odot$) than those near NGC 1850, and for these objects small to intermediate amounts of overshoot ($\lambda \lesssim 0.5$) are indicated (Sebo & Wood 1994).

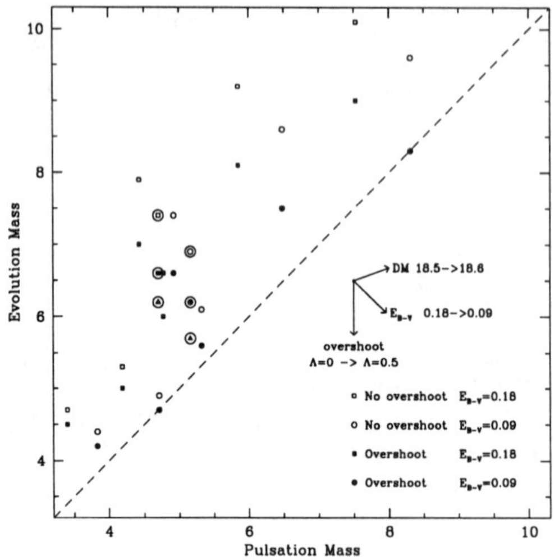


Figure 1. Evolution mass plotted against pulsation mass in $M_\odot$ for the seven Cepheids in the field centred on NGC 1850. Each Cepheid is plotted 4 times according to the amount of core overshoot and reddening. The Cepheid which is a member of NGC 1850 is circled. The triangle shows this star plotted using the evolution mass derived from the cluster age and $E_{B-V}$ =0.09 and 0.18. The arrows show the effect of changing the amount of overshoot, distance modulus and reddening.

A large number of LPVs were found in the fields around the clusters, although no cluster red giants were found to be variable. A (K,logP) diagram for the LPVs shows two distinct sequences which can be understood if the Mira variables are fundamental mode pulsators while the small amplitude variables are first overtone pulsators.

## References


Bertelli, G., Bressan, A., Chiosi, C., Fagotto, F., & Nasi, E. 1994, A&AS, 106, 275

Chiosi, C., Wood, P.R., & Capitanio, N. 1993, ApJS, 86, 541

Chiosi, C., Wood, P.R., Bertelli, G., Bressan, A., & Mateo, M. 1992, ApJ, 385, 205

Cox, A.N. 1980, ARA&A, 18, 15

Moskalik, P., Buchler, J.R., & Marom, A. 1992, ApJ, 385, 685

Sebo, K.M., & Wood, P.R. 1994, AJ, 108, 932

Sebo, K.M., & Wood, P.R. 1995, ApJ, in press

*Astrophysical Applications of Stellar Pulsation*
*ASP Conference Series, Vol. 83, 1995*
*R. S. Stobie and P. A. Whitelock, eds.*

# Iron Abundances of the Cluster RR Lyrae Variables Based on Their Light Curves

Johanna Jurcsik and Géza Kovács

*Konkoly Observatory, H-1525 Budapest XII, Box 67, Hungary*

**Abstract.** [Fe/H] values of cluster RRab variables were calculated from the Fourier parameters of their light curves using a recently published method. The average cluster metallicities obtained this way agree with direct observations within the error range.

## 1. Introduction

A previously obtained light curve → [Fe/H] calibration (Kovács & Zsoldos, 1995) was applied on independent samples of cluster variables. Good quality CCD, photoelectric and some photographic observations were used for 8 galactic and 5 LMC clusters. Iron abundances of 83 RRab stars were determined according to the 3 and 5 parameter formulae which relate the Fourier parameters of the $V$ light curves to metallicity.

Because the calibrating formulae were constructed by use of [Fe/H] values which were obtained from $\Delta S$ observations through Blanco's transformation (Blanco, 1992), our results are compared with the $\Delta S$ cluster abundances on this scale. For clusters with no $\Delta S$ measurements the metallicities derived from different methods (mostly high dispersion spectroscopy) were also transformed to the Blanco scale.

[Fe/H] values with error less than 0.5 dex are regarded as reliable. The smallest error of the individual [Fe/H] determinations is about 0.04 dex.

## 2. Results

Our results are shown in Figure 1. The following conclusions can be drawn.
• For less metal deficient clusters the calculated [Fe/H] values are in good agreement with the average cluster abundances.
• For the most metal deficient clusters our values are somewhat high, probably because of the small number of the very metal poor stars in our sample of calibration.
• The internal standard deviations of [Fe/H] for the individual clusters vary between 0.06–0.23 dex with a bias towards higher standard deviations for the lower quality light curves. The only exception is the LMC cluster NGC 2257 where the individual abundances suggest chemical inhomogeneity or more probably non-membership of some of the variables. The spectroscopic abundance determinations give two distinct values of the metallicity of this cluster. The

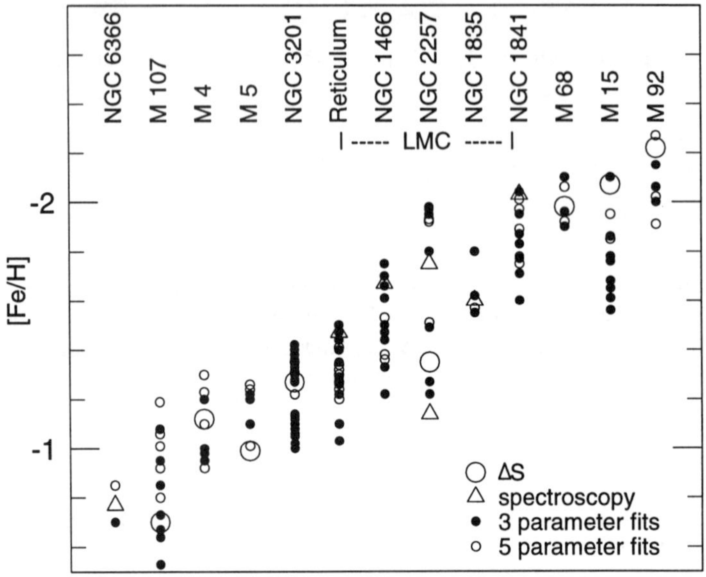


Figure 1. [Fe/H] values of individual RRab stars calculated from the Fourier parameters of their light curves (○, •). The average cluster metallicities ($\Delta S$ (◯) or spectroscopy (△), transformed to the Blanco scale) are shown for comparison.

less metal poor solution is, however, close to the $\Delta S$ abundance of the surrounding field RR Lyrae stars. Therefore, we conclude that the variables with [Fe/H] values similar to this higher metal abundance could be field variables, too.

• The present method is not applicable for the estimation of [Fe/H] in peculiar cases (e.g., in Blazhko variables, highly evolved stars...). The [Fe/H] values computed for stars with variable light curves are usually deviant. The anomalous [Fe/H] we got for M4-V2, M68-V22, NGC 1841-V6 and 47 Tuc-V9 indicate these types of peculiarities and therefore they are not included in our sample.

• The accuracy of the transformation formulae could be improved by the use of larger number of the calibrating stars with more accurate [Fe/H] values and light curves. This would enable us to derive precise metallicities on a large sample of RR Lyrae stars without resorting to the more tedious spectroscopic observations.

**Acknowledgments.** The financial support of OTKA (grant T-014183) is gratefully acknowledged.

## References

Kovács, G., & Zsoldos, E. 1995, A&A, 293, L57

Blanco, V.M. 1992, AJ, 104, 734

*Astrophysical Applications of Stellar Pulsation*
*ASP Conference Series, Vol. 83, 1995*
*R. S. Stobie and P. A. Whitelock, eds.*

# A Century in Period Changes of the Variables in M15

Katalin Barlai and B. Szeidl

*Konkoly Observatory, H-1525 Budapest XII, Box 67, Hungary*

**Abstract.** Period behaviour of 62 RR Lyrae stars in the M15 globular cluster has been investigated. About one half of the sample (30 stars) exhibited linear period change. The remaining 32 variables can be characterized by abrupt or erratic changes in their periods.

## 1. Introduction

In the year 1896 Bailey took the first photographs on the globular cluster M15 at the Boyden Station of the Harvard Observatory in Arequipa, Peru (Bailey, 1919). Since these photographs a plethora of plates (over 1000) has been taken at different observatories.

## 2. Data and results

Photographic observations have been carried out between the years 1937 - 1966 at Konkoly Observatory in Budapest with the 24-in Newton telescope (Barlai, 1989) and between 1975 - 1991 at Piszkéstető, its mountain station with the 1m RCC telescope. From the 112 variables detected so far 63 were measurable on our plates. Besides this plate material, all the observations published on M15 during this century were collected and analysed concerning the change of the periods. Period changes of 62 RR Lyraes of the cluster were followed up. From these data O–C curves were constructed considering the phase shift of the folded light curves of each single year.

If the star's period has changed linearly (either increasing or decreasing) its rate can be characterized by the quantity $\beta = dP/dt$.

The phase of the maxima has been computed from JD = 2431000 being this date roughly the middle of the 95 year long observational interval. Among the 62 RR Lyrae variables studied the period changes could have been fitted by parabolae or straight lines in 30 cases. 18 variables – 8 RRab and 10 RRc stars – exhibit (+) $\beta$ values while (–) $\beta$ values characterize the period changes of 12 variables – 6 RRab and 6 RRc stars. It means 30% increasing and 20% decreasing periods. The behaviour of the remaining 50% (32 stars) could not be characterized by linear period change.

The average values of the rates of period changes – $10^{10} 1/P(dP/dT)$ – are:
$+0.57 \pm 1.32$ for the 16 RRc stars;
$-0.47 \pm 2.86$ for the 14 RRab stars and
$+0.08 \pm 1.48$ for these 30 RR Lyraes together.

It is worth mentioning that from the 14 double mode RR Lyrae stars (Jurcsik & Barlai, 1990) in the cluster 9 exhibit erratic period change and 5 can be characterized by linear period change.

The sign of the period changes does not seem to depend on the single star's position within the RR Lyrae gap (Smith & Sandage, 1981). The slight increase in the average value of the period changes and the distribution of the $\beta$ values may support Lee's (1991) theory. The occurrence of a great number of RR Lyrae stars that show abrupt and erratic changes in their periods cannot be easily reconciled with the present theories, where period change is due to evolution alone. The small number of the stars involved and the joint handling of the fundamental and first overtone pulsators may also throw some obstacle in the way of statistical evaluation of the data.

In our view preference should be given to Sweigart & Renzini's (1979) theory. According to it the observed period changes are due to random events happening in a semiconvective zone around the convective core. Including this instability in the core's composition the size and sign of the real period changes can be explained. Still a positive mean rate of period change can be expected in course of centuries due to nuclear evolution.

The present paper covers a long observational interval and a large sample of variables in M15. This sample does not show distinct bias towards increasing periods during the century elapsed as Lee (1991) supposed in his thoroughly computed theory. Nuclear evolution of RR Lyrae stars, however, can be grasped by further sudying of period changes in the coming centuries.

**Acknowledgments.** This work has been supported by the Hungarian Research Grant OTKA I/3 – 829.

## References

Bailey, S.I. 1919, Harvard Ann., 78, Part 3

Barlai, K. 1989, Communication fron the Konkoly Observatory of the Hung. Acad. Sci., No. 92, Vol. 10, Part 6

Jurcsik, J., & Barlai, K. 1990, in: C. Cacciari & G. Clementini (eds.), Confrontation between Stellar Pulsation and Evolution, ASP Conf. Ser., 11, p. 112

Lee, Y.W. 1991, ApJ, 367, 524

Smith, H.A., & Sandage, A. 1981, ApJ, 86, 1870

Sweigart, A.V., & Renzini, A. 1979, A&A, 71, 66

*Astrophysical Applications of Stellar Pulsation*
*ASP Conference Series, Vol. 83, 1995*
*R. S. Stobie and P. A. Whitelock, eds.*

# The Distribution of the RR Lyrae Variables in $\omega$ Centauri and Background Light

J. O. Petersen[1,2] and M. Andersen[2]

[1]*Theoretical Astrophysics Center, and*
[2]*Niels Bohr Institute for Astronomy, Physics and Geophysics, Astronomical Observatory, Øster Voldgade 3, DK 1350 Copenhagen K, Denmark, (E-mail: oz@astro.ku.dk and mia@astro.ku.dk)*

**Abstract.** The distribution of the RR Lyrae stars in the globular cluster $\omega$ Centauri is studied, using the data published by Martin in 1938. Comparing the accumulated numbers of variables within the angular distance, $r$, from the centre of the cluster, it is shown that the known RRc variables seem to be less concentrated towards the centre than the RRab stars. 6–10 RRc stars are missing within $r = 2$–3 arcmin, if the two distributions are actually identical, as expected from evolution models. We conclude that the central region may contain several RRc stars that have not yet been discovered. This could be due to their relatively low amplitude, since it is very difficult to find low–amplitude variables by means of photographic photometry in the crowded, central region of $\omega$ Cen with large and varying corrections due to unresolved background light.

## 1. Introduction

According to the standard scenario for the evolution of stars in the horizontal branch in globular clusters (e.g., Lee et al. 1990) the basic physical properties of the fundamental mode RRab variables and the first overtone RRc stars are identical, except for the well known difference in effective temperature. Since most globular clusters contain less than dozen or so RR Lyrae variables, they cannot provide data that can be used to compare the distributions of RRab and RRc variables. A significant test clearly requires a large number of both RRab and RRc stars, and only a few clusters are rich in both types of variables. $\omega$ Centauri contains at least about 80 RRab variables and 60 RRc stars, and Martin (1938) published very detailed data for most of these variables. Here we use Martin's data collected in Petersen (1994) to discuss the distribution of the RR Lyrae variables in $\omega$ Cen.

## 2. Radial distribution of RRab and RRc stars

For each variable in his table, Martin (1938) gives the angular distance, $r$, of the star from the centre of the cluster. Martin analyses the effect of unresolved background light on his photographic magnitudes. He concludes that these effects are significant in a central region of $r \approx 4$ arcmin. We have studied the distribution of RRab and RRc variables including 74 RRab and 55 RRc stars with data taken from Petersen (1994). It is interesting to compare the RRc distribution with the RRab data multiplied by 55/74. We find that systematic differences are present. For $r \leq$ about 9 arcmin the number of RRc stars is 6–10 less than the number expected from the RRab distribution, whereas the number of RRc stars seems to be a little too high from $r = 11$ arcmin to about 15 arcmin. It is tempting to explain the difference between the RRc distribution and the predicted numbers by the hypothesis that 6–10 RRc stars are missing within the inner 2 arcmin of the cluster. This is precisely the crowded core region, where Martin finds large systematic corrections to his photographic magnitudes.

## 3. Conclusion

The natural conclusion from this comparison of the distribution of RRab and RRc variables in $\omega$ Cen is that about 8 RRc stars are actually present within the central 2 arcmin in $\omega$ Cen, although they have not been identified until now. Of course there could also be as yet undiscovered RRab stars present in the central region of $\omega$ Cen. If that should turn out to be the case, the expected number of "new" RRc stars should be increased with the additional number of RRab variables.

Martin (1938) could estimate the systematic effects of unresolved background light in $\omega$ Cen on his photographic magnitudes. He found that these effects are significant, typically 0.05–0.20 mag, within the central 4 arcmin. Besides the *systematic* corrections, background light will also result in additional *statistical errors* on photographic magnitudes, because the influence of the background (and crowding in the fields) will vary from one exposure to another, e.g., due to seeing. Clearly, it is more difficult to discover low–amplitude RRc variables here than in the outer regions of $\omega$ Cen. Therefore it may not be surprising, if the photographic surveys that have been used until now, have not succeeded in identifying several RR Lyrae variables — in particular low–amplitude RRc stars — in the central region of globular clusters. Modern CCD–based photometry will be able to provide much more accurate photometry in fields with background light and crowding than traditional photographic photometry.

**Acknowledgments.** We wish to thank Henning E. Jørgensen for many discussions on variable stars. Support from the Danish National Research Foundation via TAC is gratefully acknowledged.

## References

Lee, Y.–W., Demarque, P., & Zinn, R. 1990, ApJ, 350, 155

Martin, W.C. 1938, Leiden Annalen, Vol. XVII, Part 2

Petersen, J.O. 1994, A&AS, 105, 145

*Astrophysical Applications of Stellar Pulsation*
*ASP Conference Series, Vol. 83, 1995*
*R. S. Stobie and P. A. Whitelock, eds.*

# RR Lyrae Variables in NGC 7006

Amelia Wehlau, Robert W. Slawson

*Department of Astronomy, The University of Western Ontario, London, Ontario, CANADA, N6A 3K7*

James M. Nemec

*Department of Physics and Astronomy, York University, 4700 Keele Street, North York, Onatrio, CANADA, M3J 1P3*

Steve Butterworth

*Department of Physics and Astronomy, McMaster University, Hamilton, Ontario, CANADA, L8S 4M1*

**Abstract.** B and V CCD frames of the globular cluster NGC 7006 have been obtained to study the period change rates and determine photometric colours for the RR Lyr variables.

NGC 7006 is a galactic globular cluster at a heliocentric distance of 40 kpc known to contain at least 60 RR Lyrae stars. It was also one of the first clusters studied whose colour-magnitude diagram showed evidence of a "second parameter". Magnitudes of the variable stars were obtained by Rosino and his colleagues from plates taken by Hubble in the 1930's and by Sandage in the 1950's and used to determine periods for many of the variables. These earlier observations were combined with magnitudes from plates taken by Nemec in 1984 and used by Wehlau et al. (1992) to derive period change rates. Statistically significant evidence was found for a radial gradient in these rates, more negative rates being found in the outer region of the cluster.

B and V CCD frames of the cluster were obtained by Nemec in 1984, 1987, and 1989 using the Palomar 1.5 m telescope and in 1994 with the 1.8 m telescope of the Dominion Astrophysical Observatory. Most of the frames have now been reduced using magnitude calibrations based on the photometry of Buonanno et al. (1991).

These new observations made it possible to derive periods for some stars for which we were unable to obtain periods earlier. Light curves for four of these stars are shown in Fig. 1. Variables 24 ($P = 0^{\rm d}.626277$) and 27 ($P = 0^{\rm d}.567215$) are each blended with a close companion so that no periods could be obtained earlier. The light curve for V28 ($P = 0^{\rm d}.497308$) also represents a new period since the period of 0.56 day given by Pinto & Rosino (1973) does not fit our data. However this new period which is very close to half a day does fit their observations as well as ours. V76 ($P = 0^{\rm d}.549564$) was discovered on the 1984 photographic plates which were taken during a 10 day observing run.

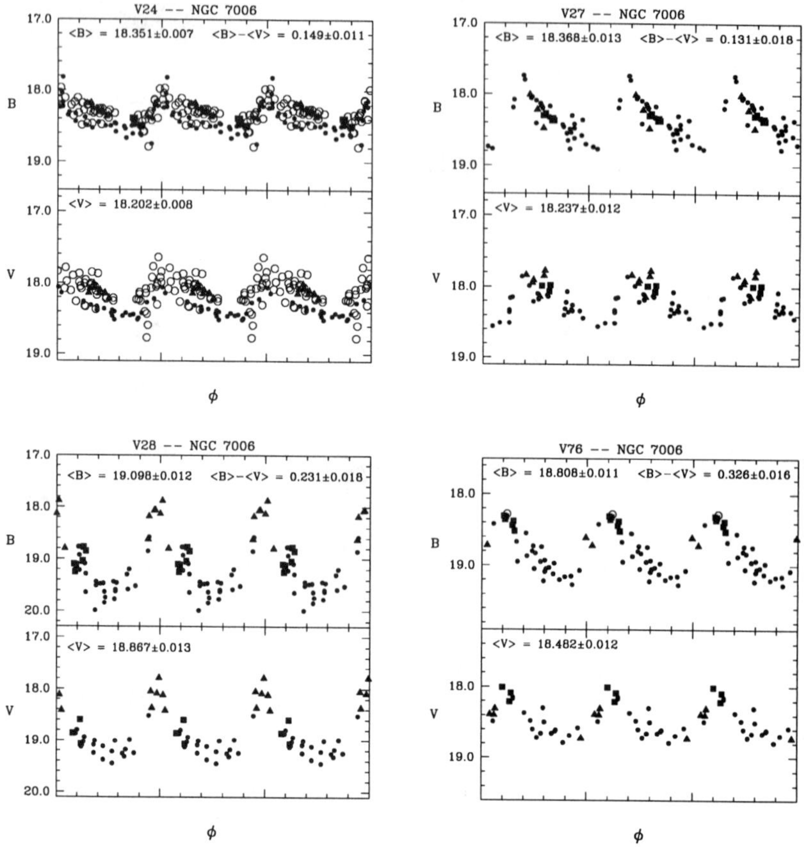


Figure 1. Light curves for V24, V27, V28, and V76 with new periods. The various symbols represent observations in different years.

## References

Buonanno, R., Fusi Pecci, F., Cappellaro, E., Ortolani, S., Richtler, T., & Geyer, E. H. 1991, AJ, 102, 1005

Pinto, G., & Rosino, L. 1973, MSAIt, 44, 255

Wehlau, A., Nemec, J. M., Hanlan, P., & Rich, R. M. 1992, AJ, 103, 1583

*Astrophysical Applications of Stellar Pulsation*
*ASP Conference Series, Vol. 83, 1995*
*R. S. Stobie and P. A. Whitelock, eds.*

# The Temperature of the Supergiant $\alpha$ Per [1]

Nancy Remage Evans

*Department of Physics and Astronomy, York University, 4700 Keele St., North York, Ontario, M3J 1P3, Canada*

Terry J. Teays [2] and Lyla L. Taylor [2]

*Computer Sciences Corp., Center for Scientific Research, Science Programs, 10000 A Aerospace Rd., Lanham-Seabrook, MD, 20706*

John B. Lester

*Department of Astronomy, Erindale College, University of Toronto, Mississauga, Ontario, L5L 1C6, Canada*

Robert B. Hindsley

*Astrometry Department, US Naval Observatory, 3450 Massachusetts Ave., NW, Washington, DC, 20392-5420*

**Abstract.** We compare observed fluxes from the ultraviolet (IUE) through J and K with recent Kurucz model atmospheres to determine a temperature for the F5 Ib supergiant $\alpha$ Per. The two most important advances in this study as compared with previous work are the use of well calibrated ultraviolet fluxes and the use of models with an appropriate microturbulence.

## 1. Introduction

There are many questions in astrophysics which require a temperature derived from the entire spectral energy distribution. As a first step in determining a temperature scale for Cepheids and non-variable supergiants, we have determined the temperature for the F5 Ib supergiant Alpha Per.

## 2. Data

Observations of $\alpha$ Per, as well as a number of other nonvariable supergiants, have been assembled including International Ultraviolet Explorer satellite (IUE) spectra and B, V, R, I, J, and K as discussed by Evans, et al. (1993). Because $\alpha$

[1] IUE Archival Research

[2] Staff Member of the International Ultraviolet Explorer Observatory

Per is located in a cluster, and because its angular diameter has been measured, its log $g$ is known observationally, as is the microturbulence in the atmosphere. We adopt a reddening E(B–V) = 0.04 mag from the stars closest to $\alpha$ Per in the cluster (Gray, 1991). These data are discussed in detail by Evans, et al. (1995), as is the process of determining the temperature.

## 3. Models

The observed energy distribution of $\alpha$ Per was compared with model atmospheres consistent with the observed values of gravity and microturbulence. Because supergiants have microturbulence higher than the standard value of 2 km $s^{-1}$, one of us (JBL) computed a series of fully line blanketed model atmospheres using the Atlas 9 code of Kurucz. For the broad-band comparisons, the model atmospheres were convolved with standard filter sensitivity functions (Evans, et al. 1995). The ultraviolet flux is sensitive to gravity and microturbulence, as well as temperature. Because we cannot resolve this degeneracy using the energy distributions, we fix log $g$ at 1.5 and microturbulence at 4 km $s^{-1}$ (as observed), and derive the temperature from the best fit to the energy distribution (6270 K).

## 4. Discussion

The rms deviation to the fit (0.065 mag) is reasonable considering that the data are assembled from many sources which are calibrated in different ways. This uncertainty corresponds to a temperature range of $\pm$ 120 K. The temperature derived from the distance to the cluster and angular diameter is within this range. Temperatures have recently been determined in several studies using the infrared flux method (Blackwell, et al. 1991, McWilliam, 1991). The differences in the temperatures between this study and the others can be attributed to the different reddenings.

**Acknowledgments.** Financial support was provided by a Natural Sciences and Engineering Research Council (Canada) grant to NRE and JBL and a NASA grant NAS5-31845 to TJT. Computing facilities were furnished by Dr. J. J. Caldwell at York University.

## References

Blackwell, D. E., Lynas-Gray, A. E., & Petford, A. D. 1991, A&A, 245, 567

Evans, N. R., Jiang, J. H., McAlary, C. W., & Campins, H. 1993, AJ, 106, 726

Evans, N. R., Teays, T. J., Taylor, L. L., Lester, J. B., & Hindsley, R. B. 1995, submitted to AJ

Gray, R. O. 1991, A&A, 252, 237

McWilliam, A. 1991, AJ, 101, 1065

*Astrophysical Applications of Stellar Pulsation*
*ASP Conference Series, Vol. 83, 1995*
*R. S. Stobie and P. A. Whitelock, eds.*

# On the Variability of Supergiant Stars

G. Burki and F. Kienzle

*Geneva Observatory, 1290 Sauverny, Switzerland*

## 1. Introduction

The Ia-type supergiants of spectral type B to G are probably all variable in radial velocity and luminosity; in addition, on the average, the amplitude of the variations increases with increasing intrinsic luminosity and are larger for the earliest and latest spectral types (Abt, 1957; Maeder & Rufener, 1972; Burki et al., 1978; Grenon, 1993). The variations are not strictly periodic, but can be frequently described by a *characteristic time* $T_c$, i.e., some kind of 'most probable period' (e.g., Sterken, 1977; Rufener et al., 1978; Percy et al., 1979; van Genderen et al., 1992).

It is possible to define a $T_c$-luminosity-colour relation for the supergiant stars (Maeder & Rufener, 1972; Burki, 1978). However, this relation exhibits a large dispersion, due to the imprecise luminosity and colour determinations for these stars and, essentially, to the fact that each supergiant can show very different $T_c$ values. Good determinations of $T_c$ require very long-term, continuous, photometric monitorings.

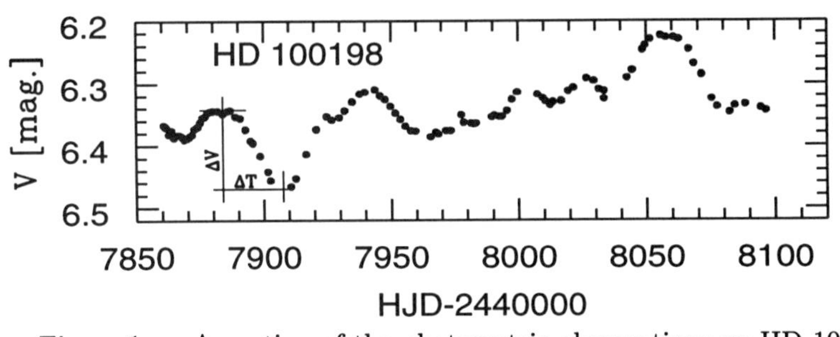


Figure 1. A portion of the photometric observations on HD 100198

## 2. A new description of the supergiant variability

A small group of bright, circumpolar, southern supergiant stars has been selected for a continuous monitoring from the Geneva station at the E.S.O. La Silla Observatory. A first result of this monitoring was the discovery that V810 Centauri (G0Ia) is a double-mode Cepheid-like pulsator, in addition to its supergiant behaviour (Burki, 1994). We present here the results of the preliminary analysis on HD 93737 (A0Ia/ab), HD 100198 (A3Ia) and HD 100261 (F7Ia/ab). It is clearly apparent in Figure 1 that, taking into account the high-precision (better than 5 millimag.) and the very good temporal coverage of the data, very faint variations can be described. These data allow us to adopt a new description of the supergiant variability : on the light curve, each increasing or decreasing portion is characterized by $\Delta V$, the luminosity variation, and by $\Delta T$, the duration of this variation. Note that this method was already used by Burki et al. (1982) to describe the variability of the extreme supergiant $\zeta^1$ Sco. In Figure 2 are plotted the data $\Delta V$ vs. $\Delta T$. On the basis of this small sample, we can tentatively postulate that **the slope $\Delta V/\Delta T$ decreases from B to G spectral type**. In addition, at a given spectral type, **the maximum value of $\Delta V$** (and, thus, also of $\Delta T$) **increases with increasing intrinsic luminosity**.

A more complete analysis, based on a larger sample of supergiants observed by us in the Geneva photometric system or by various other authors, is in progress.

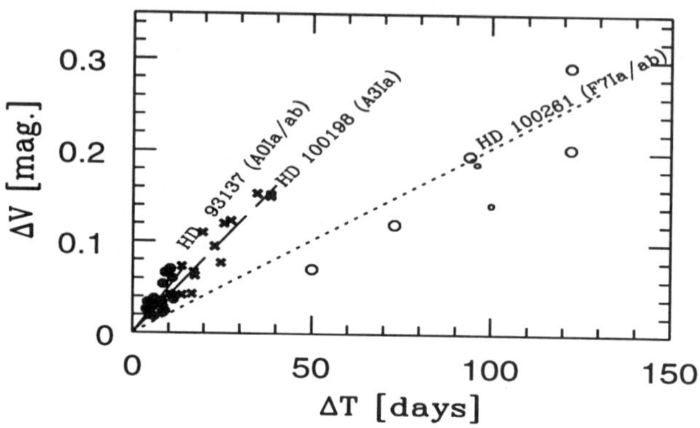


Figure 2. Relation between the luminosity variations $\Delta V$ and the corresponding time intervals $\Delta T$ for three supergiants

## References


Abt, H.A. 1957, ApJ, 126, 138

Burki, G. 1978, A&A, 65, 357

Burki, G. 1994, in: The Impact of Long-Term Monitoring on Variable Star Research, Eds. C. Sterken & M. de Groot, Kluwer, Dordrecht, 247

Burki, G., Maeder, A., & Rufener, F. 1978, A&A, 65, 363

Grenon, M. 1993, in Inside the Stars, Eds. W.W. Weiss & A. Baglin, ASP Conf. Series, 40, 693

Maeder, A., & Rufener, F. 1972, A&A, 20, 437

Percy, J.R., Barkerville, I., & Trevorrow, W. 1979, PASP, 91, 368

Rufener, F., Maeder, A., & Burki, G. 1978, A&AS, 31, 179

Sterken, C. 1977, A&A, 57, 361

van Genderen, A.M., et al. 1992, A&A, 258, 316

*Astrophysical Applications of Stellar Pulsation*
*ASP Conference Series, Vol. 83, 1995*
*R. S. Stobie and P. A. Whitelock, eds.*

# The Mass Loss and Pulsation Activity in Supernova Progenitors.

O. A. Tsiopa

*Central Astronomical Observatory of the Russian Academy of Sciences, Pulkovo 196140, St.Petersburg, Russia*

**Abstract.** The wind-formed regular structures revealed in early supernova (SN) spectra can be a result of the radial pulsation of the precursor.

## 1. Introduction

The near maximum light SN spectra contain the features, formed by the wind ejected in the last few months prior to the explosion.

Dopita et al. (1984) discovered a so called "superwind", which had been ejected with the extremly high velocities of 3000 km/s immediately prior to the SN event. The fascinating change of the velocity shifts in type II SNe 1983K (Niemela et al. 1985) and type Ia 1990M (Polcaro & Viotti 1991) are evidence of the existence of complicated, but rather regular structures in the wind at the scales of less than $10^{16}$ cm. Recently (Tsiopa 1995) the observational facts were united in a common dynamical model.

## 2. The winds of the SN precursors Type II and Ia

Though the SN envelope velocities of SNe type II are less, than in SNe type I, their winds are extremely fast (up to 2000-4000 km/s). The wind being highly variable and dense, the pronounced colliding wind-formed structures occur close to the SN progenitor. The system of interacting wind layers with different deviation of symmetry and increasing velocities can form the distinct prolate ellipsoidal shell around the SN progenitor just before the explosion.

The presupernova winds of SN type I are very difficult to detect: they are less massive, slower (hundreds km/s) and almost immediately swept up by the SN envelope expanding with very high speed. For the case of 1990M the model implies the existence of a shell formed by the matter with a maximum velocity of 1200 km/s and a minimum velocity of 600 km/s.

Judging from the SN envelope expansion velocity the wind-formed structures under consideration are situated at the radii from the explosion centre of $10^{14} - 10^{15}$ cm for SN 1983K and $3 - 6 \times 10^{15}$ cm for SN 1990M. The time of shell ejection can be estimated as $6 \times 10^6$ s for SN 1983K and $5 \times 10^7$ s for SN 1990M. Such rather small distances and short periods of the regular structure formation imply the presence of a high initial gradient of density and velocity in the wind matter.

## 3. Discussion

An extraordinary activity of the precursor can be suspected even for a single star. During the periods of interior reconstruction (preceding the SN type II event) even the outer parts of the star are likely to be in the unstable state. Strong pulsation with different modes can be evoked. Such processes as several-mode resonant coupling are probable to determine the mass-loss rates in critical periods of stellar evolution. During expansion the star throws away the very outer part of its envelope (or, in other words, the stellar wind increases dramatically) and an expanding shell is formed. In this case the precursor would be surrounded by a system of interacting shells (Tsiopa 1990)

It should be noted that the majority of the pulsating stars are located in the same region of the Hertzsprung-Russell diagram as expected for type II SN precursors. And though the modern SN theory does not provide any indications of the resonance nature of the explosion (deflagration regimes of the thermonuclear explosion of a degenerate core can provide only very short period pulsation (10 seconds), which cannot correlate with the envelope motions), there are some possible additional energy sources which can function in phase correlation with proper pulsation of the star at the last stages of its evolution. First of all, it can be the flaring up of the nuclear shell source, then the process of the neutronization can split into stages, and at last for non-Fe core (O-C, for example) the outer parts of the core can also flare up. It is possible that the explosion itself is encouraged by the resonance pulsation.

As for the SN type Ia precursors some other mechanisms are certainly to be considered. However the close systems buried in a common envelope (or one compact object inside the extended atmosphere of the other component) is in some sense indistinguishable from the rotating pulsating star. The forming circumstellar environs may occur in a similar way. Perhaps, the eccentricity of the orbit in binary system can be treated as a trigger in generating instabilities and resonance effects.

The recurrent ejections of matter with different velocities can result in formation of axial colliding wind formed structures. A very energetic individual pulsation of supernova precursor just before the explosion event produces a shell, perhaps, discovered in the early spectra of SN 1984E.

## References

Dopita, A.M., Evans, R., Cohen, M., & Schwarts, R.D. 1984, ApJ, 287, L69

Niemela, V.S., Ruiz, M.T., & Phillips, M.M. 1985, ApJ, 289, 52

Polcaro, V.F., & Viotti, R. 1991, A&A, 242, L9

Tsiopa, O.A. 1990, Astronomicheskj Tsirkular, No. 1554, 23

Tsiopa, O.A. 1995, in: K.A. van der Hucht and P.M. Williams (eds.), Wolf-Rayet Stars: Binaries, Colliding Winds, Evolution (Dordrecht:Kluwer), 425

*Astrophysical Applications of Stellar Pulsation*
*ASP Conference Series, Vol. 83, 1995*
*R. S. Stobie and P. A. Whitelock, eds.*

# Infrared Radial Velocity Results for Long Period Variables

Kenneth H. Hinkle

*Kitt Peak National Observatory*[1], *National Optical Astronomy Observatories, 950 N. Cherry Avenue, P.O. Box 26732, Tucson, Arizona 85726, U.S.A.*

**Abstract.** High-resolution time-series infrared spectra have been observed for a total of 20 Long Period Variables. The stars can be placed in four groups with distinct kinematic behaviour.

## 1. Introduction

In a series of papers (Hinkle 1978; Hinkle et al. 1982; Hinkle et al. 1984, Hinkle & Barnbaum 1995; Hinkle et al. 1995) we have explored the photospheric kinematics of long period variables (LPVs) measured in the infrared. It is difficult to observe the kinematics of LPV stellar pulsation using visible spectra. The 1.5-2.5 $\mu$m infrared has been used because it spans the opacity minimum, is relatively unblended, and offers a variety of molecular bands (especially the CO overtone bands) which make wonderful probes of the atmospheric structure.

## 2. Discussion

The 1.6-2.5 $\mu$m infrared spectra were obtained using the KPNO 4 meter telescope and Fourier transform spectrometer (FTS). The sky subtraction of this instrument (Hall et al. 1979) allows daytime observations of bright objects unaffected by the sky spectrum. As a result of being able to use daytime, a large amount of telescope time was available for monitoring projects. Nearly all observations discussed here were obtained during daylight. High-resolution time-series infrared spectra have now been observed for a total of 20 LPVs, 14 of which are Miras, 5 SRa, and 1 SRb, spanning the M, S and C abundance types.

All the Mira velocities when plotted against visual phase show good velocity repeatability from cycle-to-cycle. However, cyclic differences of up to ~2 km s$^{-1}$ from the mean occur, with the degree of scatter from the mean apparently being a function of phase. Maximum visual light roughly corresponds to the time of most negative (largest outward) radial velocity. The apparent deceleration from approximately phase 0.2 to 0.8 is approximately constant and similar in all Miras studied with periods spanning about about a factor of two in length. Comparing velocity versus phase plots for the Miras, an obvious feature that differs from star-to-star is the phase when the hot expanding material is

[1]Operated by AURA, Inc. under cooperative agreement with the National Science Foundation

first visible. In the spectrum this corresponds to doubling of the lines. In the prototypical Miras ($o$ Cet, $\chi$ Cyg) infrared line doubling is observable at phase 0.9 or shortly after. The velocity curves of R And, R Aql, R Cas, S Cep, R Leo, U Ori, and IK Tau are similar to the prototype stars. The velocity amplitude of these stars is ~25 km s$^{-1}$.

In some Miras line doubling can occur as early as 0.3 of a period before visual maximum. The appearance of the expanding gas at an early phase seems to be correlated with the conspicuous premaximum bump in the light curve. An excellent example is T Cep. Interestingly, while the lines double at an early phase they persist double until about the same phase as for the prototype Miras, i.e., doubled lines disappear shortly after maximum light. The appearance of the expanding gas must depend on the opacity of the overlying, infalling material. We have been unable to ascertain if this opacity is the result of simply a different amount of material above the shock or results from the effect of different abundances of metals or CNO. We have now identified three Miras in this group, R Aur, T Cep, and R Hya.

The third group of stars that can be discerned from the velocity measurements are Miras that have visual amplitudes near 2.5 magnitudes. SV Cas, W Hya, and X Oph are in this category. These stars have velocity curves which are essentially identical to the above Miras with the expanding gas visible at 0.3 of a period before maximum. However, line doubling is detectable only briefly with the result that infalling material disappears from the spectrum shortly after minimum light.

The SRa variables, while long considered a rather arbitrary division of the Miras by amplitude of the light curve, can be shown from the velocity results to be a separate group of variables. We have observed three members of this group, V450 Aql, RU Cyg, and SV Peg. These stars never show line doubling. The largest amplitude system we have observed, RU Cyg, has an 8 km s$^{-1}$ velocity amplitude, similar to the velocity range of the low amplitude Miras. We have also observed in some detail one SRb variable, W Cyg. W Cyg similarly does not show line doubling, has a velocity amplitude of 3.5 km s$^{-1}$, and, as might be expected, has rather irregular velocity variations. As in the Miras, in addition to regular pulsations the SR star velocities undergo non-repeatable velocity variations of a few km s$^{-1}$ amplitude. For small amplitude SR variables the non-cyclic velocities can mask the stellar pulsation.

## References


Hall, D. N. B., Ridgway, S. T., Bell, E. A., & Yarborough, J. M. 1979, Proc. Soc. Photo-Opt. Instr. Eng., 172, 121

Hinkle, K. H. 1978, ApJ, 220, 210

Hinkle, K. H., & Barnbaum, C. 1995, in preparation

Hinkle, K. H., Hall, D. N. B., & Ridgway, S. T. 1982, ApJ, 252, 697

Hinkle, K. H., Scharlach, W. W. G., & Hall, D. N. B. 1984, ApJS, 56, 1

Hinkle, K. H., Scharlach, W. W. G., & Johnson, D. 1995, in preparation.

*Astrophysical Applications of Stellar Pulsation*
*ASP Conference Series, Vol. 83, 1995*
*R. S. Stobie and P. A. Whitelock, eds.*

# Studies of Small-Amplitude Red Variables

John R. Percy, Winnie Au, Adrien Desjardins and Lawrence Yu

*Erindale College, University of Toronto, Mississauga, ON Canada L5L 1C6*

AAVSO Photoelectric Photometry Program

*25 Birch St., Cambridge MA 02138-1205 USA*

**Abstract.** We describe a survey of suspected small-amplitude red variables, and long-term (up to 10 years) monitoring of about two dozen such variables.

## 1. Introduction

Small-amplitude red variables (SARVs)' are M giants and supergiants which are pulsating with small amplitudes (up to 2.5 mag) and with periods of 20 to 200 days or more. Most SARVs are bright stars. There are 164 known and 136 suspected SARVs among the approximately 500 M giants and supergiants in the Bright Star Catalogue. SARVs are believed to be part of a sequence of red variables extending from the so-called $\sigma$ Librae stars with periods of 10 to 20 days, to the large-amplitude, long-period Mira stars. Many SARVs have a well-defined period, but the nature and cause of the irregularity (found in all of these stars) is not clear, nor is the pulsation mode. Many of these stars have a much longer secondary period, the nature of which is also not clear.

## 2. Observations

Standard photometric techniques were used, with a comparison and a check star for each variable. The internal scatter of the data ranges from 0.007 to slightly greater than 0.010.

1. We are engaged in an ongoing survey of known and suspected SARVs using: (a) the 0.4m teaching telescope of the University of Toronto; (b) the 0.25m Automatic Photometric Telescope on Mt. Hopkins AZ (Percy & Au 1994); (c) the AAVSO Photoelectric Photometry Program (Percy et al. 1994 and references therein).

2. Many SARVs have been monitored for up to a decade as part of the AAVSO's permanent photoelectric photometry program. The results presented in the table have been obtained by: (a) inspection of the light curves; (b) date-compensated discrete Fourier transform (Ferraz-Mello 1981) using a computer program kindly provided by E.P. Belserene; (c) autocorrelation analysis (Percy et al. 1993).

## 3. Results

For the stars in the survey program, the incidence and amplitude of variability increase with later spectral type. No such correlation is apparent in the monitoring program, probably because the stars in this program have been chosen for their large amplitude. Most of the stars in the monitoring program show clear periodicity. W Boo appears to be a double-mode pulsator. At least half of the stars show a long period, typically an order of magnitude longer than the "short" period. There is no clear correlation between period and spectral type. Stars of the same spectral type can have "short" periods differing by an order of magnitude. This may be partly because of the complexity of the variability, or because of luminosity differences between the stars. Or it may be due to the excitation of different modes in different stars.

| Star | Sp T | GCVS | | | Results | |
|---|---|---|---|---|---|---|
| | | Type | Range | Period | ΔV | Period(s) |
| TV Psc | M3-4 III | SR | 0.77V | 49.1 | 0.80 | 53.11, 1400: |
| RZ Ari | M6 III | SRb | 0.39V | 30 | 0.50 | 56:, 500: |
| $\rho$ Per | M4 II-III | SRb | 0.70V | 50: | 0.40 | 135, long: |
| CE Tau | M2 Iab-Ib | SRc | 0.31V | 165 | 0.65 | 272, 1200: |
| $\alpha$ Ori | M1-2 Iab | SRc | 1.3V | 2335 | 0.65 | 200:, 2000: |
| $\eta$ Gem | M3 III | SRa+ | 0.75V | 232.9 | 0.55 | 239.8 |
| V614 Mon | C4,5 (R5) | SRb | 0.35V | 60: | 0.45 | 80, long: |
| RS Cnc | M6e Ib-II | SRc: | 1.5p | 120: | 1.40 | 228, long: |
| IN Hya | M4 III | SRb | 0.60V | 65: | 0.90 | 100-150: |
| VY UMa | C6.3 (N0) | Lb | 1.13V | – | 0.50 | 200:: |
| VW UMa | M2 | SR | 0.86V | 610 | 0.5: | 50: |
| FS Com | M5 III | SRb | 0.8V | 58: | 0.55 | 55.51, long: |
| FP Vir | M4 III | SRb | 0.63V | 40: | 0.80 | 70.1, long: |
| W Boo | M2-4 III | SRb: | 0.67V | 450: | 0.40 | 24.71, 56.67 |
| ST Her | M6-7 III | SRb | 1.5p | 148.0 | 1.05 | 500: |
| R Lyr | M5 III | SRb | 1.12V | 46: | 0.60 | 53.41, long: |
| EU Del | M6.4 III | SRb | 1.11V | 59.7 | 0.95 | 62.74 |
| U Del | M5 II-III | SRb | 1.3B | 110: | 0.80 | 90, 1000 |
| W Cyg | M4-6e III | SRb | 2.1B | 131.1 | 1.20 | 120:, 260: |
| V1339 Cyg | M3-6 | SRb | 1.2V | 35: | 0.40 | 50, 800: |
| $\mu$ Cep | M2e Ia | SRc | 1.67V | 730 | 0.75 | 850 |
| $\chi$ Aqr | M3 III | Lb | 0.16V | – | 0.40 | 35, 250: |
| TX Psc | C7,2 (N0) | Lb | 0.41V | – | 0.50 | 144 and/or 278 |
| XZ Psc | M5 II | Lb | 0.36V | – | 0.45 | 52.96 |

**Acknowledgments.** We thank the AAVSO and its photoelectric observers for their participation, and NSERC Canada for a research grant.

## References

Ferraz-Mello, S. 1981, AJ, 86, 619

Percy, J.R., & Au, W. 1994, IAU IBVS No. 4114

Percy, J.R., et al. 1993, PASP, 105, 287

Percy, J.R., et al. 1994, PASP, 106, 61

*Astrophysical Applications of Stellar Pulsation*
*ASP Conference Series, Vol. 83, 1995*
*R. S. Stobie and P. A. Whitelock, eds.*

# Pulsation & Long-Periods in Three Nearby M Supergiants

M. A. Smith, T. J. Teays, L. L. Taylor, R. Wasatonic
*Computer Sciences Corp., 10000A Aerospace Rd., Lanham MD 20706*

E. F. Guinan
*Astronomy Dept, Villanova Univ., Villanova, PA 19085*

and S. Baliunas
*Center for Astrophysics, 60 Garden St., Cambridge, MA 02138*

## 1. Introduction

We report on pulsational variability found in our decade-long study of UV and optical RV, light, K-line data sets for $\alpha$ Ori, $\alpha$ Sco A, and $\alpha$ Her A. As summarized by Smith et al. (1990), our radial velocities are derived from McMath observations of five metallic lines near H$\alpha$ plus H$\alpha$ itself. Optical light curves were obtained by small photometric telescopes, and K-line emissions from the Mt. Wilson K-line project. UV continua at 3000Åand k/h emission indices were obtained with the IUE LWP camera.

## 2. "Long periods"

That $\alpha$ Ori has a "long period" (better called cycles) has been known for 80 years. Villanova photometry shows that during the last decade it has had a value of ~4 years. This modulation dominates the H$\alpha$ RVs and K-emission fluxes. The resulting correlation is that strong K-emission occurs when positive H$\alpha$ RVs and faint light do. $\alpha$ Sco shows a similar "long period" of ~6 yrs. between metallic, H$\alpha$ RVs and K-fluxes. $\alpha$ Her shows a ~4 yr. variation between H$\alpha$ and K.

The general relation between H$\alpha$ and K indices suggests an association with magnetic activity. These correlations might arise from solar-like supergranulations into which boundaries the downdrafts flow and magnetic fluxes intensify.

## 3. Pulsations

Dupree et al. (1987, 1990) discovered a 1.1-1.2 yr. period in $\alpha$ Ori from UV and optical indices. The phasing of our RV curve with their light curve confirms their interpretation of fundamental-mode radial pulsation. This period is most visible in the optical light and RV curves. A PDM diagram of our Villanova photometry for $\alpha$ Ori shows a wandering of this periodicity between 0.85 and 1.1 yrs. An examination of the long-period subtracted RVs for all three star shows that a best-fit sinusoid fits the data in the main but with important departures. The

RV curves show am/fm modulations, evidence of period-doubling, and missing cycles followed by a resumption of the previous phasing.

$\alpha$ Her shows dramatic departures from a simple oscillation. Although the RV curve shows a primary 0.9 yr. period, the optical light curve shows a complex behaviour which demands continuous coverage during the cycle even to define a mean brightness level. After a brief lull, the optical light dims, brightens, and dims a second time. During strong cycles (1994) even a secondary maximum and tertiary minimum can appear. These features are similar to the "extra bumps" that can appear in the rising or falling branch of the light curves of Mira variables and are attributable to atmospheric heating by multiple shocks.

Where do the multiple shocks come from? Bowen's (1992) HD models provide a possible answer for strong fundamental pulsations. His simulations often develop a series of three unevenly spaced shocks which are produced by a 3:1 near-resonance of pulsation energy near the atmospheric acoustic cut-off of ~90 days. The shocks propel material outward such that it returns 90 days later – which is indeed the value of the bump spacings observed in $\alpha$ Her's optical light curve! These models predict that the "short periods" are less likely to develop for stars of increasing mass. We note that $\alpha$ Her falls in between $\alpha$ Ori and the Mira variables in the H-R Diagram and seems to show hybrid features.

## 4. Pulsational shocks

The Mg II h/k lines are strong coolants in the shocks generated by strong pulsations in late stars. The phase lag between k-emission maximum and optical light maximum minimum can be used to estimate the distance the shocks transit before dissipating. The k-emission curves lag consistently $0.4 \pm 0.1$ cycles behind the RV curves (0.0 @ RV(max)).

Although the UV continuum light curves are nearly sinusoidal for $\alpha$ Ori and $\alpha$ Sco, the waveform for $\alpha$ Her shows a square-wave maximum. The brightening first appears ~50 days *before* the optical light maximum, during the first optical minimum. It attains the same level and remains constant for the next for 4 months. The light maximum ends abruptly just as the k-emission flux is reaching its maximum, signalling a transition of optical thickness to thinness.

Our observations suggest that that the onset of a pulsation cycle in $\alpha$ Her is first visible in the Balmer continuum when an optically thick shock emerges in the lower photosphere. At this time the product ${T_{eff}}^4R^2$ is not yet at its maximum so the UV shock precedes the bright star phase. Likewise, the shock is still thin in the Paschen continuum, so optical observers still "look through" the shock (Luttermoser, priv. com.). When the shock finally dissipates and becomes optically thin in the near UV, h/k emission grows to importance.

## References

Bowen, G. 1992, in: Instabilities in Evolved Supergiants, ed. C. De Jager (Amsterdam: N. Holland), p. 104

Dupree, A., et al. 1990, ASP Conf. Series 11, p. 4

Smith, M. A., Patten, B. M., & Goldberg, L. 1990, AJ, 98, 2233

*Astrophysical Applications of Stellar Pulsation*
*ASP Conference Series, Vol. 83, 1995*
*R. S. Stobie and P. A. Whitelock, eds.*

# Mode Switching in Red Semiregular Variable Stars?

J. Gál and K. Szatmáry

*Astron. Obs. and Dept. of Exp. Physics, JATE University, Dóm tér 9, H-6720, Szeged, Hungary, Internet: J.Gal@physx.u-szeged.hu*

**Abstract.** Light curve analysis of the carbon variable RY Dra and the oxygen-rich variables TX Dra and AF Cyg is summarised. In each case, two shorter periods were found (which may be connected with the radial pulsation of the stars) and also a long term variation in the mean brightness. In the case of RY Dra, the amplitudes of the shorter periods both decreased while the amplitude of the long secondary variation was found to be increasing. The light curves and frequency spectra of TX Dra and AF Cyg can be interpreted as a result of mode switching. These stars pulsate with two different periods and there are time intervals when the longer period disappears and the shorter one remains dominant, and it lasts for many cycles. We concluded that the double and multi-periodic red semiregular variables may be very important from both theoretical and observational point of view, because, based on the ratio of the periods, we can identify the pulsational modes in these variables.

Semiregular variables are red giant or supergiant stars with spectral types M, C or S. They are located on the Asymptotic Giant Branch of the H-R Diagram.

Here we present a short summary of the light curve analysis of three interesting red variables. Due to the lack of photometric data, we have used visual observations taken from different data bases. For the light curve analysis we constructed mean light curves by averaging the data in 10-day bins. The errors of these mean points were generally 0.2 mag. It is important to note that the observations from the different sources did not show systematic differences.

The periods were determined using the well-known discrete Fourier transformation. Since the periods and the amplitudes are subject to variations in all three program stars the windowed Fourier transformation was used to study the different parts of the light curves.

In the case of RY Dra the mean light curve contains 1079 points between JD 2433880 and 2449560. Unfortunately there is a long gap (JD 2437600-41930) with only a few observations. The data were divided into three segments (based on the light curve shape) and these segments were analysed separately. In the first part (JD 2433880-2437600) two short periods are detectable: 259.3 and 155.4 days. The shape of the light curve is rather different about four thousand days later which means that the amplitudes of the short periods decrease and a new slow oscillation develops; its characteristic time scale is around 900 days. After JD 2446000 the light curve is dominated by a slow variation.

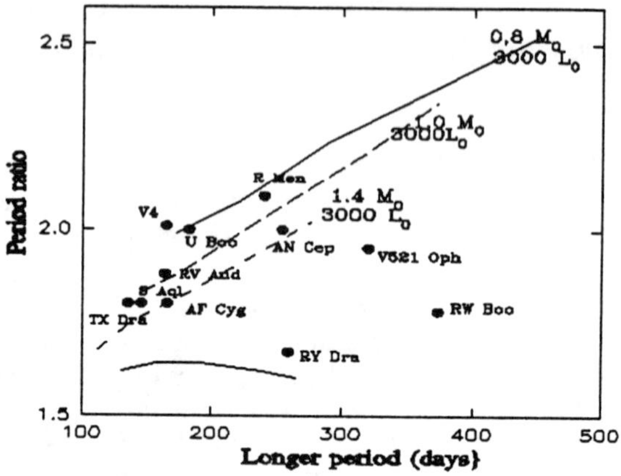


Figure 1. Longer period vs. period ratio diagram for some red semiregular stars. The lines represents the model calculations of Ostlie & Cox

The other two stars (TX Dra and AF Cyg) show a rather complex light variation. The shape of the light curves are strongly variable. In both cases there are time intervals where the brightness is approximately constant. It is important to note that TX Dra has a stable long secondary variation (700 days) which is probably not due to pulsation.

Due to the limited space the detailed description of the light curves is not presented here. The light curves are characterised by two short periods; however the amplitudes of these periods are strongly variable. In fact there are time intervals where both short periods are present and there are time intervals where the longer period disappears and the shorter one remains dominant, and it lasts during many cycles. These light variations may be connected with mode switching.

Mode switching was found in three red semiregular variables by Cadmus et al. (1991), although the general characteristic of light variation of Cadmus's stars are different from our program stars.

Since all three variables have two short periods which may connected with the radial pulsations of the stars, based on period ratios and model calculations of Ostlie & Cox (1986) we can identify these periods as pulsational modes. In Fig. 1 we plotted the period ratio as a function of the longer period. The lines represent the models of Ostlie & Cox. The figure suggests that the fundamental and the first overtone modes may be excited in these variables.

**Acknowledgments.** This work was supported by the Hungarian OTKA Grants No. T7522 and F7318.

## References

Cadmus, R.R., Willson, L.A., Sneden, C., & Mattei, J.A. 1991, AJ, 101, 1043
Ostlie, D.A., & Cox, A.N. 1986, ApJ, 311, 864

*Astrophysical Applications of Stellar Pulsation*
*ASP Conference Series, Vol. 83, 1995*
*R. S. Stobie and P. A. Whitelock, eds.*

# Blackbody Fits of Semiregular Variables

F. Kerschbaum and J. Hron
*Institut für Astronomie der Universität Wien,*
*Türkenschanzstraße 17, A-1180 Wien, Austria,*
*Offprints: kerschbaum@astro.ast.univie.ac.at*

**Abstract.** Semiregular variables of type SRa and SRb are studied by fitting combinations of blackbodies to new near infrared and IRAS data.

In Kerschbaum & Hron (1992) a systematic investigation of the basic properties of type SRa and SRb Semiregular variables (SRVs) was presented. For the O-rich subgroup a new classification scheme into 'blue', 'red' and 'Mira' SRVs was one of the main outcomes. Concerning the physical properties an independent study by Jura & Kleinmann (1992) arrived at similar conclusions, except for the long-period SRVs. New JHKL′M-photometry (Kerschbaum & Hron 1994, Paper II) supports independently the classification into 'red' and 'blue' SRVs.

As a next step, visual information is taken from the GCVS4, near infrared JHKL′M-photometry from Paper II; mid- and far-infrared measurements were extracted from the IRAS-PSC. The resulting energy distributions can be reasonably well fitted for almost all objects by a combination of two blackbody curves. The overall spectra can be roughly described by a 'photospheric' temperature $T^*$, a 'dust'-temperature $T^d$ and the 'size' of the circumstellar shell relative to the star $r = R^d/R^*$.

The resulting parameters $T^*$, $T^d$ and $r$ of course need to be related to astrophysically meaningful quantities. $T^*$ can be compared with a $T_{\text{eff}}$ derived from the synthetic H–K colours given by Bessell et al. (1989). For $T^*$ above 2100 K only a constant offset of about 400 K is found for our O-rich SRVs. For the parameters related to the circumstellar material, namely $T^d$ and $r = R^d/R^*$ the situation is more complex. A comparison of BB-Fits with the results of dust shell models by Le Bertre (1988) and Le Sidaner & Le Bertre (1993) shows that our $T^d$ and $r$ both correspond to almost the same region of the innermost part of the dust shell.

Fig. 1 shows the blackbody fit parameters for 'red' SRVs and C- and S-SRVs. The different energy distributions show up clearly. All 'blue' objects can be reasonably well approximated by only *one* blackbody whereas the 'red' and the 'Mira' SRVs need *two*. The $T^*$ values, reflecting mainly the effective temperatures for objects with only small mass loss, are typically 300 K higher in the 'blue' than in the 'red' SRV cases. With a few extreme exceptions most of the 'red' $T^d$ values are concentrated around 400 K whereas the 'Mira' like Semiregular variables show a wider distribution between 330 and 570 K. C-rich objects and S-stars differ significantly from the O-SRVs in the fit parameters. Sometimes 'unphysically' low $T^*$ are found — a result of circumstellar reddening in the high mass loss cases. Also lower values of $T^d$, accompanied by normal $T^*$

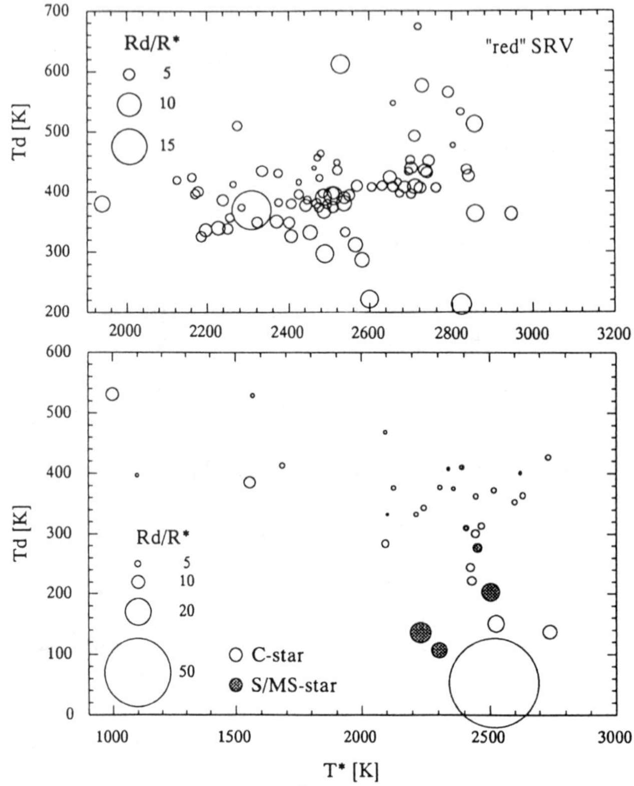


Figure 1. 'Dust'-temperature $T^d$ as a function of 'photospheric' temperature $T^*$ with the relative 'size' of the dust shell $R^d/R^*$ indicated by the diameter of the plot symbols.

and large values of $r$ are common — the most extreme example being TT Cyg with its detached envelope.

**Acknowledgments.** This work was supported by the ASF under project numbers P7252-PHY and P9638-AST.

## References

Bessell M.S., Brett J.M., Scholz M., Wood P.R. 1989, A&AS 77, 1

Jura M. 1988, ApJS 66, 33

Jura M., Kleinmann S.G. 1992, ApJS 79, 105

Kerschbaum F., Hron J. 1992, A&A 263, 97 (Paper I)

Kerschbaum F., Hron J. 1994, A&AS 106, 397 (Paper II)

Le Bertre T. 1988, A&A 190, 79

Le Sidaner P., Le Bertre T. 1993, A&A 278, 167

*Astrophysical Applications of Stellar Pulsation*
*ASP Conference Series, Vol. 83, 1995*
*R. S. Stobie and P. A. Whitelock, eds.*

# The Long-term Variation in the RV Tauri Star U Mon

Karen R. Pollard[1] and P. L. Cottrell
*Mt John University Observatory, Department of Physics & Astronomy, University of Canterbury, Christchurch, New Zealand*

## 1. Introduction

The RV Tauri stars are semiregular pulsating variables located in the brightest part of the Cepheid II instability strip. They have a characteristic light curve of alternating deep and shallow minima. A subset of the RV Tauri stars (the RVb subclass) exhibit long-term (500 to 2600 day) light and radial velocity variations. Although it is well established that the short-term variations are due to pulsations, the long-term behaviour is not well understood.

## 2. Observations and results

*BVRI* photometry and high-resolution spectra of U Mon (the brightest member of the RVb subclass) were obtained at the Mt John University Observatory (MJUO) between 1990 Aug and 1994 May. The light and colour curves obtained clearly show the long-term variation in U Mon (Fig. 1(a) and (b)). The reddest colours occur slightly later than the long-term minimum in the light curve. The short-term light and colour variations are 'damped' at the long-term minimum.

The MJUO radial velocities were analysed together with those previously published for U Mon in order to obtain an orbital solution. The derived orbital period of 2597 d is in reasonable agreement with the visual light period of 2475 d (Percy 1993). The orbit is eccentric ($e = 0.43$) with periastron at HJD 2448349.

Strong H$\alpha$ emission is seen around pulsational phases 0.15 and 0.65, corresponding to when the propagation of shock waves through RV Tauri star photospheres is observed to occur (Lèbre & Gillet 1991). Additional H$\alpha$ emission is seen during HJD 2448400 to 2448600 (Fig. 1(c)), when U Mon is faint, reddened and is closest to the earth in its orbit.

## 3. Discussion

Fokin (1994) suggests that the long-term variation in the RVb stars may be due to previously-ejected dust eclipsing the star at certain orbital phases. Using a model proposed for the HR4049 system by Waelkens & Waters (1993), Percy (1993) suggests that the long-term variation in U Mon may be due to the star being eclipsed by the inner edge of a dust torus which surrounds the binary system. The light decrease and the reddening of the colours in U Mon

[1]Present address: South African Astronomical Observatory, PO Box 9, Observatory 7935, Cape.

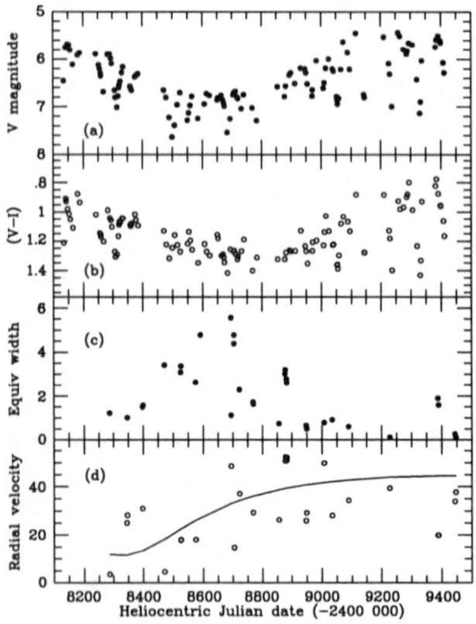


Figure 1. MJUO observations of U Mon: (a) the $V$ light curve; (b) ($V$–$I$) colour curve; (c) H$\alpha$ equivalent width; (d) radial velocity curve, showing MJUO radial velocities (∘) and the orbital solution (—).

appears consistent with both models. One effect that is difficult to explain by either model is the 'damping' of the short-term light and colour variations at long-term minimum.

An alternative explanation for the long-term variation is that the binary system involving U Mon is *interacting* at certain orbital phases. The enhanced H$\alpha$ emission is attributed to mass loss or transfer which occurs shortly after periastron due to the effect of the binary interaction. The pulsational variations will be damped at this orbital phase. U Mon appears reddened due to the presence of gas and dust from a previous or current mass loss event, the binary interaction or possibly tidal forces. It is quite possible in a system where circumstellar is known to exist, that a combination of obscuration and interaction mechanisms is present.

**Acknowledgments.** The authors wish to thank P. Kilmartin & A. Gilmore for obtaining the photometry and M. Albrow for obtaining a number of spectra.

## References

Fokin, A.B. 1994, A&A, 292, 133

Lèbre, A., & Gillet, D. 1991, A&A, 246, 490

Percy, J.R. 1993, in: Luminous High-Latitude Stars, D.D. Sasselov, ASP Conf. Series 45, 295

Waelkens, C., & Waters, L.B.F.M. 1993, in Luminous High-Latitude Stars, D.D. Sasselov (ed.), ASP Conf. Series 45, 219

*Astrophysical Applications of Stellar Pulsation*
*ASP Conference Series, Vol. 83, 1995*
*R. S. Stobie and P. A. Whitelock, eds.*

# Pulsation Properties of a Sample of Mira and OH/IR Stars

M. J. Gaylard and M. E. West

*Hartebeesthoek Radio Astronomy Observatory, P O Box 443, Krugersdorp 1740, South Africa*

**Abstract.** Well-sampled light curves have been obtained for 1612-MHz hydroxyl masers in Mira and OH/IR stars monitored with the Hartebeesthoek radio telescope. Pulsation modes are investigated via the ratios of the observed periods in the maser variations.

## 1. Introduction

The ranges of stellar masses and photospheric radii possible in Miras and OH/IR stars are such that pulsation could be occurring in the fundamental mode or the first or second overtones.

Stellar radii and masses are free parameters in theoretical pulsation models, but accurate distances are needed to derive the radii and masses of actual stars. The 1612 MHz hydroxyl masers in OH/IR stars provide a method for determining individual stellar distances. The maser intensity follows the luminosity changes of the central star, but with a phase lag across the maser profile owing to the light travel time from the back to the front of the masing shell. Measurement of the phase lag gives the linear size of the shell. The ratio of the linear size to the angular size, determined interferometrically, then gives the distance to the star (van Langevelde et al. 1990).

A programme to determine distances to a set of Miras and OH/IRs using this method was initiated in 1985. Fortnightly or weekly observations with the aim of measuring the phase lag across the OH shell are carried out with the 26-m Hartebeesthoek telescope. The MERLIN array is used to measure the angular sizes of the OH shells.

## 2. Pulsation mode determination via period ratios

The OH monitoring data can also be used in an alternative method for determining the pulsation modes which does not require the stellar distances to be found. The ratio of the observed periods $P$ of the primary and first overtone are compared to the ratios of the fundamental $P_0$ and first overtone $P_1$ and the first and second overtone $P_2$ predicted by models. The theoretical ratios form a set of curves as a function of period for each modelled mass (Ostlie & Cox 1986).

The variations of the masers from eight OH/IR stars in our sample were Fourier transformed to enable significant periodicities to be identified in the time series (Scargle 1982). No overtones were found in the power spectra of IK Tau

($P = 470$ days), IRC–20197 ($P = 640$ days) and OH358.2+0.5 ($P = 1230$ days). The periods and period ratios of the other stars are summarised in Table 1.

Table 1. Secondary periods $P_{Sn}$ with significance $\geq 2\sigma$ present in the power spectra of the OH maser variations, and their ratios to the primary period $P_P$.

| star | $P_P$ days | $P_{S1}$ days | $P_{S2}$ days | $P_P$ / $P_{S1}$ | $P_P$ / $P_{S2}$ |
|---|---|---|---|---|---|
| OH338.1+6.4 | 645 | 400 | – | 1.61 | – |
| WX Psc | 650 | 860 | 516 | 0.76 | 1.26 |
| OH1.3+1.0 | 945 | 545 | – | 1.73 | – |
| OH1.1–0.8 | 1310 | 720 | 540 | 1.82 | 2.43 |
| OH357.3–1.3 | 2000 | 1030 | – | 1.94 | – |

For OH338.1+6.4 the sidelobe levels of the window function suggest that the apparent secondary period is a sampling artifact. WX Psc has the longest span of data and is the only object showing a significant period longer than the primary but distinguishable from long term trends.

The observed ratios $P_P$ / $P_{S1}$ are smaller than predicted by the Ostlie & Cox (1986) models if $P_P$ is identified with $P_0$. The predicted $P_1$ / $P_2$ is of the order of 75 per cent of $P_0$ / $P_1$, and is more consistent with the observed period ratios. Clearly this is dependent on the applicability of the models. There is a general trend for $P_0$ / $P_1$ to increase with increasing period, in agreement with the model predictions. The period ratio for OH357.3–1.3 comes closest to the predicted $P_0$ / $P_1$ ratio for fundamental mode pulsation. Pulsation in this mode would imply that the stellar mass is at least 5 $M_\odot$ and its radius is $\geq$1250 $R_\odot$. A switch from $P_1$ to $P_0$ as the star expands would be consistent with Wood's (1974) suggestion.

Fourier transforms of the OH maser luminosity variations show significant periods other than the primary in five out of eight cases. The period ratios are more consistent with the primary pulsation mode being the first overtone rather than the fundamental, with the possible exception of the 2000-day period OH357.3–1.3.

## References

Ostlie, D. A., & Cox, A. N. 1986, ApJ, 311, 864

Scargle, J. D. 1982, ApJ, 263, 835

van Langevelde, H. J., van der Heiden, R., & van Schooneveld, C. 1990, A&A, 239, 193

Wood, P. R. 1974, ApJ, 190, 609

*Astrophysical Applications of Stellar Pulsation*
*ASP Conference Series, Vol. 83, 1995*
*R. S. Stobie and P. A. Whitelock, eds.*

# A New, Systematic Search at 1612 MHz for OH/IR Stars in the Inner Galaxy

M.N. Sevenster[1], H.J. Habing[1], J.M. Chapman[2], N.E.B. Killeen[3]

[1] *Leiden Observatory, P.O.Box 9513, 2300 RA Leiden, The Netherlands*

[2] *AAO, P.O.Box 296, Epping NSW 2121, Australia*

[3] *ATNF, P.O.Box 76, Epping NSW 2121, Australia*

## 1. Introduction

We have surveyed the inner part of the Galaxy for OH 1612 MHz maser emission from OH/IR stars. The full survey is expected to give around 2000 detections. Here we report initial results for the Bulge region between $|l| < 10°$ and $|b| < 3°$.

## 2. Data

The observations discussed here were taken with the Australia Telescope Compact Array in Oct-Nov 1993. For each field we used a total integration time of 8 min. spread over an hour-angle range of 12 hours in intervals of an hour. Interference in the 1612 MHz band is a major concern in this program - it is ubiquitous for the shorter baselines. Another concern is the beam-pattern resulting from taking short cuts spread over 12 hours, with a one dimensional array. This causes sidelobe patterns which can easily lead to spurious detections. In addition to these problems, the size of the data set calls for special reduction techniques. We have developed a non-interactive method that is not I/O intensive and is well suited for use on supercomputers. To identify the OH/IR stars, all fields are 'skimmed' at successively weaker levels, subtracting point sources at the positions of the highest peaks in the fields, until we reach the (known) 3 $\sigma$ level. Our method reduces the data reduction time by a factor of eight whilst giving comparable results to conventional cleaning methods. The disk space required is reduced by a factor of 100.

The results shown here are for detections above $4\sigma$ ($\sim$ 150 $-$ 200 mJy). The number of OH/IR stars found is 202, of which 104 have double peaks with outflow velocities between 4 km s$^{-1}$ and 30 km s$^{-1}$. The mean outflow velocity of the 104 stars is 14 km s$^{-1}$. These numbers are consistent with values obtained from other observations (e.g., te Lintel Hekkert et al. 1989; 1991 and references therein). In Fig. 1 the 202 stars are plotted in longitude versus latitude. It should be noted that most of the single-peaked stars are detected in a second peak in the next pass, since this peak is weaker. We do not show the results of the last pass here since it has not been fully reduced. The stellar velocity is only known for double-peak OH/IR stars. They form a system with solid body rotation of $\sim$ 12 km s$^{-1}$ kpc$^{-1}$ and a velocity dispersion along the line of sight of $\sim$ 100 km s$^{-1}$. We derived the same quantities for 63 double peaked stars that are closer to the plane - $|b| < 1°$. Their solid body rotation is much higher, $\sim 25$ km $s^{-1}$ $kpc^{-1}$ , and the velocity dispersion is somewhat lower, $\sim 90$ km $s^{-1}$.

Statistically, the outflow velocity is a measure of the age of an OH/IR star. There is a relation between outflow velocity, stellar luminosity and metallicity (or gas/dust ratio $\mu$ ) of the shell :

$$\mu \propto L_*^{0.5} v_{exp}^{-2}$$

On average, a lower outflow velocity will indicate an older star. Another quantity that is statistically related to the age of a star is its latitude. An older star will have more time to move out of the galactic plane by means of diffusion. So we expect that with latitude the average outflow velocity and the average luminosity of the stars decreases. This effect is indeed seen, as in previous samples of OH/IR stars (e.g., te Lintel Hekkert et al. 1989; 1991). In Fig. 2 the OH-fluxdensity is plotted as a function of latitude to indicate this. For sources in the galactic Bulge, the OH-fluxdensities are not strongly dependent on the distances, and so can be used as a statistical measure of luminosity.

The total number of double peaked sources we expect to find in the Bulge region is 400, three sources per square degree. Next to that there may be some 50 single peaked sources. With the sample of double peaked stars a dynamical analysis of the inner region of the Galaxy will be made.

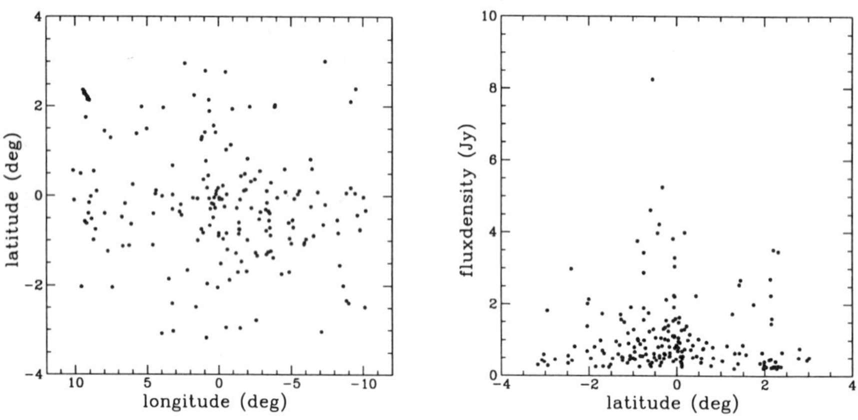


## References

te Lintel Hekkert, P., et al. 1989, A&AS, 78, 399

te Lintel Hekkert, P., et al. 1991, A&AS, 90, 327

*Astrophysical Applications of Stellar Pulsation*
*ASP Conference Series, Vol. 83, 1995*
*R. S. Stobie and P. A. Whitelock, eds.*

# Pulsations of Long-Period Variables and their Connection to Variations of Circumstellar Maser Emission

G. M. Rudnitskij

*Sternberg Astronomical Institute, Moscow State University, Moscow, 119899, Russia, gmr@sai.msk.su*

**Abstract.** I consider maser emission in the $H_2O$ spectral line $\lambda = 1.35\,\mathrm{cm}$ from long-period variable stars. Some models for $H_2O$ maser variability are suggested.

Many of long-period variables (LPVs – Miras and semiregulars) emit in maser spectral lines of molecules at microwaves. I consider in more detail $H_2O$ masers, common in variable oxygen-rich red giants.

I use data of LPV monitoring in the $H_2O$ line, obtained on the 22-meter radio telescope in Pushchino jointly with our colleagues from the Lebedev Physical Institute, Russian Academy of Sciences. For earlier results see Berulis et al. (1983). Literature data are used, too. The curves of the $H_2O$ line flux variability $F(t)$ were compared to optical light curves of the stars (Berulis et al. 1984, Rudnitskij 1987, Kudashkina & Rudnitskij 1988). In many cases (R Aql, R Leo, U Ori, U Her), there is a good correlation of the $H_2O$ line flux density with the optical brightness. There is some delay of the $F(t)$ curves with respect to the optical curves, by about $0.1 - 0.3P$ ($P$ is the period of light variations). I consider the following mechanisms that may account for such a correlation.

*a. Direct impact of a shock on the maser generation region* (Rudnitskij & Chuprikov 1990). The mechanism involves nonequilibrium conditions just behind the shock front (with gas particles heated immediately, while dust grains remain cooler for some time). Thus, a nonequilibrium post-shock gas layer is formed; conditions in it are favourable for the maser.

*b. Amplification of the free-free radio continuum emitted by postshock ionized gas* (Rudnitskij 1990). In this mechanism, $H_2O$ molecules amplify the radio continuum at $\lambda = 1.35\,\mathrm{cm}$ from the ionized post-shock gas. This gas is optically thick down to wavelengths as short as 1 cm. Thus, its brightness temperature, $T_b = (15 - 50) \times 10^3\,\mathrm{K}$, is high enough to stimulate the $H_2O$ maser, especially at visual phases $\varphi = 0.1 - 0.3$, in accordance with the above-mentioned correlation between the $H_2O$ maser intensity and the optical light curve.

*c. Amplification of the nonthermal radio continuum.* The effect itself is analogous to the previous one. It may cause strong, short-lived bursts of the $H_2O$ maser, which would repeat the time shape of the radio continuum bursts. This is a more hypothetical possibility, although strong nonthermal radio continuum was observed, at least twice, in R Aql (Woodsworth & Hughes 1973, Estalella et al. 1983).

In many stars, the $H_2O$ line profiles are simple during most time, consisting of a single emission peak at roughly the stellar radial velocity $V_*$. But, almost in every star studied (e.g., R Aql, U Ori, U Lyn, W Hya, etc.), there were episodes when the $H_2O$ line profile became more complicated, with multiple features. Most of these features were blue-shifted with respect to $V_*$, with emission extending sometimes at 10 km/s blueward. Some of the features exceeded the main peak in flux density. The effect is most prominent in LPVs classified SR (RT Vir, RX Boo, W Hya). From a comparison with optical light curves, we found that such episodes might follow the brighter visual maxima.

We present a model, in which the $H_2O$ maser is excited by a shock wave, propagating with a velocity $D_0 \sim 6 - 10$ km/s in the inner part of the circumstellar envelope during each cycle of light variability. The profile of the $H_2O$ line and the maser strength depend on the shock velocity in a given light cycle. There are indications that the episodic appearance of blue-shifted maser features is more frequent in stars with bumps on the ascending branch of their light curves (e.g, U Her, U Lyn). In our previous work (Kudashkina & Rudnitskij 1995), we connected the bumps and their duration with propagation of the shock wave in the stellar atmosphere.

Another item is the asymmetry in spatial distribution of circumstellar maser emission, implying lack of spherical symmetry of the mass loss process (bipolar maser distribution in R Aql, U Ori). This also yields characteristic structures in the $H_2O$ line profiles (triple profile or pairs of double features). We connect this with probable nonradial pulsations of the star, resulting in an equatorial torus of circumstellar matter, while masers flow out mainly in the polar directions.

**Acknowledgments.** This work was supported by the Russian Foundation for Fundamental Research (Project Code 94-02-05763). The author gratefully acknowledges also the financial assistance from the International Science Foundation (Travel Grant No. 4546/4).

## References

Berulis, I. I., Lekht, E. E., Pashchenko, M. I., & Rudnitskii G. M. 1983, Soviet Ast., 27, 179

Berulis, I. I., et al. 1984, Nauchnye Informatsii, Astron. Council, Acad. Sci. USSR, 56, 92

Estalella, R., Paredes, J. M., & Ruis, A. 1983, A&A, 124, 309

Kudashkina, L. S., & Rudnitskij, G. M. 1988, Peremennye Zvezdy, 22, 925

Kudashkina, L. S., & Rudnitskij, G. M. 1995, Odessa Astron. Publ., Vol.7 (in press)

Rudnitskij G. M. 1987, in Circumstellar Matter, Proceedings of the 122nd Symposium of the International Astronomical Union, Held in Heidelberg, F. R. Germany 23-27 June 1986, I .Appenzeller & C. Jordan, Dordrecht: Reidel, 267

Rudnitskij, G. M. 1990, in Proceedings of the International Colloquium From Miras to Planetary Nebulae: Which Path for Stellar Evolution? Montpellier, France, September 4-7, 1989, M. O. Mennessier & A. Omont, Gif sur Yvette: Editions Frontières, 268

Rudnitskii, G. M., & Chuprikov, A. A. 1990, Soviet Ast., 34, 147

Woodsworth, A. W., & Hughes, V. A. 1973, Nature, Phys. Sci., 246, 111

*Astrophysical Applications of Stellar Pulsation*
*ASP Conference Series, Vol. 83, 1995*
*R. S. Stobie and P. A. Whitelock, eds.*

# Light Curve Analysis of Long Period Pulsating Stars with the Wavelet Method: The semiregular star V Boo

K. Szatmáry, J. Gál and L. L. Kiss

*Astron. Obs. and Dept. of Exp. Physics, JATE University, Dóm tér 9, H-6720, Szeged, Hungary, e-mail: K.Szatmary@physx.u-szeged.hu*

**Abstract.** We have collected long data series of some SR and Mira stars from data banks. In many cases the light curves show strong fluctuation or variation in the period, phase and amplitude. The detection of the significant changes of these parameters is important from the point of view of pulsation theory and stellar evolution. We present here the analysis of V Boo. The used method is the Fourier and wavelet analysis in order to point out the modulations in the frequency and amplitude.

V Boo = HD 127335 (V=7.9 mag) is a SRa star of spectral type M6e similar to Mira stars, which was discovered by Dunér in 1884. This star shows a light variation due to the pulsation with $P = 258$ d. Percy et al. (1990) reported that the total amplitude decreased from 3 to 2 magnitudes between 1900 and 1975.

Mattei et al. (1990) and Cannizzo et al. (1990) searched for chaotic behaviour in the AAVSO light curve of V Boo (1905-1965). They found that there is no evidence for deterministic chaos. They concluded that the light variation can be represented by a strictly periodic (P=259 d) part and a random part whose maximal amplitude is one-half of that of the periodic part. Yanagita et al. (1993) used an AR model for the VSOLJ data, they determined 256.4 d, 135.6 d and 888.9 d periods with 1.59, 0.25 and 0.17 mag amplitudes respectively. Evidence for chaos was not found.

Our analysis based on visual measurements collected by BAA, VSOLJ, AFOEV, AAVSO and HAA between JD 2420500 and 2449600 (1913-1994). The five data sets were merged and the magnitudes were averaged if there was more than one data point in a 10 day-long bin. The final light curve contains N=2648 points (Fig. 1 left). The light change is Mira-like in the earlier part, and a strong amplitude decreasing can be seen later.

In order to study the amplitude and period variation we divided the whole data series to ten 3000 day-long segments. The amplitude spectra are in Fig. 1 (right). The next procedure was the determination of amplitude and phase for a fixed frequency component in each segment. The changes of these values are shown in Fig. 2. Assuming a linear amplitude variation the calculated rate is $dA/dt = 4 \pm 1 \times 10^{-5}\ mag/day$ or $\Delta A/P = 0.010 \pm 0.003\ mag/per$. If this tendency continues the amplitude will be zero after the elapse of 15-20 years!

The wavelet transform was calculated for 50 frequencies and 100 time-shift (Szatmáry et al. 1994). In order to detect the effect of data distribution and gaps we have used a "whitening" method. The wavelet maps (frequency-time-amplitude) show the fine structure of the amplitude change.

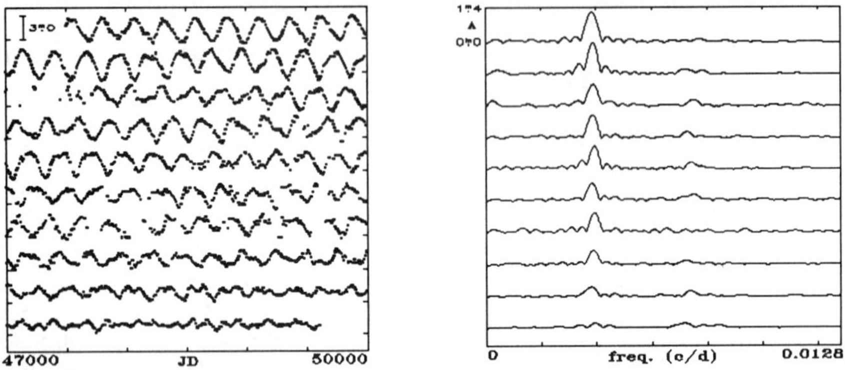


Figure 1. The light curves and the amplitude spectra of ten segments.

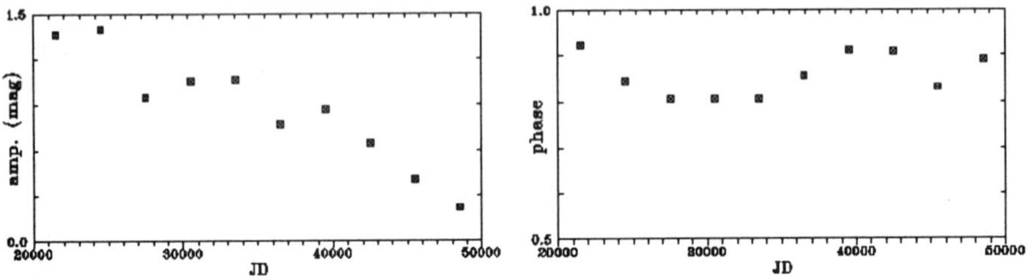


Figure 2. The amplitude and phase change of V Boo.

The O-C diagram, the plot of the instantaneous period and the CUSUM (Koen & Lombard 1993) show no evidence for significant period variation.

We have tried to identify the pulsational modes based on the models of Fox & Wood (1982) and Ostlie & Cox (1986). The possible modes for V Boo P=258 and 137 d periods are the fundamental and first overtone. Using the relations for $P_0$ and $P_1$ the mass and the radius: $M = 2.0 \pm 0.3\,M_\odot$ and $R = 290 \pm 20\,R_\odot$.

The cause of the strong decline of the amplitude of V Boo is not clear yet. The estimated duration of the short-period Mira phase is $6 \times 10^4$ yr - $5 \times 10^5$ yr. The luminosity and the period is expected to increase in time based on evolutionary consideration.

The possible interpretation of the amplitude decreasing is that the fundamental pulsational mode quickly declines, while the strength of the first overtone is relatively stable. However the reason for this phenomenon is not known yet. Perhaps we see a star evolving from a short-period Mira state to a semiregular one. Perhaps by damping of the fundamental mode the first overtone will become the dominant mode. Might V Boo stop varying in the near future? Radial velocity measurements would be useful to answer these questions.

**Acknowledgments.** This analysis based on more than 22000 observations of some hundred amateur astronomers. We are grateful to the staff of AFOEV, VSOLJ and HAA for making their data available for us. This work was supported by the Hungarian OTKA grant No.T7522.

*Astrophysical Applications of Stellar Pulsation*
*ASP Conference Series, Vol. 83, 1995*
*R. S. Stobie and P. A. Whitelock, eds.*

# On the Formation of Hydrogen Emission Lines in Miras

Patrick de Laverny

*Uppsala Astronomical Observatory, Box 515, S-75120 Uppsala, Sweden.*
*GRAAL, Université Montpellier II, France.*

Christian Magnan

*GRAAL, cc72, Université Montpellier II, 34000 Montpellier, France.*
*Collège de France.*

**Abstract.** We investigate non-LTE effects in the hydrogen recombination zone of Long-Period Variables of Mira type. We propose that the Balmer hydrogen emission lines could be formed in this deep layer. We show how the spectral cyclic variations of Miras could be mimicked with this interpretation. As a result the radius of a Mira in the Balmer transitions could be much smaller than its radius in the molecular absorptions, which are formed in higher layers.

## 1. Introduction

Long-Period variables of Mira type are giant pulsating AGB stars. Their spectrum exhibits large molecular absorptions typical of cool spectra. Balmer emission lines are also observed around the maximum luminosity of these stars. On the other hand, equilibrium departures are very important quite deep in Miras. One indeed observes shock-waves and large mass-loss rates. It is then very difficult to model these complex pulsating stars. However, the hydrodynamic effects have been investigated in some atmospheric models (see Bowen, 1988 for instance) and in some inner structure models (Tuchman et al., 1978 and Tuchman 1995). The inner structure models indicate that the hydrogen recombination zone (or H II/H I transition) is formed very high in a Mira ($R_{\mathrm{HII/HI}} \approx 80 - 90\% R_{\mathrm{Mira}}$). Its density is also too low in order to assume LTE ($N_{\mathrm{HII/HI}} < 10^{16}$ cm$^{-3}$). The aim of this work is to estimate these non-LTE effects found in layers as deep as the hydrogen recombination zone.

## 2. Simulation of the spectrum of a Mira

We determine the spectrum emerging from an optically thick hydrogen envelope where non-LTE effects are taken into account. The physical properties of this envelope are close to those of the H II/H I transition of a Mira (quite low density and large radius). We then show that emission lines are produced by recombination mechanisms in this transition zone (Magnan & de Laverny, 1994 and de Laverny & Magnan, 1995). The profiles of these lines also exhibit a large central reversal.

Since the H II/H I transition of a Mira is surrounded by an extended cool atmosphere we have then to verify that the emission lines can emerge from the star. We have thus estimated the atmospheric absorption. As a result if the hydrogen recombination zone is formed high enough in the star the emerging spectrum is close to the radiation observed near the maximum luminosity of a Mira (Balmer emission lines are superposed on a very cool continuum).

It is also possible to mimic the cyclic spectral variations of a Mira (reddening of the continua, appearance/disappearance of the Balmer lines) with this crude model. It has been shown that the hydrogen recombination zone is shifted towards the centre of the star around the minimum luminosity of a Mira. Its radius can vary by 20% during the cycle (Tuchman et al., 1978). We have shown that such a contraction of the H II/H I transition leads to larger atmospheric absorption. The emerging spectrum is then cooler without any Balmer emission lines. We thus mimic the spectrum of a Mira near its minimum luminosity.

## 3. Conclusion

We have proposed that the spectrum of a Mira could be composed of hydrogen recombination lines formed in a deep layer and cold absorption formed above (see for more details, Magnan & de Laverny, 1995).

The cycle of a Mira can also be mimicked with this schematic model. At the maximum of luminosity, the hydrogen recombination zone is formed very high in the star. Balmer lines with a central reversal are formed by fluorescence mechanisms in this transition and emerge in emission from the atmosphere. On another hand the H II/H I transition is formed deeper in the star at its minimum luminosity. The Balmer lines disappear and the star is cooler.

Finally we would like to point out a crucial observational test to this interpretation. Since hydrogen lines are formed deeper in the star than the molecular absorption the radius of a Mira could vary strongly with wavelength. We then propose that its radius in the Balmer transitions is much smaller than its radius in the molecular absorption. Future interferometric measurements of Miras diameters in narrow filters could give interesting constraints to this model.

**Acknowledgments.** This work has been partly supported by a grant from the European Space Agency. We thank the IAU and French GdR "Milieux Circumstellaires" for their financial support to attend this colloquium.

## References

Bowen, G.H. 1988, ApJ, 329, 299

de Laverny, P., & Magnan, C. 1995, A&A, submitted

Magnan, C., & de Laverny, P. 1994, Astrofizika, Vol.37

Magnan, C., & de Laverny, P. 1995, MNRAS, submitted

Tuchman, Y., Sack, N., & Barkat, Z. 1978, ApJ, 219, 183

Tuchman, Y. 1995, this volume

*Astrophysical Applications of Stellar Pulsation*
*ASP Conference Series, Vol. 83, 1995*
*R. S. Stobie and P. A. Whitelock, eds.*

# The AAVSO and its Variable Star Databank on Pulsating Stars

Janet A. Mattei

*AAVSO, 25 Birch Street, Cambridge, MA 02138, USA*

**Abstract.** The American Association of Variable Stars Observers is the world's largest organization of variable star observers. Its services to the astronomical community and database on pulsating variables are described.

## 1. Organization and services

The American Association of Variable Star Observers (AAVSO) coordinates, collects, evaluates and archives variable star observations made largely by amateur astronomers worldwide; publishes variable star data and disseminates them to researchers and educators around the world. Over 7.8 million observations have been incorporated into the AAVSO International Database since its founding in 1911. The AAVSO receives over 300,000 mostly visual observations yearly. Observations for inclusion in the database are standardized, homogeneous and subject to quality control checks to ensure the highest level of reliability.

The AAVSO receives over 200 requests each year from astronomers for its data and services to help schedule observing runs using ground-based and satellite telescopes; to provide simultaneous optical coverage of observing targets and immediate notification of their activity during particular satellite observing programs; to correlate multi-wavelength data; and to carry out collaborative research to analyze the long-term behaviour of variable stars.

## 2. Data on pulsating stars

There are over 2100 pulsating variable stars of Mira (1217), semiregular (586), Cepheid (179), RV Tauri (36), R Coronae Borealis (35), Symbiotic (34), and S Doradus (13) types in the AAVSO observing program. For many hundreds of stars, the AAVSO has almost continuous visual observations going back many decades, using consistent sets of comparison stars. Sample light curves from AAVSO International Database from 1963 to mid-1994 are shown in Figure 1. AAVSO observations and published information on maxima and minima dates and magnitudes of long period variables going back to 1900 have been used to correlate light curve parameters; to study period and pulsation mode changes; to search for chaotic behaviour; and to investigate stellar evolution. For AAVSO data and services, please contact the Director at the address above or by e-mail at aavso@cfa.harvard.edu.

I gratefully acknowledge AAVSO and IAU grants that made it possible for me to attend this well-organized, interesting and enjoyable meeting.

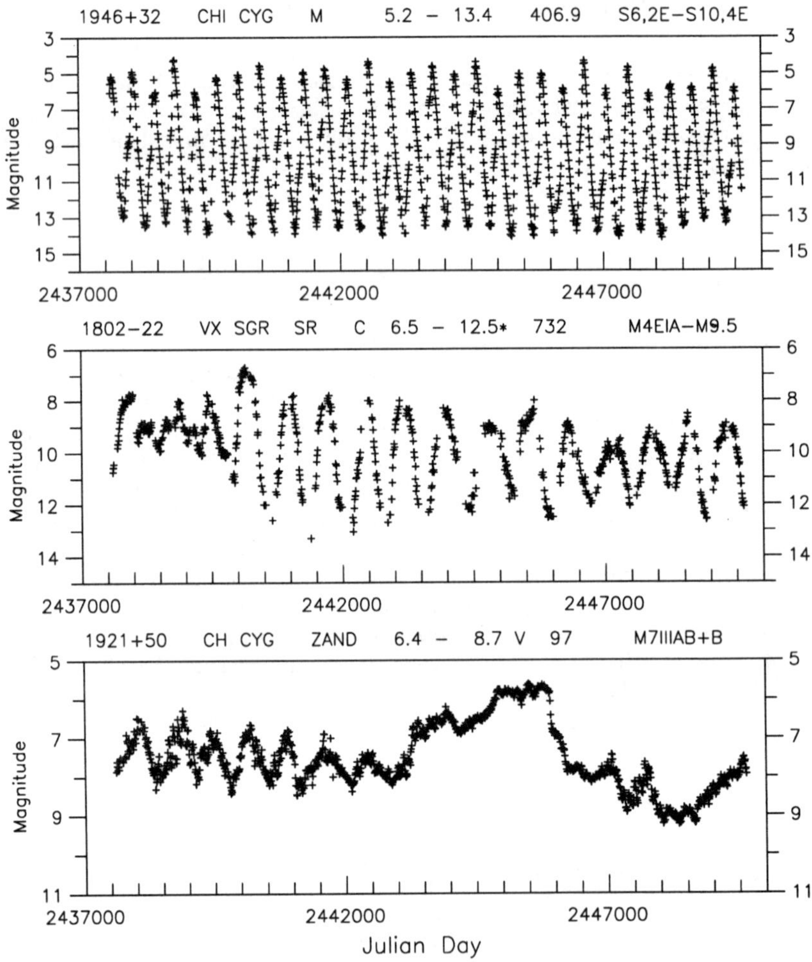


Figure 1. Sample light curves from the AAVSO International Database of (top to bottom) $\chi$ Cyg, VX Sgr, and CH Cyg. Each cross represents 10-day means of observations.

*Astrophysical Applications of Stellar Pulsation*
*ASP Conference Series, Vol. 83, 1995*
*R. S. Stobie and P. A. Whitelock, eds.*

# Miras and Galactic Populations

M.O. Mennessier

*GRAAL, Univ Montpellier II/CNRS, F-34095 Montpellier*

X. Luri, F. Figueras, J. Torra

*Departament d'Astronomia i Meteorologia, Univ Barcelona, Espagne*

**Abstract.** In order to exploit the imminent HIPPARCOS data we have developed a new Maximum-Likelihood method to determine luminosity calibrations and kinematical properties with as bias-free as possible estimates. In most cases the samples are inhomogeneous, containing stars of several types. The method includes a way to separate the different groups. We applied the method to Mira variables contained in the HIPPARCOS Input Catalogue.

## 1. Introduction

In our Maximum-Likelihood method to calibrate luminosity, the modeling of the samples has been made as realistic as possible – including the effects of the observational constraints, inhomogeneities of the sample, galactic rotation, observational errors and interstellar absorption – so the results are as bias-free as possible (Luri et al., 1993).

## 2. Sample

The sample is composed of 90 Mira variables contained in the HIPPARCOS Input Catalogue with known proper motions and radial velocities. Mean apparent magnitudes are from 75 years AAVSO data (H. Boughaleb, 1994). We take into account the process of selection used to decide the Miras to be included in the HIPPARCOS Input Catalogue – see ESA SP-111, Vol. II – basically affecting the distribution of apparent magnitudes. Selection effects in proper motions and radial velocities are not considered important after comparison with the distributions of Miras in the GCVS.

## 3. Results

Using Wilks test on likelihood, the separation of three groups of stars is decided. By the Maximum Likelihood method we get the estimates shown in Table 1. Each star can be assigned to a group. So we obtain an independent estimation of the distance of each star from observations at maximum or minimum luminosity. The agreement between both these estimates is good. The results from observations at minimum luminosity seem more reliable and agree with the estimates of velocities in the galactic plane and motion perpendicular to the galactic plane.

## 4. Conclusion

The Miras in our sample can be considered as belonging to three populations: thick disk (group 1); old disk stars (group 2) with the same velocities, aligned with the Sun: they could belong to a stream; possible halo stars (group 3), but the small value of $\sigma_M$ is probably not real (too small number of stars). However, all stars in group 3 have small period and early spectral type and seem to be very luminous. This agrees with Jura & Kleiman (1992) results and with Whitelock's review paper (these proceedings).

Table 1. Estimates of mean intrinsic V luminosity, mean velocity (km/s) and galactic disk height (pc) and percentage of stars in each group

| | group 1 | | group 2 | | group 3 | | group 1 | | group 2 | | group 3 | |
|---|---|---|---|---|---|---|---|---|---|---|---|---|
| | $\theta_{MV}$ | $\sigma$ | $\theta_{MV}$ | $\sigma$ | $\theta_{MV}$ | $\sigma$ | $\theta_{MV}$ | $\sigma$ | $\theta_{MV}$ | $\sigma$ | $\theta_{MV}$ | $\sigma$ |
| $M_0$ | 0.4 | 0.6 | -1.8 | 0.9 | -2.5 | 0.4 | 6.3 | 0.3 | 2.4 | 0.8 | 0.1 | 0.8 |
| $\sigma_M$ | 1.0 | 0.2 | 1.0 | 0.4 | 0.1 | 0.4 | 1.4 | 0.1 | 0.7 | 0.4 | 0.1 | 0.6 |
| $U_0$ | -16 | 7 | -49 | 10 | -69 | 28 | -19 | 9 | -51 | 8 | -65 | 70 |
| $\sigma_U$ | 44 | 5 | 12 | 6 | 85 | 25 | 46 | 5 | 12 | 6 | 86 | 60 |
| $V_0$ | -33 | 6 | -37 | 5 | -139 | 55 | -38 | 8 | -39 | 4 | -220 | 60 |
| $\sigma_V$ | 39 | 6 | 0 | 2 | 83 | 29 | 41 | 3 | 0 | 5 | 93 | 50 |
| $W_0$ | -5 | 6 | -22 | 12 | -9 | 32 | -7 | 3 | -23 | 9 | -4 | 160 |
| $\sigma_W$ | 23 | 7 | 12 | 8 | 205 | 39 | 31 | 2 | 16 | 7 | 300 | 130 |
| $Z_0$ | 329 | 81 | 47 | 18 | 390 | 180 | 900 | 800 | 51 | 25 | 550 | 260 |
| % | 75 | 8 | 9 | 2 | 16 | 7 | 84 | 3 | 10 | 2 | 6 | 3 |
| | apparent magnitude at maximum | | | | | | apparent magnitude at minimum | | | | | |

## References

Boughaleb, H. 1994, these Univ. Montpellier II

Jura, M. 1992, ApJS, 79, 105

Luri, X., Mennessier, M.O., Figueras, F., & Torra, J. 1994, in "New approaches in Classification and Data Analysis", Springer-Verlag, p.672

*Astrophysical Applications of Stellar Pulsation*
*ASP Conference Series, Vol. 83, 1995*
*R. S. Stobie and P. A. Whitelock, eds.*

# On the Nature of M stars with a 60 microns Excess

C. Loup[1], L.B.F.M. Waters[2], F. Kerschbaum[3], J. Hron[3], E. Josselin[1], D. Kester[4], Tj.R. Bontekoe[4], A. Zijlstra[5]

[1] *Institut d'Astrophysique de Paris, CNRS, France*
[2] *Astronomical Institute Anton Pannekoek, Amsterdam, the Netherlands*
[3] *Institut für Astronomie, Wien, Austria*
[4] *SRON Space Research Laboratory, Groningen, the Netherlands*
[5] *European Southern Observatory, Garching, Germany*

## 1. Introduction

Some years ago, Willems & de Jong (1988) noticed that many carbon stars display an excess of emission at 60 $\mu m$ and explained it by the presence of a fossil dust shell, containing only cold dust. This detached dust shell would be the result of an interruption of the mass loss, consequence of a thermal pulse. Detached shells around C stars have actually been mapped in the CO lines (Olofsson et al. 1992), and at 60 $\mu m$ (Waters et al. 1994). In 1992, Zijlstra et al. found about 100 M stars displaying an excess of emission at 60 $\mu m$, and proposed that interruptions of the mass loss due to thermal pulses is a general phenomenon on the AGB. This assumption is now supported by the theoretical calculations of Vassiliadis & Wood (1993). Here we present a detailed study of the 100 M stars of Zijlstra et al. in order to test the previous assumption.

## 2. M stars with 60 $\mu m$ excess resolved by IRAS

For these 100 M stars, we have analysed individual and averaged IRAS scans, as well as co-added maps, at 60 $\mu m$. We find that 15 sources are clearly resolved by IRAS at 60 $\mu m$, 51 objects are clean point sources, and 39 objects have an unreliable PSC 60 $\mu m$ flux. We applied maximum entropy image reconstruction techniques (Bontekoe et al. 1994) to the 15 sources resolved at 60 $\mu m$, as well as on a few point sources for comparison. 5 sources are found to be very peculiar objects. For the 10 other sources, the dust shell appears to be roughly spherical at 60 $\mu m$. These 10 stars are located in the region VIa of the IRAS two-colour diagram (van der Veen & Habing 1988) and we conclude that they have undergone a strong decrease of the mass-loss rate in the past.

## 3. Spectral and variability types

We performed optical spectroscopy of 29 M stars with a 60 $\mu m$ excess using the B&C spectrograph on the ESO 1.5m telescope. Spectra were taken from 3600 to 9200 Åwith a resolution of 1.9 Å. M classes were determined with an uncertainty of 0.5. For other sources we use previous determinations found in the literature. We also performed JHKL'M photometry of 30 M stars at the

ESO 1m telescope with its standard infrared photometer. M stars with a 60 $\mu m$ excess can be clearly separated in two groups : sources located in region VIa generally displaying a strong excess at 60 $\mu m$, and sources located in region VII and VIb with a small 60 $\mu m$ excess. Most M stars located in region VIa have spectral types earlier than M5 and are non or irregular variables. Conversely, most sources in region VII and VIb have spectral types later than M5 and are semi-regular or Mira variables. The energy distribution of one source located in region VIb clearly shows that the dust envelope is not spherical.

## 4. Discussion and conclusions

According to Vassiliadis & Wood (1993), for stars with an initial mass smaller than about 3 $M_\odot$, the consequences of a thermal pulse would be a strong and sudden decrease of the mass loss rate, the stellar luminosity, and the period, and a strong increase of the effective temperature. After typically 5000 years, $\dot{M}$, $L_\star$, and P would increase, whereas $T_{eff}$ would decrease, regularly again, up to the occurrence of a new thermal pulse.

Under the assumption of a strong decrease of the mass loss rate, M stars with a 60 $\mu m$ excess located in region VIa should have undergone a thermal pulse a few 1000 to 10,000 years ago, while M stars located in regions VII and VIb should have undergone a thermal pulse less than a few 100 years ago (Chan & Kwok 1988). Statistically, we should then find much less sources in regions VII and VIb than in region VIa. Here we find 28 sources in region VIa and 18 in regions VII and VIb. Furthermore, following the model of Vassiliadis & Wood, M stars in regions VII and VIb should have the highest effective temperatures, while we observe the opposite.

We conclude that M stars with a 60 $\mu m$ excess located in regions VII and VIb are either stars with a slowly decreasing mass loss rate, or stars enshrouded by a non–spherical dust envelope. On the other hand, we find 28 M stars located in region VIa, of which 10 are resolved by IRAS at 60 $\mu m$, which are very good candidates to be M stars with a detached envelope. For them, our observations are in qualitative agreement with Vassiliadis & Wood concerning the behaviour of the mass loss rate and the effective temperature. However, while they predict a periodic variability for these stars, we rather find them as non or irregular variables.

## References

Bontekoe, Tj.R., Koper, E., & Kester, D.J.M. 1994, A&A 284, 1037

Chan & Kwok, S. 1988, ApJ, 334, 362

Olofsson H., Carlström U., Eriksson K., & Gustafsson B. 1992, A&A 253, L17

van der Veen, W.E.C.J, & Habing, H.J. 1988, A&A, 194, 125

Vassiliadis, E., & Wood, P.R. 1993, ApJ, 413, 641

Waters, L.B.F.M., et al. 1994, A&A, 281, L1

Willems, F.J., & de Jong, T. 1988, A&A, 196, 173

Zijlstra, A.A., Loup, C., Waters, L.B.F.M., & de Jong T. 1992, A&A, 265, L5

*Astrophysical Applications of Stellar Pulsation*
*ASP Conference Series, Vol. 83, 1995*
*R. S. Stobie and P. A. Whitelock, eds.*

# Obscured AGB Stars in the LMC

Albert Zijlstra, Cecile Loup, Rens Waters, Patricia Whitelock and Francois Guglielmo

**Abstract.** We have carried out a search for obscured AGB stars in the LMC, observing fields in the vicinity of faint IRAS sources. About 25 heavily reddened AGB stars were found, with bolometric magnitudes close to the classical AGB limit of $M_{bol} = -7.1$. 10-$\mu$m photometry for 5 sources show that these are oxygen-rich; they are similar to the Galactic OH/IR stars. Mass-loss rates vary between $5\times10^{-4}$ and $2\times10^{-6}$ $M_\odot$ $yr^{-1}$. We find no evidence that the mass-loss rates vary with luminosity. Neither do the mass-loss rates differ in a clear systematic way between the SMC and the LMC. The expansion velocities appear to be slightly lower in the LMC than in the Galaxy. All of the luminous stars for which periods are available have luminosities significantly above the period-luminosity relation derived for optical LMC Miras.

## 1. Survey

Mass-losing, obscured AGB stars have only recently been found in the LMC (Reid 1991, Wood et al. 1992). Although few are known, they are very important to relate the mass-loss phase on the AGB to the stellar evolution: they are the only high-mass-loss AGB stars for which an accurate distance is known. To increase the sample, we have carried out an infrared survey of candidate IRAS sources in the LMC. We selected IRAS sources from two deep catalogues: Schwering & Israel (1989) and Reid et al. (1990), based on flux density (for an AGB star one expects typically $f_{12} \approx 0.3$ mJy) and the 12 versus 25$\mu$m colour. The SIMBAD database was used to find possible optical counterparts. Many sources had no known identification; the ones which did were mostly identified as supergiants.

## 2. Results

About 25 obscured AGB stars were discovered, some detected only in the K-band. Their bolometric magnitudes are mostly between $-6.0$ and $-7.2$. A number of reddened supergiants (significantly more luminous than the AGB stars) were also found. Together with earlier detections, there are now 39 obscured AGB stars known in the Magellanic Clouds.

We obtained 10$\mu$m photometry for five sources using ESO's TIMMI camera. Three filters were used: N-band, a narrow-band filter centred at 9.8$\mu$m (the silicate feature) and a narrow-band filter centred at 11.3$\mu$m (SiC). These three filters allow us to separate oxygen-rich from carbon-rich stars. The colours of the LMC stars indicate they are oxygen-rich, with (except for one) the silicate feature in emission. Two of the LMC stars have a very strong silicate feature: in our Galaxy this is only seen in stars with faint 12$\mu$m emission and low mass-loss rates, where the entire 12$\mu$m flux is due to silicate emission. The LMC stars mimic this behaviour while being much more luminous than these Galactic stars.

We derive dust mass-loss rates from the near-infrared–mid-infrared colours. Derived values vary from $-5.75$ to about $-8.0$ (in $\log M_\odot\,\mathrm{yr}^{-1}$). For a reasonable dust-to-gas ratio of 400, the total mass-loss rates are between a few times $10^{-4}$ and a few times $10^{-7}$. Within our sample, there is no evidence that the values depend on luminosity, nor do we find any systematic difference between the LMC and SMC (albeit with only 5 SMC stars).

Expansion velocities are available for only four of the IRAS-detected AGB stars in the LMC, in all cases based on the OH detections by Wood et al. (1992). To compare these objects to Galactic OH/IR stars we selected all Galactic OH/IR stars which have also been detected in CO. Distances are estimated by assuming a luminosity of $10^4 L_\odot$. The CO gives an independent check on the OH expansion velocity. There is a slight tendency for the LMC stars to have lower expansion velocities. The difference is about 10–20%, although there is insufficient data to draw a firm conclusion. We note that Wood et al. derive a larger difference of about 50% from the same data: the present conclusion is based on a slightly different way of determining the expansion velocity from noisy OH spectra. In either case, the lower LMC metallicity appears to affect the outflow velocity of the stellar wind.

A period–luminosity relation for LMC Miras has been derived by Feast et al. (1989), based on stars with periods less than 420 days. The relation is quite tight, and is promising for future use in distance determinations for nearby galaxies. However, the relation cannot be easily confirmed for Galactic stars because of the poorly known distances. Periods are available for a number of the IRAS-detected AGB stars in the LMC. We find that the longer-period variables fall far above the relations, which clearly are not valid at longer periods. There is in fact little indication for any PL relation at long periods. Part of this may be due to evolution in luminosity due to convective-envelope burning. The PL relation may also widen if the envelope mass evolves significantly, which can be expected for higher-mass stars which will loose most of their mass on the AGB.

## References

Feast M.W., Glass I.S., Whitelock P.A., & Catchpole R.M., 1989, MNRAS, 241, 375

Reid I.N., Tinney C.G., & Mould J.R., 1990, ApJ, 348, 98

Reid I.N., 1991, ApJ, 382, 143

Schwering P.B.W., & Israel F.P., 1990, A catalog of IRAS sources in the Magellanic Clouds (Kluwer, Dordrecht)

Wood P.R., Whiteoak J.B., Hughes M.G., Bessell M.S., Gardner F.F., & Hyland A.R., 1992, ApJ, 397, 552

*Astrophysical Applications of Stellar Pulsation*
*ASP Conference Series, Vol. 83, 1995*
*R. S. Stobie and P. A. Whitelock, eds.*

# Mid-IR Imaging of AGB Stars and Circumstellar Modelling

M. Busso, L. Origlia

*Osservatorio Astronomico di Torino, Pino Torinese, Italy*

G. Silvestro, M. Marengo[(+)]

*Istituto di Fisica Generale dell'Università, Torino, Italy*

[(+)]*International School for Advanced Studies, Trieste, Italy*

P. Persi and M. Ferrari-Toniolo

*Istituto di Astrofisica Spaziale del CNR, Frascati, Italy*

The evolution of low and intermediate mass (1-8 $M_\odot$) stars along the Asymptotic Giant Branch (AGB) is ruled by processes of mass loss, causing the whole convective envelope to be gradually ejected into space. If the stellar mass is sufficiently high ($M \geq 1.5\ M_\odot$) the envelope itself becomes enriched in nucleosynthesis products (carbon and *s*-process nuclei) and the star evolves into a C-rich phase. AGB stars are hence surrounded by O-rich or C-rich envelopes, opaque at optical wavelengths, which are best studied through direct imaging in the infrared (IR).

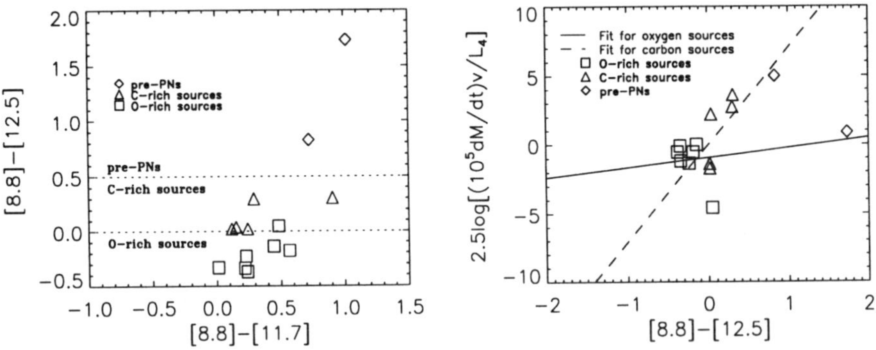


Figure 1. (a) Separation of O-rich AGB stars (squares), C-rich AGB stars (triangles) and pre-PNs (diamonds) by their mid-IR colours. (b) Correlation of the [8.8]-[12.5] colour with mass loss rates.

From the IRAS database we selected 15 bright O-rich and C-rich AGB stars, that we observed with the mid-IR camera Tircam at the TIRGO (Gornergrat, Switzerland) and the S. Pedro Martir (Mexico) IR telescopes. Filters of width $\Delta\lambda = 1\ \mu$m centered at 8.8, 9.8, 11.7 and 12.5 $\mu$m were used. Colour-colour diagrams were then constructed, to search for criteria separating C-rich from O-rich sources. Figure 1(a) allows one to see that the [8.8]-[12.5] colour clearly separates the three groups of O-rich AGBs, C-rich AGBs and pre-PNs, in different regions.

In order to investigate the correlations of IR colours with mass loss, we have adopted the mass loss rates estimated by Loup et al. (1993), together with their choice for the values of luminosity, distance and wind velocity. We hence constructed a mass-loss proportional parameter and plotted it against the mid-IR colours. As Figure 1(b) shows, O-rich and C-rich stars both display values of the ([8.8]-[12.5]) colour well correlated with their mass loss rate, but the regression lines are very different. In particular, O-rich stars (probably with a lower mass, M $\leq$ 1.3-1.4 $M_{\odot}$) display a much flatter behaviour. This is an indication that low-mass Miras evolve at lower mass loss rates than more massive red giants, which become C-stars through dredge-up of newly processed carbon. The circumstellar envelopes giving rise to the IR emission were modelled

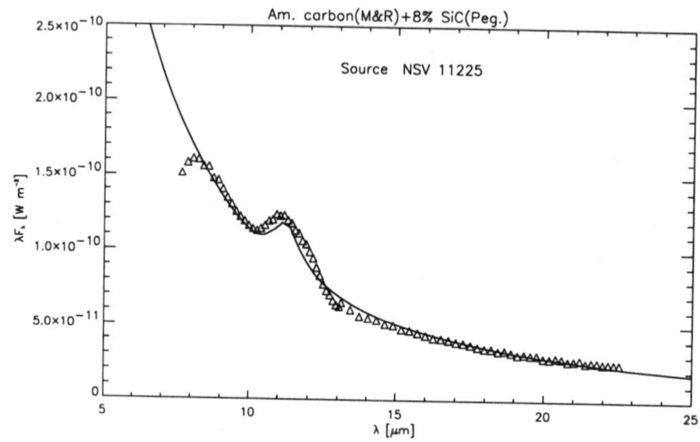


Figure 2. Comparison of the model circumstellar emission with the IRAS LRS spectrum for a C-rich star

through a numerical code, which integrates the equation of non-grey radiative transfer over a spherically symmetric dust shell. The code computes a self-consistent thermal structure for the envelope, using multiple grain components of carbon- and oxygen-rich material. The spectral energy distribution $\lambda F_{\lambda}$ is calculated taking into account the effect of non-isotropic scattering, absorption and thermal reemission by grains. Model results (Figure 2) can reproduce the IR photometric data of oxygen- and carbon-rich AGB stars with envelopes of different optical depths.

## References

Loup, C., Forveille, T., Omont, A., & Paul, J., F. 1993, A&AS, 99, 291

*Astrophysical Applications of Stellar Pulsation*
*ASP Conference Series, Vol. 83, 1995*
*R. S. Stobie and P. A. Whitelock, eds.*

# Long-term Photometric Monitoring of Cool Hydrogen-deficient Carbon Stars

P.L. Cottrell, A.C. Gilmore and P.M. Kilmartin

*Mount John University Observatory, Dept of Physics & Astronomy, University of Canterbury, Christchurch, New Zealand*

W.A. Lawson

*Dept of Physics, University College UNSW, Australian Defence Force Academy, Canberra, Australia*

## 1. Introduction and observations

Interest in the hydrogen-deficient carbon (HdC) stars, and the R Coronae Borealis (RCB) stars in particular, has been rekindled by the realisation that these stars provide the potential for testing models in a number of areas. These include: pulsation theory; grain formation; grain evolution; stellar nucleosynthesis and extended stellar atmospheres.

Observational work on these objects has increased dramatically over the last 10 years. This has included both spectroscopic and photometric projects, see Lambert & Rao (1994) and Lawson et al. (1990).

We have continued to observe a number of the cool HdC stars in the Galaxy and the LMC and now have light curves exceeding 2000 days in duration for many of these objects. In particular, we are observing the RCB stars to investigate both the low amplitude, pulsation-related variations present at maximum light and the photometric behaviour during the large amplitude declines.

With respect to the deep declines that the RCB stars undergo, we continue with the working definition of 'red' and 'blue' declines to broadly characterise this phase. This was introduced by Cottrell et al. (1990). This corresponds to the colour of the star as it initially begins to dim in light output and appears to be related to the geometry of the obscuring dust cloud. There is substantial variation in this characteristic and it is hoped that this data set will provide a significant starting point for a better definition and understanding of not only this phase, but also the rise out of these deep declines.

## 2. Comments on individual stars

Of the stars under investigation four (RY Sgr, V854 Cen, S Aps and the LMC RCB star HV 12842) have been chosen for detailed comment (see Figure 1).

RY Sgr has continued to provide distinct pulsations in both the V and colour curves and has had two distinct declines, one short, 'red' event and the other a deeper, longer and extremely 'blue' decline. The onset times of the declines are related to the 38-d pulsation period (Pugach 1977).

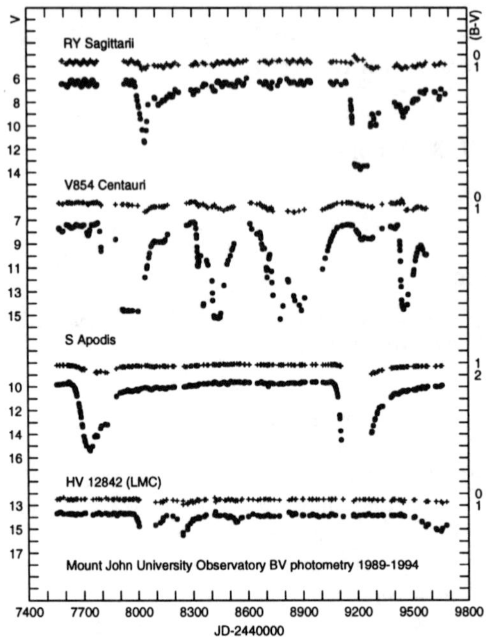


Figure 1. *V* and *(B–V) light curves for a selection of RCB stars.*

V854 Cen continues to be the most active RCB star, with declines in each of the last 5 years. There is great variety amongst the declines, with both 'red' declines and 'blue' declines. The times of onset of the declines are closely related to the 43-d pulsation period of the star (Lawson et al. 1992).

S Aps has undergone two deep declines during this period, with the one in 1989 an example of a 'red' decline, whereas the 1993 decline shows an almost constant colour for the first ten days of the decline. There are also some distinct pulsations during the rise out of these declines, which are not readily distinguishable at other phases.

HV 12842 has been the more active of the two LMC stars that we have observed. Its declines have been both short-lived and relatively shallow.

## References

Cottrell, P.L., Lawson, W.A., & Buchhorn, M. 1990, MNRAS, 244, 149

Lambert, D.L., & Rao, N.K. 1994, J. Astr. Ap., 15, 47

Lawson, W.A., et al. 1990, MNRAS, 247, 91

Lawson, W.A., et al. 1992, MNRAS, 256, 339

Pugach, A.F. 1977, IBVS, 1277

*Astrophysical Applications of Stellar Pulsation*
*ASP Conference Series, Vol. 83, 1995*
*R. S. Stobie and P. A. Whitelock, eds.*

# Pulsations and Dust Formation in R Coronae Borealis Stars

Geoffrey C. Clayton

*Center for Astrophysics and Space Astronomy, Campus Box 389, University of Colorado, Boulder, CO 80309*

Barbara A. Whitney

*Harvard-Smithsonian Center for Astrophysics, 60 Garden St., Cambridge, MA 02138*

Warrick A. Lawson

*Department of Physics, University College, University of New South Wales, Australian Defence Force Academy, Canberra ACT 2600, Australia*

The observed correlation between pulsational phase in RCB stars and the timing of their declines shows empirically that the stellar pulsations and dust formation are intimately connected (e.g., Lawson et al. 1992). However, the nature of this relationship and the process of dust formation itself are not well understood. We have shown that it is likely that dust is forming in close proximity ($< 2$ stellar radii) to the RCB star photosphere, based on time scales for acceleration of the dust, eclipse of the chromospheric region, and dispersal of the dust (Clayton et al. 1992, 1993; Whitney, Balm & Clayton 1993; Whitney et al. 1993). The temperature at which amorphous carbon forms can be as high as 4000 K, and can occur in conditions far removed from thermodynamic equilibrium, as long as a mechanism exists to contain carbon atoms within a given volume. A likely form of carbon condensate is fullerenes such as $C_{60}$ (Whitney, Balm & Clayton 1993). Shocks in the stellar atmosphere due to the pulsations may provide such a mechanism for containing the carbon. However, no spectroscopic signature related to the dust formation has been seen. We have searched unsuccessfully for IR emission bands of $C_{60}$ (Clayton et al. 1995a).

High dispersion visible and ultraviolet spectra and UBVRI photometry, covering a complete pulsation of the RCB star, RY Sgr, have been obtained (Clayton et al. 1994). The UV spectra were the first high dispersion data ever obtained for the star. The visible spectra showed the typical line splitting and radial velocity variations which have been observed previously. The simultaneous UV spectra showed much smaller, and phase shifted, velocity variations than those seen in the visible. No evidence was seen of shock induced emission at Mg II. Visible spectroscopy and polarimetry of R CrB covering 7 years has also been obtained (Clayton et al. 1995b). These spectra clearly show CN and $C_2$ band variations correlated with pulsational phase. Polarimetric variations are also seen which are related to dust formation episodes.

**References**

Clayton, G.C., Whitney, B.A., Stanford, S.A., & Drilling, J. 1992, ApJ, 397, 652

Clayton, G.C., Lawson, W.A., Whitney, B.A., & Pollacco, D.L. 1993, MNRAS, 264, L13

Clayton, G.C., Lawson, W.A., Cottrell, P.L., Whitney, B.A., Stanford, S.A., & de Ruyter, F. 1994, ApJ, 432, 785

Clayton, G.C., Kelly, D.M., Lacy, J.H., Little-Marenin, I.R., Feldman, P.A., & Bernath, P.F. 1995a, AJ, in press (May)

Clayton, G.C., Whitney, B.A., Meade, M.R., Babler, B., Bjorkman, K.S., & Nordsieck, K.H. 1995, PASP, in press

Lawson, W.A.,Cottrell, P.L., Gilmore, A.C., & Kilmartin, P.M. 1992, MNRAS, 256, 339

Whitney, B.A., Clayton, G.C., Schulte-Ladbeck, R.E., Calvet, N., Hartmann, L., & Kenyon, S.J. 1993, ApJ, 417, 687

Whitney, B.A, Balm, S.P., & Clayton, G.C. 1993, Luminous High-Latitude Stars, ASP Conference Series No. 45, ed. D. Sasselov, p. 115

*Astrophysical Applications of Stellar Pulsation*
*ASP Conference Series, Vol. 83, 1995*
*R. S. Stobie and P. A. Whitelock, eds.*

# A Nonlinear Pulsation Model for V652 Her

Yu. A. Fadeyev

*Institute for Astronomy of the Russian Academy of Sciences, Pyatnitskaya Str. 48, 109017 Moscow, Russia.*

A. E. Lynas-Gray

*Department of Physics, University of Oxford, OX1 3RH, England.*

**Abstract.** When compared with observation, stellar pulsation model predictions provide a test of stellar envelope opacities. Of specific interest, is the radially pulsating extreme helium star V652 Her for which luminosity and radial velocity observations are published. Nonlinear pulsation models of V652 Her, calculated using Opacity Project (OP) opacities, predict luminosity and radial velocity variations in close agreement with observation.

## 1. Introduction

Using radial stellar pulsation theory, linear and nonlinear calculations show V652 Her to be beyond the blue boundary of the extreme helium star pulsation instability region when envelope opacities from the Los Alamos Opacity Library (LAOL, Hübner et al. 1977) are adopted (Saio & Jeffery 1988, Fadeyev 1993). New stellar envelope opacities (OPAL, Iglesias et al. 1992) enable Saio (1993), using a linear pulsation model, to show the opacity bump at $T \sim 2 \times 10^5$K (completely absent from LAOL opacities) to be responsible for driving radial pulsations in V652 Her. OP opacities (Seaton et al. 1994), while in agreement with OPAL, also include significant contributions due to ions of chromium, manganese and nickel; this motivates the present calculation of a nonlinear radial pulsation model for V652 Her with the view to making a comparison with observed radial velocity and luminosity curves.

## 2. Nonlinear pulsation model for V652 Her

Nonlinear radial pulsation models for V652 Her were calculated using Fadeyev's (1983) code and OP opacities computed for mixtures suggested by photospheric abundances obtained by Jeffery et al. (1986). Figure 1 illustrates the best agreement with observation obtained for a 0.72 $M_\odot$ star having $T_{\rm eff}$ = 23500 K and composition by numbers of $n(H) = 0.006$, $n(He) = 0.99$, $n(N) = 0.002$ and $n(Z*) = 0.002$; $Z*$ represents all metals (except nitrogen) with abundances in proportions given by Anders & Grevesse (1989). In transforming theoretical

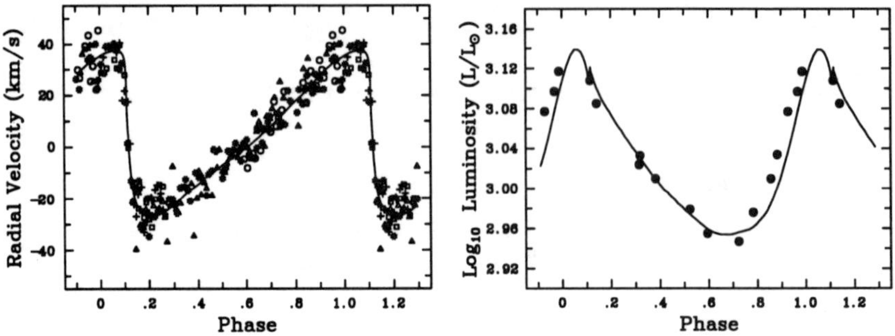


Figure 1. A comparison between observed (points) and calculated (solid line) pulsation variations in V652 Her; different symbols distinguish pulsation cycles. Radial velocity data are from Jeffery & Hill (1986); luminosities are from Lynas-Gray et al. (1984).

radial velocities from the centre-of-mass frame to the observer's rest frame, a uniform disk (no limb darkening) has been assumed.

## 3. Conclusion

In the case of V652 Her, nonlinear pulsation model calculations are limited by uncertain photospheric abundances and $T_{eff}$. Nonetheless, the agreement between observation and theory in Figure 1 is remarkably good and verifies the quality of OP opacities.

**Acknowledgments.** The Austrian Ministry of Science generously awarded a research grant to fund the HP9000/425e used for nonlinear pulsation model calculations.

## References

Anders, E., & Grevesse, N. 1989, Geochim. Cosmochim. Acta, 53, 197

Fadeyev, Yu. A. 1983, Nauchn. Inform. Astron. Council, 52, 3

Fadeyev, Yu. A. 1993, MNRAS, 262, 119

Hübner, W.F., Merts, A.L., Magee, N.H., & Argo, M.F. 1977, Los Alamos Report LA-6760-M

Iglesias, C.A., Rogers, F.J., & Wilson, B.G. 1992, ApJ, 397, 717

Jeffery, C.S., Heber, U., & Hill, P.W. 1986, IAU Coll, 87, 101

Jeffery, C.S., & Hill, P.W. 1986, MNRAS, 221, 975

Lynas-Gray, A.E., Schönberner, D., Hill, P.W., & Heber, U. 1984, MNRAS, 209, 387

Saio, H. 1993, MNRAS, 260, 465

Saio, H., & Jeffery, C. S. 1988, ApJ, 328, 714

Seaton, M.J., Yu Yan, Mihalas, D. & Pradhan, A.K. 1994, MNRAS, 266, 805

*Astrophysical Applications of Stellar Pulsation*
*ASP Conference Series, Vol. 83, 1995*
*R. S. Stobie and P. A. Whitelock, eds.*

# Pulsating Degenerate Stars in the EC Survey

R. S. Stobie, C. Koen, D. Kilkenny

*South African Astronomical Observatory, PO Box 9, Observatory 7935, Cape, South Africa*

D. O'Donoghue

*Department of Astronomy, University of Cape Town, Rondebosch 7700, Cape, South Africa*

## 1. Introduction

For asteroseismological reasons the discovery of new pulsating degenerate stars is important. Bradley (1993) lists the number of known pulsating white dwarfs at that time - 23 known DAV (ZZ Ceti) stars, 7 known DBV stars and 5 known DOV (PG1159) stars. Of these degenerate pulsators 80% (including all DBVs) are in the northern hemisphere. This illustrates the great incompleteness in the search for such objects in the southern hemisphere.

The Edinburgh-Cape Blue Object Survey is a potential source of candidate degenerate pulsators in the southern hemisphere (Stobie et al. 1992). The blue stellar objects are identified from COSMOS scans of U and B plates taken with the UK Schmidt Telescope. Follow-up photometry and spectroscopy are being obtained at the South African Astronomical Observatory (SAAO) to classify and determine the nature of each blue stellar object. Currently, over 190 DA white dwarfs have been identified, of which the majority are not in existing catalogues (O'Donoghue et al. 1993). From these white dwarfs, we have selected DA stars with B-V colours near or within the range $0.15 \leq$ B-V $\leq 0.25$ together with all the non-DA degenerate stars (DO, DB, etc.) to monitor for variability.

## 2. New pulsating degenerate stars

High speed photometry with 10s time resolution in white light has been obtained with the SAAO 1.0-m and 0.75-m telescopes at Sutherland. To detect variability at least 1.5 hours high speed photometry is obtained on two different nights and the time series photometry is frequency analysed. So far we have examined for pulsation 39 DA white dwarfs close to the instability strip with 4 new variables detected (3 DAs and 1 DAZ) and 16 DO/DB degenerates with 1 new DB variable detected. The results will be presented only for 2 DAVs and 1 DBV as the other stars require further observation and analysis.

EC23487-2424 and EC14012-1446 are two large-amplitude DAV pulsators. The data and analysis of EC23487-2424 have already been described (Stobie et al. 1993). The light curve is non-sinusoidal, with a peak-to-peak variation ranging from 0.06 to 0.24 mag in white light. Six nights of observation show a multiple-period structure, with periods between 800 and 1000s. The principal frequency at $f$ = 1.101 mHz is present on most nights observed with variable amplitude, probably caused by the presence of multiple close frequencies.

Photometry of EC14012-1446 shows the non-sinusoidal nature of the light curve with a peak-to-peak variation ranging from 0.08 to 0.30 mag in white light. The periodogram from 0 to 10 mHz of the first night of photometry exhibits at least 8 frequencies well above the noise. It is apparent that some of these frequencies are either harmonics or combination terms (e.g., $2f_1$, $f_1 + f_2$, $f_1 + f_4$) resulting in the identification of five independent frequencies at 1.067, 1.382, 1.639, 1.885 and 2.507 mHz. The frequency analyses of all four nights shows that these frequencies are present on all nights, although there is strong evidence that the amplitude of the dominant frequency (at least) is variable.

The asteroseismological understanding of individual DAV stars has long resisted interpretation for two reasons (a) the low-amplitude pulsators do not reveal sufficient modes to enable analysis (b) the large-amplitude pulsators have power spectra which change radically on many timescales. Recently, however, Clemens (1993) has shown that progress can be made by considering the DAV stars as a group rather than as individuals. He showed that the hottest DAVs have the shortest periods and smallest amplitudes. Both period and amplitude increase monotonically with decreasing temperature, consistent with theory.

A remarkable result found by Clemens (1993) was that the periods of the DAVs as a group form a highly non-random distribution in period space with strong clusterings near preferred values. EC14012-1446 (and to some extent EC23487-2424) fits this pattern. These results are interpreted as evidence that the DAV stars are remarkably homogeneous as a group with similar total mass and similar hydrogen layer mass ($M \sim 10^{-4} M_*$). As all DA stars pass through the instability strip at some stage in their evolution this conclusion holds for all DA stars.

EC20058-5234 is the eighth pulsating helium white dwarf to be discovered. More than 10 periodicities have been discovered in about 52 hours of high speed photometry, the most prominent being at 281 and 257s. Many of the periodicities appear to be permanent features of the light curve, but the amplitudes vary: the strongest mode has been observed at amplitudes between 6 and 12 mmag. EC20058-5234 is exceptional in the group of DBVs both because of its short period and its low amplitude.

## References

Bradley, P. A. 1993, in Proc. 2nd WET Workshop, E. Meistas & J. Solheim, Baltic Astronomy, 2, 559

Clemens, J. C. 1993, in Proc. 2nd WET Workshop, E. Meistas & J. Solheim, Baltic Astronomy, 2, 407

O'Donoghue, D., Chen, A., Kilkenny, D., & Stobie, R. S. 1993, in Proc. 8th European Workshop on White Dwarfs, M. Barstow, Kluwer, 39

Stobie, R. S., Chen, A., Kilkenny, D., & O'Donoghue, D. 1992, in Variable Stars and Galaxies, B. Warner, ASP Conf. Ser., 30, 87

Stobie, R. S., Chen, A., O'Donoghue, D., & Kilkenny, D. 1993, MNRAS, 263, L13

*Astrophysical Applications of Stellar Pulsation*
*ASP Conference Series, Vol. 83, 1995*
*R. S. Stobie and P. A. Whitelock, eds.*

# The Effect of Proper Motion on $\dot{P}$ in Pulsating Stars

G. Pajdosz

*Mt. Suhora Observatory, Cracow Pedagogical University, ul. Podchorążych 2, 30-084 Kraków, Poland*

**Abstract.** We show that the proper motion of a pulsating star results in a secular increase of the radial velocity of the star and consequently contributes to the observed $\dot{P}$. We found that this mechanism is very important in studying period changes in pulsating DA white dwarfs, where its effect is comparable with the expected evolutionary period change in these stars due to their cooling.

While a pulsating star is travelling in space, the variation of its line-of-sight position causes the Doppler shift of arrival of the pulsation's maxima, by changing the travel time of light between the star and the observer. The observed period $P_{\rm obs}$ of the star is coupled to the period $P$ in the star's rest frame by the formula:

$$P_{\rm obs} = P(1 + v_{\rm r}/c) \tag{1}$$

Differentiating eq.(1), under assumption that the total relative velocity of the star is constant we obtain:

$$\dot{P}_{\rm obs} = \dot{P}(1 + v_{\rm r}/c) + P\dot{v}_{\rm r}/c \simeq \dot{P} + Pv_{\rm t}^2/(cr) \tag{2}$$

The first term in eq. (2) describes the rate of pulsation period change intrinsic to the star, while the second ("proper motion") term ($\dot{P}_{\rm pm}$) results from the spatial motion of the star. Note that always $\dot{P}_{\rm pm} \geq 0$, independently of the direction of motion of the star.

The discussed mechanism was originally proposed by Shklovskii (1970) in his study of period behaviour of high-velocity pulsars. To say something about the significance of the proper motion term for other groups of pulsators, let us rewrite the proper motion term in a form more useful for direct applications:

$$\dot{P}_{\rm pm} = 1.081 \cdot 10^{-19} P[{\rm s}](v_{\rm t}[{\rm km/s}])^2 (r[{\rm pc}])^{-1} \tag{3}$$

and let us consider the following cases:

- white dwarf pulsators (DAV, DBV and DOV)

  For the DAVs the expected evolutionary values of $\dot{P}$ for g-modes range from $10^{-14}{\rm ss}^{-1}$ to $10^{-15}{\rm ss}^{-1}$ (Bradley et al. 1992). The observed rate of a secular period increase for the dominant pulsation mode in DA white dwarf G117-B15A was recently reported by (Kepler 1993); the obtained value $3.2 \cdot 10^{-15}{\rm ss}^{-1}$ is completely consistent with the value expected for secular evolution. The expected values of $\dot{P}$ for DAV stars are the lowest among all pulsating white dwarfs – for DB pulsators they are one order and for DOVs several orders of magnitude higher. Making assumption, that for white dwarfs the distance $r \simeq 50$pc, $v_t \simeq 50$km/s (Lang 1992), $P \simeq 500$ s we can roughly estimate the proper motion term: $\dot{P}_{pm} \simeq 3 \cdot 10^{-15} ss^{-1}$. We can see that the proper motion term can be very significant in the case of DAV's and can contribute to the observed value as much as the expected evolutionary term, so its effect should be subtracted from the observed period derivative for a proper evaluation of $\dot{P}$ – see Pajdosz (1995).

- $\delta$ Sct, roAp and $\beta$ Cep stars

  The stellar evolution calculations yield the expected values of rates of pulsation period changes of the order of $10^{-11} ss^{-1}$ for $\delta$ Sct variables, while the observed values are still one or two orders higher (Breger 1990). For roAp stars the predicted values are $10^{-12}$–$10^{-13} ss^{-1}$ (Heller & Kawaler 1988), but they have not been confirmed observationally yet (Kurtz 1993). $O-C$ diagrams for $\beta$ Cep stars were studied recently by Pigulski (1992). He estimated the evolutionary rates of period increase to be $10^{-10}$–$10^{-11} ss^{-1}$ in good accordance with theoretical predictions.

  We put these three groups of variables together, because their kinematical characteristics are similar; these stars are characterized by low relative space velocities (Harwit 1988). Assuming, that $r > 50$pc, $v_t \simeq 20$km/s, $P \simeq 10000$ s we get $\dot{P}_{pm} < 10^{-14} ss^{-1}$. We can see that the proper motion term is several orders of magnitude too small to be important.

- other pulsators

  For stars which do not fall into the first two categories even if we assume their velocities to be extremely high ($v_t \simeq 200$km/s), a distance greater than 50pc, pulsation period below half a day we obtain that the $\dot{P}_{pm}$ should not be higher than $4 \cdot 10^{-12} ss^{-1}$.

**Acknowledgments.** This work was partly supported via KBN 2 P03D015 08 project.

## References

Breger, M. 1990, A&A, 240, 308

Bradley, P.A., Winget, D.E., & Wood, M.A. 1992, ApJ, 391, L33

Heller, C.H., & Kawaler, S.D. 1988, ApJ, 329, L43

Harwit, M. 1988, Astrophysical Concepts, Springer-Verlag New York, 588

Kepler, S.O. 1993, Baltic Astronomy, 2, 444

Kurtz, D.W. 1993, Ap&SS, 210, 207

Lang, K.R. 1992, Astrophysical Data, Springer-Verlag, New York, 535

Pajdosz, G. 1995, A&A, in press

Pigulski, A. 1992, PhD Thesis, Univ. of Wroclaw

Shklovskii, I.S. 1970, Soviet Ast., 13, 562

*Astrophysical Applications of Stellar Pulsation*
*ASP Conference Series, Vol. 83, 1995*
*R. S. Stobie and P. A. Whitelock, eds.*

# The Pulsating White Dwarf L19-2

Denis J. Sullivan

*Department of Physics, Victoria University of Wellington, NZ*

**Abstract.** Previous studies of the southern DAV white dwarf L19-2 have revealed that it pulsates non-radially in at least five modes with periods $\sim 10^2$ s and amplitudes less than 10 mmag. All of these modes showed fine structure splitting, but the one-day aliases resulting from the gaps in the single-site data made identification of the various frequencies, at best, difficult. Recent observations of L19-2 obtained from Mt. John (NZ) are presented, along with a discussion of the prospects for a Whole Earth Telescope run on this object scheduled for 1995.

## 1. Introduction

A significant number of degenerate stars are observed to pulsate, and it appears likely that most, if not all, of these objects will develop instabilities leading to pulsation as they slowly cool and evolve. The degenerate pulsators appear to be grouped into several distinct classes: the very hot DOV ($T_e \sim 10^5$ degrees) and PNNV objects; the relatively hot helium atmosphere DBV stars ($T_e \sim 22{,}000°$); and the cooler hydrogen atmosphere DAV white dwarfs ($T_e \sim 12{,}000°$).

All of these pulsators show periods between $\sim 10^2$ s and $10^3$ s, arising from nonradial $g$-mode pulsations of the fluid sphere. The DOV and DBV pulsators, in particular, display such a rich array of frequencies, that Whole Earth Telescope (WET) multi-site observations have proven to be essential in the identification process (e.g., Winget et al. 1991, 1994). A comparison of the observational data with numerical models has been so fruitful, that the descriptive term 'asteroseismology' has increasingly been used.

Some of the DAV pulsators display a much simpler $g$-mode spectrum. L19-2 is one of these. However, in some cases the modes are closely spaced in frequency (as a result of rotational or magnetic degeneracy removal), and then extended (multi-site) data sets are also required to overcome frequency aliasing problems in the data interpretation.

Interestingly, the pulsation modes of the cool white dwarfs are remarkably stable: the amplitudes may be variable, but the pulsation frequencies remain constant (as a direct result of the rigid geometrical constraints of the nearly static white dwarf structure). However, pulsation theory for these objects does predict very small period changes, resulting from evolutionary cooling effects. The pre-white dwarf DOV star, PG1159, has revealed a definite period change (Winget et al. 1991), but only one cooler DA object has demonstrated definite indications of a change (Kepler et al. 1991).

O'Donoghue & Warner (OW) (1987) undertook extensive observations of L19-2 between 1979 and 1985 and were able to infer the presence of five coherent pulsations with periods near 350, 192, 143, 118 and 113 s, and corresponding amplitudes of 1.0, 6.0, 0.55, 1.8 and $2.2 \times 10^{-3}$ mag, respectively. Using their combined data set, OW discussed mode identification, rotational frequency splitting effects (including an inferred second-order effect), and the stability of the principal frequency.

## 2. MJUO observations

Commencing in 1994, a programme of regular high-speed photometry of L19-2 has been carried out in New Zealand at Mount John University Observatory (MJUO). A two-star photometer attached to the one metre telescope is used to obtain high-speed "white" light photometry. The essential aim of the project is to provide a supporting data base for a WET run on this object. MJUO, with a latitude of –44°, is an ideal place to observe an object as far south as –81° declination throughout the year.

The MJUO data clearly shows the dominant 192 s oscillation detected in the previous work, as well as the 118 s and 113 s oscillations. A detailed inspection of the discrete Fourier transform (DFT) spectrum of a three-night photometric data set (5, 7 and 8 Dec) reveals excess power (relative to the DFT window function) in the vicinity of these 3 obvious components. This is undoubtedly the fine structure splitting detected by OW.

## 3. A WET campaign on L19-2

L19-2 has been included as one of the objects for the first semester 1995 WET campaign. A quality (extended coverage) WET data set (even though the object will be inaccessible to northern hemisphere observers) should improve the DFT spectral window function such that the various frequency components in the light curve are unambiguously identified. Inspite of the comprehensive nature of the OW data set, aliasing problems resulting from the single-site data gaps made some of their conclusions arguable (including some frequency identifications and the apparent second-order rotational splitting effect).

The OW analysis led to an upper limit on the period change of the 192s pulsation of $3 \times 10^{-14}$ ss$^{-1}$. Given the relative dominance of this mode, and the fact that L19-2 is one of the hotter DAV objects, it is a good candidate to continue searching for secular cooling effects in these white dwarfs. A WET data set on this object will contribute significantly to this programme.

## References

Kepler, O., et al. 1991, ApJ, 378, L48

O'Donoghue, D., & Warner, B. 1987, MNRAS, 228, 949

Winget, D.E., et al. 1991, ApJ, 378, 326

Winget, D.E., et al. 1994, ApJ, 430, 839

*Astrophysical Applications of Stellar Pulsation*
*ASP Conference Series, Vol. 83, 1995*
*R. S. Stobie and P. A. Whitelock, eds.*

# Pulsations in the Nucleus of Planetary Nebula IC 418?

E. Kuczawska

*Mt. Suhora Observatory, Pedagogical University, ul. Podchorazych 2, 30-084 Cracow, Poland*

S. Zola

*Astronomical Observatory, Jagiellonian University, ul. Orla 171, 30-244 Cracow, Poland*

**Abstract.** IC 418 belongs to the group of about a dozen planetary nebulae with central stars that show irregular spectroscopic and photometric variations. Recent theoretical results provide arguments that this group may constitute a new class of pulsators. We present new photometric observations of the nucleus of IC 418 showing variations with a timescale of about 1/4 day. We discuss two possible explanations of the observed phenomena: pulsational instability and an unstable wind.

## 1. Introduction

The planetary nebula IC 418 contains the relatively bright ($V = 10^m$) nucleus HD 35914, classified as O7f. HD 35914 is a prototype star of a recently proposed class of variable PNN (planetary nebula nuclei). This group include nuclei of He 2-131, He 2-138, PB 8, NGC 2392, IC 4593, NGC 6826, NGC 6543, Hu 2-1, IC 3568, NGC 40, IC 2149. The common characteristics of these objects are:

- presence of a compact, bright nebula
- effective temperature between 27000 and 50000 K
- spectral type mostly between O4f and O7f
- presence of strong stellar wind
- short-term radial velocity variations
- photometric variability of low amplitude over timescales of several hours
- observed variations are not strictly periodic.

Photometric variations of PNN are usually due to either hemispheric heating and/or eclipses produced in binaries or nonradial g-mode pulsations present in extremely hot pre-white dwarfs. IC 418 - like PNN constitute a separate class, still waiting for a complete observational and theoretical description.

In an exploratory investigation of the linear stability of envelopes of post-AGB stars Gautschy (1993) used new stellar opacity tables and observational constraints to model the nucleus of IC 418. It turned out that instability of the fundamental radial pulsation is possible in this star. From his modal diagram one can find that the theoretical period of pulsations of HD 35914 should be in the range from $0.^d15$ to $0.^d32$.

## 2. Observations and a search for periodicity

We analyzed the differential photometry of HD 35914 obtained at Mt. Suhora Observatory in 1989 (on 2 consecutive nights) as well as observations obtained by Jerzykiewicz (1994) at the South African Astronomical Observatory in 1991 (3 consecutive nights). The maximum range of variations of the central star together with the surrounding nebula reached $0.^m053$ in the U filter.

We examined the observations for the presence of periodicities by means of Fourier analysis. For observations from 1991 the dominant spike in the power spectrum appears at frequency corresponding to the period of about $6^h$, which is consistent with a period found by Mendez et al. (1983) in their investigation of photospheric radial velocities (cf. also Handler et al., 1995). However, the results for our observations from 1989 do not confirm this frequency, although they also show variations on a timescale of several hours (Kuczawska et al. 1995).

## 3. Discussion

No single period can fit all of our observational runs, implying that we have to do with either multimodal or irregular phenomenon. Apart from pulsational instability driven by the $\kappa$-mechanism (Gautschy 1993) a key role in the observed variations may also play a fact that all IC 418 - like PNN show strong observational evidence of mass outflow via a radiatively driven stellar wind. Such an outflow can be highly unstable against even small changes in photospheric parameters, as it has been demonstrated by theoretical models (Owocki et al. 1988, Rybicki et al. 1990, Pauldrach & Puls 1990). The source of the initial fluctuation can be stellar pulsations. In accordance with such models, the photometric variations of the nucleus of IC 418 could be at least partially ascribed to the changes in optical thickness caused by the formation and propagation of dense shells or clumps above photospheric level. The analogous continuum light variations have been observed for WR stars (Drissen et al. 1989).

Longer series of continuous data are obviously needed to determine whether the nucleus of IC 418 along with similar PNN constitute a new class of pulsators.

**Acknowledgments.** We are grateful to Prof. M. Jerzykiewicz for his observations. This work was sponsored by the grant KBN 2P03D 021 08.

## References

Drissen, L. et al. 1989, ApJ, 343, 426

Gautschy, A. 1993, MNRAS, 265, 340

Handler, G. et al. 1995, these proceedings

Jerzykiewicz, M. 1994, private communication

Kuczawska, E., Wlodarczyk, K., & Zola, S. 1995, A&A, in press

Mendez, R.H., Verga, A., & Kriner, H. 1983, Rev. Mex. Astron. Astrofis., 8, 175

Owocki, S.P., Castor, J.I., & Rybicki, G.B. 1988, ApJ, 335, 914

Pauldrach, A., & Puls, J. 1990, A&A, 237, 409

Rybicki, G.B., Owocki, S.P., & Castor, J.I. 1990, ApJ, 349, 274

*Astrophysical Applications of Stellar Pulsation*
*ASP Conference Series, Vol. 83, 1995*
*R. S. Stobie and P. A. Whitelock, eds.*

# Thermonuclear Excitation of $R$-modes in Neutron Stars

Umin Lee

*Laboratoire de Physique, Ecole Normale Superieure de Lyon, 69364 Lyon Cedex 07*

Tod E. Strohmayer

*Code 666, NASA/Goddard Space Flight Center, Greenbelt, MD 20771*

**Abstract.** Neutron stars display phenomena which result from the burning of accreted nuclear fuel in their surface layers. In addition to producing X-ray bursts and other temporal phenomema, such burning can excite oscillation modes of the neutron star. Here we investigate the stability of $R$-modes in uniformly rotating neutron stars using nonadiabatic pulsation calculations of steady burning envelopes on accreting neutron stars. We find that the fundamental $R$-modes are unstable for a large range of accretion rates, and thus might manifest themselves observationally in the power spectra of accreting sources above about 1 kHz.

## 1. $R$-modes oscillations in rotating neutron stars

Sensitive searches in LMXB have yet to reveal the pulsation signal of the rapidly rotating neutron star commonly thought to reside in these systems (Wood et al. 1991). Detection of such pulsations would firmly establish the hypothesized evolutionary link between these systems and the millisecond pulsars. The ability to observe rotationally induced pulsations depends on the degree to which these systems can produce intensity variations on the neutron star, for example, an accretion hotspot, or flux asymmetry caused by the magnetic field. If the magnetic fields in these systems are incapable of producing an observable rotational modulation, then what other physical processes can produce asymmetries, or patterns, which might be modulated at small, but detectable levels by stellar rotation? One possibility is that surface waves or oscillations can be excited to sufficient amplitude to produce observable modulations. Indeed, there are several reports in the literature of periodic intensity variations which occurred either during or in the vicinity of an X-ray burst (Livio & Bath 1982). At present there is no widely accepted scenario which explains all of the reported periodicities, but one possibility is that some of these periodic variations are due to surface oscillations. A potentially interesting class of such oscillations are the $R$-modes (also called planetary waves or Rossby waves), whose frequencies are closely related to the spin frequency of the star (Saio 1982). These modes reside largely in the surface layers of a neutron star, and involve motions of a small fraction of the total stellar mass. They therefore require less energy to excite to potentially observable amplitudes than more global modes. Our nonadiabatic calculations are performed using surface envelopes which have been constructed using the thin shell formalism described by Paczynski (1970). Hydrogen burning in accreting systems occurs in the steady-state for a variety of accretion rates, so for these calculations we assume that the shell steadily burns fuel at a rate equal to the accretion rate. We have investigated a range of accretion rates, however, here we report only results for a single illustrative case with $\dot{M} = 0.8\dot{M}_{edd}$.

In the lowest order approximation, $R$-modes are toroidal modes with oscillation frequency $\sigma_0 = 2m\Omega/l(l+1)$, where $\Omega$ is the angular rotation frequency of the star and $l$ is the spherical harmonic index. We investigate the pulsational stability of $R$-modes using the nonadiabatic formulation for uniformly rotating stars described by Lee & Saio (1987). In this treatment the displacement vector of an oscillation mode is expanded as an infinite series of spherical harmonics with constant azimuthal index $m$. The results of our four term, nonadiabatic calculations are a series of complex eigenfrequencies which describe as a function of rotation period the oscillation period in the rotating reference frame and the excitation or decay timescale. In Table 1 we have compiled the results for the $m = -1$, fundamental $R$-mode (denoted $r_0$) for three values of the rotation period of the neutron star. Shown are the rotation period, the $R$-mode periods in the rotating and inertial frames (in msec), and the growth rate $\mu$ for each mode. The large, positive growth rates indicate that these modes are strongly unstable to hydrogen burning.

Table 1. Periods and Growth Rates for $R$-modes ($\dot{M} = 0.8$)

| Modes | $P_{spin}$ | $\Pi_{rot}$ | $\Pi_{inert}$ | $\mu$ |
|---|---|---|---|---|
| $r_0$ | 46.09 | 175.9 | 62.46 | 0.042 |
| $r_0$ | 9.219 | 125.6 | 9.949 | 0.070 |
| $r_0$ | 4.609 | 120.5 | 4.793 | 0.061 |

## 2. Discussion

We find that the fundamental $R$-modes are unstable to the nuclear burning which occurs in the accreted surface layers of neutron stars. The large growth rates we find suggest that these modes may be driven to sufficient amplitudes to become observationally significant. The expected mode periods indicate that searches in the X-ray by future missions such as XTE be optimized in the kHz region of the power spectrum.

## References

Lee, U., & Saio, H. 1987, MNRAS, 225, 643

Livio, M., & Bath, G. T. 1982, A&A, 116, 286

Paczynski, B. 1970, Acta Astro., 20, 287

Saio, H. 1982, ApJ, 256, 717

Wood, K. S., et al. 1991, ApJ, 379, 295

*Astrophysical Applications of Stellar Pulsation*
*ASP Conference Series, Vol. 83, 1995*
*R. S. Stobie and P. A. Whitelock, eds.*

# Near-surface Effects in Modelling Oscillations of $\eta$ Boo

Jørgen Christensen-Dalsgaard

*Teoretisk Astrofysik Center, Danmarks Grundforskningsfond*

Timothy R. Bedding

*School of Physics, University of Sydney 2006, Australia*

Günter Houdek

*Institut für Astronomie, Universität Wien, Austria*

Hans Kjeldsen, Colin Rosenthal, Regner Trampedach

*Teoretisk Astrofysik Center, Danmarks Grundforskningsfond*

Mario J.P.F.G. Monteiro

*Centro de Astrofisica, Universidade do Porto, Portugal*

Åke Nordlund

*Niels Bohr Institutet, Københavns Universitet, Denmark*

Following the report of solar-like oscillations in the G0 V star $\eta$ Boo (Kjeldsen et al. 1995), a first attempt to model the observed frequencies was made by Christensen-Dalsgaard et al. (1995). This attempt succeeded in reproducing the observed frequency separations $\Delta\nu$ and $\delta\nu_{02}$, although there remained a difference of $\sim 10\,\mu$Hz between observed and computed frequencies. In those models, the near-surface region of the star was treated rather crudely: convection was described by means of a local mixing-length theory neglecting turbulent pressure, and the oscillations were assumed to be adiabatic. These approximations are likely to affect both $\Delta\nu$ and absolute frequencies.

Here we consider more sophisticated models of the external regions:

- **NL.a** A non-local mixing-length theory (Balmforth 1992) including turbulent pressure. Frequencies computed in the adiabatic approximation.
- **NL.na** The same, but with nonadiabatic frequencies, including the Lagrangian turbulent-pressure perturbation.
- **HYD.a** Outer structure obtained from horizontal and temporal averaging of a hydrodynamical convection model (e.g., Stein & Nordlund 1991). This was extended into a deep-envelope model through continuous matching at the lower boundary of the hydrodynamical model. Frequencies computed in the adiabatic approximation.

We compared the radial oscillation frequencies computed from these sophisticated models with those of normal models whose (local) mixing length was

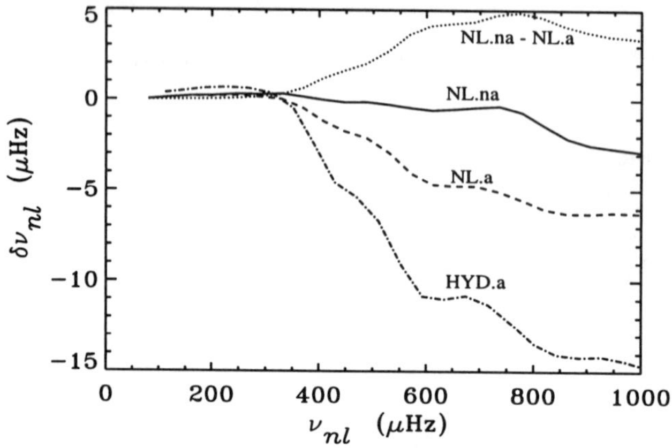


adjusted to obtain the same convection-zone depth. The comparisons are shown in the figure, in the sense (sophisticated treatment) – (normal treatment). Note that oscillations in $\eta$ Boo were observed in the frequency range 750–950 $\mu$Hz.

Two effects influence the frequencies: (i) Changes in the equilibrium model structure. It can be shown that these are dominated by the inclusion of turbulent pressure. Our results for **NL.a** and **HYD.a** imply frequency shifts of $-5$ to $-15\,\mu$Hz in the region of the observed oscillations, with the region of substantial turbulent pressure being considerably larger in the **HYD** model. (ii) The physics of the oscillations (nonadiabaticity and turbulent-pressure perturbations), whose effects are shown by the dotted line (**NL.na** – **NL.a**). Interestingly, the two influences almost cancel for the **NL** model (solid line), implying that little correction is required to the frequencies of Christensen-Dalsgaard et al. (1995). However, the **HYD.a** result suggests that the **NL** calculations underestimate the change in the equilibrium structure.

The corrections to the large frequency separation depend on the slope of the curve where the oscillations are observed. From the figure we estimate corrections to $\Delta\nu$ of up to about $-1\,\mu$Hz.

It is clear that assumptions in modelling the near-surface region introduce significant uncertainties in the computed frequencies. The present results give some hope that more detailed observations of $\eta$ Boo, the Sun and other stars will help improve our understanding of the physics of convection.

## References

Balmforth, N. J. 1992, MNRAS, 255, 632

Christensen-Dalsgaard, J., Bedding, T. R., & Kjeldsen, H. 1995, ApJ, in press

Kjeldsen, H., Bedding, T. R., Viskum, M., & Frandsen, S. 1995, AJ, 109, 1313

Stein, R. F., & Nordlund, Å. 1991, in: Challenges to theories of the structure of moderate-mass stars, Lecture Notes in Physics, 388, (eds.) D. O. Gough & J. Toomre, Springer, Heidelberg, 195

*Astrophysical Applications of Stellar Pulsation*
*ASP Conference Series, Vol. 83, 1995*
*R. S. Stobie and P. A. Whitelock, eds.*

# An Analytic Theory of Adiabatic *p*-modes in the Atmospheres of the Sun and Sun-like Stars

Namig S. Dzhalilov

*IZMIR RAN, Troitsk, Moscow Region, 142092 Russia*

Jürgen Staude

*Astrophysikalisches Institut Potsdam, Sonnenobservatorium 'Einstein-Turm', Telegrafenberg, D–14473 Potsdam, Germany*

**Abstract.** An analytic theory of resonant transmission has been used to study adiabatic, non-vertical oscillations in a realistic model of the solar or a stellar atmosphere and subphotosphere. Basic observed features of solar *p*-modes can be explained in this way.

## 1. Basic assumptions and analysis

The height profile of temperature $T$ in a realistic solar or solar-type stellar model, extending from the subphotosphere through the atmosphere up to the corona, has been piecewise approximated by various nonlinear and linear functions for the different regions. Solving the basic equations of adiabatic, non-vertical oscillations for these regions in the Cowling approximation we obtained the eigenfunctions for the *p*-modes in exact analytic form in terms of known functions. One of the advantages of this solution is an adequate matching of the complicated region around $T_{\rm min}$ — with steep gradients of $T$ on both sides — by one smooth function. The complete solution is obtained by matching to each other the partial solutions for the different layers. This solution provides the base for a detailed discussion of the reflection and transmission of waves through the acoustic barrier and above the barrier for higher frequencies $\nu$. The consideration of a lower reflecting boundary is not necessary in such a model of resonant transmission.

## 2. Results

Using the exact solution it has been demonstrated, that the mechanism of resonant transmission by itself is sufficient to explain the main features of the observed frequency spectrum of the solar *p*-modes, including the 5-min oscillations and the power peaks in the high-frequency tail between 5 and at least 10 MHz. The dependence of the power spectrum on the height $z$, on the frequency $\nu$, and on the degree $l$ has been demonstrated. The first resonant frequencies always correspond to the 5-min oscillations.

It has been shown that the upper turning point of the resonator for low frequencies $\nu \lesssim 4$ mHz does not coincide with the boundary of the acoustic potential barrier, the latter extending from the steep gradient of $T$ at the upper subphotosphere up to that above the minimum $T_{min}$. The wakes of the waves penetrate deep into the barrier (up to a height of 500 km in the solar atmosphere) and are reflected from there. High frequencies are reflected from the chromosphere-corona transition region. For not any frequency there exists a total reflection; with increasing frequency there is an increasing leakage of wave energy into the corona.

It has been shown that all observable characteristics of oscillations depend directly on the phase shift $\Delta\phi$ between upward and downward running waves. With changing height and frequency $\Delta\phi$ changes in jumps; the resonance peaks in the spectrum arise at $\Delta\phi = 0$.

## 3. Discussion

The phases of the waves can be modified by various mechanisms such as reflections, nonadiabatic effects (mainly interaction with radiation, if acoustic waves are considered), scattering at structural inhomogeneities, etc. A theoretical model consistent with the observed data should take into account all of these mechanisms or at least the first and second of them. So far our simple model considers the first effect only.

Apart from the neglect of nonadiabatic effects, the power spectra resulting from our present model turn out to be satisfactory and to reflect basic features of the observations. A conclusive comparison with ground-based and space-borne observations, however, should be postponed until the nonadiabatic effects have been taken into account. In other papers we considered nonadiabatic effects, that is the interaction between radiation, acoustic and thermal waves, in great detail, but only for a homogeneous mean model without gravitational stratification (Dzhalilov et al. 1994; Staude et al. 1994; Zhugzhda et al. 1993). In a future investigation both the gravitational stratification and the nonadiabatic effects are planned to be taken into account. That analysis will provide the basis for facing our modelling with measurements of ground-based velocity data and of luminosity oscillations planned to be measured aboard the satellites CORONAS-F and SOHO.

**Acknowledgments.** The authors gratefully acknowledge support of the present work by the German Space Agency (DARA) under grant No. 50 QL 9207 2.

## References

Dzhalilov, N.S., Zhugzhda, Y.D., & Staude, J. 1994, A&A, 291, 1001

Staude, J., Dzhalilov, N.S., & Zhugzhda, Y.D. 1994, Solar Phys., 152, 227

Zhugzhda, Y.D., Dzhalilov, N.S., & Staude, J. 1993, A&A, 278, L9

*Astrophysical Applications of Stellar Pulsation*
*ASP Conference Series, Vol. 83, 1995*
*R. S. Stobie and P. A. Whitelock, eds.*

# Visibility Functions of Solar and Stellar Irradiance Variations

Jürgen Staude and Günter Bartling

*Astrophysikalisches Institut Potsdam, Sonnenobservatorium 'Einstein-Turm', Telegrafenberg, D–14473 Potsdam, Germany*

Yuzef D. Zhugzhda and Namig S. Dzhalilov

*IZMIR RAN, Troitsk, Moscow Region, 142092 Russia*

**Abstract.** We calculated visibility functions of intensity fluctuations due to adiabatic waves in a realistic model of a non-grey, stratified atmosphere in LTE. The results for different optical wavelengths $\lambda$ and their angular dependence are discussed; moreover, a method for selecting various low-degree $p$-modes from observations of spectral irradiance fluctuations at different $\lambda$ including the UV is suggested.

## 1. Introduction

The presence of the $p$ modes of global solar oscillations has been proved in the time series not only of velocity, but also of spectral and total irradiance observations. There are hints of a non-adiabatic behaviour as well as of the influence of the gravitational stratification.

In some recent papers we investigated non-adiabatic hydrodynamic waves in a radiative, thermally conducting, homogeneous model atmosphere in LTE for the grey case (Dzhalilov et al. 1992, 1994) and for the non-grey case as well (Zhugzhda et al. 1993). Our theoretical approach took into account the non-adiabatic effects in great universality, but future investigations should consider the basic effect of the gravitational stratification as well. In the present paper we take first steps in this direction.

## 2. Intensity fluctuations due to waves in a non-grey stratified atmosphere

In order to obtain the visibility of luminosity fluctuations in a non-grey, stratified atmosphere, we computed the relative intensity fluctuations $I'/I_0$ for different $\lambda$ (or frequencies $\nu$) and positions $\mu$ on the disk. The intensity fluctuation depends on the relative fluctuations of temperature $T'/T_0$ and density $\rho'/\rho_0$, multiplied by the logarithmic derivatives of opacity with respect to $T$ and to $\rho$, respectively, describing the action of the opacity fluctuations ($f_\chi$), and of the gradient of the source function with respect to $T$ (influence of $T'$ on the source function, $f_S$). As a first step we consider the degree $l = 0$ only and normalise the relative fluctuation of $I_\nu$ to those of $T$, the latter assumed constant with depth. The absolute value of that expression will be called spectral darkening function (or visibility function for $l = 0$): $D_\lambda = |(I'_\nu / I_{0\nu})/(T'/T_0)| = |f_S + f_\chi|$. After integrating $I'_\nu$ and $I_{0\nu}$ over $\nu$ or the wavelength $\lambda$ we obtain the integrated darkening function $D$.

## 3. Calculations for a solar model

As an example we used the VAL3C model of the average solar atmosphere (Vernazza et al. 1981) in a recalculated form, taking into account the equation of state for a realistic solar plasma with 10 elements, considering molecules as well as non-LTE for H and all important opacity sources. The darkening and visibility functions have been computed for $\lambda$ from the UV up to the IR.

For the adiabatic case our calculations show small but important differences to the results of Toutain & Gouttebroze (1993) who considered a pure H plasma and an opacity by $H^-$ only: we derive a negative visibility function in the disk centre, which means opposite phases between $T'$ and $I'$. This effect is strongest in the UV and decreases with increasing $\lambda$. At shorter $\lambda$ the visibility is focussed onto a smaller sphere around the disk centre, thus suggesting a new method of mode selection by observing the spectral irradiance fluctuations at different $\lambda$. This method is applicable not only to the Sun, but also to stars.

## 4. Conclusion

In the next step our modelling will be completed by introducing information on the normal modes from a model of non-radial acoustic oscillations in a realistic solar model, to begin with the adiabatic case. However, the final step should consider non-adiabatic conditions as well. Such an analysis which is under preparation will provide the basis for confronting our modelling with measurements of ground-based velocity data and of luminosity oscillations planned to be measured aboard the satellites CORONAS-F and SOHO and for suggesting new space-borne experiments.

**Acknowledgments.** The authors gratefully acknowledge support of the present work by the German Space Agency (DARA) under grant No. 50 QL 9207 2.

## References

Dzhalilov, N.S., Zhugzhda, Y.D., & Staude, J. 1994, A&A, 291, 1001

Staude, J., Dzhalilov, N.S., & Zhugzhda, Y.D. 1994, Solar Phys., 152, 227

Toutain, T., & Gouttebroze, P. 1993, A&A, 268, 309

Vernazza, J.E., Avrett, E. H., & Loeser, R. 1981, ApJS, 45, 635

Zhugzhda, Y.D., Dzhalilov, N.S., & Staude, J. 1993, A&A, 278, L9

*Astrophysical Applications of Stellar Pulsation*
*ASP Conference Series, Vol. 83, 1995*
*R. S. Stobie and P. A. Whitelock, eds.*

# On Expected Constraints on Stellar Transport Processes from Space Seismic Missions: EVRIS, COROT

M.J. Goupil[1], E. Michel[1], S. Cassisi[1], W.A. Dziembowski[2], Y. Lebreton[1], A. Baglin[1], M. Auvergne[1], and C. Catala[3]

[1] *Observatoire de Paris, URA 335, 92190 Meudon France*

[2] *Copernicus Astronomical Center, 00-716 Warsaw, Poland*

[3] *Observatoire de Midi-Pyrénées, 31400 Toulouse, France*

## 1. Introduction

The first space seismological experiment, EVRIS (Baglin 1991) will be launched in 1996 and will observe $\sim$ 10 stars over $\sim$ 20 days each. A second one, COROT (Catala et al., 1994), should follow soon after (1999-2000), which should monitore $\sim$ 5 stars over $\sim$ 5 months each. The observed stars will be main sequence stars. For an optimum scientific outcome of EVRIS mission, one must select the best possible target stars. The first step of this selection is outlined. What would be further gained with COROT project is discussed.

## 2. $\delta$ Scuti stars

Specifications of both missions imply that among possible target stars, $\delta$ Scuti stars should present a large number of self excited modes, likely many more than presently observed from ground based networks.

Convective core overshooting : Among the detected modes of a given star, one nonradial mixed mode - a mode having amplitude in the vicinity of the convective core and in surface - can reasonably be expected. The knowledge of such a mixed mode and two radial modes, for instance, yields the overshoot distance $d_{ov}$, mass and age of the star (assuming no other uncertainties in stellar modelling). In this specific case and taking into account expected frequency observational errors ($\sim$ frequency resolution) of the mission, the current uncertainty ($\sim$ 100%) on $d_{ov}$ determination will not be much improved with EVRIS. COROT, in contrast, would yield $d_{ov}$ with a $\sim$ 20% uncertainty (Lebreton et al., 1994). Radial modes are not much sensitive to core overshooting so that higher precision should be obtained with more than one detected mixed mode. One should then be able to test more sophisticated treatment of the overshooting process, effects of core rotation and of element diffusion on the core extension which may not be negligible at this level.

Rotation profile : With COROT, one expects to detect self excited, low radial order modes of degree up to $\ell = 6$. Assuming that among these modes, all (or a subset of) opacity driven modes are detected for a given star, inversion techniques can be used to obtain its rotation profile with depth. When COROT specifications are used, the rotation profile is remarkably well recovered both at the surface and above the convective core (Goupil et al., 1995). Specifically here, local conservation of angular momentum is assumed for building an artificial rotation profile. From this profile, frequency splittings are computed for a model of the $\sim 1.7M_\odot$ $\delta$ Scuti star $\epsilon$ Cep, a suitable target star. These splittings are next contaminated with a white noise (variance $\sim$ the expected frequency uncertainty) and used as the 'observational' data. With EVRIS, one can hardly infer much more detailed information than detection of a non uniform rotation. But relative local variations of rotation as small as 10% could be detected with COROT. Noise in the data results in deviations from the real rotation rates within the error bars, which include observational uncertainties and error magnification of the inversion procedure.

## 3. Solar like oscillation

The star $\beta$ Vir is a possible candidate for EVRIS experiment (Cassisi et al., 1995). With a mass in the range $[1.25, 1.35]M_\odot$, it is expected to exhibit detectable solar like oscillations. Because of uncertainties in its HR diagram position, one must first investigate whether high order $\ell$ =1, 2 $p$-mode frequencies can discriminate between models for this star with slightly different stellar parameters. Frequencies of a model $A_*$, the 'observational' data, are compared with frequencies of two models $A_1$ and $A_2$ falling very close to $A_*$ in a HR diagram. $A_1$, $A_2$ differ from $A_*$ respectively through overshooting distance ($\delta d_{ov} = 0.1$) and chemical composition ($\delta Z = 0.01, \delta Y = 0.02$). Frequency uncertainties depend on observing time duration, signal/noise ratio and mode linewidths. With these estimations (Catala et al., 1994) and assuming a conservative $3\sigma$ error, discrepancies between $A_*$ and $A_1, A_2$ are detected with both missions. Further, COROT would detect smaller discrepancies in the second frequency differences (Gough, 1990) between model $A_*$ and $A_2$.

## 4. Conclusions

The above few examples show that EVRIS will provide more information than a mere detection of solar like oscillations. However, to test with high precision theories of transport processes, the required frequency resolution and number of detected modes make COROT an essential step in the asteroseismological field.

## References

Baglin, A., 1991, Solar Physics, 132, 155

Cassisi, S., Baglin, A., Michel, E., & Goupil, M.J., 1995 in prep.

Catala, C. et al., 1994, GONG 94 Conf., ASP Conf. Series, in press

Goupil, M.J, Diembowski, W.A., Goode, P., & Michel, E., 1995, A&A, submitted

Gough, D.O., 1990, Lecture Notes in Physics **367** , Eds Y. Osaki, H. Shibahashi, Springer Verlag, p. 283

Lebreton, Y., Michel, E., Goupil, M.J., & Baglin, A., 1994, in 'Stellar age determinations and accurate parallaxes', IAU Symp. 166

*Astrophysical Applications of Stellar Pulsation*
*ASP Conference Series, Vol. 83, 1995*
*R. S. Stobie and P. A. Whitelock, eds.*

# STARS - A Seismic Telescope for Astrophysical Research from Space

Ian W Roxburgh

*Astronomy Unit, Queen Mary and Westfield College, University of London, Mile End Road, London E1 4NS, UK.*

**Abstract.** STARS is a proposed asteroseismology (and stellar activity) space mission which is currently under phase A study in the European Space Agency as a candidate for the Medium Size Mission (M3).

STARS is a proposed asteroseismology space mission which was submitted to the European Space Agency in May 1993 by a consortium of 31 astronomers (ESA 1993b). It passed the assessment phase (ESA 1994) and is currently under phase A study as one of the short-listed candidates for the European Medium Size Mission (M3). Selection is in 1996. The primary scientific objective is to further our understanding of the evolution of stars by using asteroseismology to probe the interior structure of stars across the Hertzsprung Russell diagram. A further objective is to study stellar activity and surface rotation. STARS is a successor to an earlier proposal, PRISMA, which was one of four short-listed candidates for the M2 mission but it was not selected (ESA 1993a). The primary targets are open clusters with solar type stars brighter than $m_v = 10$. The core programme is directed towards a selected group of about 500 stars which is only a very small fraction of the total number of stars (about 20,000) that will be observed during the initial 2 year lifetime of the mission. There will be a substantial guest investigator programme.

The proposed instrumentation is a triply reflecting telescope with an effective diameter of 80cm, a focal ratio of $f/3$ with a $1^\circ$ x $1^\circ$ field of view, a white light CCD mosaic detector (1024 x 2048), slightly defocused, and operating at a temperature below -50°C (regulated to 0.1°C). This should provide a photometric precision of 0.2ppm down to $m_v = 8$ and 0.8ppm for $m_v = 11$, in 3s. The total time on each target field will be about one month, or longer, giving an accuracy in frequency determination of solar type stars of the order of 0.1 - 0.3$\mu$Hz. A photometric precision of 5ppm can be achieved for stars with $m_v = 15$. The maximum number of objects that could be observed in a field of view is about 1,000 giving a total of over 20,000 in the design mission lifetime of 2 years. The mission lifetime could be extended to 5 years. STARS will also carry a Lyman $\alpha$ Monitor, fed by a beam splitter, providing information about stellar rotation and chromospheric activity. The spacecraft will be placed in an Earth-Moon L5 orbit.

The STARS mission will include observations of members of open clusters of different ages, the four target clusters being $\alpha$ Per, the Pleiades, Coma, and the Hyades, but other clusters could be added to this list. The target list will include binaries with good mass and radius determinations, and stars observed by HIPPARCOS, for which a distances are well determined. The chosen targets should provide good coverage of the HR diagram including $\delta$ Scuti, roAp, white dwarf, RR Lyrae, $\beta$ Cephei, Am, $\lambda$ Boo and SPB stars. STARS can also be used for microlensing, the detection of stellar companions (possibly planets) and other serendipitous targets. The instrument will be capable of observing simultaneously all stars in a 1° x 1° field. STARS will remain pointed towards a given field, whether or not it belongs to an open cluster, for at least 1 month, with a duty cycle better than 95 percent.

By devoting 14 months observing the four clusters listed above, and one month on each of 10 different fields, one could observe 120 stars brighter than $m_v = 8$ (for which a photometric precision of better than 0.2 ppm will be obtained), and about 150 open cluster members. The observed fields would contain more that 2500 stars brighter than $m_v = 11$, for which the expected photometric precision of better than 0.8 ppm will be sufficient to measure solar-like oscillations to a precision of $0.3\mu$Hz, and more that 20,000 stars brighter than $m_v =$ 15, for which a photometric precision of better than 5 ppm will be achievable.

The payload is to be provided by Principal Investigator (PI) teams, funded by ESA Member States. ESA would be responsible for the overall mission and spacecraft design, integration of the payload, launch, operations and transmission of raw data. PIs would have about 30 percent of observing time with 70 percent being available to Guest Investigators (GIs). Data will be available to the general community 1 year after it has been received by the scientist who proposed the observations.

A first Call for Proposals both to PIs and to the community (GIs) would be announced 2 years before launch. A Time Allocation Committee (TAC) would draw up a list of scientific priorities and an Input Consortium would draw up a list of target fields to achieve these goals. The community (GIs) could then propose secondary objectives in each of the fields. It is expected to find between 200 and 1,000 scientifically worthwhile targets per field. Targets need to be selected well in advance (1-2 years) to enable a ground-based supporting programme to be instituted.

## References

ESA 1993a PRISMA, Probing Rotation and Interiors of Stars, Microvariability and Activity. Report on the Phase A Study, European Space Agency ESA SCI(93) 3.

ESA 1993b STARS: Seismic Telescope for Astrophysical Research from Space. Proposal for ESA M3 Mission. ESA 1993.

ESA 1994 STARS: Seismic Telescope for Astrophysical Research from Space. Report of Assessment Study. ESA SCI(94) 8.

*Astrophysical Applications of Stellar Pulsation*
*ASP Conference Series, Vol. 83, 1995*
*R. S. Stobie and P. A. Whitelock, eds.*

# Variability Studies and Asteroseismology with the XMM Optical Monitor

S. Horner

*Department of Astronomy and Astrophysics, Penn State University, University Park, PA 16802, USA*

E. Antonello

*Osservatorio Astronomico di Brera, Milano–Merate, Italy*

**Abstract.** It will be possible to make photometric variability studies from space of stars serendipitously falling in the field of view of the Optical Monitor, and to search for asteroseismological phenomena in bright target stars.

## 1. Optical Monitor

The Optical Monitor (OM) (cf. Horner et al. 1994) is one of three instruments on ESA's XMM observatory (cf. Mason et al. 1995) to be launched in 1999. The OM is a 30 cm aperture Ritchey-Chrétien telescope for UV and optical band observations that is co-aligned with the X-ray telescopes. It has two detectors for photometric and low resolution spectroscopic observations. A beam splitter separates the light into blue (160-550 nm) and a red beams (500-1100 nm) – the blue detector is a Micro-channel plate Intensified CCD (MIC), while the red detector is a frame transfer CCD.

The fields of view of the two detectors are similar, approximately 17 arcmin square. The blue detector (MIC) has a resolution of 1 arcsec, while the red detector has a resolution of 3.4 arcsec. The 11 position filter wheels in each of the beams carry broad and medium band filters, grisms and a magnifier; the red beam filter wheel also carries a defocusing lens.

It will be possible to perform variability studies from space of stars serendipitously falling in the field of view of the Optical Monitor, and to search for asteroseismological phenomena during long observations by XMM to detect the coronal X-ray spectra of bright stars. Monte Carlo simulations using the Guide Star Catalog (Jenkner et al. 1990 & Lasker et al. 1988) predict that if XMM's pointings are randomly distributed on the sky, an average 0.85 stars of $m_V \leq 12$ will fall into the field of view per pointing. If the pointings are randomly distributed within 15° of the galactic plane, an average 1.3 stars of $m_V \leq 12$ will fall into the field of view per pointing.

## 2. Photometric precision

The expected differential photometry precision and minimum detectable amplitude for bright stars observed with the red detector are reported in Table 1. The $1\sigma$-precision has been estimated for a source temperature of 5770 K. Owing to the brightness of the stars, the single exposures are assumed very short, 1 s or less, and the star image is spread over many pixels. The estimates have been made for a cumulative number of exposures of 30 s. In general, the effects of readout noise due to the very high cumulative number of exposures are not significant for very bright stars. The precisions have been estimated assuming a perfect stabilization of the images and of CCD temperature. Moreover, no contamination owing to cosmic rays and no degrading of the CCD performances due to radiation have been taken into account. It will be possible to correct for the cosmic ray contamination by means an on board frame comparison technique only for images of faint sources with relatively long exposure times; it will not be possible to apply a similar correction to images with short exposure times. Therefore, we expect an incidence of about a few percent of the time series data to be contaminated by cosmic rays in the case of bright star observations.

Table 1. Differential photometry precision and minimum detectable amplitude (see text)

| $m_V$ | precision ($1\sigma$) | amplitude |
|---|---|---|
| 4 | 9.3E–5 | 8.9E–6 |
| 5 | 1.5E–4 | 1.5E–5 |
| 6 | 2.3E–4 | 2.2E–5 |
| 7 | 3.7E–4 | 3.5E–5 |
| 8 | 5.9E–4 | 5.6E–5 |
| 9 | 9.3E–4 | 8.9E–5 |
| 10 | 1.5E–3 | 1.5E–5 |
| 11 | 2.3E–3 | 2.2E–4 |

**Acknowledgments.** Work in the US is supported by NASA contract NAS5-31787 at Penn State. In Italy, financial support is provided by ASI, in the UK by SERC, and in Belgium by the Belgian Ministry of Science.

## References

Antonello, E., & Cropper, M.S. 1992, in *Stellar Photometry - Current Techniques and Future Developments*, Butler, C.J., Elliott, I., Cambridge Univ. Press, 358

Horner, S.D., et al. 1994, in *Instrumentation in Astronomy VIII*, Crawford, D.L., Craine, E.R., SPIE, 2198, 1238

Jenkner, H., et al. 1990, ApJ, 99, 2081

Lasker, B.M., et al. 1988, ApJS, 68, 1

Mason, K.O., et al. 1995, Adv. Space Res., 16, 41

## Author Index

## Subject Index

This book was prepared with the help of editor's tools written by Chris Biemesderfer and Jeannette Barnes (NOAO).